BEAM INSTRUMENTATION WORKSHOP 2004

Previous Proceedings in the Series of Beam Instrumentation Workshops

	Year	Held in	Publisher	ISBN
10th	2002	Upton, New York	AIP Conf. Proceedings vol. 648	0-7354-0103-9
9th	2000	Cambridge, Massachusetts	AIP Conf. Proceedings vol. 546	1-56396-975-0
8th	1998	Stanford, California	AIP Conf. Proceedings vol. 451	1-56396-794-4
7th	1996	Argonne, Illinois	AIP Conf. Proceedings vol. 390	1-56396-612-3
6th	1994	Vancouver, B.C, Canada	AIP Conf. Proceedings vol. 333	1-56396-352-3
5th	1993	Santa Fe, New Mexico	AIP Conf. Proceedings vol. 319	1-56396-389-2
4th	1992	Berkeley, California	AIP Conf. Proceedings vol. 281	1-56396-190-3
3rd	1991	Newport News, Virginia	AIP Conf. Proceedings vol. 252	0-88318-934-8
2nd	1990	Batavia, Illinois	AIP Conf. Proceedings vol. 229	0-88318-832-1
1st	1989	Upton, New York	AIP Conf. Proceedings vol. 212	0-88318-645-4

The Collection of Beam Instrumentation Workshops 1-9, held 1989-2000, is also available on CD-ROM, ISBN 0-7354-0144-6.

Other Related Titles from AIP Conference Proceedings

705 Synchrotron Radiation Instrumentation: Eighth International Conference on Synchrotron Radiation Instrumentation
Edited by T. Warwick, J. Arthur, H. A. Padmore, and J. Stöhr, May 2004, 0-7354-0180-2
CD-ROM: 0-7354-0179-9

693 Beam-Halo Dynamics, Diagnostics, and Collimation: 29th ICFA Advanced Beam Dynamics Workshop on Beam-Halo Dynamics, Diagnostics, and Collimation and the Beam-Beam'03 Workshop
Edited by Jie Wei, Wolfram Fischer, and Pamela Manning, December 2003, CD-ROM included, 0-7354-0166-7

691 High Energy Density and High Power RF: 6th Workshop on High Energy Density and High Power RF
Edited by Steven H. Gold and Gregory S. Nusinovich, November 2003, CD-ROM included, 0-7354-0164-0

680 Application of Accelerators in Research and Industry: 17th International Conference on the Application of Accelerators in Research and Industry
Edited by Jerome L. Duggan and I. L. Morgan, August 2003, CD-ROM included, 0-7354-0149-7

To learn more about these titles, or the AIP Conference Proceedings Series, please visit the webpage **http://proceedings.aip.org/proceedings**

BEAM INSTRUMENTATION WORKSHOP 2004

Previous Proceedings in the Series of Beam Instrumentation Workshops

Year		Held in	Publisher	ISBN
10th	2002	Upton, New York	AIP Conf. Proceedings vol. 648	0-7354-0103-9
9th	2000	Cambridge, Massachusetts	AIP Conf. Proceedings vol. 546	1-56396-975-0
8th	1998	Stanford, California	AIP Conf. Proceedings vol. 451	1-56396-794-4
7th	1996	Argonne, Illinois	AIP Conf. Proceedings vol. 390	1-56396-612-3
6th	1994	Vancouver, B.C, Canada	AIP Conf. Proceedings vol. 333	1-56396-352-3
5th	1993	Santa Fe, New Mexico	AIP Conf. Proceedings vol. 319	1-56396-389-2
4th	1992	Berkeley, California	AIP Conf. Proceedings vol. 281	1-56396-190-3
3rd	1991	Newport News, Virginia	AIP Conf. Proceedings vol. 252	0-88318-934-8
2nd	1990	Batavia, Illinois	AIP Conf. Proceedings vol. 229	0-88318-832-1
1st	1989	Upton, New York	AIP Conf. Proceedings vol. 212	0-88318-645-4

The Collection of Beam Instrumentation Workshops 1-9, held 1989-2000, is also available on CD-ROM, ISBN 0-7354-0144-6.

Other Related Titles from AIP Conference Proceedings

705 Synchrotron Radiation Instrumentation: Eighth International Conference on Synchrotron Radiation Instrumentation
Edited by T. Warwick, J. Arthur, H. A. Padmore, and J. Stöhr, May 2004, 0-7354-0180-2
CD-ROM: 0-7354-0179-9

693 Beam-Halo Dynamics, Diagnostics, and Collimation: 29th ICFA Advanced Beam Dynamics Workshop on Beam-Halo Dynamics, Diagnostics, and Collimation and the Beam-Beam'03 Workshop
Edited by Jie Wei, Wolfram Fischer, and Pamela Manning, December 2003, CD-ROM included, 0-7354-0166-7

691 High Energy Density and High Power RF: 6th Workshop on High Energy Density and High Power RF
Edited by Steven H. Gold and Gregory S. Nusinovich, November 2003, CD-ROM included, 0-7354-0164-0

680 Application of Accelerators in Research and Industry: 17th International Conference on the Application of Accelerators in Research and Industry
Edited by Jerome L. Duggan and I. L. Morgan, August 2003, CD-ROM included, 0-7354-0149-7

To learn more about these titles, or the AIP Conference Proceedings Series, please visit the webpage **http://proceedings.aip.org/proceedings**

BEAM INSTRUMENTATION WORKSHOP 2004

Eleventh Beam Instrumentation Workshop

Knoxville, Tennessee 3 – 6 May 2004

BIW04

EDITORS
Thomas Shea
R. Coles Sibley III
Oak Ridge National Laboratory
Oak Ridge, Tennessee

SPONSORING ORGANIZATION
Spallation Neutron Source,
Oak Ridge National Laboratory

CD-ROM INCLUDED

AMERICAN INSTITUTE OF PHYSICS

Melville, New York, 2004
AIP CONFERENCE PROCEEDINGS ■ VOLUME 732

Editors:

Thomas Shea
R. Coles Sibley III

Spallation Neutron Source
Oak Ridge National Laboratory
Oak Ridge, TN 37831
USA

E-mail: shea@ornl.gov
sibley@ornl.gov

L.C. Catalog Card No. 2004114116
ISBN 0-7354-0214-0
ISSN 0094-243X
Printed in the United States of America

CONTENTS

FARADAY CUP AWARD

TUTORIALS

INVITED TALKS

CONTRIBUTED TALKS

POSTERS

APPENDICES

PREFACE

The Eleventh Beam Instrumentation Workshop, hosted by the Spallation Neutron Source (SNS), Oak Ridge National Laboratory (ORNL), Oak Ridge, Tennessee, was held in Knoxville, Tennessee, May 3 to 6, 2004, at the Knoxville Marriott Hotel. The workshop opened with an address by Dr. Thom Mason, Associate Laboratory Director, ORNL. There were more than 110 participants from 11 countries, with about 30% of the participants from overseas—a good showing. We had very good attendance from vendors, with 16 vendors supporting the workshop.

The workshop included 3 tutorials, 6 invited talks, 13 contributed talks, 1 discussion session, 1 poster session, and a tour of SNS. There were 48 posters at a very cramped poster session because of a last-minute switch by the hotel for the poster meeting room. The discussion session format was changed from previous years. This year the emphasis was on small discussion groups, six persons per table, with a variety of topics to choose from. Although the format was well received by the attendees, we learned that more than 30 minutes per session is needed for everyone at a table to participate.

There are 15 members on the BIW organizing committee. During this workshop, Mike Plum (SNS) and Roland Jung (CERN) resigned from the committee. Mike will be organizing the SNS ring installation and commissioning, and Roland will begin to relax before retirement. We would like to thank Mike and Roland for years of effort in making these workshops a success. An agreement was reached for FermiLab to host BIW06 near Chicago.

The banquet was held at the Crescent Bend Historic Home in Knoxville. The dinner was excellent, as was the outdoor setting overlooking the Tennessee River. The rare comfortable and dry evening on the river with bluegrass music was enjoyed by all.

The committee would like to thank all of the local organizing committee members for their efforts in making this workshop a success. The organizing committee is grateful to Myra Fultz, Muriel Tate, Lori Jean Lane, Angela Beach, and Debbie Weaver for all their efforts, from sending out the workshop announcements to keeping us in line and getting us ready for the workshop. We also would like to acknowledge the SNS Information Technology group for their support in setting up the computer room and equipment required for the workshop. Also, the editors would like to send a special thank you to Charlie Horak and Anna Lionberger for their efforts in getting these proceedings published.

Coles Sibley
Tom Shea
BIW04 Cochairmen

Organizing Committee

Tom Shea, ORNL/SNS, Cochair
Coles Sibley, ORNL/SNS, Cochair
Glenn Decker, ANL
Robert Hettel, SLAC
Ken Jacobs, UWSRC
Roland Jung, CERN
Geoffrey Krafft, JLab
Ralph Pasquinelli, FNAL
Mike Plum, LANL
Thomas Russo, BNL
Fernando Sannibale, LBNL
Steve Smith, SLAC
Greg Stover, LBNL
Robert Webber, FNAL
Jim Zagel, FNAL

Local Committee

Myra Fultz, Conference Coordinator
Angie Beach
Judy Benton
Willem Blokland
Charlie Horak
Alan Jones
Ben Kelmers
Lori Lane
Reneé Manning
Jeff Patton
Fern Stooksbury
Muriel Tate
David Thompson
Debbie Weaver

11th Beam Instrumentation Workshop
Hosted by Oak Ridge National Laboratory and the Spallation Neutron Source

Monday 5/3/04

Time	Session	Speaker
8:00 a.m.	Registration/Coffee	
8:30 a.m.	Welcome Session and Opening	T. Mason (ORNL/ SNS)
8:45 a.m.	Overview of the Spallation Neutron Source and its Beam Diagnostic Systems	T. Shea (ORNL/SNS, BIW04 Cochair)
9:30 a.m.	Coffee	
10:00 a.m.	Resolving EMI Issues To Optimize Accelerator Beam Diagnostic Performance	M. Thuot (LANL)
11:30 a.m.	Faraday Cup Winner Announcement	
12:00 Noon	Lunch	
1:30 p.m.	Design and Tests of a New Rest Gas Ionization Profile Monitor Installed in the SPS as a Prototype for the LHC	C. Fischer (CERN)
1:50 p.m.	Laser-Compton Scattering as a Tool for Electron Beam Diagnostics	K. Chouffani (IAC)
2:10 p.m.	Sub-Picosecond Electron Bunch Length Measurements at SLAC	P. Krejcik (SLAC)
2:30 p.m.	Measurement of Bunch Length Using Spectral Analysis of Incoherent Radiation Fluctuations	V. Sajaev (ANL)
3:30 p.m.	Coffee	
4:00 p.m.	Discussion Groups, Parallel Sessions, and Small Groups	

Tuesday 5/4/04

Time	Session	Speaker
8:00 a.m.	Coffee	
8:30 a.m.	Interceptive Beam Diagnostics–Signal Creation and Materials Interactions	M. Plum (ORNL/SNS)
10:00 a.m.	Coffee	
10:30 a.m.	Reliability of the Beam Loss Monitors System for the Large Hadron Collider	G. Guaglio (CERN)
10:50 a.m.	Measurement of the Beam Longitudinal Profile in a Storage Ring by Non-Linear Laser Mixing	S. De Santis (LBNL)
11:10 a.m.	Fast Orbit Feedback and Beam Stability at the Swiss Light Source	V. Schlott (PSI)
11:30 a.m.	Energy Spread Monitoring for the JLab Experimental Program: Synchrotron Light Interferometers, Optical Transition Radiation Monitors, and Wire Scanners	A. Freyberger (JLab)
11:50 a.m.	YO!–A Time-of-Arrival Receiver for Removal of Femtosecond Helicity-Correlated Beam Effects	J. Musson (JLab)
12:10 p.m.	BPMs with Precise Alignment for TTF2	M. Wendt (DESY)
12:30 p.m.	Lunch	
1:30 p.m.	SNS and ORNL Site Tour	Departure from hotel front lobby

Wednesday 5/5/04

Time	Session	Speaker
8:00 a.m.	Coffee	
8:30 a.m.	Instrumentation Systems from an Operations Viewpoint	R. Mau (Fermilab)
9:30 a.m.	Development of a Beam Loss Detection System for the CLIC Test Facility 3	T. Lefèvre (Northwestern University)
9:50 a.m.	Fermilab Recycler Ring BPM Upgrade Based on Digital Receiver Technology	P. Prieto (Fermilab)
10:10 a.m.	Coffee	
10:30 a.m.	Beam Diagnostics Based on AC Modulation of System Parameters	M. Tienfenback (JLab)
11:30 a.m.	DIAMOND Storage Ring Optical and X-Ray Diagnostics	C. Thomas (DIAMOND)
11:50 a.m.	Development of an Abort Gap Monitor for High-Energy Proton Rings	S. De Santis (LBNL)
12:10 p.m.	Lunch	
1:30 p.m.	Exhibitors and Poster Sessions	
6:00 p.m.	Banquet	Crescent Bend Historic Home

Thursday 5/6/04

Time	Session	Speaker
8:00 a.m.	Coffee	
8:30 a.m.	Tutorial on FPGA-Based Instrumentation and Beam Feedback	G. Foster (Fermilab)
10:00 a.m.	Coffee	
10:30 a.m.	Single-Shot Measurement of Femtosecond Electron Bunches at SLAC	A. Cavalieri (SLAC)
11:30 a.m.	Faraday Cup Talk: Twelve Years of SR Monitor Development at KEK	T. Mitsuhashi (KEK)
12:30 p.m.	Closing	Tom Shea (ORNL/SNS, BIW04 Cochair)

Toshiyuki Mitsuhashi (third from the left) receiving the 2004 Faraday Cup Award from (left to right) Coles Sibley and Tom Shea (BIW04 cochairs, ORNL/SNS) and Julien Bergoz (Faraday Cup Award sponsor, Bergoz Instrumentation, Saint Genis Pouilly, France).

2004 Faraday Cup Award

The 2004 Faraday Cup Award was presented to Toshiyuki Mitsuhashi from the High Energy Accelerator Research Organization (KEK), Tsukuba, Japan, for his work on synchrotron radiation (SR) monitor development. At the workshop, he presented his work in his talk, "Twelve Years of SR Monitor Development at KEK."

Faraday Cup Award—Information and Rules

Purpose

The Faraday Cup Award is donated by Bergoz Instrumentation, Saint Genis Pouilly, France. The award is intended to recognize and encourage innovative achievements in the field of accelerator beam instrumentation.

Award

The award consists of a U.S. $5,000 prize and a certificate to be presented at the next U.S. beam instrumentation workshop. Winners participating in the BIW will share a $1,000 travel allowance. Selection of recipients is the responsibility of the BIW Organizing Committee.

Award Criteria

The Faraday Cup Award shall be presented for outstanding contribution to the development of an innovative beam diagnostics instrument of proven workability. The prize is only awarded for demonstrated device performance and published contribution.

Criteria Interpretation

Beam Diagnostic Instrument:
A device to measure the properties of charged elementary particle, atomic or simple molecular beams during or after acceleration, or the properties of neutral particle beams produced in an intermediate state of charged particle acceleration. The device may operate by detecting secondary beams of charged, neutral, massive, or massless particles, but its purpose should be to diagnose the primary charged particle beam. The mass of primary beam particles shall be no greater than the order of 1000 atomic mass units.

Delivered Performance:
The performance of the device should have been evaluated using a charged particle beam, rather than in a "bench-top" demonstration.

Publication:
A description of the device, its operating principle, and its performance should have been published in a journal or in the proceedings of a conference or workshop that is in the public domain. Laboratory design notes, internal technical notes, etc., do not qualify but may be submitted to support other publications. Full and open disclosure is necessary to the extent that a potential user could design a similar device. More than one article may be submitted (together) to satisfy this requirement; for example, an article describing the principle plus another article describing the performance.

Eligibility:
Nominations are open to candidates of any nationality for work done at any geographical location. There are no restrictions for candidates; however, in the event of deciding between works of similar quality, preference will be given to candidates in an early stage of their beam instrumentation career. The award may be shared between persons contributing to the same accomplishment. Once accepted by the Award Committee, a nomination shall remain eligible for three successive competitions unless withdrawn by a candidate.

Disclosure

The Award Committee may release the names of entrants and a list of publications related to an entry if requested by a third party. Unpublished supporting material will not be disclosed nor will the names of persons supporting a nomination. Discussion regarding individual entries, scoring, etc., is regarded as confidential and will not be disclosed.

Nominations

The nomination package shall include the name of the candidate, relevant publications, a statement outlining his or her personal contribution and that of others, and letters from two professional accelerator physicists, engineers, or laboratory administrative personnel who are familiar with the device and its development. Two master copies of this package, suitable for copying, must be submitted not later than October 1st the year before the workshop to the conference chair for the upcoming workshop.

Previous Winners

- 1992 A.V. Feschenko—Bunch Shape Monitors Using Secondary Electron Emission
- 1993 R. B. Fiorito and D. W. Rule—Optical Transition Radiation
- 1994 E. Rossa—Real Time Single Shot Three Dimensional Measurement of Picosecond Photon Bunches
- 1996 W. Barry and H.-C. Lihn—Measurement of Subpicosecond Electron Bunch Length and Profiles Using Coherent Transition Radiation
- 1998 A. Peters—Cryogenic Current Comparator for nA beams
- 2000 Kay Wittenburg—The PIN-Diode Beam Loss Monitor System at HERA
- 2002 Andreas Jansson—Magnetic Quadrupole Pick-Ups in the CERN PS

Faraday Cup Award

11th Beam Instrumentation Workshop

May 3–6, 2004

Twelve Years of SR Monitor Development at KEK

T. Mitsuhashi

Photon Factory, KEK, Institute of Material Structure Science, High Energy Accelerator Research Organization, 1-1 Oho, Tsukuba, Ibaraki,305 Japan, E-mail: mitsuhas@kekvax.kek.ac.jp

Abstract. In order to measure transverse and longitudinal beam sizes and profiles, I have worked on the development of synchrotron radiation (SR) monitors over the past twelve years. During these years, the development of the SR interferometer has been the most significant topic, which I will review here. History, theoretical background, some experimental results, and practical notes on using the SR interferometer are described.

INTRODUCTION

The synchrotron radiation (SR) monitor based on visible optics is one of the most fundamental diagnostic tools in the high energy circular accelerator [1]. The monitor gives a visible beam profile, beam size, longitudinal profile, etc. These greatly improve the efficiency of commissioning and operation of the accelerator. In this monitor, the visible SR beam is extracted by a mirror from the circular accelerator, then the SR beam is guided into the optical diagnostics systems. I have worked on the development of SR monitors over the past twelve years at KEK. During these years, the development of the SR interferometer has been the most significant topic. The idea of the SR interferometer for measurement of beam profile and size struck me while I was performing experiments to investigate the coherence of synchrotron radiation in 1994 [2]. After basic investigation of beam profile and size measurements with the SR interferometer, this apparatus was applied to daily monitoring of beam size [3]. Nowadays, the SR interferometer is recognized as a powerful tool for easily measuring small beam sizes [3]. I review the development of the SR interferometer in this paper.

CP732, *Beam Instrumentation Workshop 2004: Eleventh Workshop*,
edited by T. Shea and R. C. Sibley III

HISTORY OF TWELVE YEARS OF DEVELOPMENT OF THE SR MONITOR AT KEK

The history of twelve years of development of the SR monitor at KEK is summarized below.

1992-1994	Renovation of SR monitor for the Photon Factory. Development of deconvolution of the beam image by means of MEM.
1993-2000	Design, construction, and commissioning of SR monitor for the B-factory. Development of beam profile measurement system by adaptive optics.
1994-1996	Development of SR interferometer.
1995-1997	Measurement of small beam sizes at the AURORA ring of Ritsumeikan-University.
1996-	Measurement of small beam sizes at the ATF, KEK. Development of SR interferometer based on reflective optics.
1998-1999	Measurement of small beam sizes in low emittance lattice at the Photon Factory.
1999-2000	Construction of profile monitor for KEKB injector linac.
2000-2002	Construction of SR monitor for the Photon Factory Advanced Ring. Development of non-aplanatic SR interferometer.
2002-2003	Development of intensity interferometry for the measurement of bunch length.
2003-2004	Development Lyott optics for the measurement of beam tail and halo.
2002-2004	Measurement of instantaneous beam profile measurement by the use of high-speed gated camera.

The following section describes the development of the SR interferometer.

BEAM PROFILE AND SIZE MEASUREMENT WITH INTERFEROMETRY

The measurements of beam profile and size are two of the most fundamental diagnostics in an electron storage ring. The most conventional method to observe the beam profile is beam profile monitoring via imaging of the visible SR beam [1]. The resolution of this monitor is generally limited by diffraction phenomena. In the usual configuration of the profile monitor, the RMS size of diffraction (1σ of the point spread function) is no smaller than 50 μm. In the last 10 years, research and development in electron storage rings (especially in reducing the beam emittance) has been very remarkable. We can realize a few nmrad beam emittance in electron storage rings. The above-mentioned profile monitor via imaging of the visible SR beam becomes ineffective in precise quantitative measurements of the beam profile and size due to the diffraction resolution limit. In visible optics, the interferometer is one of the standard methods for measuring the profile or size of very small objects. The principle of measuring the profile of an object by means of spatial coherency was first proposed by H. Fizeau [4] and is now known as the Van Cittert-Zernike theorem [5]. It is well known that A. A. Michelson measured the angular dimension (extent) of a star with this concept [6]. We developed the SR interferometer (an interferometer for SR beams) to measure the spatial coherence of the visible region of the SR beam. We demonstrated that this method is applicable to measure the beam profile and size at the KEK Photon Factory [2]. Since the SR beam from a small electron beam has good spatial coherence, this method is suitable for measuring a small beam size. The characteristics of this method are: 1) one can measure beam sizes as small as 3-4 μm with 1-μm resolution in a non-destructive manner; 2) the profile is easy to measure using visible light (typically 500 nm); 3) the measurement time is a few seconds for size measurement and a few minutes for profile measurement. Milestones in development of the SR interferometer from 1994 to 2004 are summarized below:

1994	First experiment with interferometry. Demonstration of possibility to apply the beam size measurement through interferometric method.
1995	Measurement of beam profile at the Photon Factory.
1996	Measurement of 10.5-μm beam size at the SR center of Ritsumeikan University.
1997	Measurement of 6.2-μm beam size at the ATF.
1998	2D interferometer experiment at the Photon Factory.
2003	4.7-μm beam size measurement by SR interferometer with Hasherian reflective optics.

THEORETICAL BACKGROUND OF THE INTERFEROMETRY

Simple Understanding of Beam Size Measurement With Interferometry

According to the work of H. Fizeau in 1868, the visibility (contrast) of the interference fringe taken by an interferometer is higher for a small light source and lower for a large light source. At the smallest limit of light source size, a point source gives an interferogram with visibility 1. We now interpret the point source as the single mode of a photon which gives an interferogram of visibility 1. We can represent the general light source by an ensemble of point sources. The interferograms given by such an ensemble of point sources are superimposed, and the interference fringe is smeared. Then, the visibility of the obtained interferogram will have reduced visibility, as shown in Fig. 1.

In the reverse manner, we can measure the size of the light source by measuring the visibility of the interferogram. This concept is now known as the van Cittert-Zernike theorem.

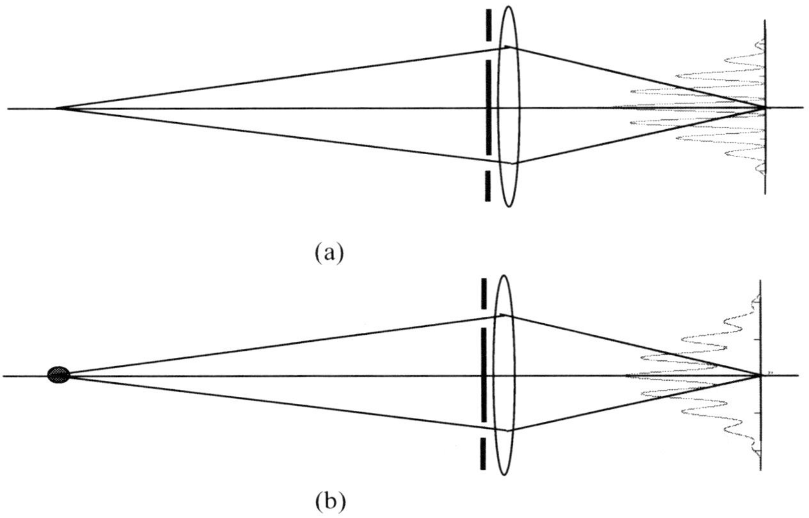

FIGURE 1. A simple representation of the visibility of an interferogram and light source size. (a) A point source yields an interferogram with a visibility of 1; (b) an ensemble of point source yields a reduced visibility.

Spatial Coherence and Beam Size

According to the van Cittert-Zernike theorem, the profile of an object is given by the Fourier transform of the complex degree of spatial coherence at longer wavelengths, as in the visible light [5][3]. Let f denote the beam profile as a function of position y, R denote the distance between the source beam and the double slit, and γ denote the complex degree of spatial coherence as a function of spatial frequency υ. Then γ is given by the Fourier transform of f(y) as follows:

$$\gamma(\upsilon) = \int f(y)\exp(-2\pi i\upsilon \cdot y)dy\,, \qquad \upsilon = \frac{2\pi D}{\lambda R}\,. \qquad (1)$$

We can measure the beam profile and the beam size via spatial coherence measurement with the interferometer.

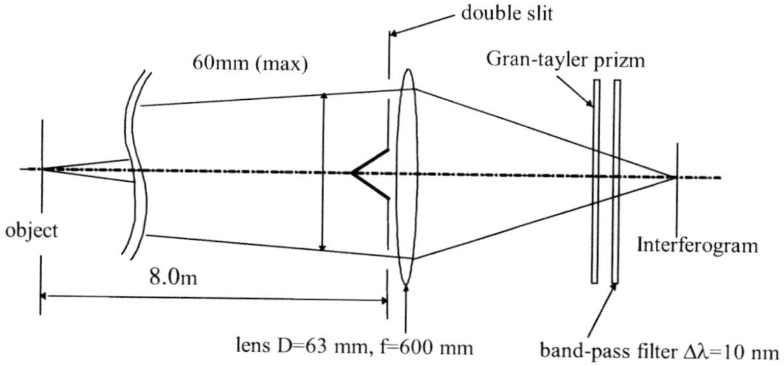

FIGURE 2. Outline of the SR interferometer.

SR INTERFEROMETER

To measure the spatial coherence of SR beams, a wavefront-division type of two-beam interferometer using polarized quasi-monochromatic rays was designed as shown in Fig. 2 [3].

In the vertical plane, the elliptical polarity of the synchrotron radiation is opposite to that in the median plane of the electron beam orbit. Therefore, there exists a π phase difference between the phases of the interferograms corresponding to the σ- and π-polarized components [2]. To eliminate the π-polarized component, we use a polarizer. With this interferometer, the intensity of the interferogram is given by

$$I(y,D)=\int\varsigma(\lambda)\cdot(I_1+I_2)\cdot\left\{\operatorname{sinc}\left(\frac{\pi\cdot a\cdot y\cdot\chi(D)}{\lambda\cdot f}\right)\right\}\cdot\left\{1+\gamma\cdot\cos\left(k\cdot D\cdot\left(\frac{y}{f}+\psi\right)\right)\right\}d\lambda$$

,

$$\gamma=\left(\frac{2\sqrt{I_1\cdot I_2}}{I_1+I_2}\right)\left(\frac{I_{max}-I_{min}}{I_{max}+I_{min}}\right),\quad \psi=\tan^{-1}\frac{S(D)}{C(D)} \tag{2}$$

where y denotes position in the interferogram, a denotes the half-height of a slit $\zeta(\lambda)$ denotes the rocking curve of the band-pass filter, and f denotes the distance between secondary principal point of the lens and the interferogram [3]. γ is the real part of the complex degree of spatial coherence (visibility of the interferogram). S(D) is the sine component and C(D) the cosine component of the Fourier transformation of the distribution function of the SR source. $\chi(D)$ in this equation represents an instrumental function of the interferometer; this term has a cosine-like dependence and comes mainly from two sources: (1) a cosine term in the Fresnel-Kirchhoff diffraction formula [3], which represents the angular dependence between the incident and diffracted light of a single slit, and (2) reduction of the effective slit height as the double slit separation D increases. This term χ is normally neglected in diffraction theory under the paraxial approximation, but we cannot neglect this term in the practical use of the interferometer. A typical interferogram observed with the SR interferometer is shown in Fig. 3.

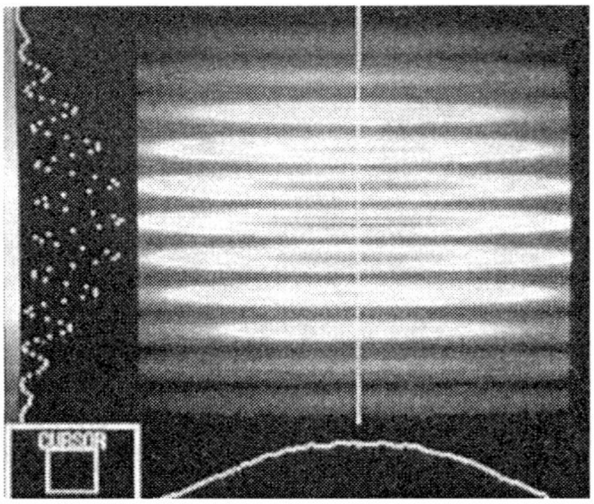

FIGURE 3. A typical interferogram observed with the SR interferometer.

THEORETICAL RESOLUTION OF THE INTERFEROMETER: COMPARISON OF THE RESOLUTION LIMITS DUE TO THE UNCERTAINTY PRINCIPLE BETWEEN IMAGING AND INTERFEROMETRY

If we assume the most restrictive class of random processes of photon statistics to be the class of ergodic random process (ensemble average is equal to temporal average), the theoretical resolutions due to uncertainty principle are as follows.

Imaging

The image is a convolution between the geometrical image and the diffraction pattern PSF. The resolution is limited by the uncertainty principle;

$$\frac{\Delta\theta}{\lambda}\cdot\Delta x \geq 1\,, \tag{3}$$

where $\Delta\theta$ and Δx are the uncertainties of the divergence of the SR and the width of the PSF. This means we need a large uncertainty $\Delta\theta$ to obtain good spatial resolution. In the case of synchrotron radiation, since the opening angle of the light is strongly limited by relativistic effects, the resolution is fully limited by this uncertainty principle. Of course, instrumental errors such as deformation of the mirror are significant too.

INTERFEROMETRY

Interferometry is a method for determining the axis of photon propagation by means of triangulation using the photon phase. Actually, an interferometer (for measuring first order coherence) is an apparatus for measuring the coherence of the photon phase. Earlier it was mentioned that the contrast of an interferogram from a single photon mode is 1. But from quantum theory, we have the uncertainty principle:

$$\Delta\phi \cdot \Delta n \geq \frac{1}{2} , \tag{4}$$

where $\Delta\phi$ and Δn are the uncertainties of the photon phase and the photon number. This means that a large phase uncertainty accompanies a small number of photons and a small phase uncertainty accompanies a large number of photons. Normally, we have a sufficient photon intensity for an interferogram measurement, and the instrumental error of the phase is much more important than the theoretical limit in the interferometry. Also, instrumental errors due to deformation of the mirror are significant in the interferometry, though we can easily correct for this particular problem as described later.

BEAM PROFILE MEASUREMENT

We can measure the beam profile by making a Fourier transform of the spatial coherence. Figure 4 shows the absolute value of the complex degree of the spatial coherence ($|\gamma|$, visibility) as measured while changing the double slit separation from 5 mm to 15 mm at the Photon Factory [2]. The resulting beam profile calculated as the Fourier transform of the spatial coherence is shown in Fig. 5.

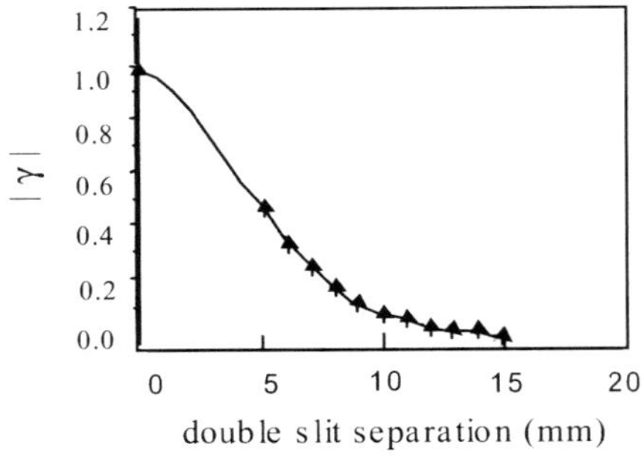

FIGURE 4. Result of $|\gamma|$ at the Photon Factory.

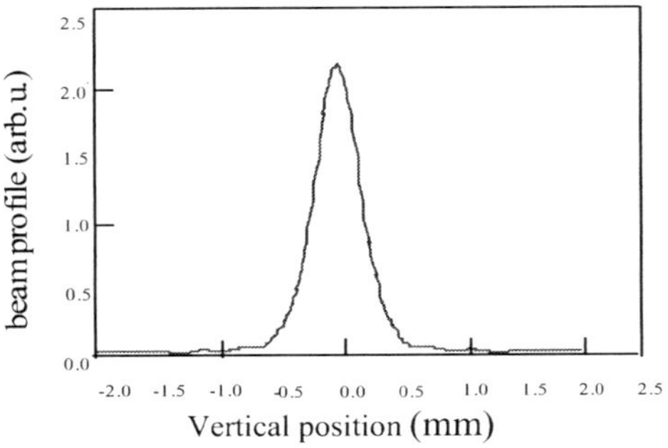

FIGURE 5. Beam profile by Fourier transformation of the spatial coherence at the Photon Factory.

TWO-DIMENSIONAL INTERFEROMETER

We can easily extend the one-dimensional SR interferometer into a two-dimensional interferometer using a 4-hole arrangement. Figure 6 shows a two-dimensional inteferogram measured at the Photon Factory.

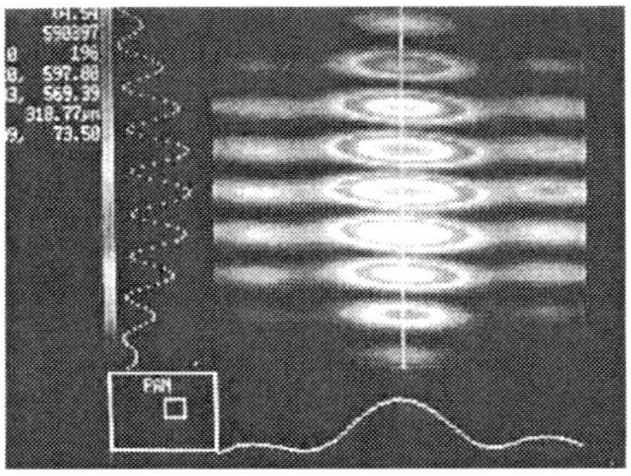

FIGURE 6. Two-dimensional interferogram measured at the Photon Factory.

With this two-dimensional interferometer, we can measure the vertical and the horizontal beam sizes simultaneously. This method also has the following demerits: (1)

we cannot easily change the parameters of interferometers, such as slit separation, in each axis independently, and (2) due to wavefront error, it is difficult to choose four good points to sample the wavefront. In conclusion, it seems convenient to use two independent interferometers.

SMALL BEAM SIZE MEASUREMENT BY MEANS OF GAUSSIAN APPROXIMATION OF BEAM PROFILE

We often approximate the beam profile with a Gaussian shape. With this approximation, we can skip any phase measurement. The Fourier transform of an even function (Gaussian) is simplified to a Fourier cosine transform. The spatial coherence is also given by a Gaussian function. We can evaluate the RMS width of the spatial coherence by using a least-squares analysis. The RMS beam size σ_{beam} is given by the RMS width of the spatial coherence curve σ_γ as follows:

$$\sigma_{beam} = \frac{\lambda \cdot R}{2 \cdot \pi \cdot \sigma_\gamma} \tag{5}$$

where R denotes the distance between the beam and the double slit. Experimentally, we must measure $|\gamma|$ (the contrast of the interferogram) as a function of slit separation D [3].

We can also measure the RMS beam size from one measurement of visibility, which is measured at a fixed separation of the double slits. The RMS beam size σ_{beam} is given by

$$\sigma_{beam} = \frac{\lambda \cdot F}{\pi \cdot D} \cdot \sqrt{\frac{1}{2} \cdot \ln\left(\frac{1}{\gamma}\right)} \tag{6}$$

where γ denotes the visibility, which is measured at a double slit separation of D [3]. With this technique, how small a beam we can measure? In Fig. 7 is shown a result of simulation. In this simulation, the measuring conditions at the ATF were assumed: the double slit separation is 50 mm, and the distance between the source point and the interferometer is 7.4 m. From this simulation, the beam size of 4 μm will give a contrast of 0.94 at a wavelength of 500 nm and 0.91 at 400 nm. A beam size of 3 μm gives a contrast of 0.97 at 500 nm and 0.95 at 400 nm. Since we can easily measure the intensity of light at 1% precision with a CCD and image processor, we can measure the difference between beam sizes of 3 μm and 4 μm with a resolution of better than 1 μm.

SMALL VERTICAL BEAM SIZE MEASUREMENT

As mentioned in the "SR Interferometer" section, if we assume a Gaussian profile, we can evaluate the beam size of the small beam based on the degree of spatial coherence. We introduce two examples of small beam size measurements in this section.

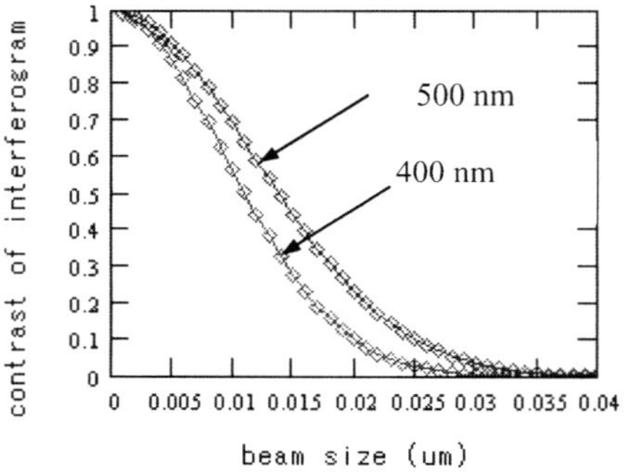

FIGURE 7. Result of simulation of contrast of interferogram as a function of beam size.

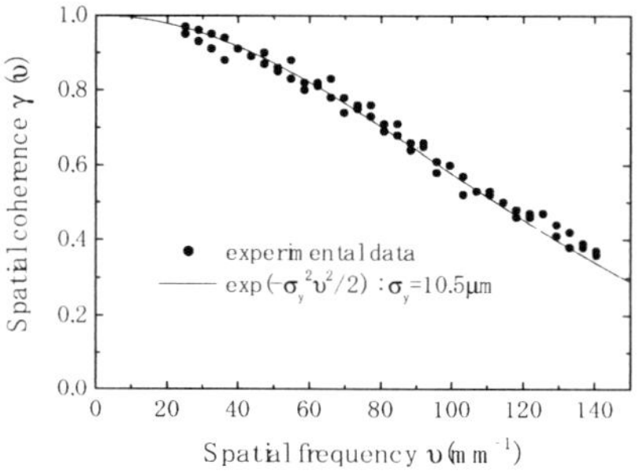

FIGURE 8. Absolute value of the complex degree of spatial coherence in the vertical direction. Dotted line denotes measured $|\gamma|$, and solid line denotes the best-fit beam size of 10.5 ± 0.6 μm.

One is the result of vertical beam size measurement at AURORA [3], and other is the vertical and the horizontal beam size measurement the ATF damping ring at KEK [3].

Figure 8 shows the result of $|\gamma|$ as a function of slit separation with a least-squares fitting by a Gaussian profile at AURORA. The beam size obtained from this fitting is 10.5 μm.

The result of measuring $|\gamma|$ as a function of slit separation in the vertical direction with a least-squares fitting by a Gaussian profile at the ATF damping ring is shown in Fig. 9. The beam size obtained from this fitting is 6.2 μm.

AUTOMATIC BEAM-SIZE MEASUREMENT AT THE KEK B-FACTORY

As described in the "SR Interferometer" section, with a Gaussan beam profile approximation again, we can estimate the RMS beam size from one measurement of the visibility of one interferogram, which is measured at a fixed separation of double slit. Using this method, we can easily measure a beam size automatically from an analysis of the interferogram taken at fixed separation of double slit D [7]. To find the visibility γ

from the interferogram, we use the standard Levenberg-Marquart method for non-linear fitting. After the image processing of the interferogram, the results are relayed to a computer in the control room for display and further analysis. Figure 10 shows an example of the display panel for the two rings of the KEK B-Factory, the HER and the LER. The interferogram, best fit curve and beam size versus beam current graphs for the vertical and horizontal directions are shown in the panel. By this automatic beam-size measurement system, we can measure the vertical and horizontal beam sizes every second, which is extremely useful for beam tuning.

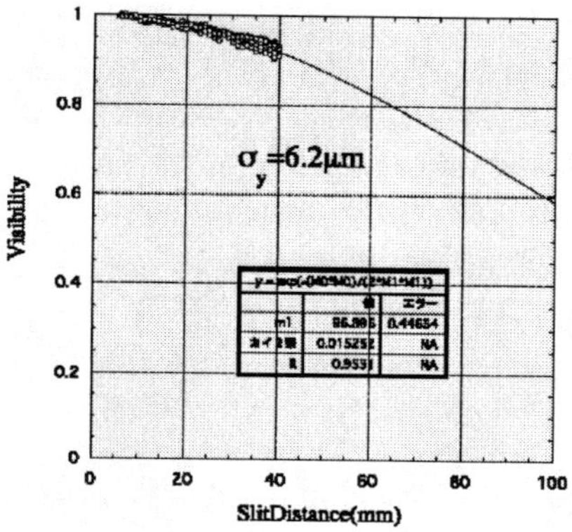

FIGURE 9. Absolute value of the complex degree of spatial coherence in the vertical direction at the ATF damping ring. The dotted line denotes the measured |γ| (visibility), and the solid line denotes the best-fit beam size of 6.2 ± 0.6 μm.

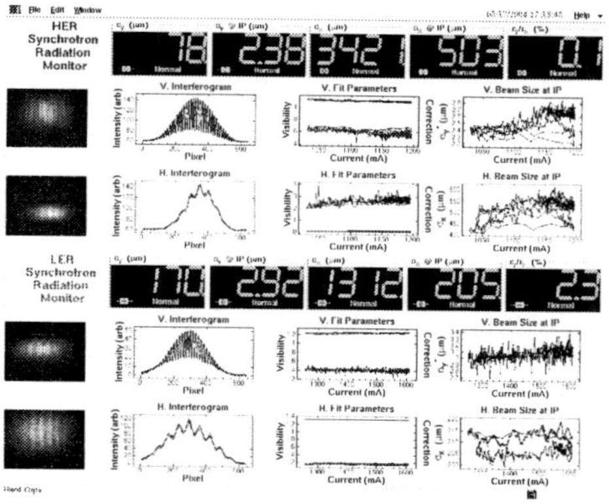

FIGURE 10. SR monitor panel in the control room, showing HER and LER vertical and horizontal beam sizes.

PRACTICAL NOTES ON THE USE OF THE INTERFEROMETER

In this section, some practical notices to use the interferometer are described. The method to correct deformation of the mirror is also described.

Linearity of the CCD Camera

Since we measure the contrast of the interferogram in the beam size measurement, the linearity of the CCD camera is very important. Here are shown some results of linearity measurements of CCD cameras. The linearity of the CCD camera is measured by using accurately calibrated neutral-density filters. The optical densities of these filters are calibrated by a spectro-densitometer to within a 0.1% error in spectrum range from 400 nm to 650 nm. Since the stability of the intensity of the SR beam is within 0.1% over a few minutes of measurement at the Photon Factory, we used an SR beam as the incident light. The measurement is performed in the intensity range from zero to the saturation level. The results of a linearity measurement for a commonly used CCD camera is shown in Fig. 11. As can be seen from this figure, the CCD camera is not always linear in the region far from saturation. The results of linearity measurement for another CCD camera are shown in Fig. 12. This camera is sold as a camera for the quantitative measurement of intensity.

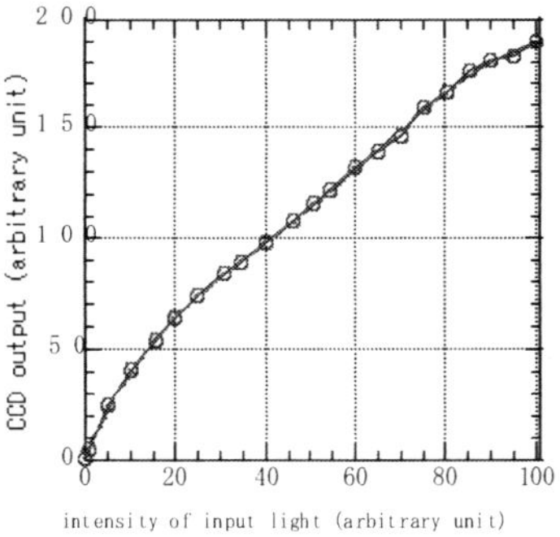

Figure 11. Linearity measurement of a commonly used CCD camera.

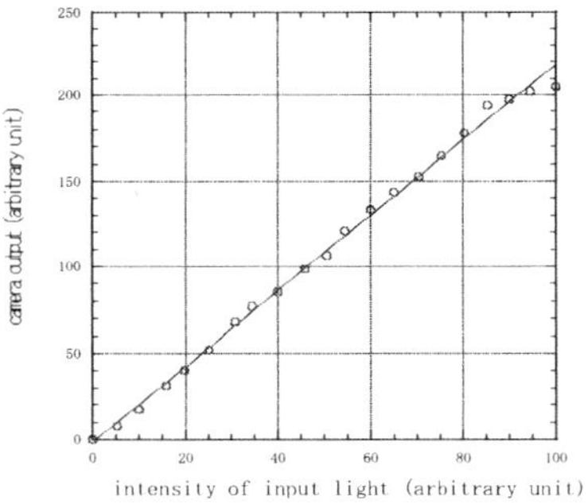

Figure 12. Linearity measurement of a CCD camera designed for quantitative measurement.

As can be seen in Fig. 11, the CCD camera which is sold for quantitative measurement has good linearity. We always check the linearity of a CCD camera before installation in the interferometer.

Effect of Floor Vibration

The vibration of the floor of the accelerator building is not negligibly small for interferometry. To see the effect of floor vibration on the interferometer, we measured the beam size as a function of exposure time of the CCD camera. The results of measurements at the Photo Factory are shown in Fig. 13.

From these data, a decrease of the beam size is observed at shorter exposure times of the CCD camera. From the view point of accelerator physics, the beam size without the effect of floor vibration is the more correct one, but an instantaneous beam size is often inconvenient for accelerator users, so we use an exposure time of 15 msec.

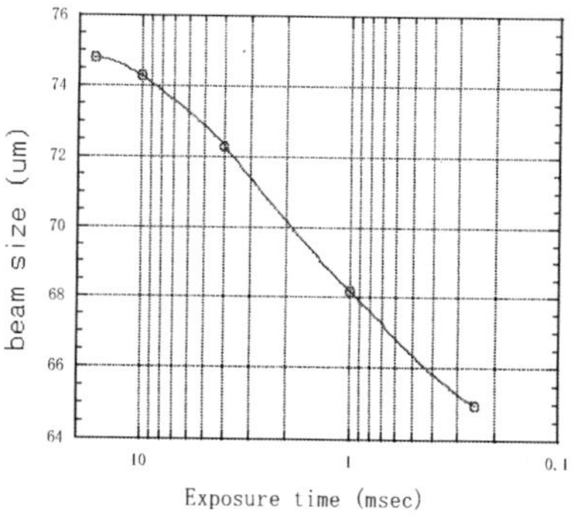

FIGURE 13. Beam size as a function of CCD exposure time.

Measurement of Deformation of SR Extraction Mirror

The mirror for extraction of the SR beam is easily deformed by the strong irradiation of SR. To measure the resulting wavefront deformation, we utilized the Hartmann screen test [8]. If the wavefront is sampled by a number of rays normal to it, the ray deviation at the observation plane can be obtained as shown in Fig. 14.

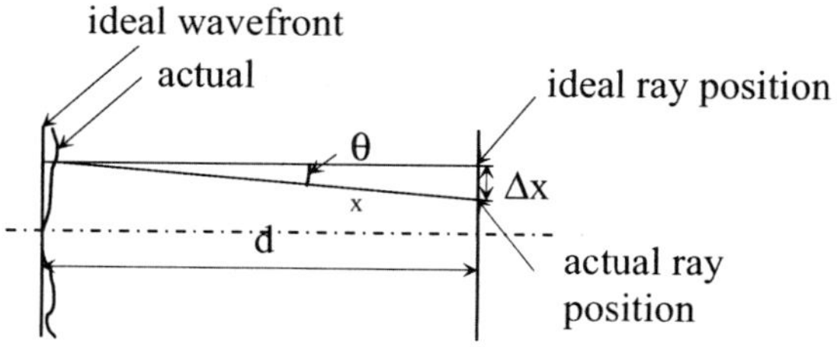

FIGURE 14. Schematics of rays and wavefronts in Hartmann test.

In Fig. 14, Δx is the value of one of the components of the ray deviation, d is the distance between the extraction mirror and the observation plane, and θ_x is the angular tilt of a portion of wavefront. The relation between the wavefront aberration W (separation between ideal wavefront and actual wavefront) and the x-component Δx of the ray deviation in the observation plane is given by

$$\frac{\partial W}{\partial x} = \frac{\Delta x}{d} \ . \tag{7}$$

Integrating this equation, we can obtain,

$$W = \frac{1}{d}\int \Delta x dx \ . \tag{8}$$

Since Δx is sampled at discrete points by the holes in a sampling screen, the integration is performed by a summation. There exist sampling screens with a number of different hole patterns [8]. We used a 100-hole square array as shown in Fig. 15. The hole diameters should be small—not so small that their diffraction patterns on the observation plane overlap each other but yet sufficiently small to permit an accurate measurement of their positions. The spacing of the holes should also be small, having a sufficient number of sampling points on the wave front but not so small that the interference fringe pattern between surrounding holes stands out. We use 1 mm for the hole diameter and 5 mm for the interval.

FIGURE 15. A 100-hole square-array screen.

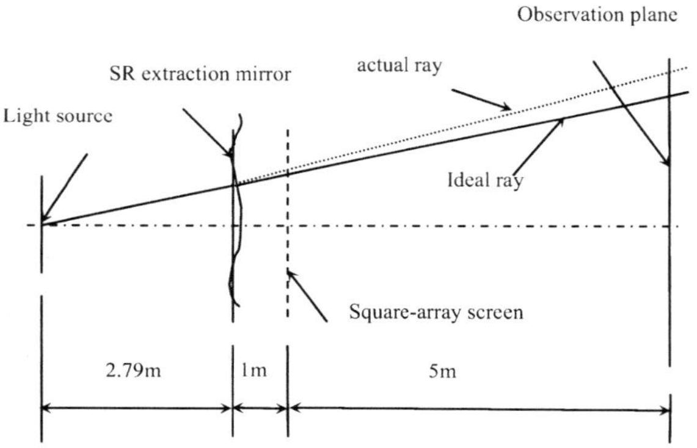

FIGURE 16. Measurement setup.

The setup of the wavefront measurement at the Photon Factory is shown in Fig. 16 [9]. Since the dimension of the stored beam is sufficiently small, we use the beam as a primary light source. Since we have a vacuum chamber system for the SR extraction mirror, we set the screen 1 m downstream from the extraction mirror. With this setup, we can measure the dot positions of the Hartmann pattern on the observation plane with 0.1-mm resolution, and so we can measure the wavefront with λ/6 (here, λ is 633 nm) precision. A typical Hartmann pattern observed at the Photon Factory is shown in Fig. 17; the reconstructed wavefront error is also shown.

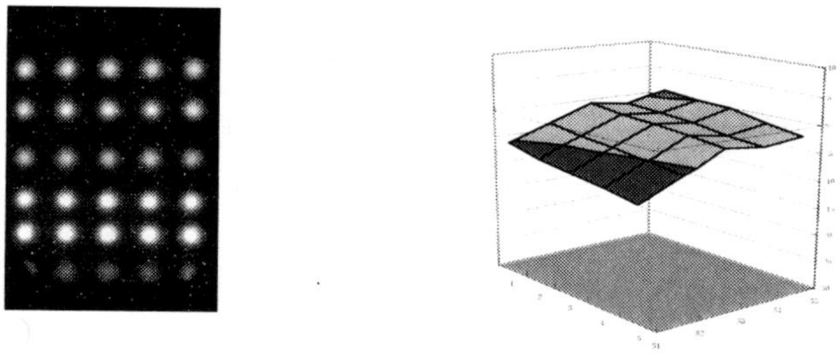

FIGURE 17. Typical Hartmann pattern observed at the Photon Factory and reconstructed wavefront. The central part of the mirror expands from irradiation by SR. The beam current is 230 mA.

Calibration of SR Interferometer Using a Single-Hole Screen

The actual separation of the double slit of the interferometer is changed by deformation of the mirror. We must find the ideal separation between the two rays at the location of the double slit for the calibration of the SR interferometer. For this purpose, we utilize a single-hole screen to trace the actual rays [9]. A setup at the Photon Factory is shown in Fig. 18.

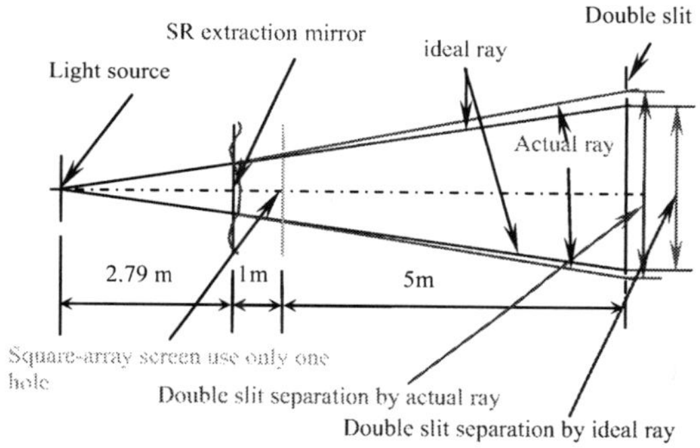

FIGURE 18. Setup of determination of true double slit separation by scanning a single-hole screen.

First, we scan the single hole screen to find the two actual rays which pass through the two slits of the SR interferometer. From this scan, we can find the separation of the two actual rays on the single-hole screen. Since the single-hole screen is located near the deformed mirror, we can easily find the ideal ray propagation by the use of geometrical optics analysis. Finally, we can find the ideal separation of the ideal rays D_{ideal} on the double slit of the interferometer. Then we can obtain the calibration factor ξ of the interferometer by using the separation of the ideal rays, D_{ideal}, and the separation of the actual rays, D_{actual} :

$$\xi = \frac{D_{actual}}{D_{ideal}} \quad . \tag{9}$$

The beam size is given by

$$\sigma_{beam} = \xi \cdot \frac{\lambda \cdot F}{\pi \cdot D_{actual}} \cdot \sqrt{\frac{1}{2} \cdot \ln\left(\frac{1}{\gamma}\right)} \tag{10}$$

CONCLUSIONS

During the twelve years of development of the SR monitor, the SR interferometer was developed to measure the spatial coherency of the visible region of the SR beam, and we have demonstrated that this method is capable of measuring the beam profile based on the van Cittert-Zernike theorem. Using a Gaussian beam profile approximation, we can measure μm-range very small beam sizes with a resolution of less than 1 μm. With an automatic analysis system, the SR interferometer is conveniently used as a beam size monitor. The measuring interval is about 1 sec. A calibration method for the SR interferometer system having extra focusing power, such as a deformed mirror, was also developed to measure the absolute value of the beam size.

ACKNOWLEDGMENTS

I would like to thank all KEK colleagues who were involved in the developments of SR monitors at the KEK. I would like to thank Prof. Emeritus Kobayakawa of KEK for strongly encouraging me to pursue the idea of beam size measurement with an interferometer. I would like to thank Prof. Iwasaki of Ritsumeikan University, who invited me the SR Center to measure the beam size. I would like to thank Professors Kurokawa and Hiramatsu of KEK, who gave me a chance to participate in the design and construction of the B-Factory SR monitors. J. W. Flanagan developed an excellent automatic analyzing system for the interferogram, and he also organized the installation of the SR monitors in the B-factory tunnel. T. Naitoh organized the installation of the SR interferometer in the ATF damping ring. Thanks are also due to M. Tadano for his technical support of the system and for writing very useful application software.

Last but not least I would like to thank the BIW2004 organizing committee for selection of this work for the Faraday Cup Award of 2004.

REFERENCES

1. T. Mitsuhashi and M. Katoh, Proc. 5th European Particle Accelerator Conference, 1669, (1996), Sitges, Barcelona.
2. T. Mitsuhashi, Proc. of 1997 Particle Accelerator Conference, 766, (1997), Vancouver, Canada.
3. T. Mitsuhashi, "Beam Profile and Size Measurement by SR Interferometer," in Beam Measurement, Ed. by S. Kurokawa et al., pp. 399-427, World Scientific (1999).
4. H. Fizeau, *C. R. Acd. Sci. Paris*, 66, 934 (1868).
5. P. H. van Cittert, Phisica, 1, 201 (1934). Also see M. Born and E. Wolf, "Principles of Optics," chapter 10, P491, Pergamon Press, 1980.
6. A. A. Michelson, *Astrophsyc.* J51, 257 (1920).
7. J. W. Flanagan, S. Hiramatsu, and T. Mitsuhashi, Proc. 7th European Particle Accelerator Conference, 1714, (2000), Vienna, Austria.
8. I. Ghozeil "Hartmann and Other Screen Tests" in *Optical Shop Testing*, Ed. by D. Malacara, pp. 367-396, Wiley-International, 1992.
9. T. Mitsuhashi and M. Tadano, Proc. 2nd Asian Particle Accelerator Conference. See also J. W. Flanagan, S. Hiramatsu, and T. Mitsuhashi, Proc. 13th Symposium on Accelerator Science and Technology, (2001) pp.381-383, Japan.

Tutorials

11th Beam Instrumentation Workshop
May 3–6, 2004

Interceptive Beam Diagnostics—Signal Creation and Materials Interactions

Michael Plum

Los Alamos National Laboratory, Los Alamos, New Mexico, USA
and
Spallation Neutron Source, Oak Ridge National Laboratory, P.O. Box 2008, Oak Ridge, TN, USA

Abstract. The focus of this tutorial will be on interceptive beam diagnostics such as wire scanners, screens, and harps. We will start with an overview of the various ways beams interact with materials to create signals useful for beam diagnostics systems. We will then discuss the errors in a harp or wire scanner profile measurement caused by errors in wire position, number of samples, and signal errors. Finally we will apply our results to two design examples–the SNS wire scanner system and the SNS target harp.

INTRODUCTION

What happens when a material is inserted into a particle beam? There are many different effects, and each one has been used in some way in a beam diagnostics system. I've tried to assemble a complete list of all possible interactions below:

1. Beam scattering. There are several categories of scattering—multiple, large-angle Coulomb, and nuclear scattering. An example beam diagnostic that takes advantage of scattering is the wire scanner profile measurement, which measures beam loss as a function wire position to determine the beam profile.
2. Beam energy loss. An example beam diagnostic that takes advantage of energy loss (and the consequent thermal heating) is a beam profile measurement utilizing wire temperature (and therefore change in resistance) vs wire position.
3. Secondary electron emission (SEM). An example is beam profile measurement using the SEM wire signal vs wire scanner position.
4. Optical transition radiation (OTR). An example is beam profile measurement by imaging the OTR.
5. Ĉerenkov radiation. An example is particle identification in particle and nuclear physics experiments.
6. Scintillation and fluorescence. An example is profile measurement utilizing light from a view screen.
7. Gas ionization (from residual or injected gas). An example is a beam profile measurement by measuring the quantity of ions collected as a function of position.
8. Beam absorption. If the beam energy is low enough, the beam may stop in the material. An example is a profile measurement using the wire signal, due to charge deposition, as a function wire scanner position.

CP732, *Beam Instrumentation Workshop 2004: Eleventh Workshop,*
edited by T. Shea and R. C. Sibley III

9. Beam stripping. If it is an ion beam, some or all of the electrons may be stripped from the beam. An example is the laser profile monitor, which measures the quantity of electrons stripped from an H^- beam as a function of laser beam position.
10. Secondary particle creation, including delta rays. An example is profile measurement using secondary flux (i.e., a loss monitor signal) as a function wire position.
11. Nuclear interactions. An example is transmutation, where a beam profile is measured utilizing foil activation.

You may have noticed that most of the examples cited above involve beam profile measurement. This is because in most cases we can use non-interceptive methods to measure the other beam parameters such as beam position and current. Usually when an interceptive monitor is chosen over a non-interceptive monitor it is because the signal levels are to low to be practical for the non-interceptive monitors. Let us now go into a little more detail for some of the more important beam-material interaction effects.

THE BETHE-BLOCH EQUATION

Practically any discussion about the interaction of charged particles heavier than electrons with matter will somehow involve the Bethe-Bloch equation [1] for the mean rate of energy loss:

$$-\frac{dE}{dx} = Kz^2 \frac{Z}{A}\frac{1}{\beta^2}\left[\frac{1}{2}\ln\frac{2m_e c^2\beta^2\gamma^2 T_{\max}}{I^2} - \beta^2 - \frac{\delta}{2}\right], \tag{1}$$

where $K = 0.3071$ MeV cm^2, z is the atomic number of the incident particle, Z and A are the atomic number and mass of the material, m_e is the mass of the electron, β and γ are the usual relativistic factors, I is the mean excitation energy, and δ is for the density effect correction. T_{max} is the maximum kinetic energy that can be imparted to a free electron in a single collision. For an incident particle of mass M,

$$T_{\max} = \frac{2m_e c^2\beta^2\gamma^2}{1 + 2\gamma m_e / M + (m_e / M)^2}, \tag{2}$$

In the low energy approximation ($2\gamma m_e/M << 1$) T_{max} simplifies to

$$T_{\max} = 2m_e c^2\beta^2\gamma^2 \text{, and} \tag{3}$$

$$-\frac{dE}{dx} = Kz^2 \frac{Z}{A}\frac{1}{\beta^2}\left[\ln\frac{2m_e c^2\beta^2\gamma^2}{I} - \beta^2 - \frac{\delta}{2}\right] \left[\frac{\text{MeV}}{\text{gm/cm}^2}\right] \tag{4}$$

A plot [1] of dE/dx for muons, pions, and protons incident on various materials is shown in Fig. 1. In the case of protons, for example, the minimum loss rate occurs at about 3 GeV/c (or 2.2 GeV kinetic energy).

An important consideration when designing an interceptive beam diagnostic is thermal heating due to energy loss from the beam. Of course we don't want the material to melt or change its properties in some way that could affect our measurement. The thermal heating depends on both the range and incident energy of the particle beam. From Fig. 1 we see that at low energies the dE/dx is high and the range is small, so the beam can deposit all its energy into the material. The highest thermal heat load for typical wire scanner and harp diagnostics is where the beam just stops in the wire. At the SNS, where we use 32-micron-diameter carbon wires, the protons just stop when their energy is about 1.5 MeV. The wire scanners we have closest to that energy are in the Medium Energy Beam Transport (MEBT), where the beam energy is 2.5 MeV. In this area we must limit the beam duty factor to 50-μs-long pulses at a few Hz to avoid damaging the wires. In the other wire scanner locations the restrictions are not as severe.

For thin materials and high beam energies, it is more accurate to use the restricted energy loss equation [1], which takes into account the energy that escapes due to knock-on electrons. This effect can substantially reduce the energy deposition and therefore the thermal heating [2,3]. For proton beam energies less than about 100 MeV, this effect is negligible; at about 1 GeV proton kinetic energy this effect will cause about a 20% reduction in heating for typical wire and foil thicknesses, and at higher beam energies the effect is increasingly greater.

MULTIPLE SCATTERING

Multiple scattering is a series of small-angle kicks to a charged particle beam caused primarily by Coulomb interactions, and to a lesser extent by nuclear interactions. The scattering angle distribution [1] is roughly Gaussian with a width

$$\theta_{space}^{rms} = \frac{\sqrt{2}\cdot 13.6\,\text{MeV}}{\beta c p}\, z \sqrt{x/X_0}\left[1 + 0.038\ln\left(x/X_0\right)\right], \tag{5}$$

where p, βc, and z are the momentum, velocity, and charge number of the incident particle, and x/X_0 is the thickness of the scattering medium in radiation lengths. The radiation length, X_0, is the distance an electron travels before it loses $1/e$ of its energy. For carbon, $X_0 = 18.8$ cm; for copper, $X_0 = 1.43$ cm; and for tungsten, $X_0 = 0.35$ cm.

The main points for our purpose is that the scattering angle increases as the incident particle velocity decreases, as the charge of the incident particle increases, and as we move to heavier elements in the periodic table. This can be an important consideration when designing a beam-intercepting diagnostic. For example, when designing a wire scanner system, a carbon wire would be preferred if it was desirable to minimize beam loss during a scan. On the other hand, if the wire scanner system relies on detecting beam loss vs wire position, a tungsten wire may be preferred.

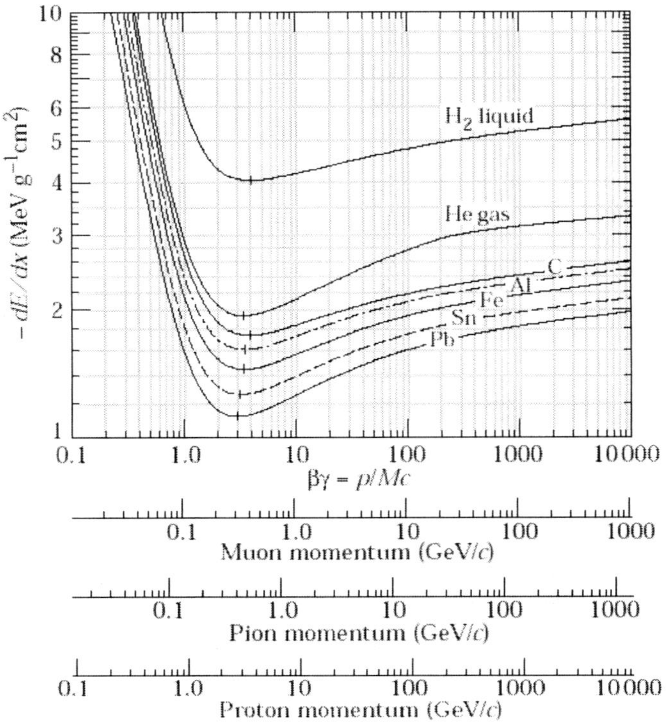

FIGURE 1. Stopping power graph for muons, pions, and protons in various materials.

OPTICAL TRANSITION RADIATION

Optical transition radiation (OTR) is created by a charged particle traveling across the boundary between two media with different dielectric properties [1]. The use of OTR for beam diagnostics is relatively new, and it has quickly grown in popularity and range of application in the last 10 years.

The OTR radiation has an angular distribution with a strong peak at angle $\theta = 1/\beta\gamma$ relative to the incoming beam velocity vector. The OTR process is an inefficient one, with an intensity proportional to γ^2, so practical applications have been limited to electron beams with kinetic energy greater than a few tens of MeV, and protons with kinetic energy greater than a few tens of GeV.

Transition radiation is also polarized. By separately measuring the polarization components parallel and perpendicular to the plane of incidence formed by the velocity vector of the incoming particle and the normal to the foil, the horizontal and vertical beam divergences can be separately measured. Together with the beam spot size measurement (also using OTR), and by focusing the beam down to a waist at the OTR foil, this provides a nice measurement of beam emittance.

An extension of OTR is OTR-interference (OTRI), obtained by placing two foils in the beam. In this case the radiation creates interference patterns that allow a more sensitive measurement of the beam divergence, and therefore a more accurate beam emittance measurement. The interference pattern fringe positions also provide a measurement of the beam energy. For example, a divergence error of 30% and a beam energy measurement error of 1% was obtained by Fiorito and Rule [4] with a 107-MeV electron beam.

Transition radiation is emitted practically exactly at the moment when the particle enters the OTR foil. This convenient property has been successfully employed to measure longitudinal profiles of sub-picosecond electron bunches [5].

OTR does of course have some limitations. These include beam scattering in the foil, which may interfere with the accuracy of the OTR-interference method or with beam operations; foil melting in high-intensity beams; and the low OTR intensity, particularly for non-electron (e.g., proton) beams with less than a few GeV kinetic energy.

SCINTILLATION AND FLUORESCENCE

Also in the optical interaction category we have scintillation or fluorescence (the difference between the two does not seem to be rigorously defined), from scintillators, view screens, and residual or injected gas, and from scintillators. To the best of my knowledge, plastic, crystal, and glass scintillators are not commonly used at accelerator facilities to interceptively measure primary beam properties. Their application is usually reserved for secondary beams or for particle detectors in the experiments that use the accelerated beam. View screens, such as phosphor-coated materials and chromium-doped aluminum oxide, however, are commonly used to measure beam profiles. The gas scintillation profile monitor is an interesting example of light-creating beam diagnostic.

In a gas scintillation profile monitor, gas molecules in the beam pipe, from either residual or injected gas, are excited by the passing particle beam. The molecules may or may not be ionized, but in any case electrons are promoted to excited states. When the electrons fall to lower energy orbits, photons are emitted. If the molecules have not traveled too far from the points where the interactions with the particle beam took place, the photons can be collected to produce an accurate measure of the beam profile. If the emission lines are in the visible spectrum, simple glass lenses can be used to focus the light onto detectors. The perfect gas will have large cross sections in the visible region and short excited-state lifetimes. If gas injection into the beam pipe is necessary, the gas should also be easily pumped by the vacuum system.

Nitrogen gas is commonly used for gas scintillation profile monitors because of its large cross sections in the visible range, the short (60 ns) excited state lifetimes, and because most vacuum systems efficiently pump N_2 gas. The intensity of the gas scintillation depends on the gas density, the beam current, and the beam energy. For the case of N_2 gas, it takes about 3.6 keV energy loss [6] to create a visible photon. From the Bethe-Bloch stopping power equation, we see that for a given set of beam parameters, the beam will deposit significantly more energy into a gas at low beam energies, and it is for this reason that most gas scintillation profile monitors are found at the low energy ends of particle accelerators.

GAS IONIZATION

Closely related to electron stripping and gas scintillation is gas ionization. Many non-intercepting profile monitors are based on measuring the ions and/or electrons created in the residual (or injected) gas by the passage of the beam. By guiding the ions and/or electrons to detectors using electric and magnetic fields, the beam profile can be measured. It takes between 25 and 35 eV of energy to ionize a gas molecule, and the number of ions and electrons created can be simply calculated using the Bethe-Bloch equation. To make a simple example and to give a feeling for the numbers, if we assume a 1-GeV, 10-mA proton beam passing through 10^{-7} Torr of N_2 gas at 20°C that has 35 eV ionization energy, and a detector length of 10 cm, the ionization current (i.e., ion pairs created per second) is

$$i_{ion} = \frac{(dE/dx)\rho t}{I} i_{beam} , \tag{6}$$

where I is the ionization energy, ρ is the mass density of the gas, t is the thickness or length of the gas in the sensitive volume of the detector, and i_{beam} is the beam current. From Fig. 1 the stopping power is 1.8 MeV-cm^2/g, so

$$\begin{aligned}
\rho &= \left(\frac{28\text{ g}}{\text{mole}}\right)\left(\frac{\text{mole}}{22.4\text{ liter}}\right)\left(\frac{\text{liter}}{1000\text{ cm}^3}\right)\left(\frac{10^{-7}\text{ Torr}}{760\text{ Torr}}\right)\left(\frac{273\ ^{o}K}{293\ ^{o}K}\right) = 1.53\text{ x }10^{-13}\text{ g/cm}^3 \\
i_{beam} &= (0.01\text{ Amp})\left(\frac{\text{coul/sec}}{\text{Amp}}\right)\left(\frac{\text{proton}}{1.602\text{ x }10^{-19}\text{ coul}}\right) = 6.24\text{ x }10^{16}\text{ proton/second} \\
i_{ion} &= \left(\frac{1.8\text{ x }10^{6}\text{ eV}}{\text{g/cm}^2}\right)\left(\frac{\text{ion pair}}{35\text{ eV}}\right)(1.53\text{ x }10^{-13}\text{ g/cm}^3)(0.1\text{ m})(6.24\text{ x }10^{16}\text{ proton/second}) \\
i_{ion} &= 4.91\text{ x }10^{7}\left(\frac{\text{ion pairs}}{\text{sec}}\right)
\end{aligned} \tag{7}$$

In some applications, one may have the option of a residual gas ionization profile monitor or a gas scintillation profile monitor. I believe that the residual gas ionization profile (IPM) monitor is better because (1) the number of ion pairs created is higher than the number of photons created. Both scale with dE/dx, but the ion pair creation mechanism is more efficient than the photon creation mechanism (35 eV/ion vs 3,600 eV/photon for nitrogen). (2) Almost all the ions are collected in an IPM, but only a small fraction (<10%) of the photons are typically collected in a gas scintillation profile monitor. (3) Ion and electron detectors are more efficient than photon detectors. In other words, for a given gas density and detector length, IPMs are at least three orders of magnitude more efficient. For most applications a gas scintillation profile monitor would be preferred over an IPM only at low beam energies where copious photons are produced.

ELECTRON STRIPPING

Electron stripping is an important issue for ion beams. It takes very little material to strip electrons from an ion beam. In synchrotrons and long linacs, just the residual gas pressure alone can be enough to cause excessive electron stripping. For example, for H^- ion beams the total stripping cross section (i.e., both electrons are stripped) for carbon, nitrogen, or oxygen is [7]

$$\sigma = \frac{7x10^{-19}\ \text{cm}^2}{\beta^2} \quad \text{per gas atom, carbon or nitrogen or oxygen.} \tag{8}$$

The number of ions stripped is then

$$n = \sigma \rho_N F L \ , \tag{9}$$

where σ is the stripping cross section, ρ_N is the numerical density, F is the flux, and L is the length of the material (or gas). At 1000 MeV, for example, it takes a carbon foil of just 400 μg/cm^2 (2.2 microns thick) to strip 98% of the H^- ions to H^+. Down at 2.5 MeV it takes just 15 nm of carbon to strip the electrons. Stripping just one electron from an H^- ion takes even less material.

The net result is that just about any material used for an interceptive beam diagnostic device will strip electrons from an ion beam. Sometimes this can be used to advantage, for example, in the SNS laser profile monitor system [8], where the beam profile is measured by scanning a laser beam through the H^- beam. The laser strips the H^- beam to H^0, and the stripped electrons are diverted into a detector. The current in the detector is plotted as a function of the laser position to get the beam profile very much like a wire scanner system.

Sometimes electron stripping can be a real problem. For example, when the CERN PS accelerates lead ion beams, introducing even a small amount of gas for a gas scintillation profile monitor strips the Pb ion beam enough to prevent operations. In high-intensity machines such as the SNS, if the residual gas pressure rises above approximately 10^{-8} Torr, the beam loss due to H^- beam stripping will rise above acceptable levels.

An often unappreciated aspect of electron stripping is the amount of energy the stripped electrons can deposit in a material. At beam energies above a couple hundred MeV the protons in an H^- beam pass through thin materials without depositing much energy. But the electrons stripped from the H^- beam can stop in the material and deposit all their energy. The electron range, i.e., the distance a stripped electron will travel in a material, can be estimated with Feather's rule [9] suitably modified [10] for materials other than aluminum:

$$R(E) = 412\left(\frac{13}{27}\right)\left(\frac{A}{Z}\right)E^n \quad \left[\frac{\text{mg}}{\text{cm}^2}\right] , \tag{10}$$

where $n = 1.265 - 0.0954 \ln(E)$, A and Z are the atomic mass and number, and E is in MeV. Shown in Fig. 2 is a calculation of energy deposition in a 25-micron-thick tungsten foil. The worst case is at about 350 MeV incident H^- beam energy, where the electrons just barely stop in the material and deposit about six times more energy than the protons. Figure 3 shows the range and stopping power [11] for electrons in carbon as a function of energy.

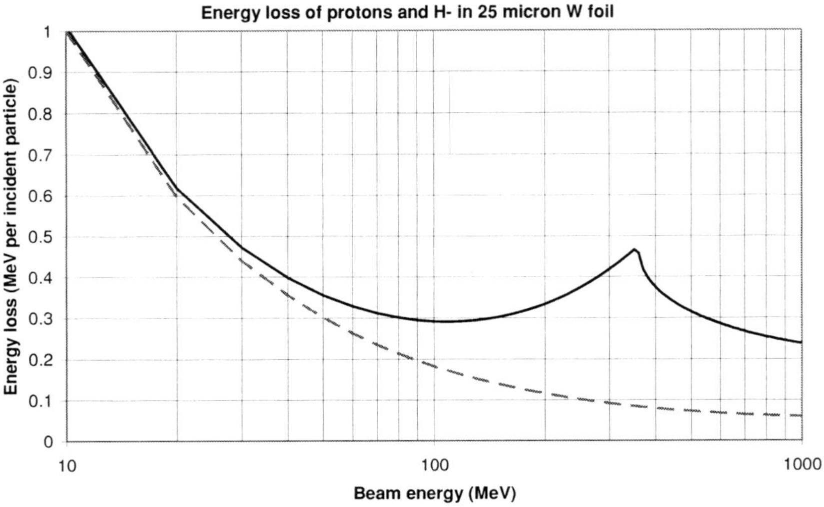

FIGURE 2. Energy loss, as a function of beam energy, per incident H^+ (dotted line) or H^- (solid line) particle incident on a 25-micron thick tungsten foil.

SECONDARY ELECTRON EMISSION

When a beam enters or exits a material, secondary electrons are created. This is a surface phenomenon, and the amount of secondary electrons created is very sensitive to the condition of the surface. Oxidation, surface cleanliness, and monolayers of contaminants can significantly change the secondary electron emission (SEM) coefficient.

A well-known theory from Sternglass [12] describes the secondary emission yield as

$$Y = \frac{P\,d_s}{E^*}\frac{dE}{dx} = (0.002\ \text{cm/MeV})\frac{dE}{dx}\ , \qquad (11)$$

where Y is the secondary yield, P is the probability of an electron escaping, d_s is the average depth from which secondaries arise, E^* is the average kinetic energy lost by the incoming particle per ionization produced in the target, and dE/dx is the stopping power of the target. Values suggested by Sternglass are $P \approx 0.5$, $d_s \approx 1$ nm, and

$E^* \approx 25$ eV. If the particle beam passes through the material then secondary electrons are emitted from both the entrance and exit surfaces. Note that for the case of a round wire the number of particles intercepted by the wire is calculated from the wire diameter, but the SEM signal is calculated from the wire circumference. Secondary electron emission is like a 1-nm-thin ion chamber fastened to the surface of the wire (or foil), with a ionization energy of 25 eV (much like a gas) and a collection efficiency of 50%.

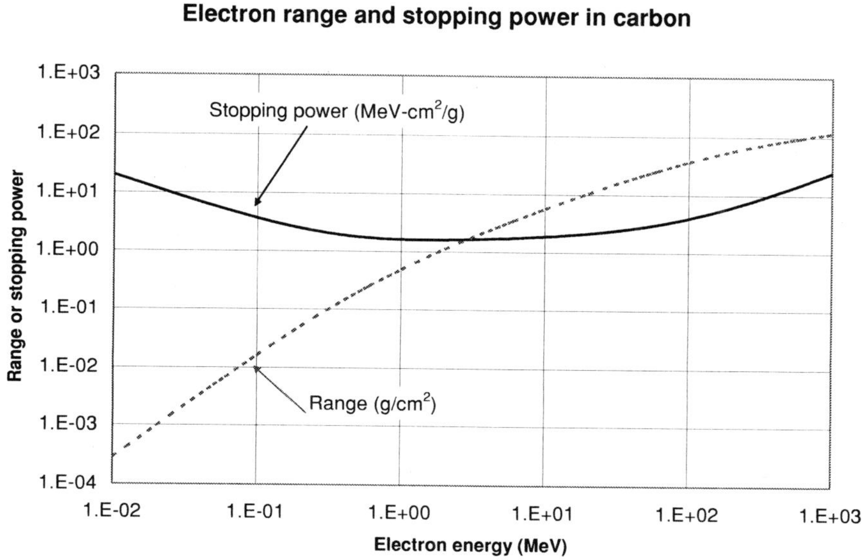

FIGURE 3. Range (dotted line) and stopping power (solid line) for electrons incident on carbon, as a function of electron energy.

The SEM signal computation therefore reduces to calculating stopping powers and ranges for the beam particles. The Bethe-Bloch equation and Feather's Rule predict precise ranges for the proton and electron, but in reality there is straggling and scattering that leads to imprecision. For the case of a wire of diameter D, the calculation can be simplified by assuming a uniform thickness equal to the average thickness $t = \pi D/4$.

For the case of an H^- beam the SEM process can be modeled by considering the H^- particle as a single particle entering the wire, whereupon it fragments into two electrons and a proton. Any electrons or protons that stop in the wire or exit the wire should be considered separately. The signal will thus comprise several sources: (1) the SEM caused by the H^- particle entering the wire, (2) any electrons or protons that stop in the wire, and (3) SEM caused by any electrons or protons exiting the wire.

A simple computer program [13] can compute and add up the various signal contributions. A sample plot of the calculated signal strength as a function of beam energy for the case of a 0.032-mm-diameter carbon wire in the center of 26 mA,

0.2 cm rms H^- beam is shown in Fig. 4. Results for other beam currents and sizes can be scaled from these values.

We have tested [14] this model of signal generation for wires in H^-, H^0, and H^+ beams by comparing model predictions to observations made at the Los Alamos Neutron Science Center (LANSCE) and other facilities. The model agrees quite well for absolute measurements made with H^+ beams and relative measurements made with H^-, H^0, and H^+ beams.

SEM coefficients have been observed [15] to change by up to 50% as target materials are exposed to beams. The values predicted by the computer program should therefore be given at least a ±50% error bar just for the degradation effect. Additional error sources are the approximations that the wire has a uniform thickness and that the H^-, proton, and electron ranges are precise. These latter effects will tend to smooth out the sharp peaks in the plots of signal strength as a function of beam energy.

WIRE SCANNERS AND HARPS—ACCURACY OF THE BEAM SIZE MEASUREMENT

The error on a beam size measurement, when measured with a wire scanner or a harp, depends [13] on several parameters: (1) the error in positioning the wire, (2) the absolute error on the signal amplitude measurement, (3) the relative error on the signal amplitude measurement, (4) the number of positions used for the scan, and (5) the beam jitter. The error also depends on the data analysis method, e.g., if a fitting function is used or if the width is directly computed from the data, as in an rms measurement. In most real-life situations it is best to fit the data with a function, such as a Gaussian, that accurately describes the data.

To get a feeling of how these factors influence the beam size measurement, we use as an example the SNS wire scanner system. We begin by simulating a wire scan by sampling a simulated 40,000 particle, 1.80-mm rms beam [16], shown in Fig. 5, with a variable number of equally spaced wire positions between ±1 cm. Typical rms beam sizes in the linac range from 0.05 to 0.25 cm. The signal, y_i, for each wire position is assigned to be all the particles within ±0.01 cm of the wire center. An error is assigned to each y value:

$$\varepsilon_{y_i} = \left[(eabs)^2 + (erel \cdot y_i)^2\right]^{1/2}, \tag{12}$$

where *eabs* is the absolute error (e.g., electronic noise), and *erel* is the relative error (e.g., non-linearity in an amplifier circuit).

The wire position error, δx, is folded into the overall error, ε_i, by computing $\varepsilon_x = f'\,\delta x$, where f' is the derivative of the fitting function, and then $\varepsilon_i = \sqrt{\varepsilon_{x_i}^2 + \varepsilon_{y_i}^2}$.

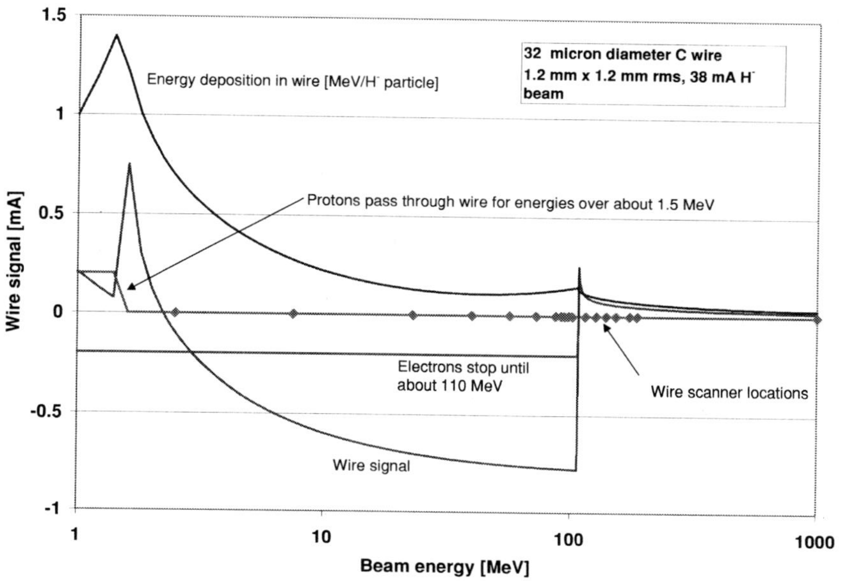

FIGURE 4. Plot of calculated wire signal vs beam energy for a 0.032 mm dia. carbon wire. The wire scanner signal is in units of mA. The line for electrons stopping, when non-zero, indicates that electrons stop in the wire (in this case for energies less than about 108 MeV). The line for protons stopping, when non-zero, indicates that protons stop in the wire (in this case for energies less than about 1.5 MeV). Energy deposition is also plotted, in units of MeV per H^- particle hitting the wire.

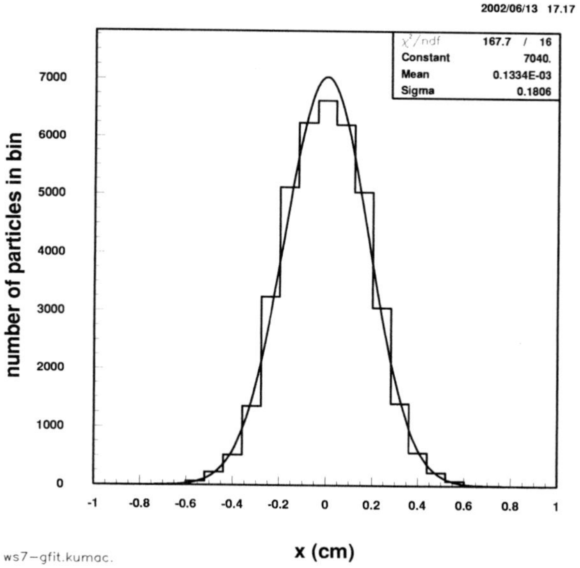

FIGURE 5. A histogram of a simulated beam the SNS DTL tank 1, from a 40,000 particle PARMILA model calculation.

We define the error in the fitted width to be *abs[(true width) – (fitted width)] + (error in fitted width)*, where *true width* is the rms beam size of 1.80 mm, and *abs* denotes the absolute value function. A data analysis package (PAW [17]) script then steps through the above procedure for a range of numbers of wire positions. Realistic values chosen for the errors are *eabs* = 2% of the peak number of counts, *erel* = 0.01, and $\delta x = 0.01$ cm. The result is shown in Fig. 6. It is evident from the figure that once the beam is sampled at more than 12 positions the error is minimized. Twelve positions across 2 cm corresponds to 0.18 cm between sample positions, or just one rms beam width. In other words, if you know in advance and have confidence in the shape of your beam, then surprisingly few samples are needed to accurately measure the beam size.

Next we scan through a range of errors on the wire position, for three different beam sizes, keeping *eabs* = 2% of the peak signal, *erel* = 0.01, and scaling the wire position spacing to be equivalent to 15 positions equally spaced across ±1 cm for a 1.80 mm rms beam. The result is shown in Fig. 7. We see that there is a roughly linear relationship between the wire position error and the resultant error in the width measurement. We also see that, as expected, a given error in the wire position affects smaller beams more than larger beams.

Finally we scan through a range of absolute and relative signal errors for the 1.80-mm rms beam, while keeping the number of wire positions fixed to 15 and assuming no error in the wire positions. These results are shown in Fig. 8. We see, for example, that once the absolute error reaches about 5% of the peak signal, relative errors up to 15% do not contribute significantly.

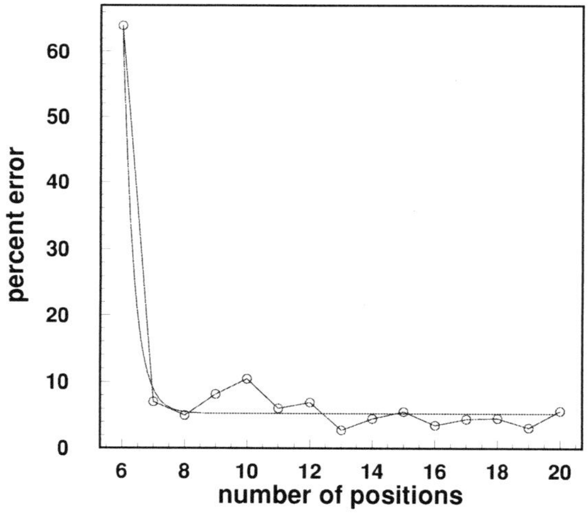

FIGURE 6. Percent error in the width measurement vs number of wire positions. The smooth solid line is a fit to the data points. The absolute error is about 2% of the peak signal, the relative error is 1% of the signal, and the wire position error is 0.01 cm.

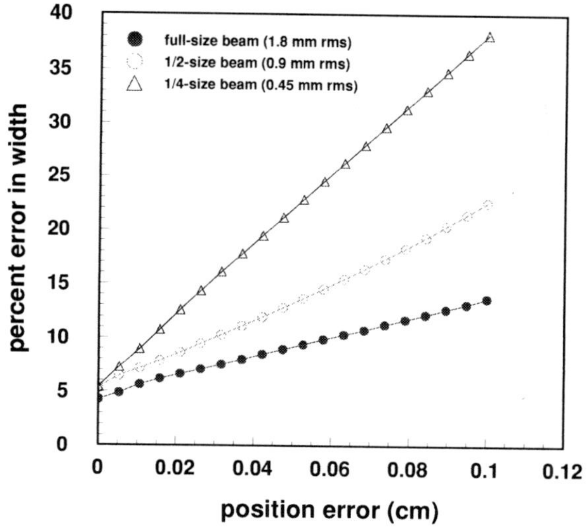

FIGURE 7. Error in the beam width measurement as a function of wire position error, for three different beam sizes, assuming an absolute error equal to 2% of the peak signal and a relative error of 1%. Full size beam = 1.8 mm rms, half size = 0.9 mm rms, quarter size = 0.45 mm rms.

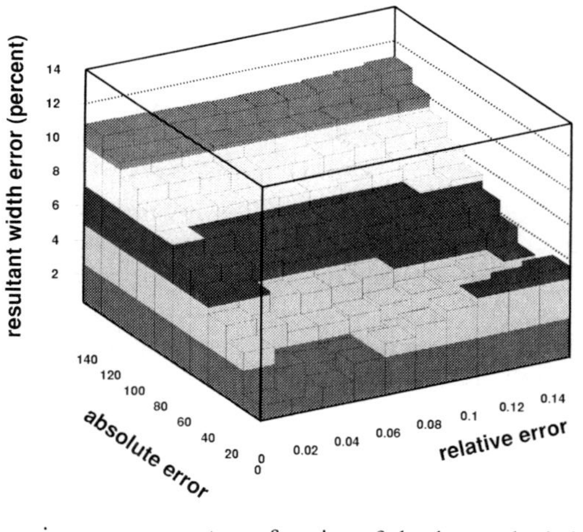

FIGURE 8. Error in the beam size measurement as a function of absolute and relative error on the wire signal. The wire position error is assumed to be negligible here. An absolute error of 150 corresponds to 10% of the peak signal.

Once an analysis such as this one has been conducted, one may simply plug in the errors for the as-built harp or wire scanner. Remember that these simulations do not include the effects of beam position and beam width jitter, which if severe enough could add additional error. Also, the work presented here assumes a gaussian fit to the data. It is of course reasonable to use other functions as well. However, as long as the fitted function does a good job at describing the data, the results will be the same, since the wire position error enters the overall beam width error through the slope of the function at a given wire position, and the slope will barely change from one function to the next given the constraint that the function must describe the data.

DESIGN EXAMPLE—THE SNS WIRE SCANNER SYSTEM

Up to now our discussions have been somewhat abstract. Let us now apply ourselves to a real-life example—the design of the SNS linac wire scanner system. Before any design work can begin, we must of course first define the requirements, shown in Table 1.

TABLE 1. SNS linac wire scanner requirements.

Parameter	**Value**
Beam type	H^-
Beam energy	2.5 to 1300 MeV,
Beam current	5 to 38 mA peak, <1.4 mA avg.
Beam size	0.7-mm to 4.3-mm rms
Beam size measurement accuracy	10%
Beam pulse length	50–1000 μs
Rep rate	Up to 60 Hz
Beam pipe aperture (diameter)	2.5 to 20 cm
Actuator cycles over lifetime	30,000
Machine protect interface?	Yes
Measure position too?	Preferred
Radiation levels	Up to 7 Mrad in 30 years

Signal from Wire or from Loss Monitor?

Our first requirements from the above table involve the beam energy, intensity, and pulse structure. This already tells us a lot about what our wire scanner system will look like. At beam energies less than about 10 MeV, the radiation caused by the beam interacting with the wire does not pass through the vacuum chamber, so we must use the option of detecting the wire signals. Measuring beam loss as a function of wire position in not an option. This means each signal wire must have a connection to the outside world through a vacuum feedthrough. If we are designing a wire scanner system that will be measuring profiles in more than one plane, we now also have the option to cross our wires to reduce the stroke requirement. (Remember that when the secondary radiation option is used, only one wire can be in the beam at a time.) Drawbacks to crossing the wires are cross talk and increased beam loss. At SNS, to mitigate these drawbacks we chose to offset the wires so that only one wire at a time is within ±2 rms of the beam center. We also chose to mount the actuator at a 45° angle so that with one actuator we can have three wires pass through the beam to measure the horizontal, vertical, and diagonal profiles.

Once the beam energy gets above 10 or 20 MeV we no longer need to rely on just the secondary emission signal. But to simplify the wire scanner system at the SNS we chose to stay with the secondary emission option throughout the facility, except in the ring, where we are concerned about the rf structure of the beam interfering with the measurement. Of course the SNS also has an extensive loss monitor system, so we always have the option of tying the two systems together to augment our profile measurements.

Choice of Signal Wire

From the beam energy, beam current, and beam size we see that the SNS beam is very intense. At low beam energies, where the beam fully stops inside the wire, the energy deposition is very high, and it is impossible to find a wire that will survive a full pulse length. As a compromise, we chose a wire that will maximize our pulse length. Detailed thermal analysis shows that carbon wire has the best thermal performance. We chose the 32-micron-diameter wire—the largest diameter carbon wire on the market. An additional benefit of carbon is its low atomic number, so that only a minimal amount of beam loss is caused by multiple scattering.

From Fig. 4 we see that for a 32-micron-diameter carbon wire in an H^- beam, the worst beam energy is about 1.5 MeV. The SNS has wire scanners in the 2.5 MeV Medium Energy Beam Transport, and it is these wire scanners that suffer the greatest thermal effects. In this portion of the linac we must limit the beam duty factor to 50 μs at 1 or 2 Hz. The wire scanners in other portions of the linac can be used at higher duty factors.

Fork Design

The fork should ideally be large enough that it never interferes with the beam aperture. The simplest way to do this is to make the space between the tines slightly larger than the inside diameter of the upstream beam pipe. Of course every rule of thumb has its exceptions, and sometimes it will be okay to make the tine spacing smaller, depending on the application. The same argument is used to determine the length of the tines—i.e., they should be long enough that the signal wires pass completely through the beam aperture.

The method used to mount the wires to the fork also depends on the application. Some forks can be as simple as a piece of G-10 circuit board with the signal wires soldered on. However, if the wires will get hot enough that thermal expansion can cause them to sag, it may be desirable to mount them under tension, as we do at SNS. Also, if we will be measuring the wire signals they must be electrically isolated. Figure 9 shows an SNS wire scanner fork with a nice mounting mechanism (Fig. 10) developed at LANL and used at SNS that allows small wires to be both mounted under tension and electrically isolated. At SNS we bring both ends of each signal wire out through a vacuum feedthrough for a self-test feature that will be explained later. Of course the materials used should also be rad hard and be compatible with the vacuum system.

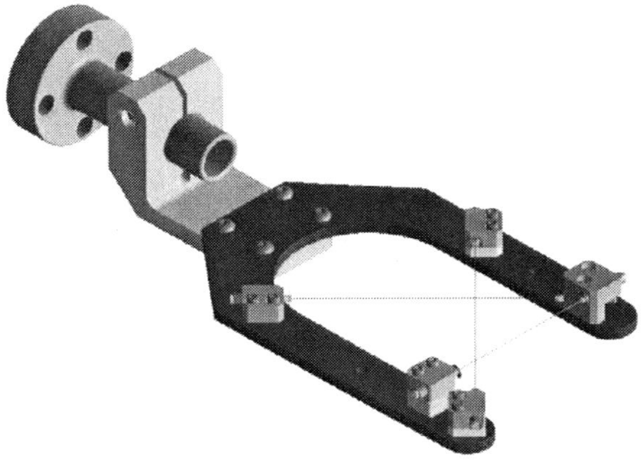

FIGURE 9. An isometric CAD image on an SNS linac wire scanner fork.

Stroke and Other Actuator Details

We now have our wire type (32-micron carbon), our wire spacing (4 x rms beam size), our fork size, and our wire mount. The design of the business end of the actuator is essentially complete. For the actuator itself, the design should allow a stroke large enough to pass all the signal wires completely through the beam aperture. Almost all the wire scanner actuators at the SNS have a stroke long enough to satisfy this criterion. Each actuator has three signal wires, and each end of each signal wire is connected to the outside world through a vacuum feedthrough, and this means we must have a six-pin (or equivalent) vacuum feedthrough on our actuator. This is a complication that can be avoided in wire scanner systems that measure secondary radiation rather than wire signals. The actuator should also have limit switches to control its motion and interact with the machine protection system if necessary. The SNS actuators have one limit switch at each end for motion control, and one in-limit switch and two out-limit switches (for redundancy) for the machine protection system. Since running the wire scanner during high-power beam operations can destroy the wire, we have redundant out-limit switches for extra reliability. Finally, wire scanner actuators can usually benefit from some sort of position read-back device. Optical encoders, potentiometers, and linear variable differential transformers (LVDTs) are commonly used. At SNS we chose to avoid the optical encoders because of our high-radiation environment and instead used the LVDTs. Potentiometers are typically less accurate than LVDTs but are also less expensive. In many applications it is also desirable to use brakes on the actuators, as we do on our SNS actuators, so that the forks will not be sucked in by the vacuum when the stepper motors are not energized.

We must also take into account radiation levels in the design of our actuator. We anticipate up to 7 Mrad over 30 years of operations in the linac. We chose to use rad-hard BNC connectors made with Rexolite (good to about 1000 Mrad) and circular multi-pin connectors made with polycholoprene (good to about 100 Mrad). The LVDT is a rad-hard model that does not use any Teflon wiring. The signal cables are Kapton-coated solid-core wire inside the vacuum and Kapton-insulated coax outside the vacuum. For the limit switch and stepper motor wiring we avoided using Teflon. Probably the most radiation-sensitive component is the stepper motor. We investigated rad-hard stepper motors but determined that they were too expensive. Fortunately the design allows straightforward replacement of the stepper motors if necessary.

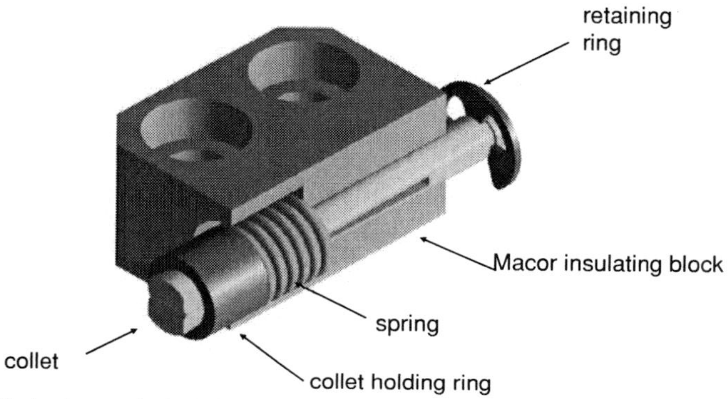

FIGURE 10. An isometric CAD image of the wire mount used for the SNS wire scanner system.

Positioning Accuracy

To get an accurate beam profile measurement we must of course have an actuator that can travel in well-defined steps through the beam. The accuracy of the step size directly impacts the accuracy of the beam profile measurement. Using the plots shown earlier in this paper, to meet the SNS requirement of a 10% or less error on the rms beam size (assuming an absolute signal error of 2% of the peak, a 1% relative signal error, and a minimum beam size of 0.7 mm), our actuator should have a positioning accuracy of at least 0.2 mm. We have measured [13] the positioning accuracy of our prototype actuators to be ±0.15 mm or better, so our actuators meet the positioning accuracy requirement.

Signal Processing

Our wire scanner system detects signals from the wires, so our signal processor will have three channels (one for each signal wire). The signals are weak, so the signal cables should be well shielded. At SNS we use RG-223 coax cables run in cable trays with solid metal shielding. To minimize our susceptibility to electromagnetic noise pick up we chose to make the bandwidth of the signal processor front end as low as we could and still meet requirements. The minimum beam pulse length is 50 μs (30 μs to ramp up to full current and 20 μs to stabilize for a profile measurement), so we chose a bandwidth of 36 kHz. There is a wide range of signal strengths in the SNS wire scanner system, so we chose to design a signal processor [18] with three gain ranges. Some of the other relevant processor specifications are shown in Table 2. The noise floor on the SNS wire scanner electronics has been measured to be less than the least significant bit on the 12-bit digitizer, so it is essentially zero. The dominant noise source will be due to the accelerator environment. Assuming a reasonable quality cable plant, the noise level should not exceed about 2% of the peak signal level. The electronics linearity is better than 0.1%, so the electronics design is consistent with the assumptions we made on the signal level errors when we analyzed the overall accuracy of the system. Of course we will have to wait until the system is fully installed before we know the true level of electromagnetic noise pick up.

TABLE 2. Some relevant parameters for the SNS linac wire scanner signal processor.

Parameter	Value
Minimum signal	16 nA at high gain
Maximum signal	2.1 mA at low gain
No. gain ranges	3, separated by 18 dB
Linearity	<0.1%
Rise/fall time	8 μs

Self-Test

Every beam diagnostic system needs a diagnostic. This is especially important in a facility where a large number of units are deployed, and this is where the idea of bringing both ends of each signal wire out through the vacuum comes in. If the signal wire breaks it can be easily detected with a continuity test. One can also build a self-test feature into the electronics. At SNS we chose to add a 180-kΩ resistor to ground at one end of each signal wire, down at the actuator. The signal processor circuit has the ability to apply a high-voltage bias to the signal wire, and if the wire breaks the read back voltage increases to the supply voltage and indicates a fault. Other solutions include simulating a beam profile by injecting pulses onto the far end of the signal wire (e.g., as done at LANSCE). To test the motion control we use the position read back and the limit switches. As we drive the actuator from the out limit to the in limit, we should see the switches actuate and a nice linear dependence of the position read back when plotted as a function of the pulses issued to the stepper motor.

Motion Control

Every wire scanner system needs motion control. Typical solutions range from flying wires swept through the beam using servomotors to "crawling" wires using stepper motors. At SNS we chose the stepper motor solution and off-the-shelf motion control based on National Instruments units. There seems to be some debate in the diagnostics community about linear drivers vs stepper drivers. Linear drivers are inherently electrically quieter but also run hotter. However, a well-designed chopper driver can meet electrical noise requirements. For SNS we tried each type and finally settled on the chopper driver solution since the electrical noise was below detectable limits. In the design of our motion control system we must also decide if we want to use position feedback. A properly functioning stepper motor actuator is more accurate than an LVDT or potentiometer position read back, so at SNS we chose not to use position feedback. We use the position read back as a check to make sure the device is functioning properly. Some photographs of the SNS wire scanner system are shown in Fig. 11.

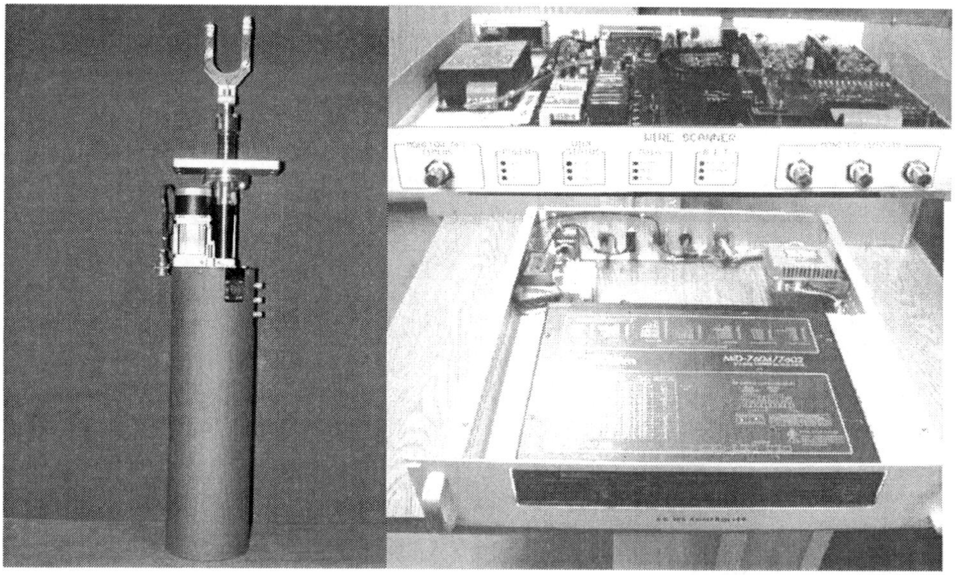

Figure 11. Some photos of the SNS wire scanner system. Left: an actuator. Top right: a signal processor chassis. Bottom right: a driver module chassis.

HOW TO DESIGN A HARP

Let us now consider another real-life design example, this time of the SNS target harp system. Designing a harp system is a lot like designing a wire scanner system. The main difference is that the wires are fixed and that many signal processing channels are required. There are some more subtle differences, including increased beam loss due to all the wires in the beam, high-voltage bias planes, channel-to-channel gain differences, wire plane spacing, wire plane fabrication, etc.

Given all these complications, why use a harp instead of a wire scanner? The main reason is to get fast profile acquisition in a single beam pulse. Harps also have the advantage that the motion control is simpler. Usually harps are either fully inserted or fully retracted, or sometimes they are permanently positioned in the beam with no motion control at all. Wire scanners have the advantage that the signal processing is simpler (just one processor per signal wire, or if the loss monitor method is used just one loss monitor for all the signal wires on the actuator), gain variation from one wire to the next is not an issue, beam loss is less, construction is simpler, and we have the option to use a loss monitor signal to measure the profile.

As in the case of the wire scanner system just discussed, before any design work can begin, we must first define our requirements, shown in Table 3.

How Many Wires?

The first question to answer is how many wires should be used? Obviously the more wires, the better the beam profile measurement. But how many is enough? As in the case of the wire scanner, we can run a simple simulation to determine how the wire spacing affects the beam profile measurement. An example of such a simulation is shown in Fig. 12. From this figure we see that a once we have about 10 to 15 wires per signal plane, the error in the measurement is minimized. For the SNS harp we chose to use 30 wires, since we also are interested in the shape of the distribution, and we must also measure the size of off-nominal beams.

TABLE 3. SNS target harp requirements.

Parameter	Value
Beam type	H^+
Beam energy	1.0 to 1.3 GeV
Beam current	10 mA to 24 A peak, <1.4 mA avg.
Beam pulse length	700 ns
Rep rate	Up to 60 Hz
Radiation levels	Up to 6,000 rad/h when inserted
Beam profile accuracy	Measures tails down to <4% of peak
Beam size (FWHM)	17 cm (horizontal) x 4.8 cm (vertical)
Beam size at base	20 cm (horizontal) x 7 cm (vertical)
Active area	34.2 x 17.1 cm^2
Signal planes	Horizontal, vertical, and diagonal (15 deg.)
High-voltage planes	4, to sandwich each signal plane.
Machine protect interface?	Yes, needed for insertion and retraction
Measure position too?	Yes, ±2 mm

The horizontal wire plane will use 30 wires spaced 1.18 cm apart to measure a horizontal beam size of 17 cm FWHM, and the vertical wire plane will use 30 wires spaced 0.589 cm apart to measure a vertical beam size of 4.8 cm FWHM. It is simplest to interpret the beam profile measurement if the wires are uniformly spaced, but this is not necessary. The 33-cm, 56-wire harps at LANSCE had three different wire spacings. If the wire count becomes excessive, one may consider spacing the wires further apart out at the tails of the beam and/or near the edges of the beam aperture. Ideally the wires will span the entire beam aperture so that the profile can be measured even if the beam is off center, but again this is not necessary and it may be the victim of a design optimization.

Wire Type and Temperature

The next question to ask ourselves is what type of wire shall we use? As we saw in our wire scanner example, carbon is best for high thermal loads, but in this case simulations show that, as far as wire temperatures are concerned, metal wires will be fine. An example of a thermal performance model calculation is shown in Figs. 13 and 14. The model includes thermal radiation, thermal conduction along the length of the wire, energy deposition by convoy electrons (e.g., in an H^- beam) and temperature-dependent thermal emissivity and specific heat. Figure 13 shows the case for the harp wire in the middle of the beam and with the wire ends held at a constant temperature. Figure 14 shows the case for the harp wire in the middle of the beam and with no heat allowed to flow out of the ends of the wire. Reality is somewhere in between, since we will have poor thermal contact with the circuit board that the harp wires are mounted to. As we see from the figures, the peak temperature of the wire is actually the same whether we have the ends of the wire in a thermal bath or not.

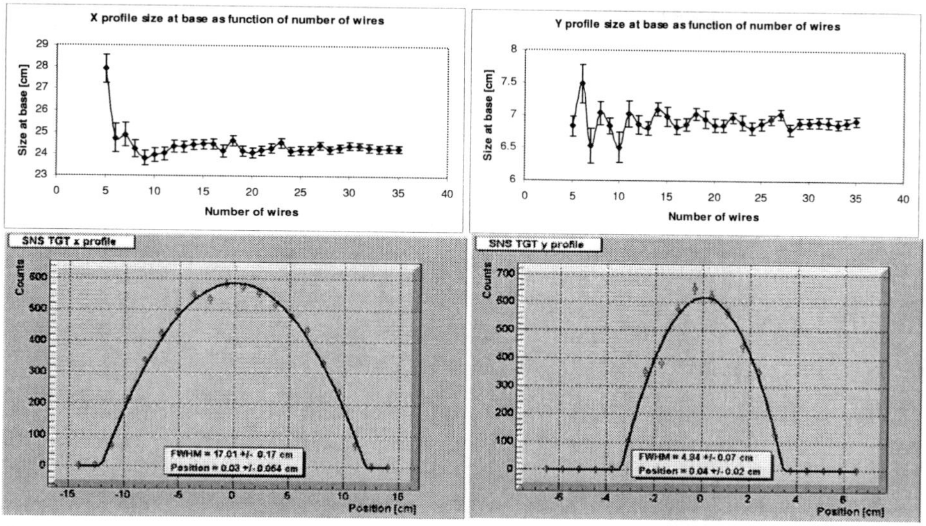

FIGURE 12. Top: Plots of the error in the measurement of the size of simulated beam distribution at the SNS target harp as a function of number of wires on the harp, for the horizontal (left) and vertical (right) profiles. Bottom: the simulated beam distributions.

Another consideration is beam scattering. If we use carbon or SiC wires, the beam will be scattered less and we will incur less beam loss. However, this is not an issue for our target harp application, since shortly after our harp we have a vacuum window which will create much more scattering and beam loss than a harp. Yet another consideration is that carbon and SiC wires are more difficult to handle and mount. In my opinion one should use metal wires unless there is a clear reason to use carbon or SiC wires. For the SNS harp we chose 100-micron-diameter tungsten wires.

WIRE MOUNTING

The next consideration is how to mount the wires. As in the case of the wire scanners, if the wire gets hot, and thermal expansion and contraction becomes an issue, one may wish to mount the wires using springs. This is the method used for the SNS harp and all the large (≥10-cm-diameter) harps at LANSCE. It is often advantageous to use printed circuit board technology to mount the wires and to run the signals over to the edge of the board and possibly to a mass termination connector. For the SNS harp we chose a ceramic substrate because of the high radiation environment. We will also use platinum or molybdenum-manganese traces and lead-free solder to avoid mixed waste issues.

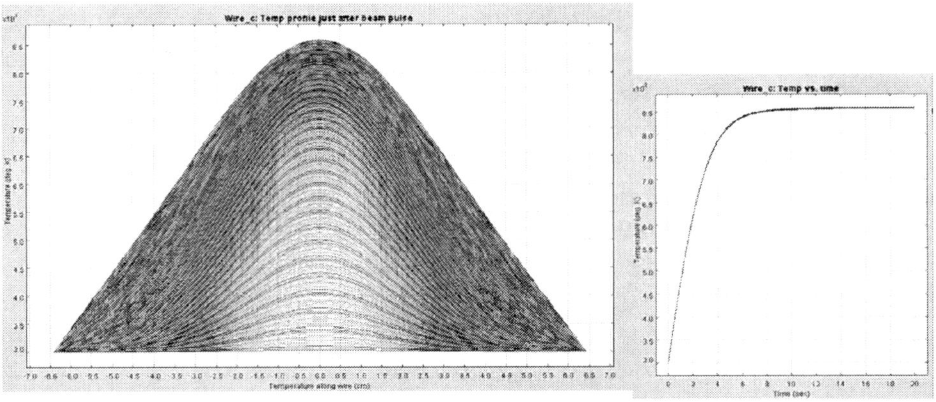

FIGURE 13. Example calculation of wire heating for a wire in the middle of the beam. Left: the wire temperature along the length of the wire, assuming each end of the wire is held at a fixed temperature. The individual lines on the plot show the temperature distributions for successive beam pulses. Right: a plot of the temperature in the middle of the central wire as a function of time. The maximum temperature is 586 °C.

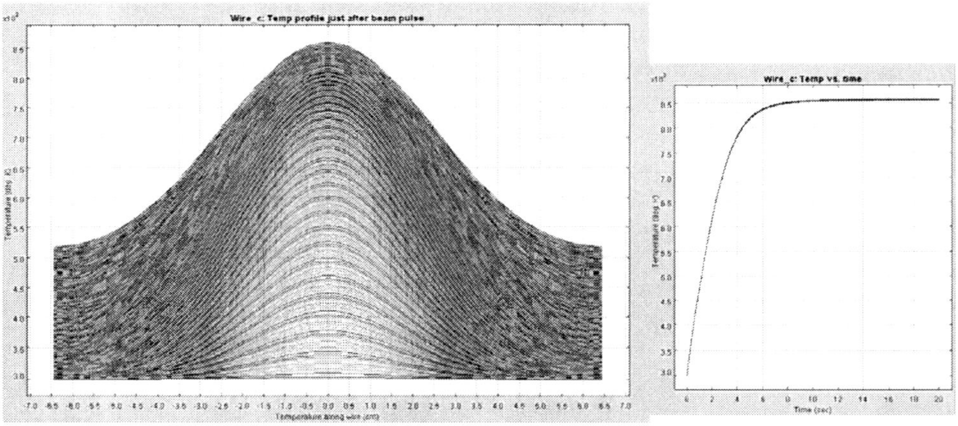

FIGURE 14. Example calculations of wire heating for a wire in the middle of the beam. Left: the wire temperature along the length of the wire, assuming no heat is allowed to flow out of the ends of the wire. The individual lines on the plot show the temperature distributions for successive beam pulses. Right: a plot of the temperature in the middle of the central wire as a function of time. The maximum temperature is 586 °C.

Most harps are built with all the signal wires in a flat plane. But this can cause problems if a wire breaks in either the signal plane or in an adjacent high-voltage plane. The broken wire can fall across several signal wires and short them out. If the broken wire is from the high-voltage plane it can saturate the signal levels on the signal wires. To mitigate this problem we have chosen to stagger the wires on the signal wire plane, so that the wires are on two different levels. Then only one level of wires is likely to short out, leaving the other signal wires fully functional.

Harp insertion mechanisms may be the same or different from the wire scanner mechanisms. First of all, in some cases no insertion mechanism may be required; i.e., as long as wire survival and beam scattering is not important to the application the harp may be in the beam all the time. At SNS we are concerned about wire survival in the intense beam, so the harp will be insertable. Unlike wire scanners, insertion mechanisms can be either air actuators or stepper motors, since it is not always necessary to move the harp during the beam measurement.

Signal Processing

Harp signal acquisition systems can be of two basic types—the wire signals can be sampled simultaneously or sequentially. The cost per channel is often low enough that simultaneous sampling is not a problem. It is usually desirable to use a calibration method to tweak the gain of each channel so they will all have the same gain. At SNS we chose to use simultaneous sampling with a custom-designed analog front end followed by off-the-shelf analog-to-digital signal acquisition.

Self-Test

As we discussed in the wire scanner example, every beam diagnostic system needs a diagnostic. This is especially important for our harp which is buried under tons of shielding and is very difficult to access. For the harp self-test we chose to mount 1-pF capacitors to the harp signal wire board, on the ends of the signal wires opposite to the ends where we take our beam profile signals. Each signal wire has its own capacitor, and the other ends of the capacitors are bussed together to a separate cable onto which we can inject a pulse. This pulse will be capacitively coupled onto the signal wires and then transmitted up to the signal processor, thus checking for broken signal wires and providing a test of the signal processor circuits. Some isometric CAD images of the SNS target harp system are shown in Fig. 15.

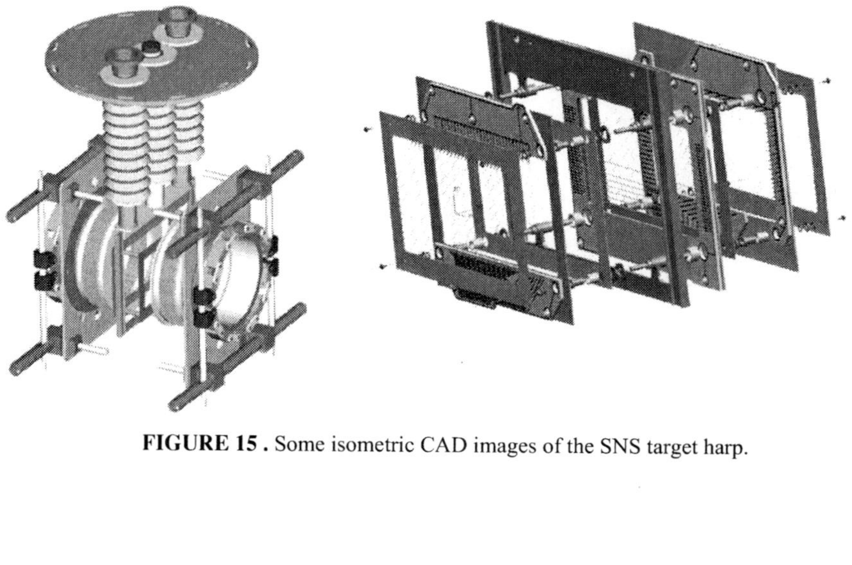

FIGURE 15 . Some isometric CAD images of the SNS target harp.

REFERENCES

1. K. Hagiwara et al., "The Review of Particle Physics," *Phys. Rev. D* **66**, 010001 (2002).
2. C. Fischer, "Ionization Losses and Wire Scanner Heating: Evaluation, Possible Solutions, Application to the LHC," Proceedings of the 1999 DIPAC Conference, Chester, UK, May 16-18, 1999.
3. T. Osiecki, S. Kopp, and Z. Pavlovic, "Studies of Beam Heating of NuMI Proton Beam Profile Monitors," NuMI Tech Note NuMI-B-926, April 1, 2004.
4. R. B. Fiorito and D. W. Rule, "Optical Transition Radiation Beam Emittance Diagnostics," in *Beam Instrumentation Workshop*, edited by R. Shafer, AIP Conference Proceedings 319, New York: American Institute of Physics, 1994, pp. 21-37.
5. W. Barry, "Measurement of Subpicosecond Bunch Profiles Using Coherent Transition Radiation," in *Beam Instrumentation, Proceedings of the Seventh Workshop*, edited by A. Lumpkin and C. Eyberger., AIP Conference Proceedings 390, New York: American Institute of Physics, 1997, pp. 173-185.
6. M. A. Plum, J. Bosser, E. Bravin, and R. Maccaferri, *Nucl. Instr. & Meth.* A492, 74-90 (2002).
7. S. Assadi et al., "The SNS Laser Profile Monitor: Design and Implementation," Proceedings of the 2003 Particle Accelerator Conference, May 12–16, 2003, Portland, Oregon.
8. Evans, *The Atomic Nucleus*, McGraw-Hill, 1965.
9. R. E. Shafer, "Comparison of H-minus and Proton Beam Heating in Thin Foils," SNS Tech Note SNS-104050000-TD0001, June 16, 2000.
10. http://physics.nist.gov/.
11. E. J. Sternglass, *Phys Rev* **108**, 1 (1957). Also J. E. Borovsky and D. M. Suszeynsky, *Phys. Rev.* **A43**, 1416 (1991).
12. M. A. Plum et al., "SNS Linac Wire Scanner System: Signal Levels and Accuracy," Proceedings of the 2002 Linear Accelerator Conference, Gyeongju, South Korea, August 19-23, 2002.
13. M. A. Plum, "H^+, H^0, and H^- Secondary Electron Emission Coefficients for 800 MeV Beams," PSR Tech Note PSR-99-001, March 23, 1999.
14. G. Ferioli and R. Jung, "Evolution of the Secondary Emission Efficiencies of Various Materials Measured in the CERN SPS Secondary Beam Lines," Proceedings of the DIPAC 1997 Conference, Frascati, Italy, October 12–1997.
15. D. Jeon, PARMILA Simulation of a Typical SNS Linac Beam at the Exit of DTL Tank 1, at 7.5 MeV, private communication.
16. Physics Analysis Workstation, ©CERN, Geneva 1992-1999.
17. M. A. Plum et al., "The SNS Wire Scanner System," Proceedings of the 2003 Particle Accelerator Conference, Portland, Oregon, USA, May 12-16, 2003.
18. G. Gillespie, *Phys. Rev.* A16 (1996) 3201.

Resolving EMI Issues To Optimize Accelerator Beam Diagnostic Performance

Michael Thuot

Los Alamos National Laboratory
LANSCE Division
Los Alamos, New Mexico

Abstract. If you have struggled to get the last bit of performance from a beam diagnostic only to find your dynamic range limited by external sources of electromagnetic interference (EMI) once the system is installed, then you will find this tutorial on electromagnetic compatibility and grounding useful. The tutorial will provide some simple, direct methods to analyze, understand and mitigate the impact of EMI on beam diagnostic systems. Several common and unique accelerator EMI sources will be characterized. The dependencies of source frequency and distance to the source on the optimal choice of grounding and shielding methods will be illustrated. The emphasis is on a stepwise process that leads to understanding and cost-effective resolution of EMI impacts on beam diagnostic systems.

INTRODUCTION

Many methods of reducing electromagnetic interference (EMI) in beam diagnostic systems are known and widely recommended to those who encounter EMI in the process of fielding diagnostic instruments or systems. EMI is the degradation of the performance of a device, equipment or system by an electromagnetic disturbance. However, the benefits derived from many EMI-reducing design modifications vary widely in practice: some design changes or additions effectively mitigate the offending EMI, while others seem to have either no effect or even a deleterious effect on the measurement system performance.

Over the last few decades, an engineering process that aids in the selection of effective EMI mitigation strategies out of the myriad of possible mitigation approaches has emerged.[1][2] In this paper, the process will be called an electromagnetic compatibility system design process. Electromagnetic compatibility, EMC, is a term created to recognize that EMI problems can be resolved by controlling the source of the interfering energy, the medium that couples the energy, and the electronic device whose performance is being degraded by the interfering electromagnetic energy. The objective of EMC is compatible operation, not total elimination of the interfering energy transfer. The use of the term "system design" is to remind the diagnostic or accelerator facility designer that it is far more cost-effective to design for electromagnetic compatibility early, during the system design or prototyping stage, than it is to revise designs during a crisis re-design after EMI issues have been shown to adversely impact the performance of an installed diagnostic

CP732, *Beam Instrumentation Workshop 2004: Eleventh Workshop*,
edited by T. Shea and R. C. Sibley III

system.[1] This tutorial paper will describe this EMC design process that results in the mitigation of EMI and optimal performance of beam diagnostic systems in an adverse EMI environment.

This paper is organized around a six step EMC system design process. The goals of this six-step process are outlined below, then each step is expanded to include some of the knowledge and techniques required to successfully complete each step. Beyond these six steps, separate sections treat the basics of shielding, grounding and EMI control in signal cables. Because of limits on the length of the paper, useful EMI control techniques like balancing, filtering, reducing bandwidth, isolation, cancellation, etc. are not covered in any detail. Those interested in these topics will find a useful treatment of these topics in the referenced sources.[1][2][3]

AN ELECTROMAGNETIC COMPATIBILITY SYSTEM DESIGN PROCESS

The primary goal in EMC system design is to prevent or limit the transfer of energy from a source of an electromagnetic disturbance to a diagnostic system receptor by modifying the source to limit the disturbance, or by preventing coupling to the environment, or by reducing the susceptibility of the diagnostic system. The reduction of emissions from an EMI source is accomplished primarily by reducing di/dt in the source and/or by reducing the area enclosed by the path of EMI currents. A reduction in the area of EMI radiating loops is accomplished by returning the noise currents to the source locally. Reducing or preventing coupling to the environment is usually accomplished by interposing various shielding materials that limit coupling through radiation, or by making improvements in grounding that limit coupling by impedance or conduction. Reducing the susceptibility of a diagnostic system can be done in a myriad of ways: by balancing, filtering, isolation, orientation, separation, cancellation, impedance matching, grounding, bonding, shielding, bandwidth reduction, gain reduction, conductor geometry or component placement.

As detailed below, the first two steps in the EMC design process result in a characterization of the radiated fields and conduction paths from the source and focus on how to reduce the effect of the EMI source on the electromagnetic environment. The next two steps in the EMC design process result in a characterization of the EM radiation and conduction paths to the receptor. The focus here is on modifying the coupling medium to reduce radiated EMI by shielding, or to reduce conducted EMI by grounding, isolating or filtering. Characteristics of the coupling medium and the distance from the source to the receptor modify the radiated EMI field and therefore modify the noise power spectrum at the receptor. The last two steps of the EMC design process focus on reducing the susceptibly of the diagnostic system receptors to the previously characterized EMI coupling paths. By determining the EMI frequency spectra, the field impedance and the field magnitude at the receptors, a cost-effective EMC solution for a diagnostic system can be selected from the large number of possible mitigation strategies.

A Summary of the Electromagnetic Compatibility System Design Process

1. Identify and characterize EMI sources
2. Apply EMC methods to limit the disturbance at the source and/or to minimize EMI coupling to the environment
3. Identify and characterize EMI coupling mechanisms
4. Apply EMC methods (shielding and grounding) to minimize coupling through radiation or conduction paths
5. Identify and characterize receptors
6. Minimize receptor EMI susceptibility by shielding, grounding, isolation, filtering, balancing, orientation, separation, impedance, etc.

To successfully complete the EMC process, it helps to think of electronic systems from the viewpoint of high frequency energy transfer, visualizing energy transport through stray and common mode paths, locating conductive and parasitic ground loops, and reviewing effects on all sub-systems and components.

EMC Design Step 1: Identification and Characterization of EMI Sources by Measurement and/or Modeling

The first step in the EMC design process, identifying and characterizing EMI sources, is an essential step for understanding an EMI environment and subsequently creating a cost-effective EMC solution. There are hundreds of EMI sources in a modern particle accelerator facility. Most are the traditional components of accelerator systems: high power rf systems, magnet power supplies, beam kickers, beam injectors, etc., and there are a few newer ones with interesting characteristics: high voltage converter/modulator power supplies and switching mode DC supplies for high power magnets. In this section we will look at some of the more troublesome sources of EMI for beam diagnostic systems. Because of the complexity of accelerator systems it is sometimes difficult to locate an offending source and to identify its coupling mechanisms.

Following a good systems engineering approach, all the significant local sources should be evaluated, otherwise one will find that the far more costly crisis approach to EMC will need to be employed after the diagnostic has been installed in the facility. One should pay particular attention to any equipment located near the diagnostics that operates at high power levels and/or has the capacity to produce a high di/dt discharge. Low-frequency, high-power sources, from 60 hertz to a few megahertz, can be especially troublesome since it is likely that most of the facility is in the low-impedance induction field, the near field, generated by this type of source. Low-impedance emissions are notoriously difficult to shield, as we shall see in the section on shielding that follows. Extremely high di/dt sources such as vacuum breakdown in a beam injector; rf system HVDC diverters or beam kicker power supplies often are significant sources of disruption for diagnostics located nearby. A key step in the EMC system design process is to characterize radiated and conducted emissions from each important EMI source, along with the distance from each source to the susceptible circuit or system, to generate an estimate of the field impedance at a shield or receptor.

What Is Electromagnetic Field Impedance?

The impedance of a radiated electromagnetic field can characterized by the ratio of the electric component of the field (E) to the corresponding magnetic component of the field (H), or field impedance Z = E/H. The wave or field impedance of a radiating EM wave is a function of the characteristics of the source and the medium, and the distance from the source.[1][3]

A radiated electromagnetic field is said to be "high impedance" when the electric field component dominates, as it does in radiation from a dipole antenna, and the resulting field impedance, E/H, is greater than the intrinsic impedance of the medium. Conversely, when an electromagnetic field is radiated from a current source/loop antenna and the magnetic field component dominates, the E/H ratio will be less than 377 ohms and the radiated field is called "low-impedance".

As electromagnetic radiation propagates through a medium, the impedance of the wave, E/H, approaches the intrinsic impedance of the medium (Fig. 1). The distance the wave must travel in the medium for this transition in impedance to occur is $\sim\lambda/2\pi$. Close to the source, in the near field, the electric and magnetic fields must be considered separately since the ratio E/H is initially determined by the source and is not constant. Most of the troublesome accelerator EMI sources radiate noise at frequencies below a few megahertz, so most receptors are in the near field of these radiation sources. For example, the near field of the 60 kHz noise radiated from the SNS high voltage converter modulator extends out to about half a mile from the source.

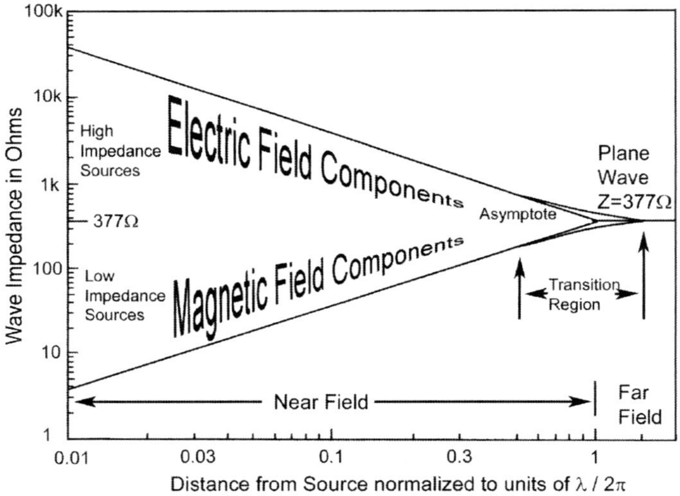

FIGURE 1. Wave or field impedance depends on the source circuit impedance and the distance from the source.

Sources of High Impedance EMI Fields

The nature of a radiated EMI field depends on the characteristics of the energy source and the coupling means, which is usually a parasitic path that creates an effective antenna for noise generated by the source. If the noise source is a voltage source and the coupling mechanism is a dipole structure, the local radiated field impedance will be greater than the free space field impedance. Any ungrounded moderately sized (>1/10 λ wavelength) structure, or even a grounded large structure whose dimensions approach resonance at multiples of ½ λ, that is charged by rf frequencies or by any high dv/dt voltage source can act as a dipole antenna and radiate significant energy into the local environment. The defense against this type of EMI threat is relatively simple; just ground the radiating structure in a way that returns the interfering current to the source as locally as possible or in a way that disrupts the resonance condition on large structures. As we will see when we review shielding, a high impedance field is quite easy to shield effectively. As the distance from a high impedance EMI source increases, the field impedance of the radiated field *decreases* asymptotically towards the impedance of free space, 377 ohms, at a distance of the EMI wavelength(s)/2π from the source.

Sources of Low Impedance EMI Fields

Much like high impedance field sources, a low impedance EMI source usually couples energy to the local environment through parasitic paths. A low impedance field, one much less than 377 ohms, can be produced by any low impedance high di/dt current source and one or more current-loop coupling structures. Any moderately large (>1/20 λ) double grounded or loop structure that carries rf current or high di/dt current from the EMI current source can act as an effective loop antenna and radiate significant energy into the local environment. Defense against a low-impedance radiated field is far more difficult than a high-impedance field, since as the field impedance approaches a shield's impedance, the ability of the shield to reflect the EMI approaches zero. Also, low-impedance fields can induce significant currents into grounded cable shields that can couple into the signal path. As the distance from a low-impedance EMI source increases, the field impedance of the radiated field *increases* asymptotically towards the impedance of free space at a distance of λ/2π from the source. This means that at moderate distances from a low-impedance high-frequency source, the radiated EMI fields become easier to shield. Therefore, the preferred location for a shield, in a low impedance field, is surrounding the receptor.

EMC design Step 1, Example 1: Characterizing A High-Power Low-Impedance EMI Source

The Spallation Neutron Source (SNS) project employs a new type of high voltage power supply/modulator to provide pulsed HVDC power for the klystrons.[4] This 20/60 kHz poly-phase high voltage converter-modulator (HVCM) power supply also acts as an EMI generator of a type of which there was originally little information or experience. EMI emission measurements were made at Los Alamos, under prototype HVCM operating conditions a few years before final installation, and again at SNS, when the first unit was operational, to assess the noise coupled from this high voltage power supply into local electronics and cables.

A Direct EMI Source Measurement Technique

At the prototype, the EMI coupling was measured directly to unshielded and shielded twisted pair cables installed in grounded cable trays about 2 meters above the HVCM. Since the orientation of the EMI fields were unknown, the cables were run in orthogonal directions above the HVCM with about 15 meters of cable exposed to the EMI field. The twisted pair cables were terminated on both ends, as would be the case in a normal installation. Induced voltage measurements were made while the HVCM was operating between 95 kV and 130 kV output, at 300 kW to 1,900 kW average power with about a 1.1 ms output pulse at a 60 Hz pulse repetition rate. At the full 11 MW power level, the HVCM can operate at 140 kV with a 1.2 ms pulse. It was expected that the EMI levels from an HVCM installed at SNS would be comparable to or only slightly more severe than those measured on the prototype. The measurements were made with a Tektronics digital storage oscilloscope with a 60 MHz bandwidth, used in a differential mode. The structure of the noise pulse observed on the 75 ohm terminated pair was that of a noise burst, occurring every 16.7 ms with an ~100 microsecond wide initial peak as the modulator and klystron turned on, a lower level for ~1 ms in the middle of the HV output pulse and a higher-magnitude ~30 microsecond wide final peak as the modulator turned off. The initial and final peaks consisted of 4 MHz converter switching noise, while 20 kHz and 60 kHz inverter noise frequencies dominated the middle portion of the pulse.

The HVCM and high power klystron rf system proved to be a significant source of low-impedance radiated EMI even though many EMI reducing measures had been designed into the device. Nevertheless, we measured induced normal mode noise spikes >4 volts peak to peak on twisted pair cable terminated with 75 ohms to ground, and we mounted an effort to reduce these emissions; see step 2, example 1, following.

While it is possible to calculate the field impedance of this source, the calculation is complex. A far more convenient method is to use the graphs in Don White's book "Shielding".[5] These graphs link circuit impedance, frequency and distance from a source to make a close estimate of the field impedance. In the case of the HVCM, the source circuit impedance is about 3 ohms. Therefore, at the instrumentation racks about 10 meters away from the HVCM, the field impedance would be about 6 ohms for the 60 kHz field components and about 280 ohms for the 4 MHz components.

EMI measurements were made again when the first HVCM was operating at SNS. EMI disturbances from HVCM operation had already been observed in the timing and other data systems. By tracking HVCM noise currents with a small single turn magnetic field probe coil and a scope, a significant low impedance EMI field source generated by 60 kHz currents flowing in the unshielded DC supply cables connecting the HVCM to the SCR DC power regulator was located. This source of EMI was not recognized on the prototype since, at LANL, these 4/0 cables were twisted together with the current return path, which canceled much of the radiated field. Near these cables at SNS, the field induced more than 2 volts into a 4" one turn coil terminated in

50 ohms. Measurements made near the center of the ~ 15 X 25 foot cable loop showed 20 kHz, 60 kHz and 4 MHz transients in a low impedance field. This EMI field extended to the nearby control racks and the surrounding cable trays. This radiative coupling path was the most likely source of the interference that adversely affected the accelerator operations, and a decision was made to shield the DC supply cables with grounded conduit; see step 3, example 1, following.

EMC Step 1, Example 2: Characterizing High di/dt Transient EMI Source

The Accelerator Production of Tritium (APT) project required the development of a 75 kV, 130 mA CW proton source.[6] During the injector prototype development program, several ion source magnet power supplies were repetitively blown out by EMI generated by spark-down transients. Measurements indicated an ~500 V displacement of the local ground during the transient and there were concerns that beam current probes and other diagnostic instrumentation would suffer a similar fate. The injector developers, who had little previous experience with high current CW injectors, wanted to understand how to preserve the electronics and diagnostics that would someday be deployed around this beam injector.

Determination of EMI Characteristics Through EMI Source Modeling

In this case, modeling of the circuits in an EMI source was used to characterize the EMI generated by spark-down in the beam injector, since accurate direct measurements of very high di/dt intermittent transients is difficult. A spark-down circuit model of the injector and the associated high voltage power supply (HVPS) was constructed to estimate the peak current, di/dt, dv/dt and the frequency spectrum of a spark-down transient in the injector. The values of the HVPS output circuit components in the model were set to the actual circuit values of the HVPS energy storage capacitors along with the series resistors and inductors. The spark discharge arc was modeled. The ion source stray capacitance was measured from the actual injector, but the distribution of this stray capacitance to various ground paths was estimated from the physical arrangement of the conductors and reflected in the model.

The inductances of the two major discharge current loops involved in the spark-down control the natural frequency of the discharge current. These two loops consist of a large loop formed by the coaxial HV cable connecting the HVPS to the injector above the facility floor, and a much smaller loop formed by the ion source and the source drive connections above the grounded ion source table and beamline. The inductances of these loops were estimated using a formula for the inductance of a conductor above a ground plane, from Ott.[1] These calculated inductance values were adjusted to correlate with oscilloscope traces taken during actual spark-down transients. The HV coaxial cable that connects the HVPS to the injector was modeled as a transmission line with approximately one lumped element per foot. Since this transmission line is not impedance matched at the load and source ends, it had a large effect on the rate-of-change of the voltage and the current in the circuit. The model of the injector spark-down circuit produced waveform records through SPICE analysis performed by a commercial software package, Electronic Workbench, running on a

PC. Electronic Workbench represents the circuit as a schematic rather than a SPICE node list, which simplifies the modification of circuit values and connections.

The waveforms and power spectra generated by the SPICE model of the spark-down were analyzed to extract the peak transient current, ~1300 A, the transient current natural frequencies, 3 to 6 MHz, and the discharge di/dt, ~2.3E+10 A/s and dv/dt, ~1.6E+12 V/s. These parameters quantify the level of the source of EMI that the diagnostic system must suppress. At the injector the EM field impedance is almost equal to the discharge circuit impedance, ~60 ohms. But at the instrumentation rack 3 meters away, the field impedance of the 6 MHz components of the field has increased to ~180 ohms, making it easier to shield effectively. If the EMI coupling from this strong EMI source exceeds the transient suppression capability of the diagnostic system interface, then the transient is a threat to the proper operation of the instrumentation. In step two of the EMC process, source EMI mitigation strategies appreciably reduced the threat from this EMI source.

These two examples demonstrate that a variety of methods can be applied to characterize an EMI source. If the source is complex, like the HVCM, measurements on a prototype dependably provide a characterization of the source. However, as we discovered on the HVCM, slight differences in the installation of a strong EMI source can have important negative consequences. If the EMI source is an intermittent transient, modeling can conveniently provide a far more complete analysis of the power spectrum of the source than trying to capture fleeting transients in a noisy environment. The primary shortcoming of the modeling approach is that it is difficult to visualize and include the important parasitic coupling paths in the EMI source evaluation.

EMC Design Step 2: Apply EMC Methods To Limit the Disturbance or Minimize EMI Coupling to the Environment from the Source

The second step in the EMC design process includes two activities that are generally very productive and often overlooked steps in resolving EMI problems: (1) applying EMC methods to limit the disturbance at the source and (2) minimizing EMI coupling from the source to the environment. The reduction of EM disturbances at the source often provides the greatest benefit to the EMI environment for the lowest cost. Usually the number of receptors far outnumbers the EMI sources in a facility, so reducing the disturbance produced by few sources provides benefits for all receptors. EMI source mitigation can easily decrease the number of EMI crises during commissioning. Often, mitigating source emission has a side effect of improving the reliability of source operation, since the source will stop interfering with itself. The generalized approach to this problem is to first, look at ways to reduce di/dt in the source, and second, look for ways to reduce coupling from the source enclosure and cables by shielding or grounding. Reducing the return-current loop area by installing cables on a ground plane can reduce EMI radiated from power cables. Another way to prevent radiation of a low impedance field from a conductor connecting systems that are grounded at both ends is to shield the conductor with the shield grounded at both ends. This allows a shield current equal and opposite to that in the center conductor to flow, which cancels the external magnetic field at high frequencies.[1]

EMC Step 2, Example 1: Mitigation of the HVCM EMI Source

As detailed in step one above, initial measurements indicated very significant EMI emissions from the HVCM, which resulted in modifications to the HVCM enclosure, the cabling and the shielding. Gaps in the metal HVCM enclosures were closed with conductive panels and more closely spaced fasteners were added to access hatches to reduce magnetic field leakage from internal currents. Improvements were made to control and instrumentation cable routing to reduce the area of the multiple coupling loops formed by these grounded cables. Cables were routed on a ground plane to reduce the loop area and unused cables were disconnected and removed.

The largest magnitude transient measured on the early configuration of the HVCM was the transient produced when the modulator shut off the ~100 kV, 50 A output current at the end of the 1.1 ms pulse. This transient was generated by the 900 Hz, 1.1 ms modulator pulse current penetrating the output cable shield that was only <2 skin depths thick. To shield this very low frequency, a >10 skin depths thick ferrous conduit was installed around the output cable. The addition of a grounded rigid conduit enclosing the output coax cable eliminated this turnoff transient as a significant component of the radiated EMI. These modifications of the HVCM reduced the level of EMI coupled to the environment by >10 dB.

After the measurements at SNS located the additional EMI threat from the 60 kHz currents radiating from the DC supply cables, routing the DC supply and return current cables together inside a grounded ferrous conduit further reduced the EMI from the HVCM. This reduced the size of the noise current loop from 15 X 25 feet to < 1 X 25 feet, and reduced this source of radiated noise.

EMC Step 2, Example 2: Mitigation of the Spark-down EMI Source

After modeling in step one above had characterized the injector spark-down EMI source, several defenses were included in the design to reduce the threat from this source. Capacitive coupling from the HVPS/injector high dv/dt through a local high-impedance field to the ion source magnets proved easy to shield. Proper grounding and closing gaps in enclosures surrounding the HVPS, the ion source and the shields on signal and magnet cables eliminated almost all of the dv/dt driven interference.

To control low impedance coupling from the high di/dt spark-down current, the best defense was to reduce the area of all source and signal loops. The primary source of this coupling was the large loop formed by the HVPS output coax cable connecting the grounded HVPS to the grounded injector. We changed the grounding of the coaxial shield on this cable by adding a ground on both ends. The resulting low inductance path, formed by the coaxial current flow, forced the HVPS discharge return current to flow back along the cable shield rather than through the facility grounds. This configuration reduced the area of the large loop, formed by the HV cable above the floor, to a small area formed by the HV coax with the current returning through the shield. Measurements made on the injector prototype demonstrated a greater than 14 dB reduction in the EMI from the injector after grounding the high voltage cable shield at both ends. Using these EMI source mitigation strategies, injector operational reliability and the prospects for obtaining reliable beam diagnostic signals from the injector were greatly improved.

EMC Step 3: Identify and Characterize EMI Coupling Mechanisms

There are two primary EMI coupling mechanisms: conduction and electromagnetic field radiation. Coupling by conduction usually occurs through common mode paths, either common ground impedances or shared power supply impedances, or through improperly grounded cable shields. Shared ground bus impedances are a common occurrence in almost all ac power system grounds. Radiative coupling usually occurs through the components located closest to the EMI source or through the components of the system with the largest physical dimensions, the cables. Therefore, we will treat noise induced in cables in detail in step six of the ECM process. While either conduction or radiation may be the primary EMI coupling path, some combination of both may act to transmit the source EMI energy to the receptor. Since the offending noise usually couples through stray capacitance or stray mutual inductance and along common mode paths, and these types of circuit elements do not appear in a circuit analysis, the EMI coupling mechanism usually avoids detection and direct circuit analysis by the designer.

EMC Step 3, Example 1: EMI Coupling Paths from the Injector

The coupling between the injector spark-down and the instrumentation follows several paths, but the most significant path is through local low-impedance field coupling through any mutual inductance between the discharge circuit and the instrumentation. Anywhere near the discharge, significant currents were induced in cable shields that had been grounded on both ends, since these formed grounded loops in the induction field. Just as adding a missing shield ground may reduce EMI coupling in a plane wave or high-impedance field, so in this case, adding an extra shield ground increased the EMI coupling in this low impedance field. The solution was to ground the shields at one point or to add signal cable isolators to break local ground loops. By using an isolator, the voltage induced in the cable loop appears as a common mode voltage on the isolator, rather than a voltage drop on the cable shield that is coupled into the signal.

EMC Step 4: Apply EMC Methods (Grounding and Shielding) To Minimize Coupling Through Radiation or Conduction

The primary way to minimize coupling from EMI sources to diagnostic system receptors is by grounding and shielding. Shielding mitigates radiative coupling, but for most cable shields to perform effectively, they must also be grounded. The process of grounding always relates a diagnostic subsystem to a larger entity, the facility. Where, and to what should the ground be connected? How many ground connections are needed? The following two major sections provide information on shielding and grounding concepts and methods used to reduce coupling between EMI sources and receptors.

Shielding To Minimize Radiative EMI Coupling

Shielding is the principal way to reduce radiative coupling from a noise source to a diagnostic system receptor. A shield is a metallic partition used to control the propagation of electric and magnetic fields by reflecting and/or absorbing the energy. [1] In use, a shield is interposed between the EMI source and the receptor. It may be placed around the EMI source, or the receptor, or both, since a shield may be used to confine the radiated field from a noise source or it may be used to exclude radiated noise from a receptor.

Shield performance, shielding effectiveness, or shield loss is a measure of the reduction in magnetic and/or electric field strength caused by a shield. It is measured by the ratio of the field strength on the source side of the shield to the diminished field strength on the receptor side of the shield, expressed in dB. The total shielding effectiveness is the sum of the shielding effects of two different processes: reflection loss and absorption loss. When a metallic shield is placed between an EMI source and a receptor, the incident wave from an EMI source is partially reflected from the metal barrier and partially transmitted into the metal. As the wave passes through the metal it is partially absorbed. When it reaches the other surface of the shield, it is once more partially reflected back into the metal and partially transmitted into the air and travels on to the receptor. The shielding effectiveness is the sum of the losses from these processes.

Shield Reflection Loss

The reflection loss at a metal shield is dependent on the type of field (electric or magnetic), the type of metal, the frequency and the wave impedance. The reflection coefficient at the air-to-metal interface depends on the ratio of the incident wave field impedance to the metal impedance. As previously noted, this is the basis of how shield performance is related to EMI field impedance. If the incident field impedance is high, then the ratio to the metal impedance is large and the wave is mostly reflected, i.e. the reflection loss is high. But if the incident field is low impedance, like when a low-impedance source is located close to a shield, the ratio of the field impedance to the metal impedance is not large, and a larger portion of the wave is transmitted into the shield. The reflection loss can be calculated, but the calculation is complicated. For the rest of us, there are graphs published by White [5] or nomographs [7] that solve for the reflection loss for various combinations of shield materials, EMI frequency, and field impedance or distance to the source.

From the above referenced graphs on reflection loss, we see that in copper, reflection loss for electric fields is much greater than for magnetic fields in the near field. Far field (plane wave) reflection loss is greatest at low frequencies and for high conductivity materials, like copper. Shield impedance is minimized and reflection loss is increased by using shield materials with high conductivity and low permeability, so steel has much less reflection loss than copper.

Shield Absorption Loss

An EM wave passing through an absorbing medium is attenuated exponentially by ohmic losses arising from induced currents.[1][2][5] Once an EM wave is traveling inside a metal shield, the transmitted field is attenuated at a rate of ~9 dB per skin depth. This means that a shield's absorption loss increases with the frequency of the wave and with shield permeability, conductivity and thickness. For low impedance, low frequency fields, where shield absorption dominates, skin depth in steel is much smaller than in copper; therefore a reasonably thick steel shield is more effective than a comparable one made of copper, and less expensive too. As EMI frequencies increase above 100 kHz and the wave impedance increases towards 377 ohms, copper shields become more effective than steel.

Holes, Joints, and Conductors Compromise a Shield's Effectiveness

In real life applications, a shield rarely performs as well as the shielding effectiveness of an uninterrupted sheet of the material would indicate.[2] A completely unbroken shielded box is, of course, not very useful since most electronic systems require external power, signal cables and ventilation. These lead to compromising the shielding effectiveness by routing cables through holes in the shield, by attaching current carrying cables to the shield and by diverting the noise-field-induced currents flowing in the surface of a shield. One of the common shield leakage paths is through ventilation slots or holes or other joints in a shield. The amount of shield leakage from a shield discontinuity depends on how much the hole or joint obstructs and diverts the shield currents needed to maintain the shield reflectivity to the EMI field.[1] Therefore the leakage from a shield discontinuity depends on the maximum linear dimension, not the area, of the opening. This means a large number of small holes in a shield will degrade the shielding effectiveness much less than a large hole of the same total area. Much better (>100 dB) shield performance can be obtained from ventilation holes if they are shaped to form a waveguide beyond cutoff structure. This is accomplished if the shield wall is extended to form one or more open tubes, where the diameter of the tube is much less than the cutoff frequency, [1][2] and the tube length is about 3 times the diameter.

Discontinuities in shields can allow fields to radiate inside the shielded space. If the discontinuities are small, less than $\lambda/100$, compared to the EMI wavelengths, this effect is small. An effective way to think about this is that a slot in a metal shield produces a radiated field through the shield just as if it was an antenna of the same dimensions as the slot, driven with the incident wave's power.[2] Narrow slots, longer than 1/20 λ, can cause significant leakage. Maximum radiation occurs when the slot length is ½ λ. Fasteners at a seam may form slot antennas; fasteners should be installed <1/50 λ apart, or a conductive gasket should be used.

A similar degrading effect comes from a cable routed through a hole in a shield. An unshielded wire will act as an antenna on the outside of the shield, picking up the EM field, and conducting the noise inside, to re-radiate it into the shielded volume if the exposed lengths of the wire exceed 1/20 λ. This is a very common way to degrade a shield and ac power entrances are the usual culprits. However, if the wire is shielded by an extension of the outside shield on the outside of the shielded enclosure, and by a similar shield on the inside, then the wire may be brought through the shield with minimal degradation to the shielding effectiveness. Alternatively, for ac power entrances that preserve the shield integrity, the power leads should enter the enclosure near the ground point; the power leads should be fully enclosed in enclosures several skin depths thick; the power system ground should not be allowed to penetrate the shielded space; and a filter should be used on every power lead in high impedance EMI fields, or an isolation transformer for low impedance fields.[8] These steps will minimize shielding degradation from the ac power entrance.

Grounding To Reduce Conductive Coupling of EMI

An effective rule-of-thumb for ground system design is to think of ground as a low impedance conductor rather than a zero-potential plane. Inadequate grounding is an example of a failure to think of the performance of conductors at high frequencies. We usually think of a ground as a zero-impedance, equipotential surface, like a perfect conductor. Currents with frequency components from dc up into 100s of MHz typically pass through ground conductors. At frequencies in the MHz range, resistance of the conductor, even including skin effects, is negligible compared with the impedance of the ground conductor inductance. It is important to think of "ground" as a path for current to flow, instead of an equipotential surface to design ground systems that are effective in mitigating EMI.

Choice of a Grounding System

Several grounding systems are available for an accelerator facility: bus, single-point, multipoint (distributed), and hybrid ground. The bus grounding system, commonly used in ac power systems, is adequate for safety; but due to its nature, noise currents produced by electrical devices connected in series to a common ground bus couple directly into other series connected devices through common ground impedances.

A single-point grounding system, where each device has a separate ground conductor, mitigates this direct coupling problem. Usually a ground near a control center is chosen as the single grounding point and all devices are connected in parallel, in a star structure, to that single point. The single-point approach is a good grounding method for low frequency EMI, but the system suffers when interconnection must be made between devices at the 'points' of the star structure, or when the distances between the 'points' and the single ground point exceed $\sim\lambda/10$ of the noise field. Ground currents in conductors longer than $\sim\lambda/10$ give rise to significant ground potential differences and EMI radiating from the ground conductors. The first issue may be avoided by using isolators in all connections between devices connected to separate ground conductors, but the second is a definite problem in a facility the size of a typical accelerator.

Multipoint (distributed) grounding systems operate well where high-frequency interference is the problem: rf noise is reflected/absorbed at the many ground connections and is attenuated before affecting the equipment. However, at low frequency, 60 Hz to 100 kHz, ground loops abound and low-impedance EMI sources (like klystron modulators) can cause high currents to flow in the ground connections.

This current causes interfering potential drops and unsatisfactory instrumentation grounding when used on moderate to large-scale systems.

Hybrid grounding systems have some of the advantages of single point and multipoint systems, without some of the disadvantages. The hybrid system reduces to a single-point grounding system at low frequencies and has few problems with low-frequency grounding loops. With the addition of an underlying ground plane, normal stray or externally added capacitance to this plane makes the system behave like a multipoint grounding system at rf frequencies, overcoming the limitations of ground impedance in long connections to a single ground point. The hybrid system thus has lower ground impedance at higher frequencies and is more capable of causing rf reflection on cables than the single-point system. The only limitation of the hybrid system, which is ideal for systems with a large bandwidth, is overall size, since the ground plane voltage drops and large parasitic capacitance that can cause ground loop currents become significant when the size of the plane exceeds $\sim\lambda/10$.

For most large accelerator facilities, a distributed hybrid grounding system would perform better than other grounding systems because of large facility size, wide noise bandwidths, and mechanical restrictions. Local hybrid ground areas could be defined in the facility, perhaps centered around each instance of an rf station or each group of instrumentation racks, with each area respecting the $\sim\lambda/10$ size limitation. The distributed ground reference planes for the hybrid system could consist of a conductive mesh embedded in the facility floors, along with connections to the grounded building steel and rebar. This mesh would provide an approximation to a local low-potential grounding plane that could be built within cost and mechanical limitations. The local "single-point" ground references for the distributed hybrid ground system could be extensions of the facility-wide common ground provided by the accelerator structure in the tunnel. This would enable many diagnostic cables to follow a reference ground plane from the accelerator to the racks. Metal conduits, rf wave-guides or 8" to 12" wide copper sheet ground reference conductors, connected to the accelerator/beam-line structure, could provide the reference for the single-point ground for each local hybrid ground area. Employing non-conductive instrumentation interconnections between the local ground areas would mitigate the effects of the potential differences in the large reference plane arising from the large facility size.

Connection of Diagnostic System Equipment to Ground

Electrical safety requires that all equipment accessible to an operator be grounded in a manner that limits human accessible potentials to less than 50 volts, even during transient faults. The manner of grounding, the design of the equipment, the interconnections between equipment, and inadvertent ground connections caused by power or cable entrances will all affect EMI performance.

Design of Equipment Grounds

The reference potential in each piece of equipment, a chassis, a rack, a power source, or a screen-room ground stud, should be considered as a separate subsystem ground. This subsystem ground should be extended to form a shield around all components, cables, sensors, and outputs directly connected to that subsystem. This shield will provide a common reference potential for the subsystem within the following constraints:

1. The distance to the local ground point for any subsystem should be less than λ/10 (some sources give a conservative λ/20).[1] λ/10 is ~10 m in a 3 MHz EMI field.
2. No appreciable currents should flow in any EMI shield.
3. The shield should not be interconnected with other subsystems or their shields at more than one point to avoid creating ground loops.

These three design constraints can be met. The first criterion puts a limit on the size of a subsystem and thereby requires some isolation means, like fiber optics or signal isolators, to connect distant (>λ/10) grounded devices to a subsystem, e.g. connections to remote transducers, facility data networks and site-wide timing systems.

The second criterion, no significant current flow in shields, can be met in several ways. It requires the inclusion of a current return path for every signal and supply within the subsystem shield. It also may require the use of shielded twisted-pair coax with electrically thick shields or triax cables for signals, as determined by the fields around each cable. Because strong low impedance field coupling can cause significant currents to flow in subsystem shields, the second criterion also requires application of magnetic field control methods around the subsystem shields where the field strength is high or the loops are large. This constraint can be met by enclosing the shielded signal cables in thick enclosed cable trays or in rigid metal conduits. Routing the cables on/in an extension of the local ground plane to reduce loop area can also reduce magnetic field coupling. A third approach to reducing currents in shields is to replace the conductive signal cables with a nonconductive medium, fiber optics, or to insert signal isolators in the conductive paths.

The third criterion requires that subsystem shields connect to each other only near the local area single-point ground. Because connection to the designated ground point is required for safety reasons, and because additional connections to grounded equipment more than ~λ/10 distant in a low impedance noise field form ground loops, other conductive interconnections cannot be allowed. If such interconnections are required, they should be made through an isolating medium, e.g., isolating devices, shielded transformers or fiber optics.

The third criterion causes difficulty with the connection of the subsystem to a power source. The direct connection to a line power source ground would constitute an additional ground connection creating a ground loop, and isolation with a transformer could leave a fairly low-impedance ground path through the transformer stray capacitance. The isolation of this ground path can be improved with a shielded transformer, which can also reject common mode and transverse mode noise present on the line. Power line filters are not desirable for use in a low impedance EMI environment because they convert the noise induced currents into significant ground currents that can generate an additional source of near-field low impedance EMI that can then penetrate shields. Filters can also cause appreciable current flow in grounds and shields, in violation of criteria 2.

From the above discussion, an ideally grounded and shielded subsystem would be one located near the local area single-point ground and connected to it with a low-impedance conductor. It would be totally enclosed by a reference-potential shield with a properly grounded shielded transformer power supply and would be interconnected to other distant grounded devices by isolated signal cables or fiber optics.

12 Rules for Grounding and Shielding Accelerator Diagnostics

This section condenses the previous sections into rules for equipment, interconnection, and ground-connection design. These rules are meant to guide a designer in providing diagnostics compatible with an accelerator EMI environment.

1. Mutually ground all operator-accessible equipment to the designated local ground point.
2. Minimize magnetic flux coupling to/from cables: route cables on the ground plane, use enclosed raceways or conduit for cabling and keep raceways close to the ground plane unless they are constructed of material >10 skin depths thick.
3. Use single-point grounding practices with respect to each local area ground point. Connect each of the subsystem grounds to the local ground point with a low-impedance conductor.
4. Contain each subsystem in an EMI shielded enclosure connected to the subsystem reference. The shielding effectiveness of the enclosure should be scaled to the local EMI field strength and field impedance and the receptor susceptibility.
5. Keep currents through shields and ground connections to a minimum. Provide current return paths in the same cable or tray for every source or supply. Do not use thin shields for current return; use balanced sources and cable if possible. Signal cable shields from ungrounded signal sources should be grounded only on one end.
6. No conductive cables should be allowed to interconnect subsystems connected to separate single-point grounds.
7. Shields in any subsystem, including cable shields, should be shorter than $\sim\lambda/10$. Cables inside modulators or other high power equipment shields should be isolated and/or be shorter than $\sim\lambda/50$ and be provided with common mode chokes where they exit the shield.
8. Eliminate or minimize conductive cables entering shielded enclosures. Position cable entrances near the ground point/power entrance. Use fiber optics or isolators where possible. If not possible, cables should meet restrictions of Rules 5, 6 and 7.
9. Power entrances to subsystems often define the ground point. Respect them; group all required ground connections together to minimize currents in shields.
10. Use shielded transformers, properly connected, for power supplies. Avoid the use of line filters with low impedance EMI; use filters only with respect to the ground point they define.
11. Completely shield all high di/dt power sources. Noise radiating cables should be shielded and grounded at both ends, run high power cables in metal conduit. Avoid openings or discontinuities in shields whose maximum dimension exceeds $\sim\lambda/50$. Include trigger sources and rf amplifiers inside the shield if possible.

12. Establish and mark local subsystem ground points that do not violate ground rules or impair system integrity for casual use by technicians and experimenters. Show grounds on the system prints.

EMC Step 4, Example 1: Grounding/Isolating The Injector AC Power

Proper ac power grounding can reduce the coupling of spark-down transients into the instrumentation system. There are three grounding systems of particular interest on the APT injector: the AC power system, the HVPS system and the diagnostics signal cable system. After analysis of the circuit estimated the transient frequencies, in step 1 above, an elegant solution to the issue of ac power conducted interference was indicated. The >75 mil thickness of the steel used in enclosures and conduits of the electrical power distribution systems are >10 skin depths thick at the 3 MHz to 6 MHz transient frequencies, which allows transient EMI grounds to act as if they were electrically independent from the ac grounds. This fact allows the use of conduits and junction boxes for barrier shields and permits effective safety and transient current grounds to be easily made. One three-phase shielded isolation transformer was employed to isolate the ac power to the HVPS, thereby limiting the coupling of the spark transients into the AC power lines. The transient-carrying secondary wiring of this transformer was enclosed in rigid conduit to prevent coupling to the diagnostics environment. A second shielded three-phase transformer provided three isolated single-phase 115 V power sources: clean power for the diagnostics and computer, "semi-clean" power for the vacuum controls, and "dirty" power for the injector auxiliary power supplies, each wired in separate conduits and grounded at separate distribution panels. By employing this ground/isolation plan, the spark-down caused interference was mitigated and the injector and the instrumentation operated reliably.

EMC Steps 5 & 6: Identify and Characterize Receptors Then Minimize the Receptors EMI Susceptibility

Once the diagnostic system designer arrives at these last two steps, the source emissions have been characterized and should be mitigated; a grounding system has been chosen and your system should be properly grounded; and shielding has been selected and installed to isolate the diagnostic electronics from the EMI environment. That should leave us with no further EMI problems. Well, perhaps, but more likely the mitigation of the sources left an EMI field that does not destroy the diagnostics, but merely wipes out your noise margins. And, correct grounding and shielding certainly helped, but somehow noise is still coupling into your system. Now where do you look? *Look at your cables.* Since the cables have the largest physical dimensions of any component in a diagnostic system, they form the largest antenna for coupling EMI, so they are likely the most susceptible component to noise. By proper grounding and shielding of cables the designer can mitigate a dominant EMI coupling mechanism.

Electric fields, or capacitive coupling, can transfer noise energy to an unshielded conductor. The electric field coupled noise voltage increases with frequency, coupling capacitance and receptor impedance to ground.[1] A shield around the conductor may reduce this capacitive coupling, but grounding the interposed shield can reduce the noise voltage to zero. For effective electric field shielding, use shields that provide ~100% coverage, minimize the lengths of unshielded portions of conductors, and provide a good ground on the shields. A single ground connection makes a good shield ground if the cable is less than ~$\lambda/10$ long. On longer cables, multiple grounds may be required to avoid resonance effects as the cable length approaches $\lambda/4$.

Magnetic fields, or inductive coupling, can transfer noise energy through the mutual inductance between circuits. The magnetic field coupled noise voltage increases with frequency, source current and coupling mutual inductance, which is proportional to the area enclosed by the disturbed circuit and is independent of receptor impedance to ground. To reduce the noise voltage, reduce the receptor circuit loop area, increase the distance to the source or otherwise decrease the flux, or change the orientation of the loop to link less flux. An ungrounded or single point grounded non-magnetic shield around a conductor has *no effect* on the magnetically induced noise voltage since the shield does not reduce the linked flux in any way.

The magnetic coupling, the mutual inductance, between a shield and an inner conductor is equal to the self-inductance of the shield.[1] This curious fact can be visualized by considering a current carrying tubular shield conductor that generates a magnetic field outside, but not inside. All the flux produced by this conductor would encircle an inner conductor placed inside the tubular conductor and the flux from their mutual inductance would be identical to that from the shield self-inductance, so the mutual inductance is equal to the shield self-inductance. The mutual inductance equivalence leads to another useful result, at high frequencies: *the noise voltage induced in the center conductor of a coax equals the voltage drop on the shield*, if shield current is allowed to flow. The noise voltage induced into the center conductor is zero at dc and increases to equal the shield voltage at a frequency of ~5 times the shield resistance/shield self-inductance radians/s. This frequency is called the shield cutoff frequency and is typically in the range of 0.6 kHz to 2 kHz for coax cables.[1]

Noise levels induced in signal cables in a low impedance field are very sensitive to grounding. The best way to protect against induced noise is to decrease the area of the signal loop above the ground plane. Connecting the cable shield so that all of the signal current will return on the shield is the best way to reduce the coupling area. If the signal circuit is grounded on both ends, then some degree of protection from the low impedance field can be gained by grounding the shield on both ends and allowing the low inductance path created by coaxial current flow to force the return current to follow the shield at frequencies above cut-off. However, if a circuit is grounded at both ends, only a limited amount of magnetic field protection is possible because ground loops induce significant noise currents into the shield and these produce a shield voltage drop that couples into the signal.[1] For maximum low impedance field noise protection at low frequencies, the cable shield should not be one of the signal conductors, that is, one end of the signal circuit should be isolated from ground. If the signal circuit is grounded at both ends, minimize the loop area and ground potential differences and use a cable with a shield that is many skin depths thick at the noise frequencies.

Solutions to the Quandary About Current Flow in Cable Shields

As we have seen, a coax cable shield grounded at one point provides no shielding effect against low impedance (magnetic) fields. A cable shield grounded at both ends provides some magnetic field shielding, but the resulting current in the shield generates a shield voltage drop, a noise source that couples into the signal path through the mutual inductance. Is there a way to let shield currents provide some magnetic field shielding for cables, but keep the shield current from coupling noise into the signal conductors?

Tri-axial cable can be used to provide a separate shield conductor for the shield current by grounding both ends of the outer shield while isolating the inner shield for the signal return current, but tri-axial is expensive and clumsy to terminate properly. Also, the braided outer shield is not many skin depths thick and may couple low frequency noise onto the inner shield. As an alternative, twin-axial cable also keeps the shield current off the signal conductors and any noise coupled into the inner twisted conductors is canceled by the twisted structure. But, termination of twin-axial is somewhat more difficult than coax and many types of twisted shielded pair cable have high shield impedance leading to higher shield voltage drop and poor performance at very high frequencies.

A third solution is to make the coax shield many skin depths thick at the noise frequencies to force the EMI induced currents to flow on the outside surfaces of the shield, leaving the inside surface with low shield current generated voltage drop. This solution, often used in rf system rigid coax, could work well at lower frequencies too, if sufficiently thick shields could be obtained. However, the cable shield does not have to be part of the actual signal carrying coax for this solution to be effective – the shield can be a metal conduit, fully enclosed cable duct or any structure that fully encloses the signal cables in a conductive material that is 6 to 10 or more skin depths thick. One of the most cost-effective ways to provide this shield in a new facility is to install rigid metal conduit between the beam-line and the diagnostic electronics racks. The threaded joints provide excellent continuity and the thick ferrous walls are many skin depths thick even at low frequencies. Since the shield wall is many skin depths thick, it does not matter if the conduit is grounded at multiple points, as it would be if installed according to standard practice. EMT conduit can work as well for many applications where the slightly lower shielding effectiveness at low frequencies is not an issue, as long as the joints are assembled correctly and tightly to insure good continuity.

EMC Step 5&6, Example 1: EMI Coupling to Cables from the HVCM

After the HVCM began operating at SNS, disruptions in the previously functioning facility-wide timing system were detected. By placing a wideband current transformer around some of the timing signal distribution cables nearby, a 100 to 400 mA common mode current could be measured that correlated to HVCM operation. Shield currents like these generate a voltage drop along the shield that couples to and interferes with the signal, if the shield impedance is high or if the cable is long. The low impedance field from the HVCM coupling to these cables that had the shield grounded on both ends was the most likely coupling path for the observed EMI. To reduce the susceptibly of these cables, signal isolators could be used to break the shield current path; some of the terminal equipment could be relocated to reduce the cable lengths; the cables could be routed closer to the facility ground plane to reduce the loop area; or the cables could be installed in an enclosed raceway or conduit. All of these alternatives can reduce the shield current induced by the HVCM low impedance field, so implementing any one of these could resolve the timing system disruption.

CONCLUDING REMARKS

Each particular EMI problem will occur in an environment determined by characteristics of the EMI sources, a range of EMI frequencies, the distances to the sources, and the dimensions and other characteristics of the receptors. These characteristics determine the impedance of the EMI field emitted from the source, the change in that impedance during transit to the receptors, the coupling mechanisms at the receptors, and the resulting effectiveness of grounding, shielding or other countermeasures. The EMC assessment process starts with the sources of EMI, then identifies coupling mechanisms, and finally evaluates EMI mitigation measures for the receptors based on the characteristics of the conducted noise currents or radiated fields. If the source and coupling mechanisms are not understood, various EMI control techniques may be effective, ineffective or counterproductive when applied to a particular EMI problem. We have seen that a good EMC system design is the result of a multi-step process that begins with characterizing the sources of EMI and includes analysis of coupling paths. The resulting information about the field impedance, frequency and magnitude of the EMI helps to select cost-effective measures to reduce the susceptibly of diagnostic system receptors.

ACKNOWLEDGMENTS

The author acknowledges Henry Ott for his most useful book "Noise Reduction Techniques in Electronic Systems", which has been a guide for 27 years, containing many simple, clear explanations of the concepts and complexities of EMI. Also, I gratefully acknowledge my colleagues at LANL, Gary Allen, Lloyd Gordon, Jorg Jansen, Bill Reass, Joe Sherman, Dave Warren and Tom Zaugg; and at SNS, Bill DeVan, Alan Jones, and Coles Sibley with whom I had the pleasure of working on the projects mentioned in this paper and who contributed their time and support to understanding many EMI issues.

REFERENCES

1. H. W. Ott, *Noise Reduction Techniques in Electronic Systems*, 2nd ed., Wiley-Interscience, New York, 1988.
2. C. R. Paul, *Introduction to Electromagnetic Compatibility*, Wiley-Interscience, New York, 1992.
3. D. R. White, *Electromagnetic Interference and Compatibility*, Vol. 3, D. W. Consultants, Germantown, Maryland, 1973.
4. W. A. Reass et al., "Design and Status of the Polyphase Resonant Converter Modulator System for the Spallation Neutron Source Linear Accelerator," LINAC 2002 Proceedings, Gyeongju, Korea, 2002, pp. 546-550.
5. D. R. White, *Electromagnetic Shielding Materials and Performance*, D. W. Consultants, Germantown, Maryland, 1975.
6. J. D. Sherman and M. E. Thuot et al., "Status Report on a dc 130-mA, 75-keV Proton Injector," *Rev. Sci. Instrum.* **69**, 1003-1008 (1998).
7. R. B. Cowell, "Nomograms Simplify Calculations of Magnetic Shielding Effectiveness," *EDN* Magazine (September 1972).
8. E. F. Vance, "Electromagnetic-Interference Control," in *IEEE Transactions on Electromagnetic Compatibility*, Vol. 22, no. 37.

Invited Talks

11th Beam Instrumentation Workshop
May 3–6, 2004

Results

Measurement of Bunch Length Using Spectral Analysis of Incoherent Radiation Fluctuations*

V. Sajaev

Argonne National Laboratory, Argonne, Illinois 60439

Abstract. A measurement of the longitudinal beam profile of a relativistic charged particle beam is an important tool in modern accelerators. For bunch lengths in the range of picoseconds, such measurements can be performed by means of a streak camera. Shorter bunches usually require special techniques. In this paper we describe a novel technique that allows obtaining properties of a bunch of charged particles through measurement of the fluctuations of incoherent radiation from the bunch. Due to shot-noise fluctuations in the longitudinal beam density, this incoherent radiation has a spectrum, which consists of random spikes with width inversely proportional to the bunch length. The convolution of the beam current can also be obtained from the radiation spectrum. After the convolution function is found, the phase retrieval technique can be applied to recover the bunch shape. This technique has been used to analyze the shape of the 4-ps-long bunches at the Advanced Photon Source self-amplified spontaneous emission free-electron laser (SASE FEL) experiment.

INTRODUCTION

Recent developments in physics and technology open an exciting world of very short sub-picosecond bunches. The length of electron bunches that are going to be used in the next linear colliders is of the order of 1 picosecond.[1] The projects of X-ray free-electron lasers (FELs) require even shorter bunches, down to 100 femtoseconds or less. The Linac Coherent Light Source (LCLS) [2] under construction at SLAC utilizes electron bunches as short as 80 femtoseconds to produce self-amplified spontaneous emission (SASE) radiation in an FEL. The new Subpicosecond Pulse Particle Source (SPPS) [3] already can produce electron pulses as short as 30 femtoseconds rms.

The production and tuning of these short bunches are crucial to the performance of these colliders and FELs, but the measurement of such ultra short bunches is an interesting challenge in itself. For bunch lengths in the range of picoseconds, such measurements can be performed with a streak camera. But shorter bunches require special techniques. They include high-power rf transverse deflecting structures that streak the beam in the accelerator allowing the bunch length to be observed on a profile monitor.[4][5] Electro-optic crystal diagnostics use the electric field of the electron bunch to modulate the light emitted by high-bandwidth, femtosecond visible lasers thereby allowing one to recover the bunch length.[6] Coherent synchrotron

* Work supported by U.S. Department of Energy, Office of Basic Sciences, under Contract No. W-31-109-ENG-38.

CP732, *Beam Instrumentation Workshop 2004: Eleventh Workshop*,
edited by T. Shea and R. C. Sibley III

radiation from dipole magnets can also be detected in a terahertz band at wavelengths comparable to the bunch length.[7]

In this paper we describe a method proposed previously [8, 9, 10], which allows obtaining properties of a bunch of charged particles through measurement of the fluctuations of incoherent emission from the bunch. Emission can be produced by any kind of incoherent radiation: generated in a bend or wiggler, transition or Cerenkov radiation, etc. And, unlike the techniques using coherent radiation, this method does not set any conditions on the bandwidth of the radiation. These conditions are rather set by available detectors.

This paper begins with a theoretical description of the method. Then it describes the instrumentation used for this experiment. The rest of the paper is devoted to the experimental results on single-shot spectral measurements and on recovery of the bunch profile. Simulation results are also presented to check and confirm the technique used.

THEORY

Let us consider a microscopic picture of the bunch, where each particle radiates an electromagnetic pulse with the electric field given by a function $e(t)$. (The specific expression for this function depends on how the radiation was generated: in the dipole, or in the wiggler, etc.) The total radiated field $E(t)$ of all particles is

$$E(t)=\sum_{k=1}^{N} e(t-t_k),$$

where N is the total number of particles in the bunch and t_k defines the longitudinal position of the k-th particle within the bunch. We assume that t_k are random numbers, with the probability to find t_k between t and $t + dt$ equal to $f(t)dt$, where $f(t)$ is the bunch distribution function normalized to 1.

The spectral properties of the radiation can be obtained from the Fourier transform of the radiated field:

$$\hat{E}(\omega)=\int_{-\infty}^{\infty} E(t)e^{i\omega t}dt=\int_{-\infty}^{\infty} e^{i\omega t}dt\sum_{k=1}^{N} e(t-t_k)=\hat{e}(\omega)\sum_{k=1}^{N} e^{i\omega t_k},$$

where $\hat{e}(\omega)=\int_{-\infty}^{\infty} e(t)e^{i\omega t}dt$. In the real experiment, one measures the power spectrum of the radiation $P(\omega)$, which is proportional to $\left|\hat{E}(\omega)\right|^2$ (for simplicity we take $P(\omega)=\left|\hat{E}(\omega)\right|^2$):

$$P(\omega)=\hat{E}(\omega)\cdot\hat{E}^*(\omega)=\hat{e}(\omega)\hat{e}^*(\omega)\sum_{k=1}^{N} e^{i\omega t_k}\sum_{m=1}^{N} e^{-i\omega t_m}=\left|\hat{e}(\omega)\right|^2\sum_{k,m=1}^{N} e^{i\omega(t_k-t_m)}. \quad (1)$$

Averaging this equation, we find

$$\langle P(\omega)\rangle = |\hat{e}(\omega)|^2 \sum_{k,m=1}^{N} \int_{-\infty}^{\infty}\int_{-\infty}^{\infty} dt_k dt_m f(t_k) f(t_m) e^{i\omega(t_k - t_m)}$$

$$= |\hat{e}(\omega)|^2 \sum_{k=m}^{N} \int_{-\infty}^{\infty}\int_{-\infty}^{\infty} dt_k dt_m f(t_k) f(t_m) e^{i\omega(t_k - t_m)} + |\hat{e}(\omega)|^2 \sum_{k\neq m}^{N} \int_{-\infty}^{\infty}\int_{-\infty}^{\infty} f(t_k) e^{i\omega t_k} dt_k f(t_m) e^{-i\omega t_m} dt_m$$

$$= |\hat{e}(\omega)|^2 (N + N^2 |\hat{f}(\omega)|^2), \qquad (2)$$

where $|\hat{f}(\omega)|$ is the Fourier transform of the distribution function, and we used the approximation $N - 1 \approx N$. The first term in Equation (2) is incoherent radiation proportional to the number of particles in the bunch. The second term is the coherent radiation that scales quadratically with N. The coherent radiation term carries information about the distribution function of the beam but only at low frequencies of the order of $\omega \cong \sigma_t^{-1}$, where $\hat{f}(\omega)$ is not zero. At high frequencies $N|\hat{f}(\omega)|^2 << 1$, and the coherent radiation is negligible in comparison with the incoherent one.

To illustrate the difference between coherent and incoherent radiation intensities, let us consider a numerical example for the case of a Gaussian beam with $\sigma_t = 1$ ps and total charge of 1 nC (approximately 10^{10} electrons). The Gaussian distribution function and its Fourier transform have the forms

$$f(t) = \frac{1}{\sqrt{2\pi}\sigma_t} e^{-\frac{t^2}{2\sigma_t^2}} \quad \text{and} \quad \hat{f}(\omega) = e^{-\frac{\omega^2\sigma_t^2}{2}}. \qquad (3)$$

At 1 THz frequency, where $\omega\sigma_t \approx 1$, the coherent term in Equation (2) is about 10^{10} times stronger. At $\omega\sigma_t = 10$, the coherent term is 10^{34} times weaker.

However, the original, not averaged expression for the spectral power [Equation (1)] shows that the properties of the radiation, even at high frequencies, carry information about the distribution function. Considered separately, each term of the summation $e^{i\omega(t_k - t_m)}$ oscillates as a function of frequency with the period $\Delta\omega = 2\pi/(t_k - t_m) \sim 2\pi/\sigma_t$. Because of the random distribution of particles in the bunch, the sum in Equation (1) fluctuates randomly as a function of frequency ω, and statistical properties of these fluctuations depend on the properties of the distribution function of the bunch.

To obtain quantitative characteristics of the radiation, let us calculate the average value of the product $P(\omega)P(\omega')$, which later can be used for calculation of the power spectrum autocorrelation:

$$\langle P(\omega)P(\omega')\rangle = |\hat{e}(\omega)|^2 |\hat{e}(\omega')|^2 \left\langle \sum_{k,m,p,q=1}^{N} e^{i\omega(t_k - t_m) + i\omega'(t_p - t_q)} \right\rangle.$$

The N^4 terms in this summation can be placed in 15 different classes.[11] For the incoherent radiation, only two classes are important: $k = m$, $p = q$, $k \neq p$ and $k = q$, $m = p$, $k \neq m$. Thus it can be found that

$$\langle P(\omega)P(\omega')\rangle = N^2 |\hat{e}(\omega)|^2 |\hat{e}(\omega')|^2 \left(1 + \left|\hat{f}(\omega - \omega')\right|^2\right). \tag{4}$$

We can now define the following second-order correlation function:

$$g(\omega - \omega') = \frac{\langle P(\omega)P(\omega')\rangle}{2\langle P(\omega)\rangle\langle P(\omega')\rangle},$$

and, using Equations (2) and (4), it follows that

$$g(\Omega) = \frac{1}{2}\left(1 + \left|\hat{f}(\Omega)\right|^2\right). \tag{5}$$

This important equation relates the spectrum autocorrelation function and the Fourier transform of the bunch shape. It can be used to find the bunch length.

Another useful quantity was introduced previously.[9] It is the Fourier transform of the spectrum.

$$G(\tau) = \int_{-\infty}^{\infty} P(\omega)e^{i\omega\tau} d\omega .$$

It can be shown that the variance, $D(\tau) = \left\langle \left|G(\tau) - \langle G(\tau)\rangle\right|^2 \right\rangle$, is proportional to the convolution function of the distribution function $f(t)$:

$$D(\tau) = A \int_{-\infty}^{\infty} f(t) f(t - \tau) dt . \tag{6}$$

Strictly speaking, knowledge of the spectral function $\left|\hat{f}(\omega)\right|$ (or convolution function) does not allow a unique restoration of $f(t)$, because the information about the phase of $\left|\hat{f}(\omega)\right|$ is lost. However, as has been shown [12], using a phase-retrieval technique, one can recover the beam profile in many practical cases. This technique will be discussed later.

The above consideration assumes a filament beam, with a negligibly small transverse size. In this case, the fluctuations of the spectral intensity $P(\omega)$ are 100%, and the correlation function g in Equation (5) varies from 1 at $\Omega = 0$ to 1/2 at $\Omega \to \infty$. This regime requires that the transverse beam size be smaller than the transverse coherence size of the radiation, which is of the order of $\lambda/2\pi\sigma_\theta$, where σ_θ is the angular spread of the radiation. In the opposite case, the fluctuations become less pronounced.

We also used a classical description of the radiation process and neglected quantum effects. This approach is justified if the quantum fluctuations of the number of radiated photons are negligible. One can derive the condition for the beam intensity when this requirement is met. If the source of the radiation is synchrotron light from a dipole, the number of photons that reach the detector in the frequency interval $\Delta\omega$ (they are radiated from $(1/\gamma)(\omega_c/\omega)^{1/3}$ radians of the beam trajectory, where ω_c is the critical frequency of the synchrotron radiation) can be estimated as [13]

$$\langle n_{ph} \rangle \approx \frac{1}{2}\alpha N_e \frac{\Delta\omega}{\omega},$$

where α is the fine structure constant, and N_e is the number of particles in the bunch. Assuming $\Delta\omega \sim \sigma_t^{-1}$ and visible light range, we can rewrite this equation as $\langle n_{ph} \rangle \approx 3 \cdot 10^5 (I/I_A)$, where I is the peak current in the beam and I_A = 17 kA is the Alfven current. If the number of photons is much larger than one, the quantum effects can be neglected.

For wiggler radiation, the estimate for the number of radiated photons is

$$\langle n_{ph} \rangle \approx 4\alpha N_e M \frac{K^2}{1+K^2}\frac{\Delta\omega}{\omega},$$

where K is the wiggler parameter and M is the number of periods in the wiggler.

If condition $\langle n_{ph} \rangle >> 1$ is not met, quantum fluctuations superimpose on the random fluctuations described above. However, the information about the pulse shape is still present in the measured signal, although larger statistics would be required to suppress the additional noise introduced by quantum fluctuations.

Finally, we would like to emphasize that measurements of the beam profile using spectral fluctuations can be done in one shot, if the width of the measured spectrum is wide enough. In this case, the averaging of different quantities can be done over the spectrum, rather than over multiple shots.

MEASUREMENTS

Recently, a number of measurements were reported that confirm the fluctuational nature of the spontaneous incoherent radiation. Most of these measurements were conducted at FEL-related facilities. First measurements of the single-shot spectra of the undulator spontaneous emission with resolution sufficient to demonstrate 100% fluctuations of intensity were performed at ATF/BNL.[14] Bunch length was extracted from the fluctuations, and emittance was estimated in that experiment. More extensive analysis of the spectral fluctuations was conducted at the Low-Energy Undulator Test Line (LEUTL) at the Advanced Photon Source at ANL.[15] Later, there were a number of reports demonstrating spectral fluctuations of synchrotron [16], Cerenkov [17], and transition radiation [18].

LEUTL Facility Description

In this part, we will describe in detail the results of single-shot spectrum measurements obtained at LEUTL.

The LEUTL system was built as an extension of the existing APS linac (Fig. 1). It consists of a high-brightness photocathode rf electron gun coupled to the 650 MeV APS linac and a long undulator system. The gun is driven by a picosecond Nd:glass laser and can generate a few picosecond electron pulses at 6 Hz frequency. A bunch compressor is installed after two accelerating sections of the linac to further reduce the electron bunch length. It is capable of compressing the beam by a factor of ten. The undulator system consists of nine identical 2.4-m-long sections separated by drift spaces where visible light diagnostics (VLD) stations are installed. Each VLD can deflect the undulator light and send it to the end station located at the downstream end of the undulator line.

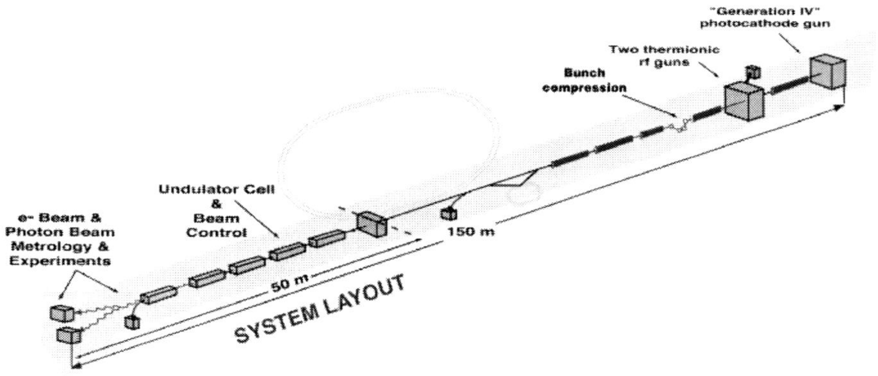

FIGURE 1. Schematic of the LEUTL system.

All spectral measurements reported in this paper have been obtained with a high-resolution spectrometer located at the downstream end of the undulator line in the end station. A mirror at each diagnostic station could direct the SASE light towards the spectrometer through a hole in the shielding wall, thus allowing one to measure spectral characteristics of the SASE light at different longitudinal locations along the undulator line.

A schematic of the spectrometer is shown in Fig. 2. It utilizes a Paschen-Runge mount. This design was chosen because of its great flexibility: it provides independence on the angle of the incoming light, it can be tuned for a wide range of wavelengths, and it is easy to modify. The spectrometer consists of three main elements all located on the Rowland circle: a vertical entrance slit, a spherical grating, and a CCD camera. The light coming from the undulator hall is focused on the entrance slit with a concave mirror. All optical elements are reflective with metal coatings. This allows the system to work over a wide range of wavelengths. The gated CCD camera can measure the radiation of each electron bunch separately. To reduce the dark current and to improve the signal-to-noise ratio, the CCD camera is cooled.

The spectrometer was calibrated with hollow cathode discharge lamps, and the designed resolution was checked on different wavelengths. The main parameters of the spectrometer are presented in Table 1.

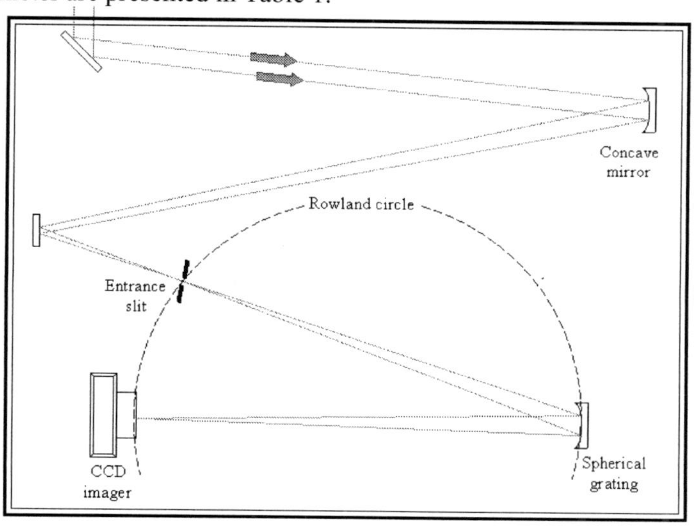

FIGURE 2. A top view of the Paschen-Runge type spectrometer for the analysis of the SASE FEL light.

TABLE 1. Main parameters of the high-resolution spectrometer

Grating	Grooves/mm	600
	Curvature radius [mm]	1000
	Blaze wavelength [nm]	482
CCD camera	Number of pixels	1100×330
	Pixel size [μm]	24
Concave mirror curvature radius [mm]		4000
Spectral resolution [Å]		0.4
Bandpass [nm]		44
Resolving power at 530 nm		10000
Wavelength range [nm]		250 – 1100

Spectrum Measurements and Bunch Length Extraction

Measurement of the typical single-shot spectrum is shown in Fig. 3 (top). The spectrum is composed of spikes of random amplitude and frequency that have a characteristic width $\Delta\omega \sim 1/\sigma_t$ and intensity fluctuation of almost 100%. The shape of the individual spectra changes randomly from shot to shot, but the average of many shots approaches the familiar wiggler spectrum, Fig. 3 (bottom).

To illustrate the spike width dependence on the bunch length, the bunch was compressed by the bunch compressor. Figure 4 presents the spectrum radiated by this much shorter bunch.

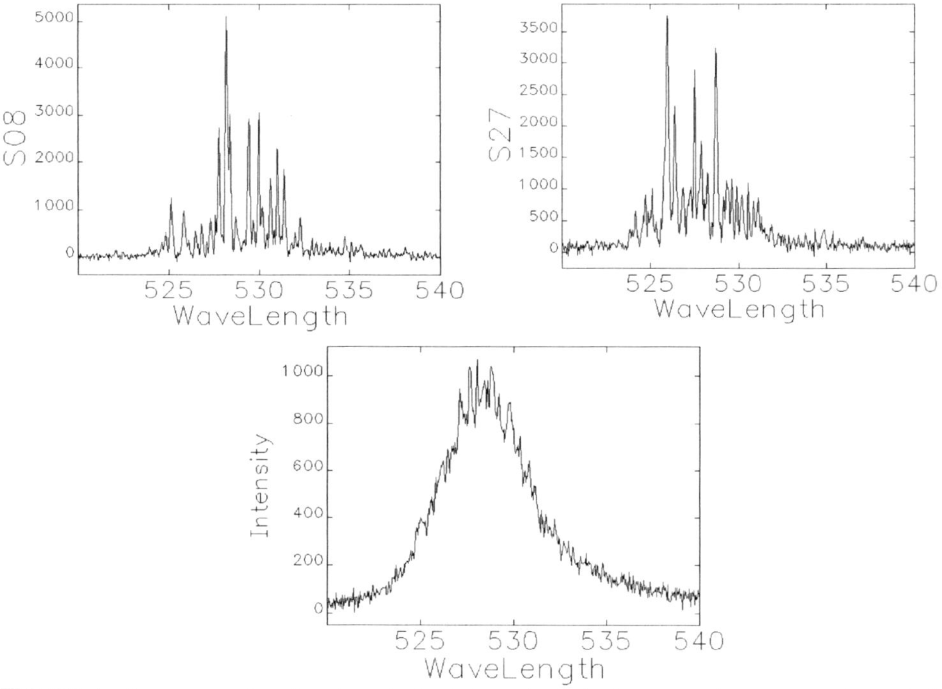

FIGURE 3. Top – examples of typical single-shot spectrum; bottom – spectrum averaged over 100 single shots.

To extract the bunch length from the spectral data, Equation (5) for the second-order spectrum correlation is used. The normalized correlation of the spectral intensity averaged over many shots was calculated from the spectrum measurements:

$$C_n = \left\langle \sum_i P(\omega_i) P(\omega_{i+n}) \right\rangle / \left\langle \sum_i P(\omega_i)^2 \right\rangle,$$

where $P(\omega_i)$ is the signal in the i-th CCD channel and n is the shift in channels. The correlation averaged over 100 shots is plotted in Fig. 5. To illustrate how the instrument resolution can affect the measurements, different sizes of the entrance slit of the spectrometer were used.

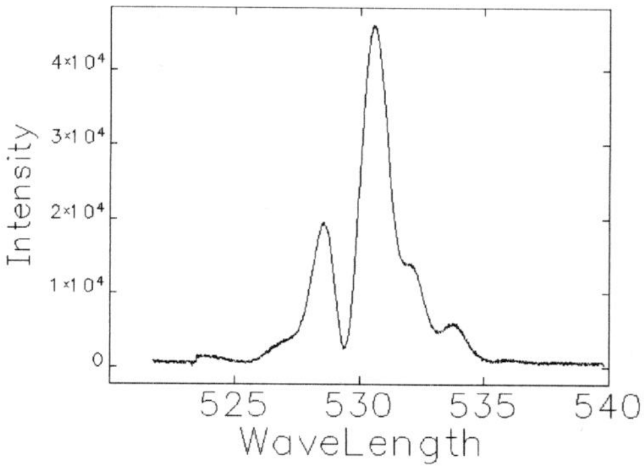

FIGURE 4. Single-shot spectrum for a very short bunch. The bunch was compressed by about a factor of 5 compared to the previous figure.

For the ideal case of a zero-emittance beam and a diagnostics with sufficient spectral resolution, 100% fluctuation of the spectral intensity will occur. However, when the beam size is large or the detector spectral resolution is poor, the spectrum will be similar to a spectrum emitted by several independent sources. The fluctuation level will be reduced, and a nonzero pedestal will appear in the spectrum.

In general, to achieve the best resolution, the entrance slit of the spectrometer should be less than the pixel size of the detector. But in our case the radiation intensity was not enough to work with slit openings less than 25 μm. Low resolution can lead to a situation where one detector channel sees more than one spike, which hampers the spike width measurements. The asymptotic level of the correlation curves at large n shows how large the pedestal was; for 100% intensity fluctuations, this level should be equal to 0.5 [see Equation (5)]. The value of this level can be used to characterize how many independent modes contributed to the radiation measured in one channel. To be able to directly apply Equation (5), the curve with the smallest pedestal should be chosen.

The spike width on half maximum according to Fig. 5 is about two pixels. The frequency step corresponding to one pixel is $\delta\omega = 2.4 \cdot 10^{11}$ rad/s. Therefore, assuming the beam to be Gaussian and using Equations (3) and (5), the sigma of the Gaussian distribution is

$$\tau_b \approx \frac{1}{n \cdot \delta\omega} \approx 2ps\,.$$

This gives the FWHM length of the bunch to be equal to 4.5 ps.

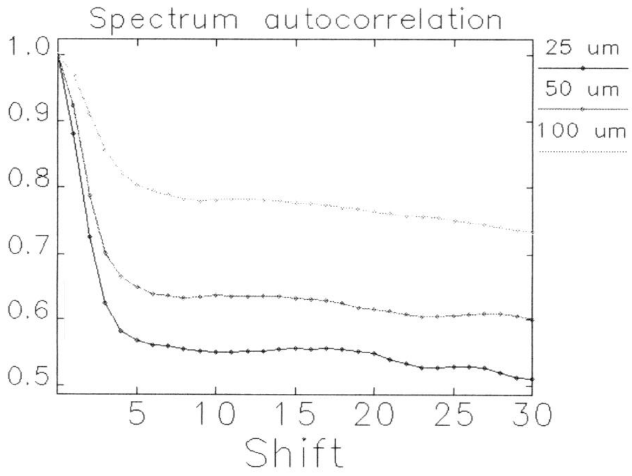

FIGURE 5. Spectrum autocorrelation for different entrance slit sizes (top = 100, middle = 50, bottom = 25). This plot illustrates how the instrument resolution can affect the measurements.

The accuracy of determination of the spike width is not so great because its size is only a few pixels. This means the resolution of the spectrometer is not enough for these bunch lengths. But the shorter the bunch, the wider the spikes, and the higher the accuracy; see, for example, Fig. 4.

Calculation of the Convolution Function of the Bunch Profile

As mentioned before, measurements of the spectral intensity fluctuations can be used not only to determine the bunch length but also to recover a longitudinal bunch profile. It has been shown [9] that the variance of the Fourier transform of the spectrum is proportional to the convolution function of the beam current. After the convolution function is found, a phase retrieval technique can be used to recover the shape of the pulse in many practical cases. These two steps are described below.

First, to calculate the convolution of the bunch current, we calculate the Fourier transform of the measured spectrum for measurement n:

$$G_{k,n} = \sum_{m=1}^{N_{ch}} P_{m,n} e^{2\pi imk/N_{ch}} ,$$

where m and N_{ch} are the channel number and the number of channels in the detector (CCD), respectively, and P is the detector signal. After accumulation of N_p number of pulses large enough for statistical analysis, the variance of the Fourier transformed spectrum is calculated the following way:

$$D_k = \sum_{m=1}^{N_p} \left| G_{k,m} - \frac{1}{N_p} \sum_{n=1}^{N_p} G_{k,n} \right|^2 .$$

It can be described in other words as the average deviation of the signal in the k-th channel from its average. As was mentioned above [see Equation (6)], the quantity D_k gives the convolution function of the particle density in the bunch averaged over N_p bunches.

Computer simulations have been performed to check this statement and confirmed it for bunches with Gaussian and step-function profiles. Figure 6 shows results of the simulation.

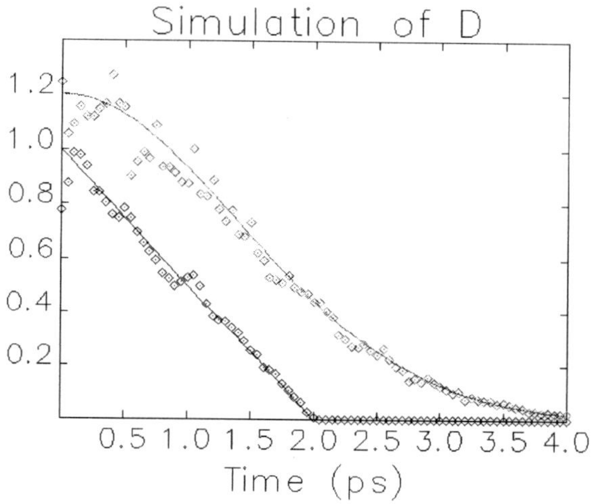

FIGURE 6. Convolutions of longitudinal bunch shape extracted from simulated single-shot spectra for Gaussian (top) and step-function (bottom) distributions. The number of particles used for the simulations was 100. The number of spectra for averaging was 200. Solid lines are analytical convolutions.

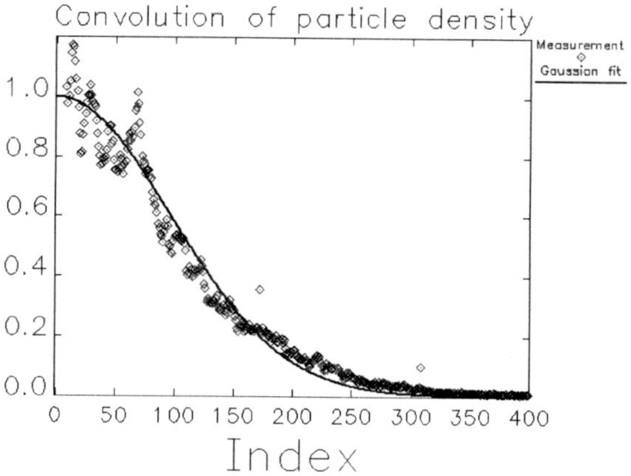

FIGURE 7. Convolution of longitudinal particle density calculated using measured spectra. Solid line is a fitted Gaussian function.

Figure 7 represents the convolution function of the longitudinal bunch distribution extracted from real spectral measurements. The number of spectrum measurements used for analysis is 100. This plot can also be used to determine the bunch length. If we assume that the bunch has a Gaussian shape, then its convolution is also a Gaussian with $\sigma = \sqrt{2} \cdot \sigma_t$. A fitted Gaussian function is also shown in Fig. 7 as a solid line. This plot gives the bunch length equal to $\tau_b = 1.8 ps$.

As mentioned above, the convolution function of the bunch profile does not allow for a strict restoration of the bunch profile itself. But, in practice, real beams often look like some non-symmetric combination of flattop and "Gaussian-like" functions. For these cases, using the Gaussian function alone to fit the convolution gives a good enough answer for the bunch length.

Reconstruction of the Bunch Shape from the Convolution Function

If the measurement of the bunch length using the convolution techniques described above does not provide enough information, one can apply a more sophisticated analysis of the convolution function. It is possible to extract both the amplitude and the phase information of the radiation source by applying a Kramers-Kronig relation to the convolution function. This technique of phase extraction was well developed in the optics of solids for the problem of reflectivity. We will briefly describe the technique following Ref. [12].

We denote the longitudinal particle density as $S(z)$ and its Fourier transform as $S(\omega)$ and write it in the following form:

$$S(\omega) = \rho(\omega)e^{i\psi(\omega)}. \tag{7}$$

Here $\rho(\omega)$ corresponds to the Fourier transform of our convolution function. Phase ψ can be extracted using the expression

$$\psi_m(\omega) + \psi_{Blaschke}(\omega) = -\frac{2\omega}{\pi} P \int_0^\infty dx \frac{\ln \rho(x)}{x^2 - \omega^2} + \sum \arg\left(\frac{\omega - \omega_j}{\omega - \omega_j^*}\right),$$

where ψ_m is the minimal phase and the ω_j's are the zeros of $S(\omega)$ in the upper half of the complex frequency plane. If $S(\omega)$ has no zeros, the contribution from $\psi_{Blaschke}(\omega)$ equals zero, and the expression above gives the minimal phase. This minimal phase is a good approximation to the actual phase in cases where the bunch density has no nearby zeros in the upper half of the complex frequency plane. The final expression for calculating the minimal phase is

$$\psi_m(\omega) = -\frac{2\omega}{\pi} P \int_0^\infty dx \frac{\ln[\rho(x)/\rho(\omega)]}{x^2 - \omega^2}.$$

Introduction of $\rho(\omega)$ in the numerator of the integral does not change the principal value of the integral and removes the singularity at $x = \omega$. The density distribution function can now be obtained from the inverse Fourier transform of Equation (7):

$$S(z) = \frac{1}{\pi c}\int_0^\infty d\omega \cdot \rho(\omega) \cdot \cos\left(\psi_m(\omega) - \frac{\omega z}{c}\right).$$

Figure 8 shows the simulated results of applying the above technique to restore different bunch shapes. The solid line is the original bunch shape, and the dots are distribution functions calculated from the convolution function with minimal phase approximation. One can see that, for some bunch shapes, this technique gives really good agreement, while for others it is not as good.

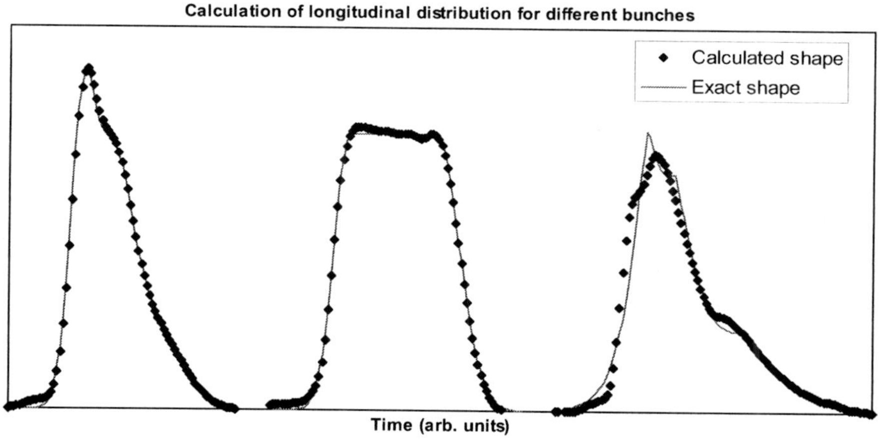

FIGURE 8. Simulation: exact longitudinal bunch distribution function compared with that calculated from the convolution function. First – sum of three Gaussian functions; second – step function combined with Gaussians; third – sum of three Gaussian-like functions $e^{-\frac{z^{1.3}}{2\cdot\sigma^{1.3}}}$.

Next, the technique described above is applied to the measured convolution function of the longitudinal bunch distribution (Fig. 7). One can see that the plot of the convolution function is still very noisy despite analyzing 100 single shots. This noise produces high-frequency content in the Fourier transformation. Therefore, before taking the Fourier transform of the convolution function, it has been smoothed by filtering out the high frequencies.

Results of calculations of the bunch shape for two different measurements (two sets of 100 single-shot spectra) are presented in Fig. 9. They were taken on the same day with an interval of several hours. The electron beam parameters were supposed to be the same, but small variations were possible. The two curves demonstrate almost the same profile of the beam with a FWHM length of approximately 4 ps.

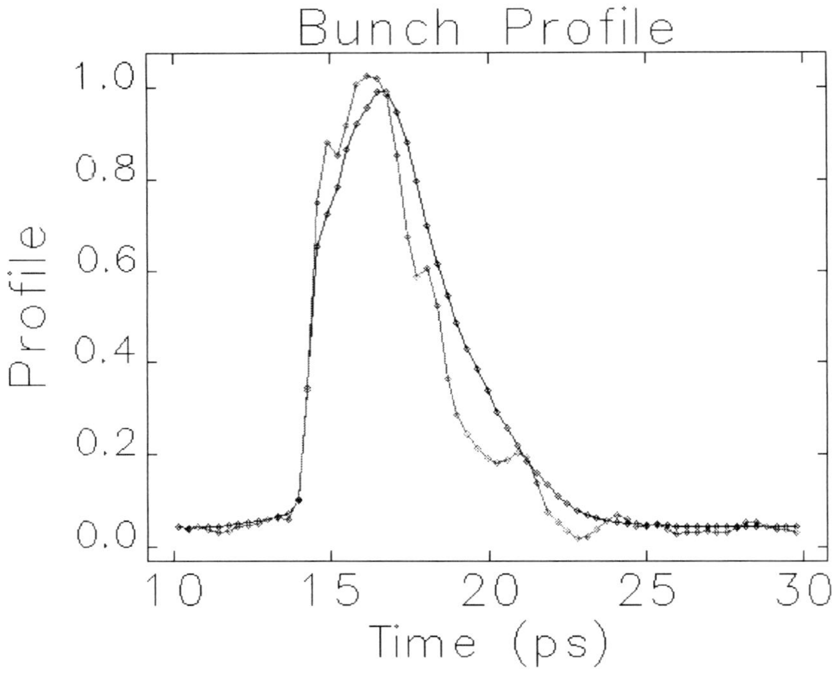

FIGURE 9. Bunch shape calculated from two different sets of measurements. The beam parameters were approximately the same, and the calculations show similar bunch profile for both measurements.

CONCLUSIONS

A technique for recovering a longitudinal bunch profile from spectral fluctuations of incoherent radiation has been implemented, and a bunch profile of a relatively short (4-ps) beam has been measured. Although we used synchrotron radiation above, the nature of the radiation is not important; it can be transition radiation, Cerenkov, etc. Typically, averaging over many (of the order of 10^2) single shots is required, however a modification of the method is possible in which one can perform averaging over wide spectral intervals in a single pulse.

To reconstruct the bunch shape from the convolution function we used the technique suggested in Ref. [12], but we used a different approach to build the convolution. Using spectrum fluctuations to construct the convolution allows us to avoid measuring the spectrum of coherent far infrared radiation of the bunch and making any assumptions about the asymptotic behavior of this spectrum.

An important feature of the method is that it can be used for bunches with lengths varying from a centimeter to tens of microns. However, there are several important conditions for this technique. In order to be able to measure a bunch of length σ_t, the spectral resolution of the spectrometer should be better than $1/\sigma_t$. Also, the spectral width of the radiation and the spectrometer must be much larger than the inverse bunch length. For example, Fig. 4 presents the situation when the radiation bandwidth (wiggler radiation) is not large enough; therefore one can observe only a few spectral spikes.

ACKNOWLEDGMENTS

The author would like to thank Max Zolotorev and Gennady Stupakov for very useful discussions on the method, Steve Milton and Efim Gluskin for stimulating and supporting this work, and Oleg Makarov for discussions and help with the measurements.

REFERENCES

1. "Zeroth-Order Design Report for the Next Linear Collider," LBNL-PUB-5424, SLAC Report 474 (May 1996).
2. "LCLS CDR," SLAC-R-593, April 2002.
3. M. Cornacchia et al., "A Subpicosecond Photon Pulse Facility for SLAC," SLAC-PUB-8950, LCLS-TN-01-7, Aug. 2001.
4. R.H. Miller, R.F. Koontz and D.D. Tsang, "The SLAC Injector," IEEE Trans. Nucl. Sci., June 1965, p 804.
5. X.-J. Wang, "Producing and Measuring Small Electron Bunches," Proc. of the PAC 1999.
6. X. Yan et al., Phys. Rev. Lett. 85, 3404 (2000); I. Wilke et al., Phys. Rev. Lett. 88, 124801 (2002).
7. T. Nakazato et al., Phys. Rev. Lett. 63, 1245 (1989); Y. Shibata et al., Nucl. Instrum. Methods Phys. Res. Sect. A 301, 161 (1991); E. B. Blum, U. Happek and A. Sievers, Nucl. Instrum. Methods Phys. Rec. Sect. A 307, 568 (1991).
8. M.S. Zolotorev and G.V. Stupakov, "Fluctuational Interferometry for Measurements of Short Pulses of Incoherent Radiation," SLAC Rep. SLAC-PUB-7132, 1996.
9. M.S. Zolotorev and G.V. Stupakov, "Spectral Fluctuations of Incoherent Radiation and Measurement of Longitudinal Bunch Profile," Proc. of the 1997 Particle Accelerator Conference.
10. J. Krzywinski, E. Saldin, E. Schneidmiller and M. Yurkov, "A New method for Ultra-Short Electron Pulse-Shape Measurement Using Synchrotron Radiation from a Bending Magnet," DESY Report DESY-TESLA-FEL 97-03.
11. J. Goodman, *Statistical Optics*, J. Wiley and Sons, New York, 1985.
12. R. Lai and A.J. Sievers, Micro Bunches Workshop, AIP Conference Proceedings 367 (NY), 1995, p. 312.
13. M. Sands, SLAC Rep. SLAC-121, 1970.
14. P. Catravas et al., Phys. Rev. Lett., **82**, 5261-5264 (1999).
15. V. Sajaev, "Determination of Longitudinal Bunch Profile using Spectral Fluctuations of Incoherent Radiation," Proc. of EPAC 2000, p. 1806.
16. L. DiMauro et al., "First SASE and seeded FEL lasing of the NSLS DUV FEL at 266 and 400 nm"; A. Murokh et al., "Results of the VISA SASE FEL Experiment at 840 nm"; V. Ayvazyan et al., "Study of the Statistical Properties of the Radiation from a VUV SASE FEL Operating in the Femtosecond Regime", all in Proc. of FEL Conference 2002.
17. T. Wanabe et al., "Diagnostics of Subpicosecond Electron Pulse by the Fluctuation Method," Proc. of PAC 2001, p. 2362.
18. A. Lumpkin et al., "Evidence for Micro-Bunching in a Saturated Free-Electron Laser Using Coherent Optical Transition Radiation," Phys. Rev. Lett., 88, 234801 (2002).

Beam Diagnostics Based on AC Modulation of System Parameters

Michael Tiefenback

Thomas Jefferson National Accelerator Facility, 12000 Jefferson Avenue, Newport News, VA 23606, USA

Abstract. To improve the performance of operating accelerators, quantities such as lattice functions, beam transfer functions, betatron frequencies, etc, can be measured turn by turn with beam position monitors or from difference measurements using step changes in system parameters. Spectral measurements in closed orbit machines provide accurate values for some properties. But for open-ended systems and some measurements in closed-orbit machines, periodic modulation can be very useful for obtaining information about the beam line. Using examples from existing machines, we compare and contrast beam based modulation techniques and step function or passive measurements. For example, large amplitude dipole modulation in rings [1] can be used in dedicated exploration of nonlinear optical properties without beam degradation, even allowing for tune spread effects. Low-level modulation can provide real-time system monitoring with no adverse effect on beam users. Examples considered include fully resonant dipole modulation in storage rings such as RHIC (hadrons) and PEP-II (electrons), and the continuous low-level modulation used in the CEBAF recirculating electron linac for real-time feedback to improve availability.

INTRODUCTION

There are many ways to measure certain physical quantities. One can measure the height of a tree by cutting it down and using a tape measure from top to root. One can measure the current or charge in a beam by intercepting it on a suitably configured Faraday cup. Both of these measurements come at the price of destroying the item of interest. Non-invasive measurements often exist. The purpose of this paper is to discuss examples of measurements which involve AC modulation of beam or beam transport parameters, as opposed to simple step function changes, pointing out why the chosen technique is used and what tradeoffs are made in ease of use, avoiding degradation of beam, etc. This type of manipulation of beam or transport parameters is often called "dithering." The reasons for choosing such a system might be diverse: to provide a speedy update rate to facilitate initial tuning of system, to increase precision through averaging to allow for fine optimization, or to provide the ability to be used in a non-invasive manner to allow feedback compensation against drifts while users are making full use of the beam from the machine.

The mechanism is often to use the beam to sample properties of the beam transport system, diagnosing hardware behavior and configuration. However, the beam itself can be part of the transport system, with self- or beam-beam effects an intrinsic part of the beam environment. We include examples of both below.

CP732, *Beam Instrumentation Workshop 2004: Eleventh Workshop*,
edited by T. Shea and R. C. Sibley III

This technique opens many opportunities for improving measurements. By selecting a particularly "clean" portion (with respect to noise) of the frequency spectrum, and by using lock-in averaging techniques in signal analysis, one can enhance measurement precision by orders of magnitude. Or one can use the same enhancement of precision to reduce the perturbation amplitude until it becomes inconsequential to the users of the beam while still providing a useful monitor of system properties. One can sometimes use the same hardware to drive the beam at invasive levels, reducing the required signal averaging and resulting in essentially real-time updates for control room use, either for initial tuning or for troubleshooting.

TYPICAL OUTLINE OF AC MODULATION MEASUREMENT

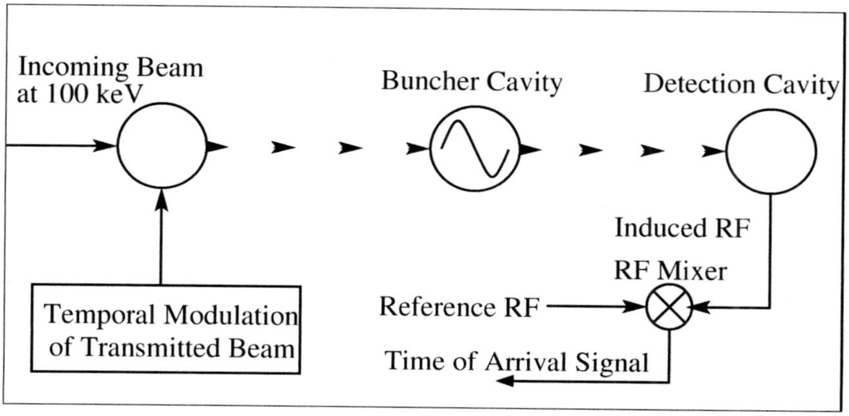

FIGURE 1. Schematic layout of modulation based diagnostic for a beam bunching region in a beam line. This system supports part of the tuning process for the CEBAF injector at Jefferson Lab. It is intrinsically destructive of the beam in that it relies upon ballistic propagation of longitudinal time-slices of the beam (longitudinal "rays") without the space-charge interactions present in the complete beam. See text for further discussion.

The basic layout of a modulation measurement is shown in Fig. 1 The figure shows a beam entering an RF deflection device, through which it is transmitted only for a selectable RF phase window[2]. The transmitted phase is modulated at acoustic frequencies. The induced RF in the detection cavity is mixed in quadrature against a master oscillator signal. The low-pass component of the mixed-down signal is proportional to variation in time of arrival of the beam particles. The bunching RF parameters are adjusted until the relationship between the beam time of arrival and the injection phase is as desired. Such a measurement can readily characterize the hardware configuration, aiding in-situ calibration and commissioning. However, this particular system is blind to longitudinal self-field effects within the beam. The measurement is made on longitudinal slices cut from the beam, each propagating in the absence of the others.

TYPICAL APPLICATIONS AND CONSIDERATIONS

It is easy to incorporate AC modulated measurements when the quantities involved depend upon beam location or timing. Existing beam position monitors or relatively simple RF phase monitors can be used in conjunction with suitable reference sources for the parameter modulation. Off-the-shelf lock-in amplifiers or dedicated DSP hardware can be used to derive accurate correlation coefficients for the repetitive signals involved. Sometimes it is possible to adapt existing control and analysis hardware to do such tasks[3], taking better advantage of capabilities of the system.

Single- and multiple-particle oscillation frequencies, optical transfer functions, detection and localization of beam loss, relative timing between the beam and RF fields, calibration of those RF fields, and differential energy gain (vs. RF phase) are ready candidates for measurement. As is often the case, tuning to null points in response can provide a very precise reference for calibration of hardware either for periodic checks or for recalibration after maintenance or other replacement of components.

As with any system, the relative capital cost of installation is one of the factors determining what actually gets installed. But other issues must also be included in the discussion. Everything costs money, and some tradeoffs are subtle and affect the installation for the long haul. Ease of use and ease of interpretation of diagnostics are worth considering. Difficult-to-use diagnostics are not used as often as easy ones. Signals which require care in interpretation are often mis-interpreted and cause unnecessary down time for the accelerator installation. The performance of modulation-based diagnostics can be adjusted within certain limits through control of the perturbation amplitude and of the integration time used to filter noise. The effective update rate in the control room can be enhanced for manual tuning or troubleshooting activities.

Another consideration is degradation of the beam. For open-ended systems, like linacs, beam degradation during setup or troubleshooting exercises may be acceptable, particularly at reduced duty factor. However, for hadron storage rings (such as RHIC), degradation of the beam emittance is not reversible and necessitates refilling the ring. Some optics measurements which are difficult to do even in this case are enabled by RF dipole techniques [1, 4].

A diagnostic device which requires dedicated beam time consumes time which could otherwise be dedicated to user experiments. One system discussed below for which this is clearly an area of benefit is the on-line transit time adjustment system at Jefferson Lab [5]. Beam time of roughly 1 hr per week was returned to the users after this system became fully operational, replacing an invasive tuning procedure.

Table 1 outlines some techniques of measuring transverse beam transfer functions and related parameters in rings and linacs. The differences in behavior between open- and closed-orbit machines affect the techniques of choice. Linac transfer functions can be easy to measure when the beam is not required to be simultaneously delivered to users. When it is advantageous for measurements to be made during routine beam delivery, small perturbations and feedback at the target or interaction region are very helpful. Use of correlation/lock-in techniques with low-level modulation of the beam steering can give highly accurate values for transfer functions when, for instance, some aspect of the transfer function is required to remain fixed. Closed-orbit systems require extra care as the perturbation does not convect out of the machine as it does for open systems.

TABLE 1. Miscellaneous transfer function measurements in rings and linacs

Non-repetitive: Standard Technique	**Ring**	**Linac**
Impulse kick: transfer functions via difference orbit	Turn by turn bpm measurements Disturbance disperses with time	Transient bpm measurements Disturbance convects out of system
Static corrector kick: transfer functions via difference orbits	Kick can affect global system: e.g., must compensate any change in circumference	Difference orbits can be made large and precisely measured from bpm averages
AC-Modulated		
Phase locked loop excitation/detection	Low amplitude modulation locked onto betatron resonances [6, 7, 8]	No direct linac analog (no betatron resonance) Perhaps sub-threshold Beam Break-Up gain measurement?
AC Dipole	Special treatment for rings Kick near betatron resonance in ring can provide stable harmonic orbit excursion Synchronous detection may make nonlinear ring dynamics accessible [1, 4]	Harmonic corrector perturbation convects out of linac with no closed orbit complications Used at JLab for optics configuration and monitoring [9, 3, 10]

BETATRON TUNE MONITORING IN RINGS

Because of the central importance of resonances in rings, the betatron tune is a critical parameter. Tune-shift effects from various sources can drive the beam to cross resonances and degrade the beam, decreasing the performance of storage rings. Measurement of the betatron tune in rings has been improved in recent years by application of lock-in techniques involving driving the beam directly on the resonance and measuring essentially an S_{21} transfer function [6, 7, 8]. Locking to the frequency for peak S_{21}, RHIC with its hadron beam and no significant synchrotron light damping, and PEP-II with its high-current, feedback stabilized beam of electrons with heavy synchrotron light damping are both able to make similar measurements.

This beam-based measurement perturbs the beam particles without otherwise perturbing the environment of the beam. The effects of space-charge, wall and wake-field interactions, beam-beam tune shift, and so forth are all involved in the measured betatron tune. As schematically illustrated in Fig. 2, an RF transverse kick is applied to the beam somewhere in the ring and a pick-up elsewhere in the machine senses the spectral component of the beam motion at the kick frequency. Narrow band, off-the-shelf lock-in hardware and special-purpose signal processors all have their place in such systems. Lower noise levels in the pick-up circuit and increased sensitivity in the detector both contribute to allowing for reductions in the perturbing kicker signal, and restricting the spectral breadth of the perturbation also contributes to reducing the total perturbation energy available to degrade the beam. The PEP-II example was chosen to illustrate its

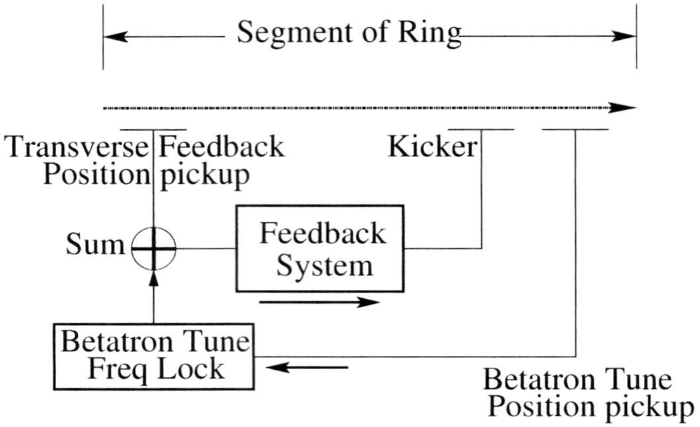

FIGURE 2. The PEP-II betatron tune system, schematically illustrated. The transverse feedback system is used to drive the betatron frequency perturbation on the beam, as it already exists and can be incorporated without degrading its own performance.

re-use of the existing transverse stabilization feedback kicker for diagnostic purposes, simultaneously with its original use.

SYSTEM SETUP EXAMPLES AT JEFFERSON LAB

In addition to the example in Fig. 1, other setup tools at Jefferson Lab are modulation-based. The setup tools are used periodically and their operation is generally inconsistent with simultaneous beam delivery. The buncher setup system is used to gather a stream of correlation data between the injection timing and time of arrival at the downstream detector. An example correlation plot is shown in Fig. 3. The plot shows a roughly 1 picosecond variation in arrival time on the horizontal axis associated with a roughly 110 picosecond variation in injection time on the vertical axis. The phase compression measured here is not a direct measure of the bunch length, but of the configuration of the bunching RF system parameters and the transmission window set to center on the longitudinal acceptance of the buncher. Particles arriving either too early or too late fall outside of the acceptance and end up as late arrivals at the detection cavity. This presentation gives operators an idea of what the system is doing. Sampling a wide range of injection phase and showing the results for the slices provides a picture of the performance of the system much better than any single scalar measurement.

A similar time of arrival detection system is in place at the end of each of the two CEBAF linacs. During pulsed beam setup work, part of the pulse structure is a short post-pulse of duration equal to the electron transit time around the accelerator, about four microseconds. The detection cavities are installed on the common portion of the beam line occupied by the beam on each circuit of the accelerator. The signal from each of the cavities provides a relative time-of-flight signal for the successive passes of the

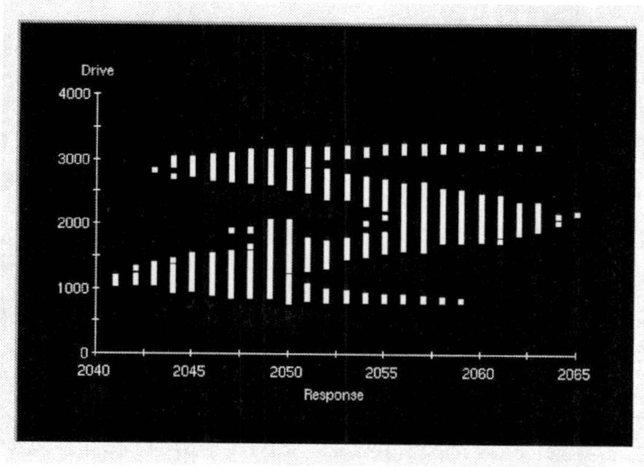

FIGURE 3. Correlation plot of time of arrival (horizontal axis; full scale approximately 1 picosecond) vs. time of injection (vertical axis; full extent of data approximately 110 picoseconds). The residual distortions from a straight line are because of aberrations in the longitudinal focus and some intentional over-focusing.

beam. This static setup tool is also used with a modulation of beam energy to detect excessive time-of-flight variations (R_{56} or M_{56} transfer matrix element).

Tuning of the optics of the machine is the largest single activity in machine setup at Jefferson Lab. Because the machine is essentially a folded linac, open-ended, resonance effects are much milder than the betatron resonance issues in closed-orbit machines. Phase advance requirements on transfer functions are very relaxed as a result. Control of the peak values of the envelope functions is very important, however. At defined locations, correctors are excited with 30 Hz signals locked to the 60 Hz frequency of the AC power grid. The resulting 30 Hz differential orbit variations are detected by the beam position monitor (BPM) system.

These synchronous position variations are fitted in groups of from 3 to 5 bpms to determine (x,x',y,y') values for the beam centroid along the machine relative to the unperturbed beam [9]. The actual number of oscillations executed by a particle on its path is not of great importance. What is important is control of the Twiss functions β and α along the accelerator. From the differential orbit data and beam momentum, one derives the emittance-like constant of the motion for each of the two linearly independent orbit perturbations in each plane. An example of the resulting plots is shown in Fig. 4. The vertical scale is arbitrary, and the horizontal dimension is simply distance of propagation through the accelerator. The yellow lines should, in the ideal case, all be featureless horizontal lines. Deviations from ideal are tuned back toward the desired value using quadrupoles defined for this purpose in the optics. This tuning is done with millimeter scale orbit perturbations, and in pulsed beam mode so that accidental scrapage of beam will not damage hardware. The update interval of the diagnostic as presently implemented is long: about 15-20 seconds to fully cycle through all four orbit perturbations.

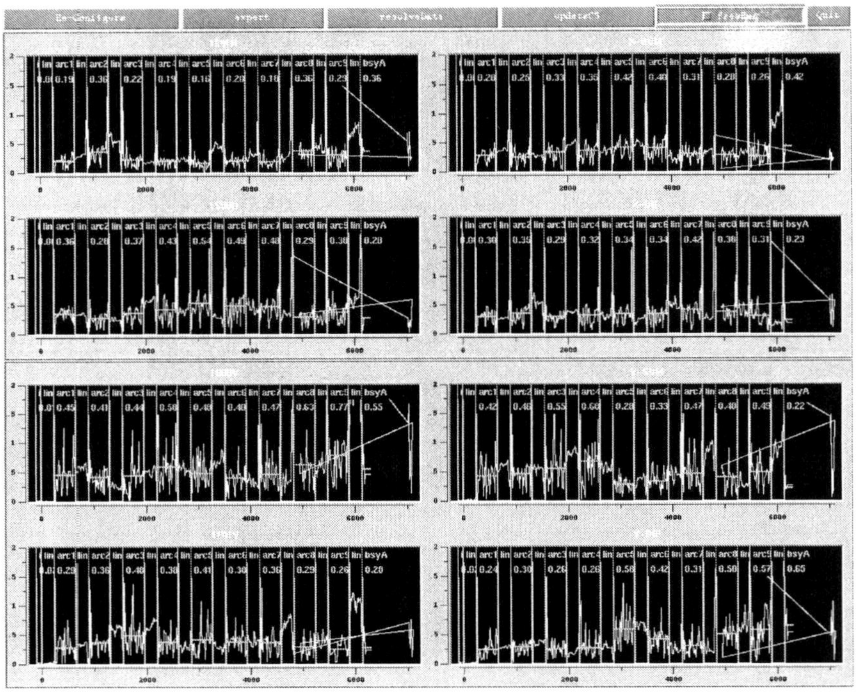

FIGURE 4. The emittance-like Courant-Snyder constant for each of four perturbed orbits through the CEBAF accelerator. The left-hand column shows the results for independent orbits, while the right-hand column shows numerically combined sums and differences for the respective planes. The vertical scale is arbitrary, and the horizontal scale is distance of propagation through the accelerator. In the ideal case, the yellow lines would be flat for each orbit.

ON-LINE EXAMPLES AT JEFFERSON LAB

Beam Scraping Monitor

In addition to the setup mode diagnostics, there are several related systems which are designed for constant monitoring of the accelerator during beam delivery. The Beam Scraping Monitor [10] uses low-level modulation of beam position (tens of microns) to search for chronic beam scraping on an obstacle. This could occur from an insertable beam line element failing to retract properly or from drifts in beam position allowing the beam to be too close to a material object like a septum. The absolute sensitivity of the beam current detectors is at the 100 nanoampere level, but drifts of that level also occasionally are found. The incremental sensitivity of the beam current detectors should be quite reliable. We have supplemented the CW beam loss protection system, which requires the total current to all users to match the current injected into the linacs, with a fundamentally different system which searches for beam current modulation at any Hall that is synchronous with the applied steering perturbation(s). A cartoon of the principle is shown in Fig. 5. The beam is traveling to the right. The portion of the beam that passed the obstruction earlier, which it was deflected toward the obstruction, lost some of its charge on the obstruction. The modulation in beam current is synchronous with the steering modulation.

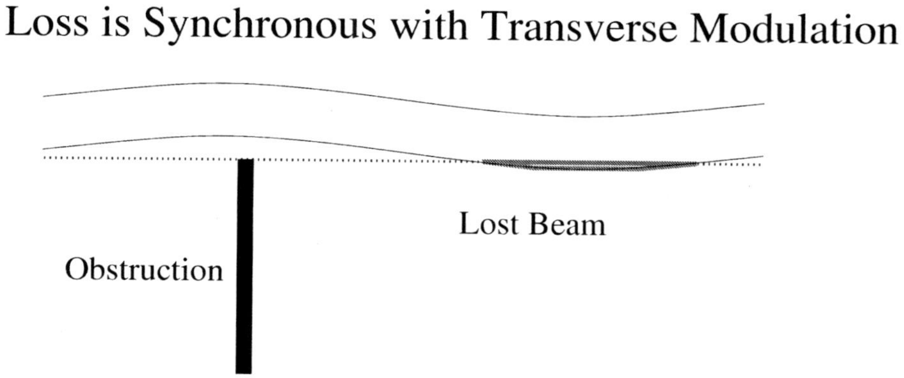

FIGURE 5. The Beam Scraping Monitor: the beam position is modulated at the scale of tens of microns. Should the beam be near an obstacle which is intercepting a significant portion of its current, an AC signal is generated in the beam current monitor hardware. A sufficiently large beam loss correlated with the steering perturbations raises a warning to the operations crew that a portion of the beam is probably being scraped by an obstruction.

Four independent modulation sources are used in this system, each at its own unique frequency for easy determination of which corrector perturbation is driving the beam loss. In this instance, the detector need only have an accurate independent time base; only the frequency is used for correlation. The relative phase of the beam loss and the perturbation are systematically ignored. This choice meant that it was not necessary to provide an analog connection to link the reference signal to the lock-in amplifier.

Transfer Function Monitor

The BSM monitors the consequences of its perturbations of the beam on the transmitted beam current. These perturbations also result in beam position correlations in the final beam transport lines to the experiments. Monitoring of the differential orbits induced by the BSM provides a measure of the overall transverse transfer function between the correctors and the final bpms. Although the Fast FeedBack (FFB) system [11] by design has negligible gain at the frequencies chosen for the BSM (600-800 Hz), it does provide a good monitoring and correlation tool. The purpose of this monitor is to provide early warning of changes in the optics along the accelerator, notifying operations when and whether a change has occurred, rather than where the change occurred. Such a monitor is useful because no other accelerator diagnostic can diagnose distributed changes in the machine optics such as are caused by transient faults in the power supplies and occasionally in the electrical utility grid. It is also a test application for the idea of distribution correlation capability generically in portions of the BPM system. If the

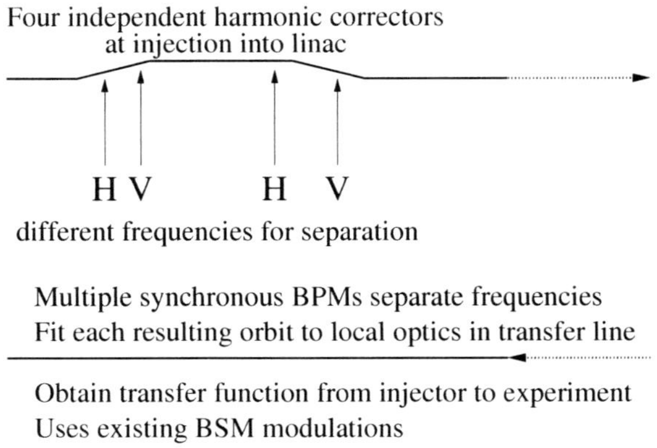

FIGURE 6. The Transfer Function Monitor: the same perturbations imposed by the Beam Scraping Monitor are detected in the experimental lines by the FFB system.

BPM system were able to extract differential orbit data accurately during beam delivery for experiments, the optics of the machine could be monitored for early detection of transient magnet power supply faults or other difficult to find faults.

Global RF Phase and Transit Time Correction

In a phase-stable accelerator, the RF systems provide a timing reference to which the beam is dynamically self-locked. The CEBAF electron linac uses on-crest acceleration with a short injected bunch to minimize energy spread for the users. The electrical master oscillator distribution system was found empirically to be insufficiently stable over the span of minutes to guarantee the required energy spread and energy stability. The Master Oscillater Modulator (MOMod) system [12] was installed to allow measurement and correction of the relative timing of beam and RF in each linac as a whole, not requiring phase correction of each cavity individually.

A small ($\pm 0.05^\circ$ peak-peak) phase modulation, considerably smaller than the typical full bunch length of 0.5°, induces a small energy modulation on the beam, sufficient to be detectable using lock-in amplifier techniques but otherwise inconsequential. This system is shown schematically in Fig. 7. The phase difference ψ between the beam and the RF and the energy modulation δE synchronous with the phase modulation are related through

$$\delta E \propto \sin(\psi).$$

The lock-in amplifier can monitor not only the magnitude of the energy modulation but also its sign relative to the RF phase modulation. Regulation of phase errors to $\pm 0.12^\circ$ is readily achieved, as shown in Fig. 8, tracking through seasonal drifts of tens of degrees. Typical short-term drifts in the absence of regulation are at the $\pm 1^\circ$ level.

In addition, ground motion through the calendar year, as well as thermal expansion (particularly after extended maintenance periods), caused variability in the pass-to-pass

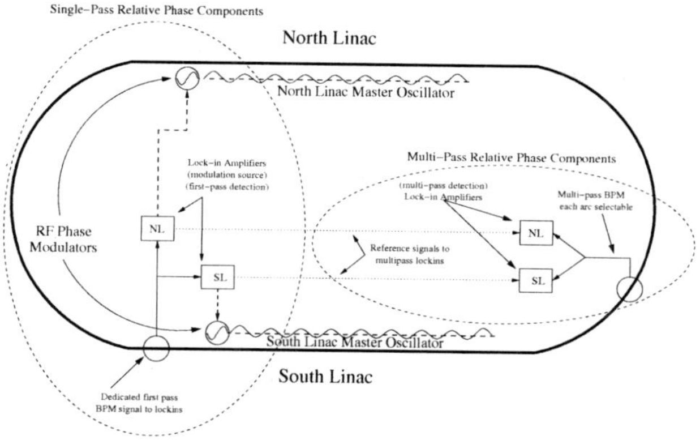

FIGURE 7. The overall layout of beam to RF timing system (MOMod) used to maintain the RF synchronous with the first pass beam and the higher pass system used to monitor variations in the transit time of the beam around the accelerator.

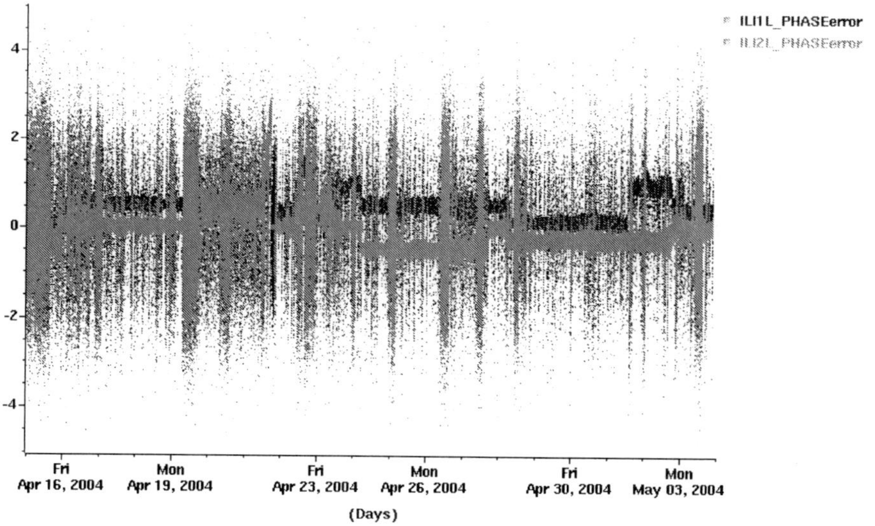

FIGURE 8. Display of regulated linac phase error. Large indicated errors are a combination of real errors measured after beam restoration and before regulation is enabled, as well as noise during the absence of beam. The discrete steps in the measured phase were due to manual setpoint changes made in attempts to trade off RF phase against transit time errors and maintain energy spread. The vertical scale is degrees of phase at 1497 MHz.

transit time. The setup tools used in the early running of the accelerator were not usable during experimental beam delivery, and they required more beam time for correction of these drifts than was desirable. A non-invasive monitor for beam transit time was devised using the same RF phase modulation applied to monitor the first-pass phase relationship

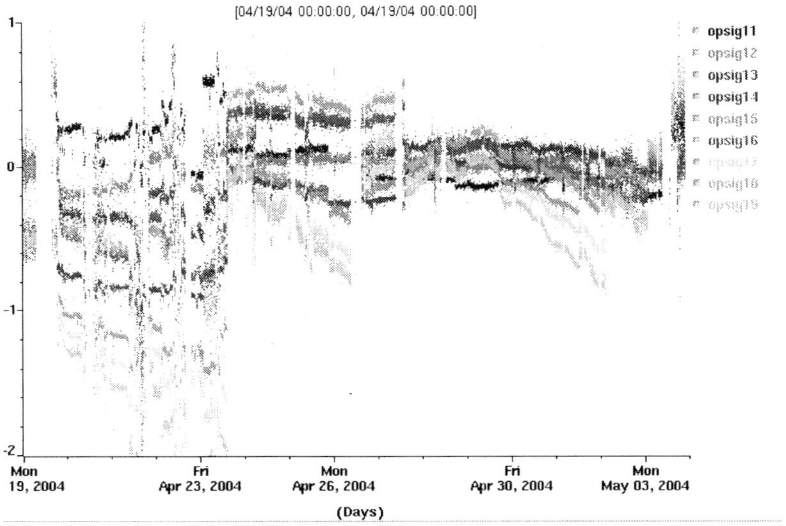

FIGURE 9. Display of the relative phase of the beam on various recirculation passes during a two-week interval, showing several discrete correction events. The vertical scale is again degrees of phase at 1497 MHz. The accelerator is expanding, so the beam arrives after the RF peak.

between the RF and beam. This phase modulation continues to affect the energy of the beam through the remainder of its acceleration. With additional pickup locations (at dispersive points) on the higher passes, it is possible to measure transit time slippage of the beam on subsequent recirculations. Such slippage is the result of change in effective machine circumference. Real-time, non-invasive measurement allows operations staff to compensate for changes in the machine circumference using the chicanes placed along the machine. Examples of this are shown in Fig. 9. These systems are part of the supporting diagnostics which are required to deliver reliably a relative energy spread under 2×10^{-5}.

UPGRADES AND POTENTIAL ADDITIONS

CEBAF Individual Cavity Phases

The field of each of the individual cavities should be at peak when the beam transits the cavity. The cavity cresting protocol which had been in routine use in the CEBAF accelerator induced too much transient energy deviation to satisfy the users as their requirements became more stringent. An upgraded AC-modulated system can eliminate this interference and provide improved performance at the same time. The former protocol [13] used discrete changes in cavity phase and measured the induced beam energy shift to determine optimum phase. By using a sinusoidal modulation instead of step-function changes, high frequency noise in the system is reduced and the FFB system can compensate for the variation in beam energy at the experimental stations. Ten Hz would be a convenient frequency for this phase modulation. The FFB error signal provides the data from which the beam phase error would be derived. This involves essentially the same algorithm as is used for setting each linac global RF phase, but implementing the phase modulation on a per-cavity basis. At the infrequent times when time is taken to correct the individual cavity phases, the linac commonly increases its energy gain by 1/2% and the overall gradients in the cavities must be reduced by this amount. Of somewhat greater concern are the occasional large phase errors found in such activities, which would be found much more rapidly by routine non-invasive monitoring.

Proton Linac Time of Flight Example

In an electron linac, dispersive regions are the obvious choice for detecting momentum changes of the beam. In mildly relativistic conditions, the same pattern of cavity phase modulation mentioned in the CEBAF upgrade scenario can be coupled with a time of arrival monitor of the kind used in the bunch compression example. This would provide a time of flight monitor which could be used to determine the phase of RF cavities. The example we will consider is patterned after the Spallation Neutron Source (SNS) parameters.

The roughly 100 femtosecond resolution of CEBAF cavities at 10 microamperes of beam current should be improved at higher beam current. Presuming a drift of roughly 15 meters from the acceleration cavity to a sensing cavity downstream, an energy gain from each cavity between 6 MeV and 10 MeV, and allowing up to $\pm 5^\circ$ of phase modulation, one should readily be able to measure phase errors of 1° from crest phase. As such linacs generally operate well away from crest phase in phase stable operation, and the signal is proportional to the sine of the off-crest phase, the sensitivity will be large. This system would target a specific value of the differential time of flight instead of the null point at peak energy gain. Commissioning experience at CEBAF indicates that use of a detection system with little or no positional dependence is wise, as beam steering effects at low beam energy associated with the cavity phase modulation were difficult to separate from real time of flight variations. For a CW machine with short bunches, the Fourier amplitudes of the RF harmonics are essentially as high as that of the fundamental frequency. Using overmoded pickup cavities can improve the timing sensitivity. The 6th harmonic 6 GHz cavity used in some applications at Jefferson Lab is very close to the 8th harmonic of many of the SNS cavities (805 MHz), for instance.

CONCLUSION

AC modulation techniques can provide readily interpretable pictures of system performance for use in the control room. The same system hardware can often be configurable for setup purposes to speed update rates or for high precision, lower bandwidth measurements for accurate feedback control during user delivery.

REFERENCES

1. Peggs, S., and Tang, C., Nonlinear Diagnostics using an AC Dipole, Tech. Rep. RHIC-AP-159, Brookhaven National Laboratory, Upton, NY, USA (1998).
2. Abbott, R. et al., "The CEBAF Injector RF Distribution and Bunch Length Measurement System," in *Proceedings of the 1992 International Linac Conference, AECL-10728, 24-28 Aug, 1992*, edited by C. R. Hoffmann, AECL, Chalk River, Ontario, Canada, 1992, p. 232.
3. Dickson, R., "Development of a BPM Lock-In Diagnostic System," in *Proceedings of the 2003 Particle Accelerator Conference, 12-16 May, 2003, Portland, Oregon, USA*, edited by J. Chew, P. Lucas, and S. Webber, IEEE, 2003, pp. 2563–2564.
4. Bai, M. et al., "Measuring Beta Function and Phase Advance in RHIC with an AC Dipole," in *Proceedings of the 2003 Particle Accelerator Conference, 12-16 May, 2003, Portland, Oregon, USA*, edited by J. Chew, P. Lucas, and S. Webber, IEEE, 2003, pp. 2204–2206.
5. Tiefenback, M., "On-Line Measurement and Tuning of Multi-Pass Recirculation Time in the CEBAF Linacs," in *Proceedings of the 2001 Particle Accelerator Conference, 18-22 June, 2001, Chicago, Illinois, USA*, edited by P. Lucas and S. Webber, IEEE, 2001, pp. 553–555.
6. Brennan, M. et al., "Tune Measurement in RHIC," in *Beam Instrumentation Workshop 2002: Tenth Workshop*, edited by G. A. Smith and T. Russo, AIP Conference Proceedings 648, American Institute of Physics, New York, 2002, p. 134.
7. Cameron, P. et al., "RHIC Third Generation PLL Tune System," in *Proceedings of the 2003 Particle Accelerator Conference, 12-16 May, 2003, Portland, Oregon, USA*, edited by J. Chew, P. Lucas, and S. Webber, IEEE, 2003, pp. 524–526.
8. Fisher, A. S. et al., "Upgrades to PEP-II Tune Measurements," in *Beam Instrumentation Workshop 2002: Tenth Workshop*, edited by G. A. Smith and T. Russo, AIP Conference Proceedings 648, American Institute of Physics, New York, 2002, p. 267.
9. Lebedev, V. A. et al., "Linear Optics Correction in the CEBAF Accelerator," in *Proceedings of the 1997 Particle Accelerator Conference, 12-16 May, 1999, Vancouver, B. C., Canada*, edited by M. C. et al., IEEE, 1997, pp. 2128–2130.
10. Allison, T. et al., "The CEBAF Beam Scraping Monitor," in *Proceedings of the 2001 Particle Accelerator Conference, 18-22 June, 2001, Chicago, Illinois, USA*, edited by S. W. P. Lucas, IEEE, 2001, pp. 2398–2391.
11. Dickson, R. A., and Lebedev, V. A., "Fast Digital Feedback System for Energy and Beam Position Stabilization," in *Proceedings of the 1999 Particle Accelerator Conference, 27 March - 2 April, 1999, Chicago, Illinois, USA*, edited by A. Luccio and W. MacKay, IEEE, 1999, pp. 646–648.
12. Lebedev, V. et al., "High-Precision Beam-Based RF Phase Stabilization at Jefferson Lab," in *Proceedings of the 1999 Particle Accelerator Conference, 27 March - 2 April, 1999, Chicago, Illinois, USA*, edited by W. M. A. Luccio, IEEE, 1999, pp. 1183–1185.
13. Tiefenback, M. G., and Brown, K., "Beam-Based Phase Monitoring and Gradient Calibration of Jefferson Laboratory RF Systems," in *Proceedings of the 1997 Particle Accelerator Conference, 12-16 May, 1999, Vancouver, B. C., Canada*, edited by M. C. et al., IEEE, 1997, pp. 2271–2273.

Contributed Talks

11th Beam Instrumentation Workshop
May 3–6, 2004

Cafe

Tektroni

Development of an Abort Gap Monitor for High-Energy Proton Rings*

J.-F. Beche, J. Byrd, S. De Santis, P. Denes, M. Placidi, W. Turner, and M. Zolotorev

Lawrence Berkeley National Laboratory, Berkeley, California, USA

Abstract. The fill pattern in proton synchrotrons usually features an empty gap, longer than the abort kicker raise time, for machine protection. This gap is referred to as the "abort gap," and any particles, which may accumulate in it due to injection errors and diffusion between RF buckets, would be lost inside the ring, rather than in the beam dump, during the kicker firing. In large proton rings, due to the high energies involved, it is vital to monitor the build up of charges in the abort gap with a high sensitivity. We present a study of an abort gap monitor based on a photomultiplier with a gated microchannel plate, which would allow for detecting low charge densities by monitoring the synchrotron radiation emitted. We show results of beam test experiments at the Advanced Light Source using a Hamamatsu 5916U MCP-PMT and compare them to the specifications for the Large Hadron Collider.

INTRODUCTION

Large proton synchrotron rings are characterized by very high energies stored in their beams and by extensive use of superconducting magnets. Such magnets, as well as other delicate components, can be easily damaged by even a very small fraction of the beam if hit. Special beam dumps, capable of absorbing such high stored energies (~350 MJ/beam in the LHC, for example) are placed along the ring together with abort kicker magnets which can steer the entire beam to be lost in the dump in case of irregular operations. It is standard practice to leave a gap, longer than the abort kicker raise time, in the ring fill pattern to avoid dispersing particles on other machine components during the kicker firing. It is vital to ensure that no particles accumulate in the gap since those would receive only a kick of reduced strength, when the dump system is activated, and would hit machine components. The maximum charge density admissible in the gap can be calculated taking into account the machine energy and the damage threshold of the most sensitive components (usually quenching in superconducting magnets sets the lower limit). The abort gap needs then to be monitored continuously in order to dump the beam or activate gap clearing procedures when the safety threshold is reached.

Several phenomena can cause the gap to become populated: timing errors at injection, drifting of unbunched particles, diffusion of particles between RF buckets, and debunching due to glitches in the RF. The time scale of such phenomena is varied,

* This work was supported by the U.S. Department of Energy under Contract No. DE-AC03-76SF00098.

CP732, *Beam Instrumentation Workshop 2004: Eleventh Workshop*, edited by T. Shea and R. C. Sibley III

going from practically instantaneous for injection errors to tens of thousands of turns for slow diffusion mechanisms.

In this paper we present a preliminary study of a simple instrument, based on a gated photomultiplier, which is suitable to be included in a ring's interlock chain. Experimental tests have been carried out at the Advanced Light Source, and their results are reported here. Though our instrument is not designed with a particular machine in mind, we will often make reference to the Large Hadron Collider for which an exhaustive study of the abort gap monitoring requirements is available [1]. The LHC also makes for an interesting case study due to the necessity of monitoring the abort gap at two different energies (450 GeV at injection and 7 TeV at final energy) with the relative consequences on the photon flux available and different safety thresholds.

ABORT GAP MONITOR OUTLINE

The fundamental component of our instrument is a gateable microchannel plate photomultiplier tube (MCP-PMT), which we use for measuring the synchrotron radiation emitted by the circulating particles in a suitable diagnostic station (dipole or undulator). By selecting an appropriately long observation window and shifting it along the abort gap, it is possible to monitor the presence of particles continuously and to estimate their longitudinal density. Commercially available MCP-PMTs [2] are the ideal instrument for such a task due to their high gain, fast gate, low noise, and variety of peak wavelength of their spectral response. Also, the low voltage required for gating and their reduced dimensions (Fig. 1) are other beneficial features that make the use of such a device particularly simple.

The optimal time window used has to be determined on the basis of the particular configuration of the dumping kicker and dump line. Other aspects to take into account are the MCP-PMT maximum duty factor, typically of the order of a few percent, the abort gap length, compared to the total length of the ring, and the maximum allowable integration time, compared to the expected photon flux. As an example, the requirements for the LHC specify maximum number of protons of $4 \cdot 10^9$ (450 GeV) and $6 \cdot 10^6$ (7 Tev) over 100 ns intervals. With a 1% duty factor, it is possible to take 100-ns-long samples only at 10-μs intervals, which means one sample per turn given that the abort gap is 3 μs and the revolution period ~89 μs. Mapping the entire abort gap requires 30 of such samples and, therefore, 30 turns. CERN specifications allow for a 100-ms integration time, which means that it is possible to average over 37 distinct measurements for every 100-ns sample. This has to be compared to the expected [3] intensity in the longitudinal diagnostic station, which is ~40 photons/0.1% bandwidth/turn (450 GeV) and ~5 photons/0.1% bandwidth/turn (7 TeV) produced by the maximum allowable number of protons at an 800-nm wavelength. The typical quantum efficiency of an MCP-PMT is in the range of a few percents, but they have a fairly large bandwidth so that it is possible to integrate photon counts over a 10% range, which gives a very good signal to noise ratio.

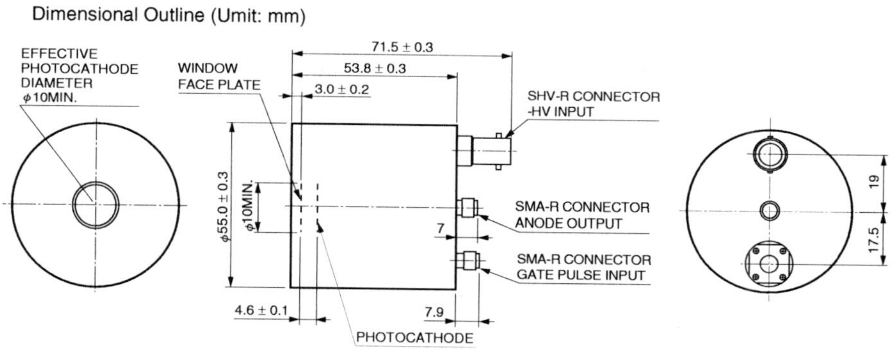

FIGURE 1. Hamamatsu gateable microchannel plate photomultiplier R5916U-50 Series.

Monitoring the beginning and the end of the abort gap presents an extra challenge due to the presence of regular bunches, having a current higher by many orders of magnitude, right next to these regions. The 1-ns gate raise time in our Hamamatsu MCP-PMT is fast enough to ensure that the filled bunches can be gated out.

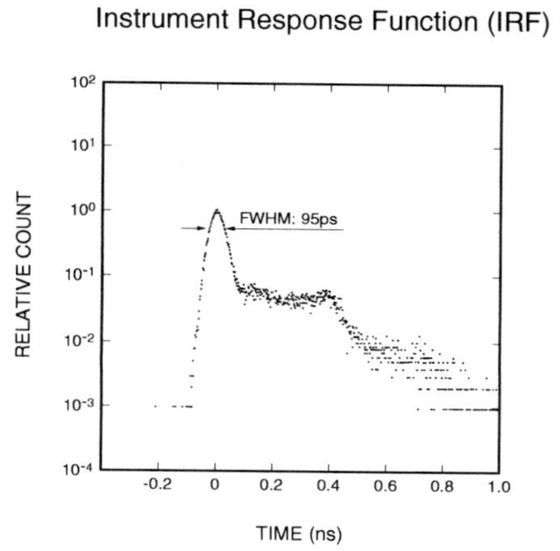

FIGURE 2. Instrument response function or the Hamamatsu R5916U-50 MCP-PMT.

On the other hand, the instrument response function (roughly equivalent to the impulse response) of the MCP-PMT decays three order of magnitude in about 1 ns so that the question arises on whether the large filled bunches could saturate our instrument, keeping in mind that the separation between RF buckets is only a few nanoseconds in many rings (2.5 ns in the LHC and 19 ns in the Tevatron, for example). We also wanted to verify that the intense radiation from the filled bunches would not saturate the photocatode, even with the gate signal off. In that case, a fast optical (Pockels cell) switch in front of the instrument would have been required, adding to the instrument's complexity. To this end we conducted experimental tests at the Advanced Light Source, which features a ~100-ns-long gap in its fill and a bunch separation of 2 ns.

Further tests on the instrument's response to unbunched beam will be shortly carried out at the Tevatron.

EXPERIMENTAL RESULTS

Some parameters of the ALS storage ring make it a good place for testing a possible utilization of an MCP-PMT as an abort gap monitor for the LHC. We mentioned the similar distance between RF buckets and a gap length equal to the length of an LHC sample in the previous paragraph. The ALS ring also features a special high-current bunch (referred to as *camshaft*) in the middle of its fill gap, which, after a while, is followed by very low current parasitic bunches due to diffusion of electrons from the camshaft into the following RF buckets. This is particularly interesting from the point of view of studying eventual saturation phenomena in the MCP-PMT, which could not allow detection of a low charge density next to a much larger one.

Our experimental station is at the ALS diagnostic beamline (BL 3.1), which uses the synchrotron radiation generated in a bending dipole (1.3 T).

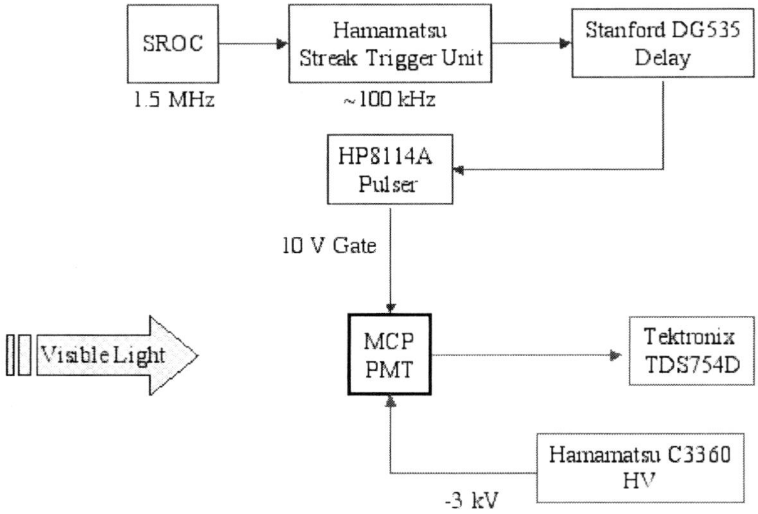

FIGURE 3. Block diagram of the experimental setup.

Figure 3 shows the block diagram of our experimental setup at BL 3.1. The 10-V gating pulse, synchronized to the ring orbit clock (SROC), can be shifted around using a delay box and the internal pulser capabilities.

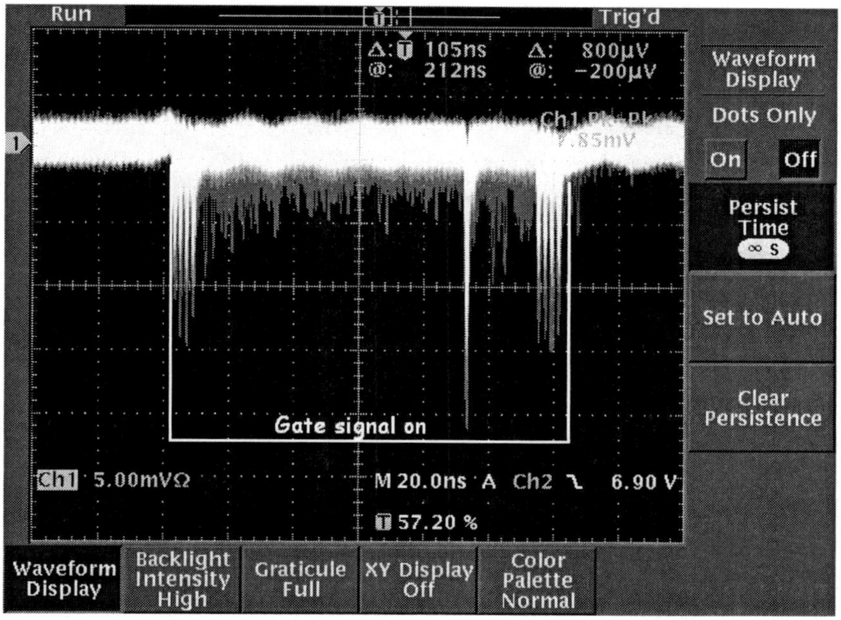

FIGURE 4. The ALS fill gap with camshaft as measured by an MCP-PMT.

Figure 4 shows an image of the entire ALS fill gap. Four regularly filled bunches can be seen at the beginning and at the end of the gap, and the camshaft is the larger signal between them. This picture was obtained setting the oscilloscope on an infinite persistence time so that we are indeed observing data accumulated over a very large number of turns. This was done because, at the beginning of the experiment, we had placed a neutral density filter in front of the MCP-PMT to be sure not to expose it to excessive light.

After this additional attenuation is removed, Fig. 5 gives an idea of the instrument time resolution. This image was taken with no averaging and shows the good signal-to-noise characteristics of our instrument.

In Fig. 6 we show how it is possible to detect very low intensity bunches simply by gating out nearby large signals. The two parasitic bunches after the first one (which is also visible in Fig. 6a) shown in Fig.6b have an estimated population of roughly 10^7 electrons, about 100 times less than the regular bunches, showing that photocatode saturation is not an issue.

Figure 7 further demonstrates that transit time spread in the MCP-PMT has only limited effects on the measurement. Comparing the left side of the trace (with the gate off) to the rightmost side (gate on, no charges) shows the noise level of the instrument. Bunched particles are still easily detectable, even at low densities, since one knows where to look for them. Tests are being planned at the Tevatron to check the performance with unbunched beam.

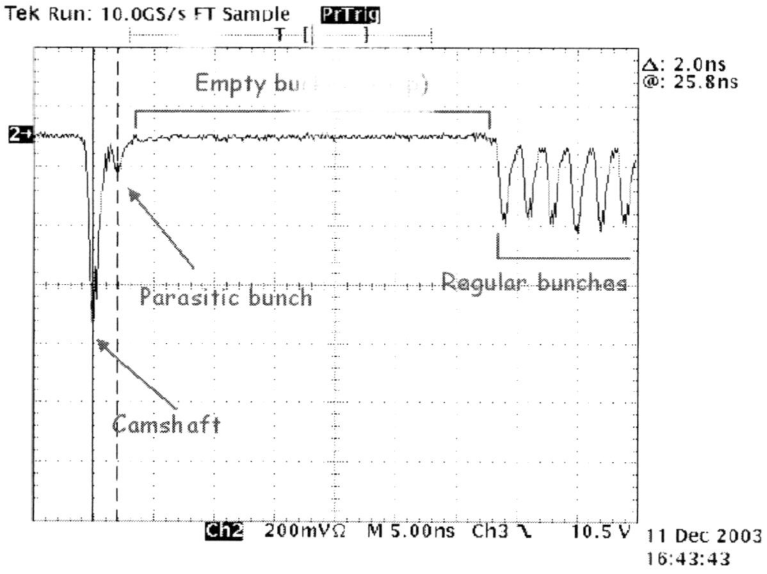

FIGURE 5. Image of the camshaft, trailing parasitic bunch, and first 6 bunches of the ALS fill train. No averages.

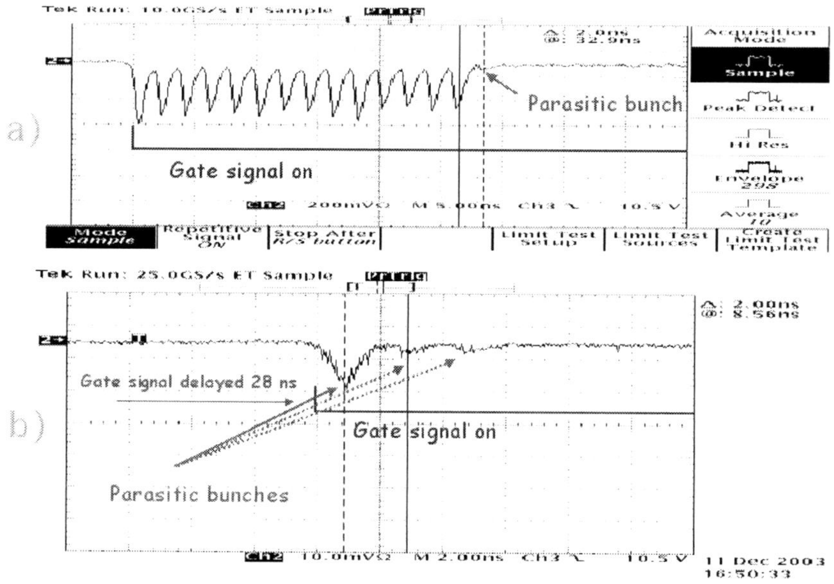

FIGURE 6. a) Beginning of the ALS gap and the last 14 filled RF buckets; b) filled RF buckets are gated out by shifting the gate signal. Vertical scale is enlarged to show parasitic bunches.

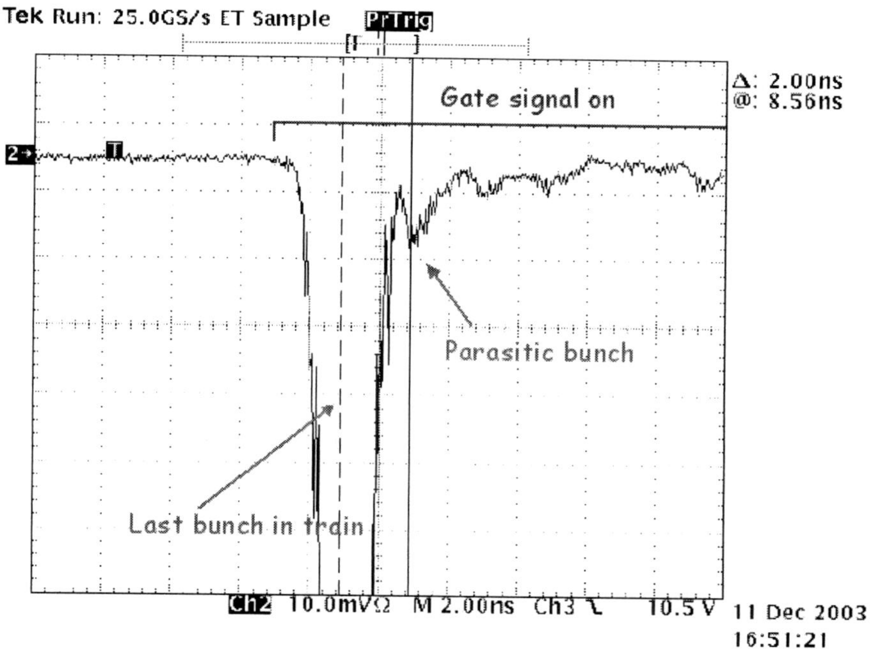

FIGURE 7. Closeup view of the last bunch of the ALS train and the nearby parasitic bunch.

IN CONCLUSION

We have presented what we believe is a promising technique for monitoring the presence of low charge densities in the abort gap of storage rings. We use a gated microchannel plate photomultiplier for measuring the synchrotron radiation emitted and it is then possible to integrate the detected signal over the desired observation window to estimate the number of particles therein. The simplicity of the scheme proposed makes it suitable to be included in a machine interlock chain, where robustness and high MTBF are paramount.

Estimates of the signal levels obtainable in the LHC and experimental results at the ALS have also been shown.

REFERENCES

1. C. Fischer, *High Sensitivity Measurement of the Longitudinal Distribution of the LHC Beams*, LHC-B-ES-0005, 2003.
2. http://www.hamamatsu.com.
3. M. Facchini and R. Jung, private communication.

Measurement of the Beam Longitudinal Profile in a Storage Ring by Non-Linear Laser Mixing*

J.-F. Beche, J. Byrd, S. De Santis, P. Denes, M. Placidi, W. Turner, and M. Zolotorev

Lawrence Berkeley National Laboratory, Berkeley, California, USA

Abstract. We report on the development of a new technique for the measurement of the longitudinal beam profile in storage rings. This technique, which has been successfully demonstrated at the Advanced Light Source, mixes the synchrotron radiation with the light from a mode-locked solid-state laser oscillator in a non-linear crystal. The up-converted radiation is then detected with a photomultiplier and processed to extract, store, and display the required information. The available choices of laser repetition frequency, pulse width, and phase modulation give a wide range of options for matching the bunch configuration of a particular storage ring. Besides the dynamic measurement of the longitudinal profile of each bunch, the instrument can monitor the evolution of the bunch tails, the presence of untrapped particles, and their diffusion into nominally empty RF buckets ("ghost bunches").

INTRODUCTION

We report on the development of a flexible beam instrument for the measurement of the longitudinal bunch distribution of beam particles in storage rings. The approach that is being developed is indicated schematically in Fig. 1 and utilizes the synchrotron radiation emitted by highly relativistic particles in a strong magnetic field. This radiation is cross correlated with the light from a solid-state CW laser that is phase locked to the storage ring radio frequency system. The cross correlation is done by mixing the laser and beam radiation in a non-linear crystal and detecting the up-converted radiation at the sum frequency on a photomultiplier tube. When the synchrotron radiation is in the near infrared, this allows shifting the detection wavelength to the optical region where the detectors quantum efficiency is higher. A prototype system has been assembled and initial data taken in Beamline 5.3.1 at the ALS electron storage ring at LBNL.

Owing to the large range of laser pulse lengths and repetition frequencies available it is possible to design a system that can accurately measure bunch shapes with ~1 ps to ~50 ps resolution and with rms lengths from ~10 ps to ~1 ns for any modern day storage ring; Tevatron, RHIC, ALS, LHC, APS, etc. A longitudinal density monitor would find a wide range of applications that provide essential information for machine operation and longitudinal beam dynamics studies, including measurements of the untrapped beam fraction; the longitudinal bunch shape; and coherent multi-bunch

* Work supported by the U.S. Department of Energy under Contract No. DE-AC03-76SF00098.

CP732, *Beam Instrumentation Workshop 2004: Eleventh Workshop*, edited by T. Shea and R. C. Sibley III

modes. Depending on the particular machine, the time required to measure these beam parameters can be substantially shorter than the synchrotron period. The concept was originally proposed and is being developed at LBNL for the Large Hadron Collider (LHC) [1] under construction at CERN. In this case the stored beam energy is very high ~350 MJ/beam and can cause extensive damage to machine components if some of it diffuses into the beam abort gap. It is therefore clearly necessary for LHC operations to have a good understanding of the longitudinal distribution of protons. In addition, it is desirable to minimize the untrapped beam fraction prior to the energy ramp and to limit the beam population in the ghost bunches which cause background in the detectors. In the LHC context our goal is to develop an instrument with four orders of magnitude dynamic range so diffusion of beam particles into unwanted regions can be detected well before reaching a level that could damage equipment. The idea of building such a system on ALS was to demonstrate the quality of information available and the practicality of such a system for routine use in an accelerator control room.

NON-LINEAR MIXING TECHNIQUE OUTLINE

The instrument can be seen as an optical sampling scope, with an important difference that we will illustrate later. Figure 1 gives a schematic view of the fundamental functional parts involved.

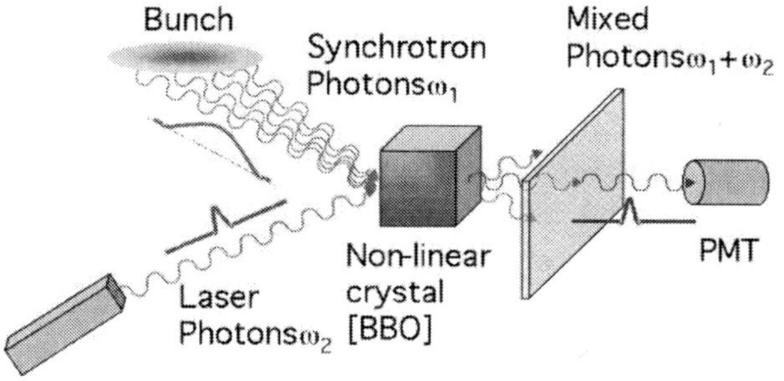

FIGURE 1. Schematics of the experimental technique for longitudinal beam profile measurement by up-conversion of synchrotron radiation with a CW laser.

The synchrotron radiation emitted by a bunch is mixed in a non-linear optical crystal with the light from a laser pulse. The sum frequency is filtered and detected on a photomultiplier (PMT). The efficiency of the up-conversion process is linearly proportional to the laser power density [2], which we can assume is constant from pulse to pulse. Therefore the number of photons detected on the PMT gives a measurement of the number of charged particles in the bunch contained in the duration of the laser pulse. Choosing a laser pulse much shorter than the bunch and sweeping it across the bunch length, we can measure the longitudinal distribution of particles in the bunch. The laser pulse repetition rate is chosen in such a way as to be synchronized with the bunch frequency or a sub-multiple of it (i.e., sampling every n-th bunch). This is done by tuning the laser cavity length. The timing of each laser pulse is then finely changed, at a slower rate, so that the laser pulse samples different longitudinal slices of a given bunch at each bunch revolution. This can be realized by mounting one of the laser cavity mirrors on a piezo-ceramic positioner, which is excited by a sinusoidal signal. This dual sweep procedure allows for concentrating data samples at high time resolution in the region of interest, for instance monitoring the bunch core at a resolution comparable to the laser pulse length, as opposed for instance to a more traditional oscilloscope. The fastest oscilloscopes available today allow for taking data at the 10-ps level, which is still slower than what is achievable with our technique anyway, but have to map the entire ring uniformly, thus generating a huge amount of unnecessary data for many applications.

The non-linear crystal length and cut angle with respect to its optical axis are optimized at a design wavelength, which should be chosen together with the laser wavelength in order to put the detection wavelength at a value where the PMT quantum efficiency is maximum. The physics of the non-linear mixing process naturally selects a bandwidth (usually of the order of a few percent) outside of which the efficiency of the up-conversion process falls sharply.

Depending on the synchrotron photon flux in the bandwidth of interest, the instrument can work in "single-photon counting" or "accumulation" modes. In single-photon counting mode, the probability of two up-converted photons to be generated in a single sample is negligible. The detection circuit, following the photomultiplier, is a simple two-level discriminator, and the analysis of a single bunch slice has to be integrated over many passes in order to build a meaningful statistics of the particle distribution in the bunch. In machines with a higher photon flux, where multiple photons can be generated by the mixing process at once, the digitizing of the PMT detected signal has to be carried out on a multiple level basis. In this case, it is possible to measure the bunch distribution taking just one sample per slice, depending on the signal-to-noise and accuracy requirements.

INSTRUMENT SETUP AT THE ALS

In order to test this technique validity, a prototype system has been assembled at Lawrence Berkeley National Laboratory, and data were taken during several experimental sessions in Beamline 5.3.1 at the Advanced Light Source (Fig. 2).

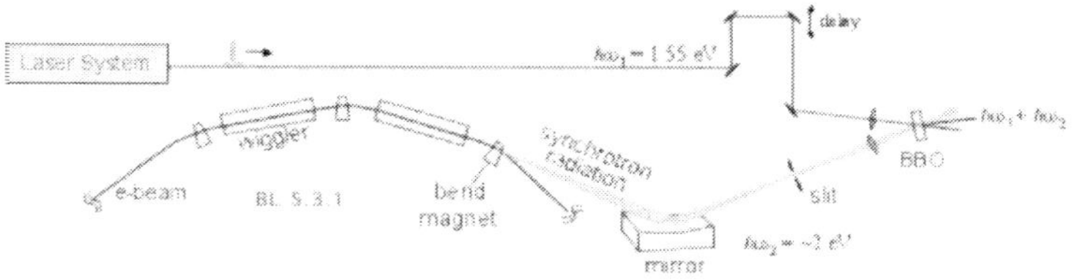

FIGURE 2. Optical layout at ALS's Beamline 5.3.1.

The laser used is a mode-locked Ti:Al_2O_3 oscillator, phase locked to a sub-harmonic of the storage ring's main RF system. The laser operates at 71.4 MHz (14 ns repetition period), with a pulse length of 50 fs, wavelength of 800 nm, and a 200-mW power.

The 197-m-long ALS storage ring has 328 RF buckets separated by 2 ns so that, after 7 revolutions, each RF bucket has been sampled once ($7 \times 47 = 328 + 1$).

The laser is phase modulated at 10 Hz to sample the longitudinal profile of each RF bucket with 32 time slices (or ***bins***) separated by approximately 4.8 ps.

The particular non-linear crystal used is a β-Barium Borate which, as a result of its large thermal acceptance bandwidth, high damage threshold, and low absorption, is of common use with laser systems like ours.

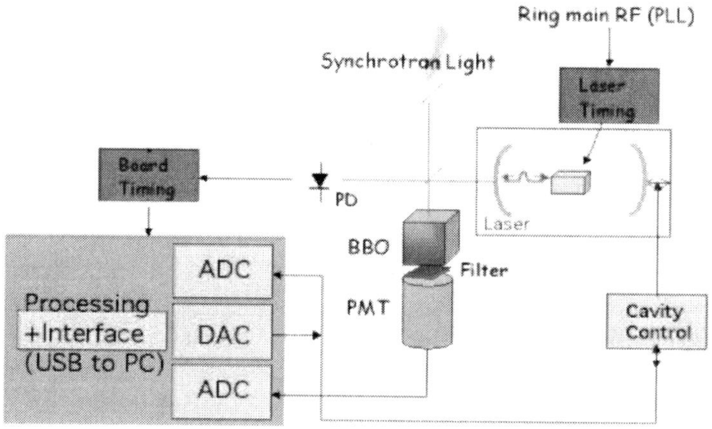

FIGURE 3. Timing scheme for the laser mixing technique used at the Advanced Light Source.

Figure 3 shows the functional units composing the instrument prototype used at the Advanced Light Source. The laser pulse is detected on a photodiode, which triggers the data acquisition board. The time delay between laser and enable signal for the ADC is adjusted at the beginning of each experimental session. We are currently investigating triggering the board directly from the ring RF so that data taking would be synchronized with the beam rather than with the laser. The cavity control signal is a 10-Hz sinusoid, which is slow enough to avoid any problem connected with mechanical vibrations induced in the mirror by faster modulating signals. This sets to 50 ms the theoretically minimum time to sample the entire bunch profile (i.e., the time it takes for a full excursion movement of the mirror). In fact, this is consistent with the photon flux out of the ring which, under typical conditions, yields about 10^6 photons per second detected by the PMT. Simple arithmetic tells us that this is equivalent to some 180 photons from each bunch, spread over the 32 bins. To obtain a better signal-to-noise ratio, especially for the bunch tails, we actually integrate for about 10 s so that our bunch profile is composed of 3×10^4 accumulated counts. The non-linearity of the

10-Hz sinusoid can be further used to our advantage: opportunely timed (i.e., with its zero-crossings coinciding with the bunch centre) it allows for taking more data points in correspondence to the bunch tails, where a lower photon count, and therefore a lower S/N, is expected.

Our in-house developed electronics board converts the PMT output into digital format and stores it in a local memory. Each bin has a separate memory register that accumulates the relative photon count. The 10-Hz waveform is also digitized, and its value at data acquisition time is stored in a corresponding register so that the count number for a particular bin is normalized to the number of samples taken.

The accumulated data are eventually dumped into a personal computer via a USB interface for data postprocessing and visualization.

EXPERIMENTAL RESULTS

A prototype of the instrument described in the previous section has been successfully tested during several experimental sessions, which allowed for the pointing out and correcting of bugs in the software for data display and minor problems in the board timing and RF noise rejection.

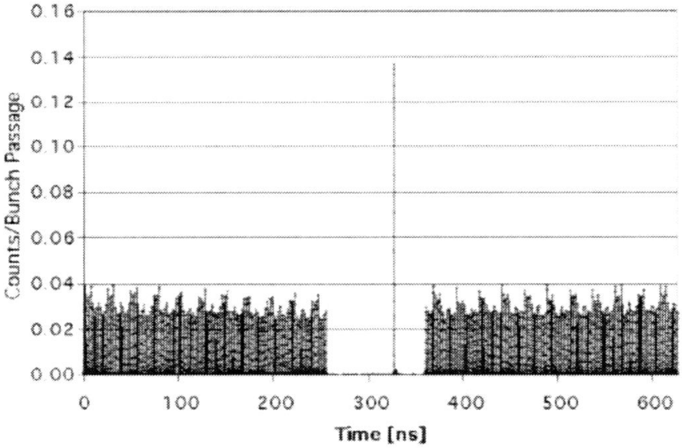

FIGURE 4. Complete ALS fill as measured using the laser mixing technique.

Figure 4 shows the data taken on such an occasion. The 657-ns-long ALS fill consists of a single train of 276 bunches, with the remaining 52 RF buckets left empty (referred to as the *gap*). One of such buckets, called *camshaft*, is usually filled separately with a higher than normal current. The total current is in the range of 200-400 mA. The periodic structure seen in the bunch train in Fig. 4 is caused by the particular injection process used at the ALS, which our technique is able to document.

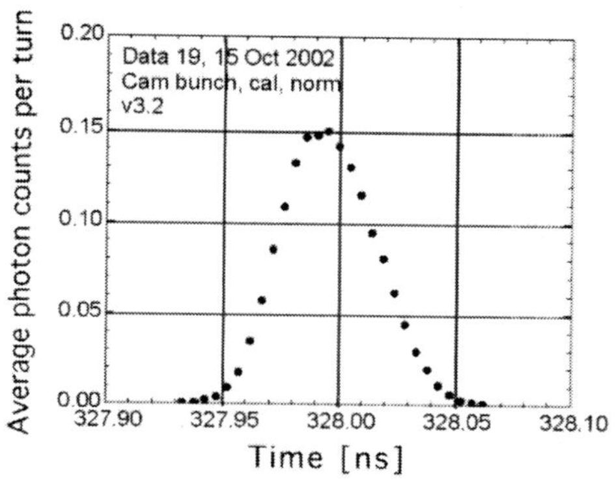

FIGURE 5. Detail of the camshaft longitudinal profile.

Thanks to the very high data throughput, expanding the horizontal time scale is all it takes for examining one of the bunches shown in Fig. 4 in much greater detail (Fig. 5). As a comparison, a streak camera can generate similar figures after a somewhat lengthier postprocessing. The fundamental difference is that a streak camera cannot take data at the picosecond level resolution necessary to analyze a single bunch when taking data over a time span of hundreds of nanoseconds. Therefore, generating Figs. 4 and 5 would require taking two separate images.

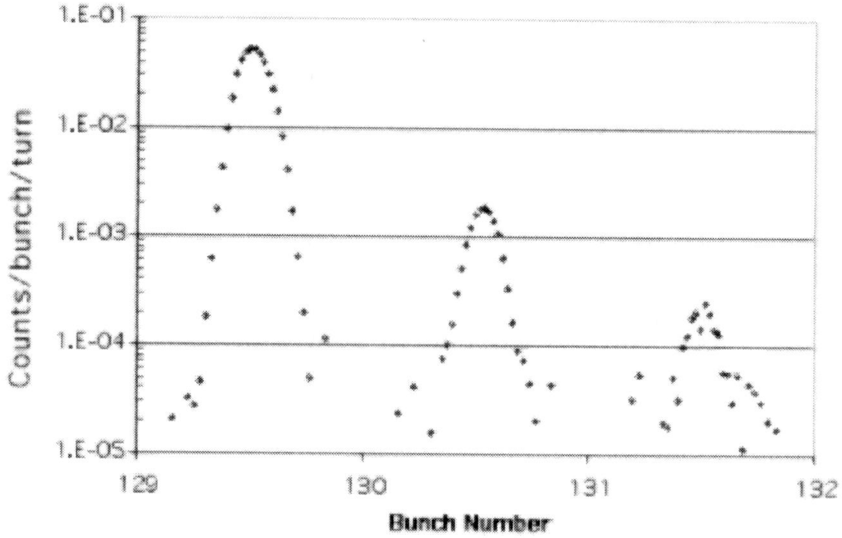

FIGURE 6. Dynamic range and sensitivity: the camshaft and ghost bunches can be measured at the same time. The instrument's noise floor is about three orders of magnitude below the camshaft signal.

Electrons lost from the camshaft can be recaptured by the following RF buckets, which are nominally empty, generating the so-called ***ghost bunches***. Figure 6 shows such a situation: the second ghost bunch has a current of about 1/500-th of the camshaft's (6 mA), which gives an idea of the sensitivity obtainable with this technique.

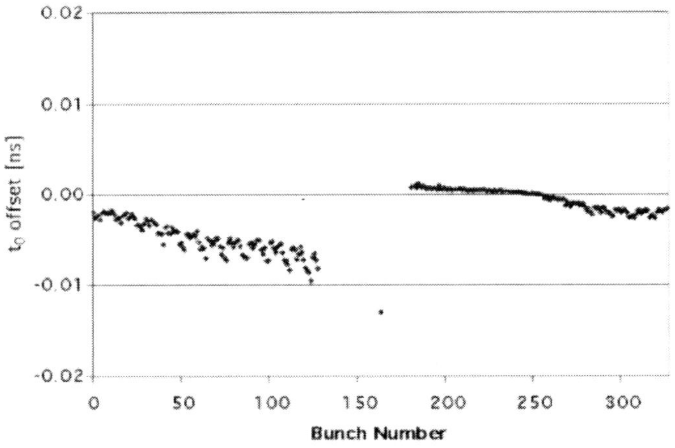

FIGURE 7. Synchronous phase transient along the ALS bunch train.

With a simple postprocessing of the accumulated data, we were also able to plot others parameters of interest: Fig. 7 shows the variation of the synchronous phase along the bunch train by plotting the position of each bunch centre. Figure 8 shows the bunch length instead. These effects in the ALS are caused by beam loading in the third harmonic cavities system installed to increase the lifetime.

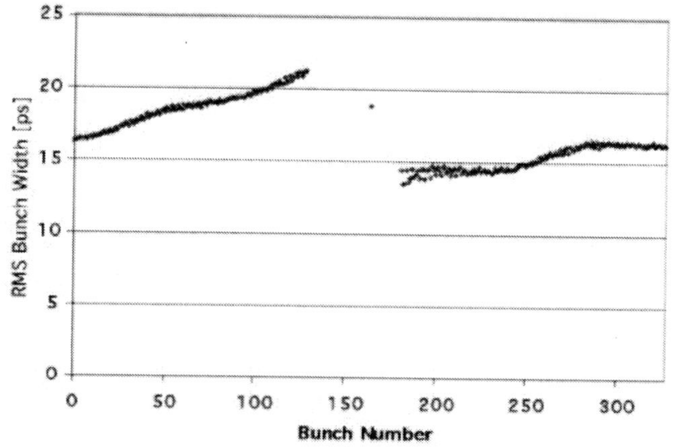

FIGURE 8. Bunch length variation along the ALS bunch train.

CONCLUSIONS

We believe that the laser mixing technique presented in this paper offers a number of exclusive features, which makes for a powerful instrument for longitudinal dynamics studies in storage rings. In particular, the maximum time resolution achievable is much higher than with other techniques available, and the combination of frequency up-conversion and choice of sampling rate make it a very flexible device. Our experimental results at the ALS document well the range of possible applications. Where a higher photon flux is available, it is possible to study the evolution of several beam parameters in a time scale substantially shorter than the synchrotron period.

REFERENCES

1. M. Placidi, W. C. Turner, and M. Zolotorev, "Notes on the Measurement of Longitudinal Beam Density in LHC Using Optical Methods," CBP Tech Note-307 (2003).
2. M. Zolotorev, J.-F. Beche et al., "An Optical Sampling Technique for Measuring Longitudinal Beam Density in a Storage Ring," *Nucl. Instr. And Methods in Phys Res. A*, submitted (2004).

Energy Spread Monitoring for the JLAB Experimental Program: Synchrotron Light Interferometers, Optical Transition Radiation Monitors, and Wire Scanners[1]

Y.-C. Chao*, P. Chevtsov*, A. Day*, A. P. Freyberger*, R. Hicks*, M. Joyce* and J.-C. Denard†

*Jefferson Lab, 12000 Jefferson Avenue, Newport News, VA 23606
†Synchrotron SOLEIL, 91192 CEDEX, France

Abstract. The hypernuclear physics program at JLAB requires an electron beam with small transverse size ($\sigma \sim 100\ \mu m$) and an upper limit on the RMS energy spread of $\frac{\delta E}{E} < 3 \times 10^{-5}$. To measure and monitor these parameters, a beam size and energy spread measurement system has been created. The system consists of a set of wire scanners, Optical Transition Radiation (OTR) detectors, and Synchrotron Light Interferometers (SLI). The energy spread is measured via a set of wire scans performed at specific locations in the transport line, which is an invasive process. During physics operation the energy spread is monitored continuously with the OTR and/or the SLI. These devices are non-invasive [or nearly non-invasive in the case of OTR] and operate over a very wide range of beam energies (1-6 GeV) and currents ($\sim$100 μA down to few μA). All components of this system are automated in an EPICS accelerator control environment. The paper presents our operational experience with the beam size and energy spread measurement system and its maintenance.

INTRODUCTION

The hypernuclear physics program at JLAB is attempting to measure very small mass splittings and widths of hypernuclear states. A 280 KeV mass resolution is needed for the experimenters to observe the closely spaced mass states. The mass of the hypernuclear state is measured by the "missing mass" technique, where the four vectors of the incoming and scattered electron are measured and used to determine the mass of the hypernuclear state. The equation for the missing mass squared (M^2_X) is:

$$M_X^2 = m_p^2 + 2m_p[E_0 - E^{'}] + 2E_0E^{'}cos(\theta)$$

where E_0 is the energy of the beam, and $E^{'}$ is the energy of the e^- after scattering off the proton target, m_p. Energy spread of the incident electron beam will broaden/smear the M_X^2 distribution. Of the 280 KeV resolution budget, 120 KeV is allocated to energy

[1] This work was supported by the U.S. Department of Energy under Contract No DE-AC05-84ER40150.

CP732, *Beam Instrumentation Workshop 2004: Eleventh Workshop*,
edited by T. Shea and R. C. Sibley III

TABLE 1. Tabulation of the operating range and capabilities of the three devices used to measure the transverse beam width in the end station transport lines.

device	current range	beam interaction	measures
wire scanners/quad scan	$<20\mu A$	invasive	$\sigma_E/E, \alpha, \beta, \gamma$
OTR	$8 \rightarrow 180\mu A$	slightly invasive	σ_{beam}
SLI	$20 \rightarrow \infty\mu A$	non-invasive	σ_{beam}

spread, which with $E_0 = 4$ GeV results in a requirement of

$$\frac{\sigma_E}{E_0} < 3 \times 10^{-5}.$$

This paper describes the instrumentation that has been installed in the end station transport beam-lines to measure and monitor the energy spread during these experiments. The instrumentation includes wire scanners, optical transition radiation monitors and synchrotron light interferometers.

MEASURING ENERGY SPREAD

The transverse beam size, σ_{beam}, measured in a dispersive location has two sources:

$$\sigma_{beam} = \sqrt{\sigma_\beta^2 + \sigma_\delta^2},$$

where $\sigma_\beta = \sqrt{\varepsilon\beta}$ is the betatron size and σ_δ is the size due to dispersion [1]. The energy spread is $\frac{\sigma_E}{E_0} = \frac{\sigma_\delta}{D}$, where D is the local dispersion. Ignoring the betatron contribution (which is safe to do when $\frac{\sigma_\beta}{\sigma_\delta} << 1$) the upper limit on the energy spread is

$$\frac{\sigma_E}{E_0} < \frac{\sigma_{beam}}{D}.$$

When the beam has a small energy spread, as with the JLAB beam, σ_β can be comparable to σ_δ and the betatron contribution to the transverse beam size must be taken into account to determine the central value of the energy spread.

In order to minimize the betatron contribution to the beam size, as well as the effect of incoming dispersion, a special optics was devised for the end station transport line to increase the dispersion from 4 m (nominal) to 8 m (high dispersion).

The transverse beam size is measured by wire scanners, optical transition radiation monitors and synchrotron light interferometers. The operating ranges and capabilities of the devices are listed in Table 1 and discussed in more detail in the following sections.

Wire Scanners

Wire scanners, being the originally installed devices to measure transverse beam size, are numerous [there are 10 wire scanners in the end station A and C transport lines and five in the end station B transport line] and well understood devices [2]. The JLAB wire scanner design uses three 25μm W97Re3 wires in vertical, horizontal and 45° orientations. The wire position is correlated with the current induced on the wire from secondary emission or with the photomultiplier tube response due to the resulting shower from the electron scattering off the wire. Measuring the beam parameters with a wire scanner is limited to beam currents below 20 μA and is invasive to the beam delivery.

The ten installed wire scanners[2] allow the extraction of the beam Twiss parameters, $\alpha,\beta,\gamma,\varepsilon$, without having to adjust any quadrupole strengths [i.e. quad scan]. The software for obtaining and analyzing the wire scan data is broken into three steps:

1. Cycle individual wire scanners and fit the processed data to a Gaussian distribution with constant background. A 'best in show' wire scan is shown in Figure 1. All wire scan data are written to disk for further analysis and comparison with previous history.
2. Cycle all or a subset of wire scanners sequentially, verify the data integrity and summarize the results along with the quadrupole magnetic fields during the scans. A complete scan of all the wire scanners in the transport line takes about 30 minutes.
3. Take the accumulated wire scan data from the previous step and extract the Twiss parameters. Details of the Twiss parameter extraction can be found in Ref. [3].

Optical Transition Radiation [OTR] Monitors

Optical transition radiation[4] monitors have been installed at selective locations in the transport line to complement the wire scanners. The OTR monitors measure the transverse beam size by imaging the OTR on CID camera and analyzing the horizontal and vertical projections. A carbon foil of 1 cm × 1 cm area and 0.2 μm thickness[3] presents such a negligible material to the beam that beam can be delivered to the end station while the OTR foil is inserted in the beam line. Neutral density filters are used to adjust the light intensity within the optimum range for the camera. The OTR operates with CW beam with current from 10 μA up to 180 μA. See Ref. [5] for details.

The very thin carbon foils are mounted on a three sided frame. OTR emitted in the forward direction is used to form the beam image. Forward OTR distribution is not affected by local shape of the foil as in the case of backward OTR. Figure 2 shows the mechanical layout of the OTR.

---

[2] Only four non-degenerate measurements of the transverse beam size are needed to extract the Twiss parameters.

[3] Foils are obtained from Goodfellow Corp., www.goodfellow.com.

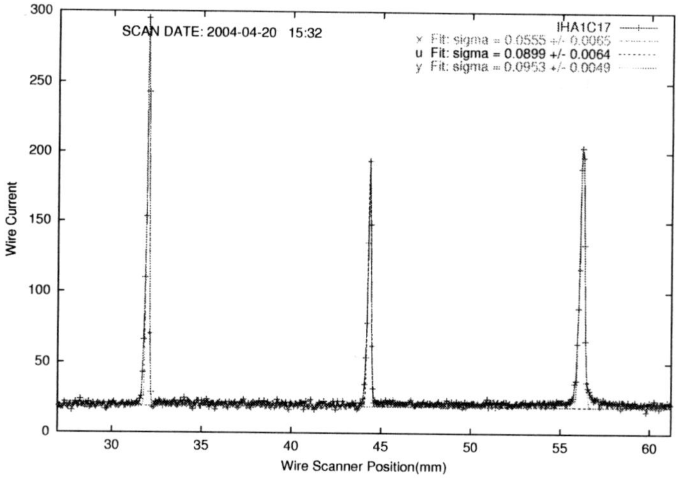

FIGURE 1. Wire Scan data and fit results. The three peaks correspond to the three wires, horizontal, 45°, and vertical orientations. Each peak is fit to a Gaussian distribution with a constant background.

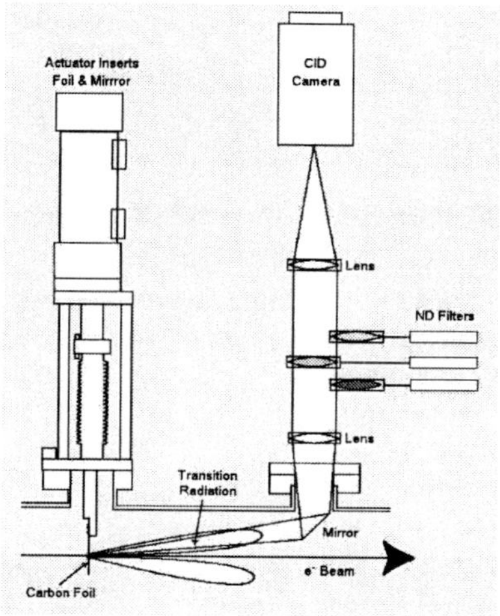

FIGURE 2. Mechanical schematic of the forward optical transition radiation monitor. The 0.2μm carbon foil and mirror are mechanically coupled.

The beam size as measured by the OTR is calibrated by comparison with the beam size as measured by the wire scanners. This calibration is done at 8μA of beam current

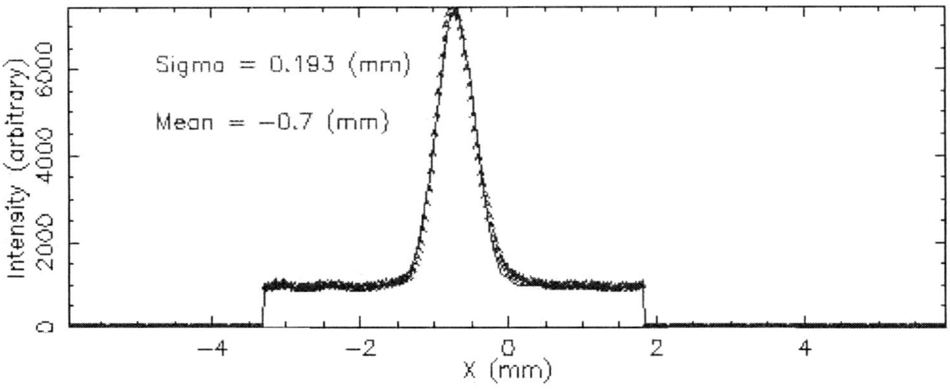

FIGURE 3. Typical OTR horizontal projection. The points are the data from the CID camera; the blue line and values for σ and position are the result of a fit to the data.

and different beam sizes are obtained by varying quadrupole magnet field strengths[4].

Figure 3 shows the horizontal beam profile as measured by the OTR. The OTR images are analyzed in real time. Horizontal and vertical projections are fit to a Gaussian distribution with a constant background.

Synchrotron Light Interferometer [SLI]

The use of synchrotron light interferometry makes it possible to measure the very small transverse beam size in a completely non-invasive manner. SLI installations at storage rings have an abundant amount of synchrotron light due to the large stored beam current. The lower JLAB beam current is advantageous in that beam heating of the SLI optical components is no longer an issue as it is at storage rings. However, at these lower beam currents the intensity of the light is so diminished that a cooled CCD camera[5] is necessary to image the interferogram.

The difference in light intensity and the choice of CCD camera is the major difference between the JLAB SLI and that of T. Mitsuhashi at KEK [6]. A sketch of the components used to make a SLI are shown in Fig. 4 and details of the design are found in Ref. [7].

---

[4] We have found that the optical components get dusty which introduces a changing point source term [offset]. This requires an OTR calibration and/or cleaning to be performed on a periodic basis. The source of the dust is a mystery since the optical components are in a closed system.

[5] Santa Barbara Instrument Group, Astronomical CCD cameras, www.sbig.com.

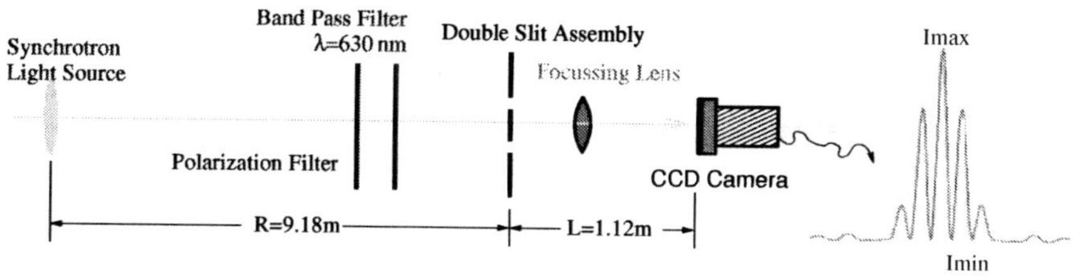

FIGURE 4. Cartoon schematic of the synchrotron light interferometer.

The beam size is a function of the visibility, $\mathcal{V}$, of the interference pattern

$$\mathcal{V} = \frac{I_{max} - I_{min}}{I_{max} + I_{min}},$$

where I_{max} (I_{min}) is the maximum (minimum) of the interference pattern. Since $\mathcal{V}$ is a ratio, most systematics involved in digitization cancel. However, noise terms contribute to the sum [denominator] leading to a reduced $\mathcal{V}$, or larger calculated beam size and therefore must be kept small.

For a Gaussian beam distribution the beam size is found to be

$$\sigma_{beam} = \frac{\lambda_0 R}{\pi d}\sqrt{0.5\ln(1/\mathcal{V})},$$

where λ_0 is the wavelength of the bandpass filter, R is the distance from the light source to the double slit assembly, d is the slit separation and $\mathcal{V}$ is the measured visibility.

The JLAB SLI uses a grid pattern for the double slit assembly. This allows the simultaneous measurement of horizontal and vertical beam sizes. Figure 5 shows the real-time display of the interferogram data and fits to the data. The data update rate is a function of the beam current; as fast as 0.5 seconds for 120μA and tens of seconds for beam currents less than 20μA. Figure 6 is a plot of the beam sizes as measured by the OTR and SLI. There is an discrepancy with the larger beam sizes that is presently being investigated.

CONCLUSIONS

A suite of instrumentation to measure and continuously monitor the energy spread at the end station transport lines has been presented. Two transport lines have been equipped with the full compliment of instrumentation. Figure 7 shows the stability of the energy for several days of operation. Initially the energy spread was near the specification of $\frac{\sigma_E}{E} < 3 \times 10^{-5}$. Improving the operation of the RF cavities and general machine setup resulted in a dramatic improvement of energy spread, seen on the plot as the drop in energy spread mid-day on April 23. The continuous operation of the SLI and OTR have proved to be useful in identifying components contributing to the measured energy spread.

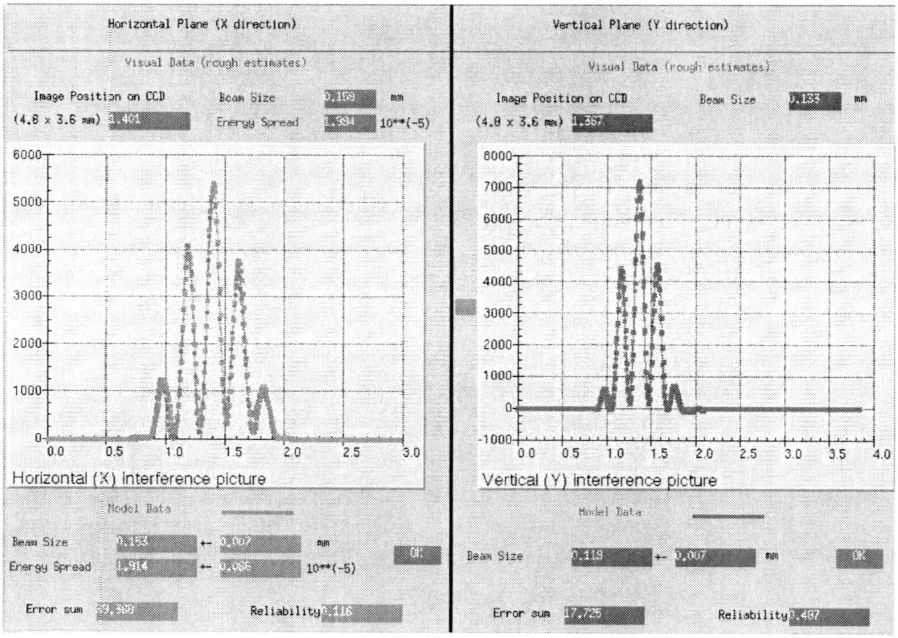

FIGURE 5. Control screen for the SLI showing the horizontal and vertical interferograms. The results for the beam size above the interferogram are the result of a simple search algorithm to determine I_{max} and I_{min}. The values below the interferogram result from a fit to the expected functional form.

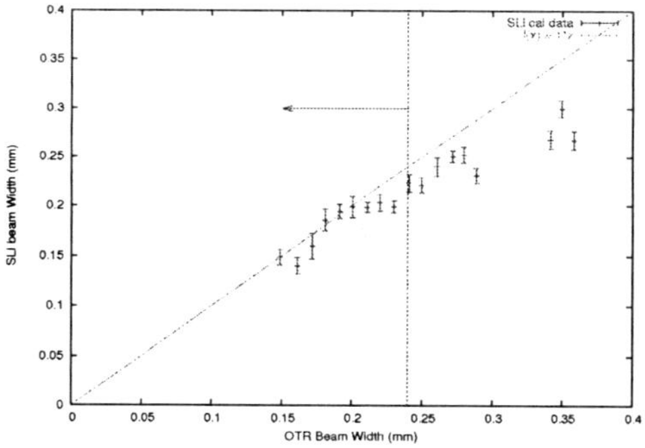

FIGURE 6. Plot of the beam width as measured by the SLI vs OTR. The dashed curve is a line with slope of one and intercept of zero.

ACKNOWLEDGMENTS

The authors thank the JLAB operations staff for support during commissioning the devices.

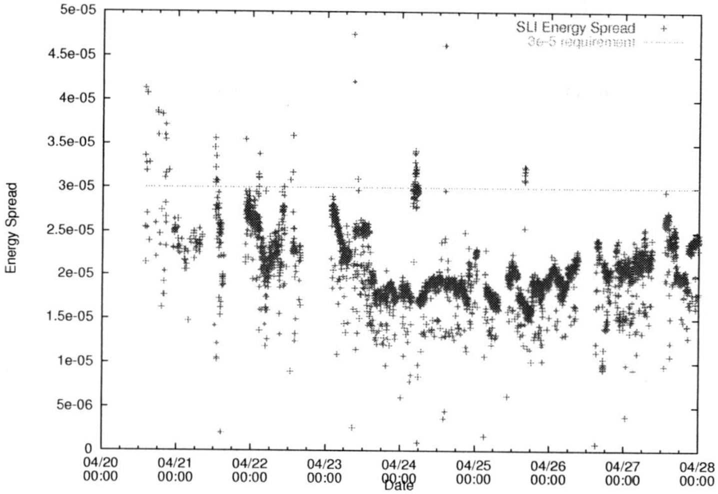

FIGURE 7. Plot of the SLI measurement of energy spread as a function of time.

REFERENCES

1. M.G. Minty and F. Zimmerman, *Measurement and Control of Charged Particle Beams*, Springer-Verlag, Berlin, 2003, pp. 163.
2. M. Plum, "Interceptive Beam Diagnostics - Signal Creation and Materials Interactions," in *these proceedings*.
3. Y-C. Chao, "Measuring and Matching Transport Optics at Jefferson Lab," in *Proceedings of 2003 PAC*, http://accelconf.web.cern.ch/accelconf/p03/PAPRS/TPB01 1.PDF
4. V.L. Ginzberg and V.N. Tsytovich, *Transition Radiation and Transition Scattering*, Adam Hilger, Bristol, 1990, p. 448.
5. J-C. Denard *et al.*, "High Power Beam Profile Monitor with Optical Transition Radiation," in *Proceedings of the 1997 PAC*, http://accelconf.web.cern.ch/accelconf/pac97/papers/pdf/5p062.pdf.
6. T. Mitsuhashi, "Faraday Cup Award Talk: 12 Years of SR Monitor Development at KEK," in *these proceedings*.
7. P. Chevtsov *et al.*, "Synchrotron Light Interferometer at Jefferson Lab," in *Proceedings of the 2003 PAC*, http://accelconf.web.cern.ch/accelconf/p03/PAPERS/WPPB06 8.PDF.

Diagnostics at the ELBE Bremsstrahlung Facility

P. Evthushenko, U. Lehnert, P. Michel, K. D. Schilling, R. Schurig, and J. Teichert

Forschungszentrum Rossendorf (FZR), Postfach 510119, 01314 Dresden, Germany

Abstract. The ELBE (Electron Linear accelerator with high Brilliance and low Emittance) machine is currently under transition from commissioning to regular user operation. The linac produces electron beams up to 40 MeV and 1 mA (CW), which are used to generate various kinds of secondary radiation. While the IR-FEL and X-ray facilities are still under construction, comprehensive nuclear physics experiments have been conducted at the Bremsstrahlung facility. Both, to meet the experimenter's demands on beam quality, stability and reproducibility and to ensure a safe operation of the machine, several diagnostic elements were developed. These devices, including backward OTR viewers, radiator temperature diagnostics, stripline detectors to monitor the beam incidence angle on the radiator, and a four-quadrant loss monitor, were developed, installed, and tested. These devices are described and results are presented in this paper.

INTRODUCTION

The Forschungszentrum Rossendorf is constructing a superconducting Electron Linac with high Brilliance and low Emittance (ELBE), which can deliver a 1-mA CW beam for 40 MeV [1]. The electron beam is used to generate infrared light (Free Electron Lasers), X-rays (electron channelling), MeV-Bremsstrahlung, fast neutrons and positrons [1]. Table 1 gives an overview on the secondary beams at the ELBE facility and the associated fields of science.

TABLE 1. Secondary Radiation at ELBE and scientific application.

0..20 MeV γ-radiation	- nuclear physics - astrophysics
10..100 keV X-rays	- radiation damage of cells - study of phase transitions in liquid metals
5..150 µm infrared FEL	- semiconductor physics - radiochemical and biological experiments
0..30 MeV neutrons	- material studies for fusion reactors
0..30 keV positrons	- defect studies in solid states

CP732, *Beam Instrumentation Workshop 2004: Eleventh Workshop*,
edited by T. Shea and R. C. Sibley III

Figure 1 shows the layout of the ELBE building, the beamlines and the experimental caves. After commissioning of the injector and the first linac stages in 2001, the nuclear physics beamline and the Bremsstrahlung facility were assembled. It has been designed for nuclear-structure experiments using photon-induced reactions such as (γ, γ^{1}), (nuclear resonance fluorescence), (γ, u), (γ, p), (nuclear photo effect) and photon–induced fission. The main components such as the beam-diagnostic elements, the water-cooled radiator holder with three radiator positions, the beam shutter with radiation hardener, the electron beam dump, and the conical collimator for the Bremsstrahlung are made of pure aluminium and are shown in Fig. 2 [2]. In 2002 the ELBE Bremsstrahlung facility went into regular operation. There are various kinds of diagnostic elements that ensure precise, stable and safe operations. These include stripline beam position monitors (BPMs), optical transition radiation (OTR), beam viewers, four-quadrant beam loss monitors and a pyrometer to measure the local spot temperature on the radiator.

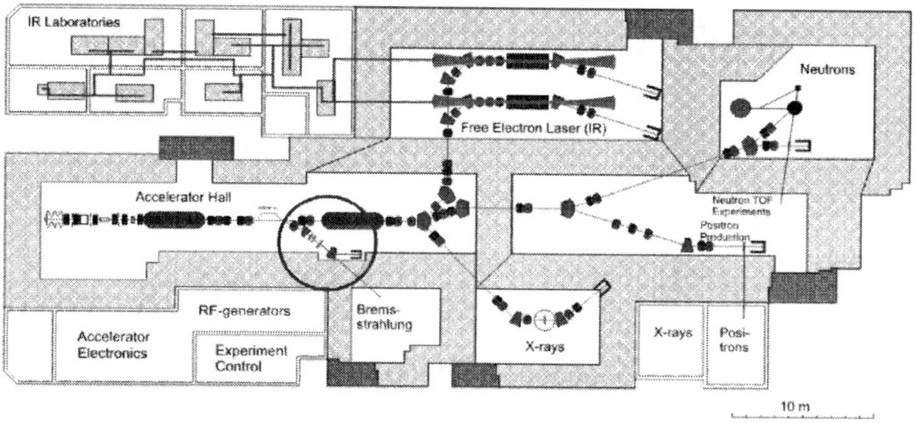

FIGURE 1. Layout of the Radiation Source ELBE.

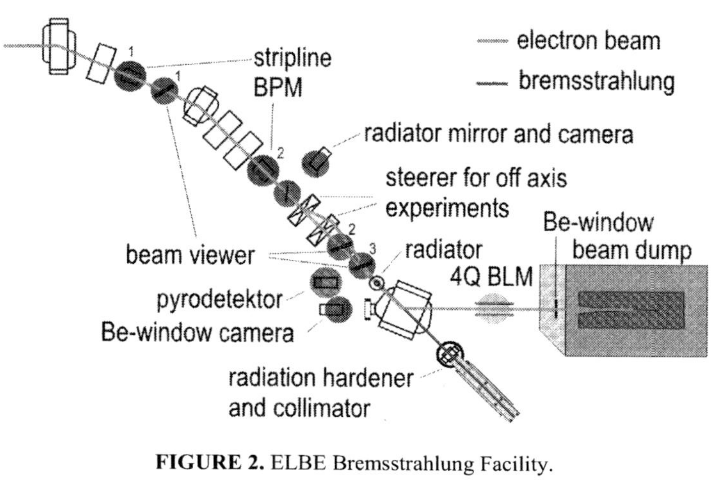

FIGURE 2. ELBE Bremsstrahlung Facility.

$^{\lambda}/_{4}$ Stripline Beam Position Monitors

Starting from the 1.5-GHz $^{3}/_{4}\lambda$ Stripline design made in JLAB, new 1.3-GHz $^{\lambda}/_{4}$ BPMs were developed at ELBE (Fig. 3) [3, 4]. The resolution of the BPM system including electronics is better than 20 μm at nominal bunch charge at ELBE of 77 pC. BPM 1 (Fig. 2) is placed in the dispersive area between the first two dipole magnets to monitor the beam energy. The achieved energy resolution is better than 0.1%. BPM 2 and the OTR viewer at the radiator are used to keep the beam and its incident angle in stable position.

FIGURE 3. $^{\lambda}/_{4}$ Stripline Beam Position Monitor.

Optical Transition Radiation Beam Viewers

Beam position and profile on the nuclear physics radiator are monitored using a camera sensitive to OTR. A mirror deflects the backward OTR from the radiator outside the beam-axis to a vidicon camera. At higher currents (> 300 μA) thermal radiation becomes visible (Fig. 4). Beam position and profile on the radiator are observable during full beam power operation. There are three viewscreens along the nuclear physics beamline. Backward OTR is produced on 18-μm thick aluminium foils and detected with Perkin-Elmer KH 500 vidicon cameras [5]. Downstream from the target, the electron beam is terminated in a water-cooled beam dump. A beryllium window separates the beamline from the dump vacuum. The beam spot on the beryllium window is also monitored.

Radiator Temperature Monitoring

The temperature of the radiator is monitored to ensure that the radiator material is not melted or evaporated by the electron beam. This could otherwise contaminate the surface of the superconducting RF-cavities. A Raytek pyro detector (RAYMA2SCCFL) is used. It measures the beam spot temperature up to 2000 °C in a 1-mm window. Figure 5 shows the temperature depending on the average curve at 12-MeV beam energy.

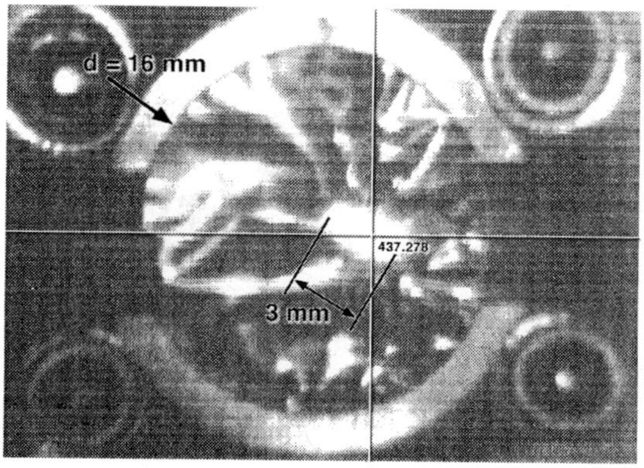

FIGURE 4. Beam spot on the radiator.

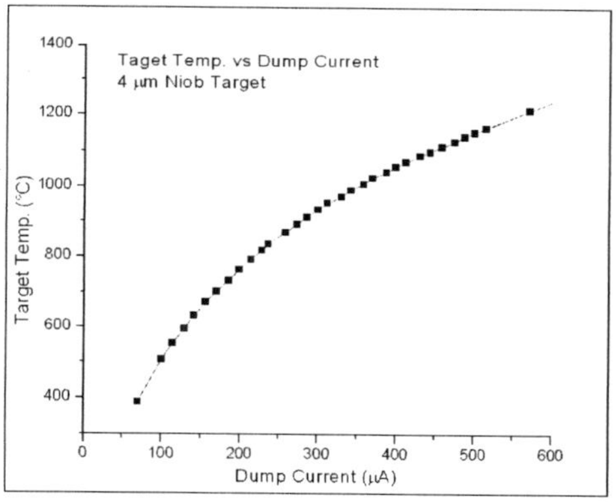

FIGURE 5. Radiator temperature at 12-MeV beam energy.

Beam Loss Monitor

Beam loss due to scattering in the radiator depends on energy of the electron beam and radiator thickness. Also factors like electron beam position, divergence and changes in energy spread contribute. A four-quadrant beam loss monitor was designed and installed around the beamline. It consists of 30-cm long pieces of air filled Heliax coax cable (Andrews HJ4-50). Figure 6 shows a simple scheme of the four detectors and their readout.

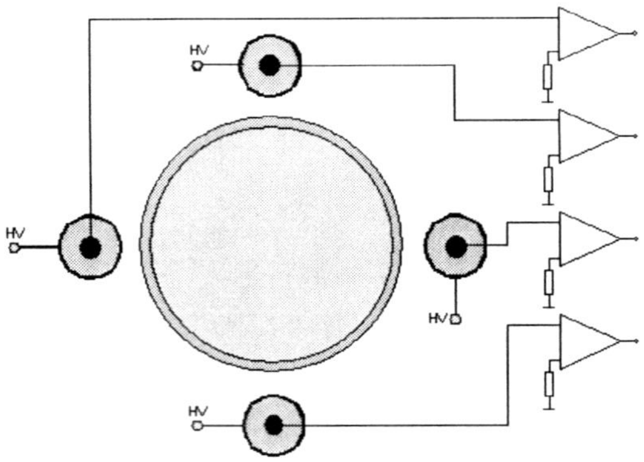

FIGURE 6. Four-quadrant beam loss monitor system.

They are operated as ionization chambers detecting radiation produced by scattered electrons hitting the beam pipe. These kinds of beam loss monitors based on Heliax cables at ELBE are described in detail in [6].

REFERENCES

1. F. Gabriel et al., NIM B 161 – 163, 2000, pp. 1143 – 1147.
2. R. Schwengner et al., "First Experiments with the Bremsstrahlung Facility," ELBE Annual Report 2002, p. 23.
3. P. Evtushenko et al., "New Design of the ELBE Beam Position Monitors," ELBE Annual Report 2001, p. 13.
4. P. Evtushenko et al., "Performance of the BPM Electronics, ELBE," ELBE Annual Report 2002, p. 17.
5. R. Jainsch et al., "The Video System for Beam Monitoring and Diagnostics at ELBE," ELBE Annual Report 2002, p. 11.
6. P. Michel et al., "Beam Loss Detection at Radiation Source ELBE," Proceedings DIPAC 2003, Mainz, Germany, p. 65.

Design and Tests of a New Rest Gas Ionization Profile Monitor Installed in the SPS as a Prototype for the LHC

C. Fischer, J. Koopman, D. Kramer, R. Perret, M. Sillanoli

CERN, Geneva, Switzerland

Abstract. Based on the encouraging results obtained with a Rest Gas Ionization Profile Monitor of a first generation, a new monitor was designed and then installed in the SPS at the beginning of 2002. Its design fulfills all the requirements for a future installation in the LHC where four such monitors are foreseen. After the initial tests performed during the run of 2002, a few upgrading steps appeared necessary mainly in order to cope with the nominal LHC beam characteristics. They were implemented during the subsequent winter stop and the operation of the monitor was resumed in 2003 under various conditions of beam, ranging from an LHC pilot bunch up to beams having nominal distributions in bunch number, intensity and energy in the SPS for injection into the LHC. After a description of the monitor design, the measurements performed with the instrument during these last two years are discussed with the difficulties encountered and the corresponding implemented cures. Data acquired in 2003 on the whole spectrum of LHC beam characteristics are presented and the modifications made to prepare the 2004 campaign with a view to still improve the performance are also discussed.

INTRODUCTION

A prototype gas ionization beam profile monitor was installed at the SPS in 1999 and has been under test since then. The device, provided by DESY, was initially designed to exploit the ion signal produced by the ionization of the residual gas on the beam passage. It was adapted and inserted within a magnet with increased gap, and the electron signal used instead, the resolution obtained with the electron signal generally being better. Encouraging results were reported on several occasions.[1] In particular, it was verified that useful information can be supplied in fast acquisition mode by measuring on consecutive turns, for example to perform injection matching tuning, as well as by integrating the signal on several beam passages in order to measure beam transverse distributions with a high precision.

Based on these results, it has been decided to equip the SPS and the LHC with such monitors for continuous transverse distribution measurements. A new monitor, working in the vertical plane, was then developed and installed in the SPS before the 2002 run. Its innovative design is compatible with incorporation in the LHC, for the start-up of 2007. Such issues as compactness, easier maintainability and optical tuning are tackled. The monitor commissioning in 2002, with the resulting steps taken, is reported, and some measurements made in 2003 are mentioned and discussed. The perspectives for the year 2004 are also presented.

CP732, *Beam Instrumentation Workshop 2004: Eleventh Workshop*,
edited by T. Shea and R. C. Sibley III

THE NEW MONITOR

In the LHC most of the beam instrumentation, apart from the Beam Position Monitors, will be incorporated on ambient temperature vacuum chambers sections, where transverse space contingencies are severe in the horizontal plane due to the proximity of the two beam vacuum chambers. A location has been reserved where 360 mm is available for the monitor section half dimension.

Another requirement of the LHC is to keep the vacuum chamber impedance budget to a minimum. This is achieved by avoiding abrupt steps in the vacuum chamber section, adding small angle transition pieces to connect parts having different cross sections.

The Magnet

For the production of the magnetic field necessary to avoid the electron signal divergence, conventional magnets are used. They are recuperated from the former LEP wigglers and match nicely the space requirements with a section of 681 x 646 mm^2. These magnets produce a magnetic field B = 0.240 T at maximum current (I_{MAX} = 55 A). Figure 1 gives a sketch of such a magnet with the basic geometrical parameters.

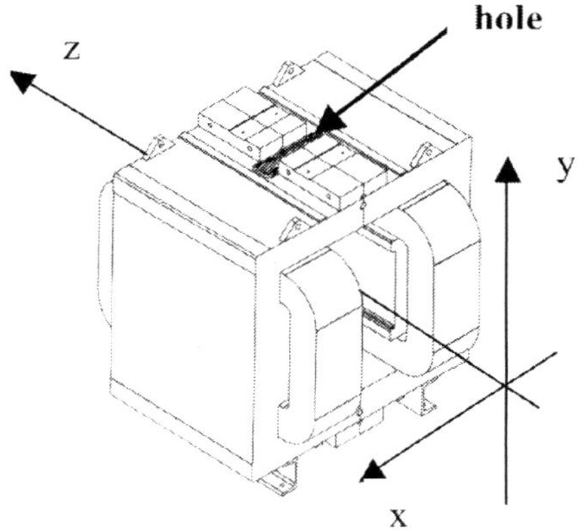

I_{MAX}	55	A
B_{MAX}	0.240	T
Strength	0.14	Tm
Gap width	200 x 200	mm^2
Section	681 x 646	mm^2
Overall length	680	mm

FIGURE 1. Sketch of the magnet used for the Gas Monitor with some basic parameters. The location of the hole through the yoke is also indicated.

A gap of 200 x 200 mm^2 is available between the yokes for incorporation of the monitor tank. As represented in Fig. 1, a racetrack shaped hole has been drilled through the magnet yoke to vertically extract the light signal coming from the monitor. The shape of the hole is such that while extracting on the left-hand side the signal coming from the ionization of the rest gas related to the vertical beam size, as explained hereafter, it is also possible to look, on the right-hand side, at the signal coming from the luminescence of the gas, providing the horizontal beam size.

Magnetic measurements performed after the hole implementation have shown that the magnetic field quality in the useful part of the magnet is only marginally affected by the presence of the hole: the linearity remains very good even up to 60 A, as shown in Fig. 2, and the field relative variation at ± 40 mm from the magnet center along the three directions is less than 1 per mil as represented for the z direction in Fig. 3.

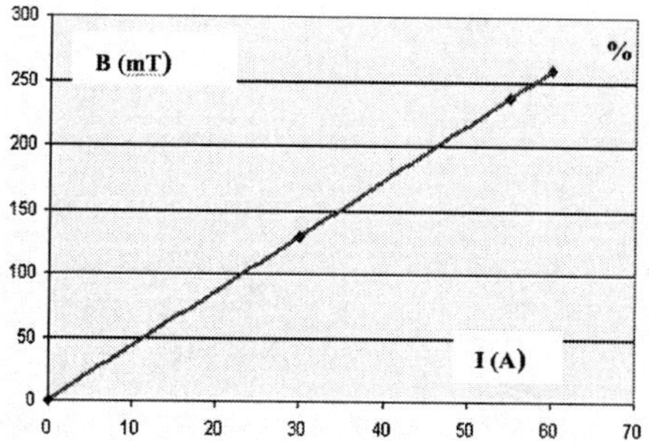

FIGURE 2. Magnet linearity in the presence of the hole.

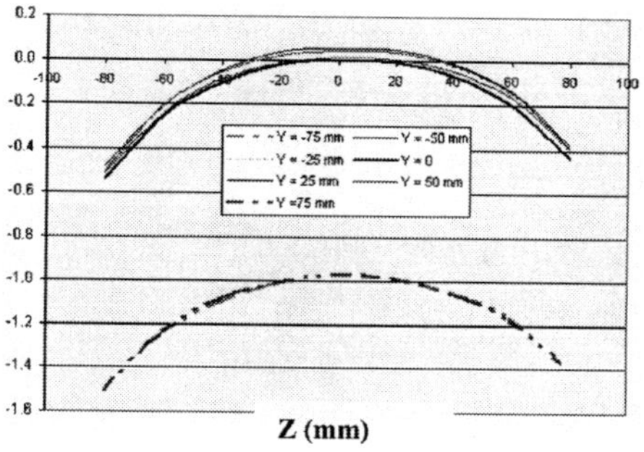

FIGURE 3. Relative variation of B along z at x = 0, for different y values.

The Central Detector

In Fig. 4 a), a cross section of the central detector is sketched. The two HV electrodes, normally set at ± 1 kV during operation, are spaced by 84 mm and are in the shadow of the 83 mm (internal diameter) vacuum chamber. The electrons are attracted to the input face of a Micro Channel Plate (MCP), which is set at the positive voltage, are multiplied within the MCP, and finally sent, at the MCP output, to the entrance face of a phosphor-coated prism set at a high positive voltage (anode).

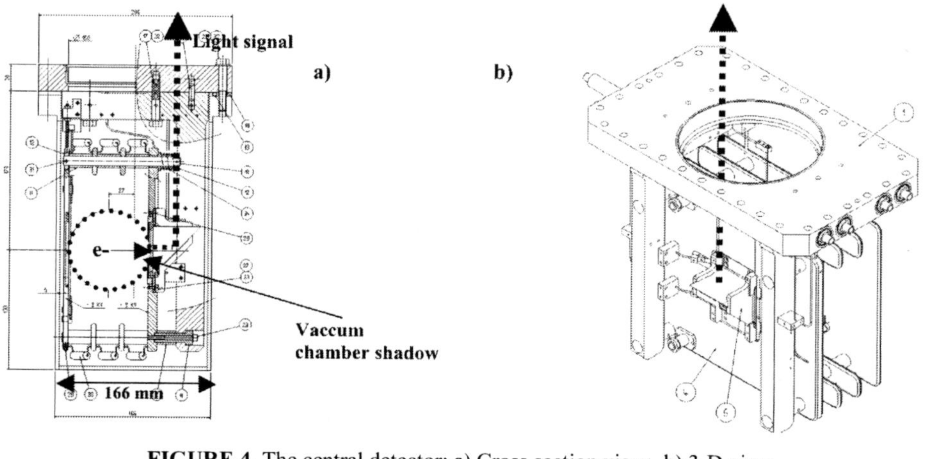

FIGURE 4. The central detector: a) Cross section view; b) 3-D view.

The light produced traverses the prism and, onto its back face, is reflected vertically to the outside.[Fig. 4 a)] It then leaves the vacuum chamber through a large fused silica viewing port, incorporated within a specially designed rectangular vacuum flange which supports the whole detector assembly [Fig. 4 b)]. The electron signal conversion and light extraction assembly transversally occupies 55 mm.

The purpose of the two small lateral field shaping electrodes appearing at the top and at the bottom on the central detector [see 3-D view in Fig. 4 b)] is to avoid electric field distortion.

The Monitor Tank

The monitor tank, made out of 316 LN stainless steel, fulfills two major requirements: it must become integrated within the magnet gap and, to avoid deteriorating the machine impedance budget, it should not present any discontinuity with respect to the normal vacuum chamber section. For this reason, its central part is terminated on each side by a fifteen-degree transition piece, which permits smoothly recuperating the normal section of 83 mm.[Fig. 5 a)] An extra 500 mm vacuum chamber length, with normal aperture, is left available on the right-hand side.[Fig. 5 a) and b)] This allows the magnet to slide along two rails [Fig. 5 b)], when access is needed to the supporting flange in order to remove the detector from the tank. The whole assembly occupies a longitudinal space of 1600 mm.

A valve and a vacuum gauge feed-through are visible on each side on the two transition ends. They have been implemented in order to control the injection of nitrogen and hence increase the pressure inside the detector. This is needed in the present SPS monitor to obtain a signal from luminescence, to measure the radial beam dimension, and could be required in the LHC where the rest gas pressure is expected to be only around 10^{-11} hPa at the monitor location.

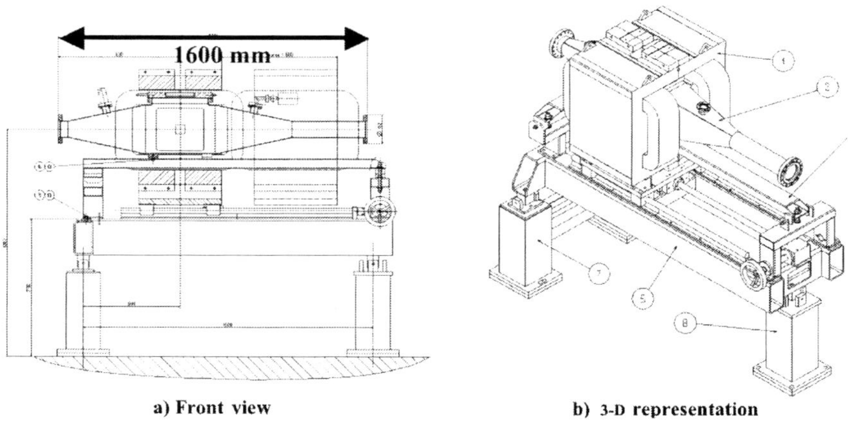

FIGURE 5. Sketch of the Monitor tank: a) front view; b) 3-D representation of the whole monitor assembly.

COMMISSIONING IN 2002

The first tests started during the SPS operation period in the summer of 2002, with two main problems generating important perturbations and a high noise level on the detector signal.

The first of these parasitic effects was electron cloud generation within the detector [2], a mechanism which starts to show up in the SPS machine when the beam structure approaches the nominal parameters required for the LHC, and which is still under investigation. Via this mechanism, secondary electrons are produced within the chamber. Under given conditions of bunch charge and bunch spacing this signal is steadily amplified and generates beam instabilities. In the SPS, the effect usually appears during the injection of the third batch, out of the four composing the nominal LHC beam. The vacuum pressure is seen to increase. The gas monitor signal exhibited the same behavior, (Fig. 8), and large fluctuations of the electrode DC voltage, and hence of the signal, were observed.

Other perturbations were induced by the high frequency modes generated by the beam structure inside the detector; the resulting high voltage instabilities could lead to complete signal loss due to vanishing gain.

During the subsequent stop several actions were taken. The high voltage electrodes were coated with a Non Evaporable Getter (NEG) layer, a treatment which is applied for vacuum getter pumping but which also reduces the secondary emission yield of the treated components, and hence their ability to generate electron clouds.

RF absorbers were also added close to the high voltage power supplies to filter the high frequencies.

Finally a conducting cover plate, leaving a reduced window, was installed on the view port for better conduction of the beam image current.

THE 2003 CAMPAIGN

Problems and Cures

In order to activate the NEG layer coating applied on the electrodes, a bake out to 200 degrees is necessary. This was performed after the monitor re-installation in 2003 with the result that electrical contacts were lost due to materials thermal expansion, in particular on the phosphor screen. In order to cope with these effects, thin CuBe frames, equipped with peripheral flexible contacts, were inserted within the detector assembly for better elasticity.

A fast but less sensitive phosphor was installed to try to test the feasibility of individual bunch measurement at 40 MHz. As a consequence, the MCP was used with high gain and premature aging of its central part, corresponding to the beam area, was observed. To overcome this problem, an image intensifier was added to the camera and a local orbit bump was applied in order to use another part of the MCP.

However, the addition of the intensifier generated other drawbacks. First, the resolution of the optical system was somewhat deteriorated. Reducing the camera lens diaphragm opening by 50% brought some improvement. Secondly, the intensifier appeared to be sensitive to the stray-field escaping from the magnet hole where the light is extracted. Hence, the intensifier was shielded with several layers of μ metal. Furthermore, the magnet was used at only 50 % of its maximum excitation.

Some parasitic lightening was still observed on the phosphor screen, at the end of the acceleration cycle when the bunch length was shrunk. This effect was overcome by removing the series resistors added close the monitor HV feed-throughs, which probably reflected back to the inside the RF modes generated by the beam.

Results

In spite of the above listed difficulties, interesting results were obtained last year. Some of them are discussed here but a more comprehensive study will be published soon.[3] Vertical profiles were measured over the whole intensity and energy ranges of LHC type beams, i.e. from single pilot bunches of 5 x 10^9 protons, to nominal beams of 3.5 x 10^{13} protons, (288 bunches of 1.15 x 10^{11} protons each), and from the injection energy into the SPS of 26 GeV up to 450 GeV which will be the injection energy into the LHC. On Fig. 6, profiles and 2-D pictures of a pilot bunch recorded at 26 GeV and at 450 GeV are shown. The data quality is improved, as seen on the lower one of the two profile curves, by subtraction of the fixed background pattern.

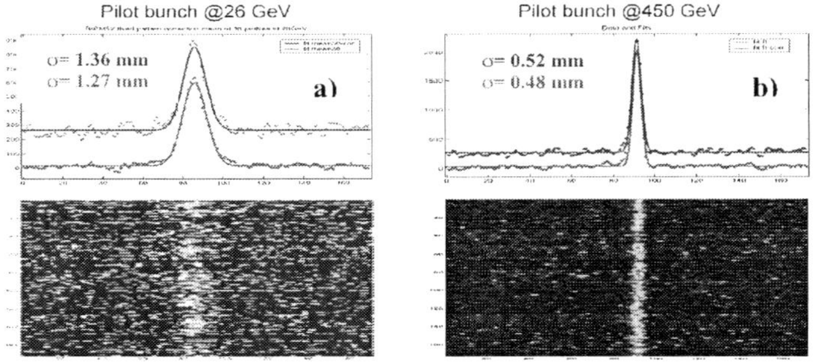

FIGURE 6. Vertical profiles and 2-D pictures of 5 x 10^9 protons: a) 26 GeV; b) 450 GeV. Removing the background pattern improves the data (lower curve).

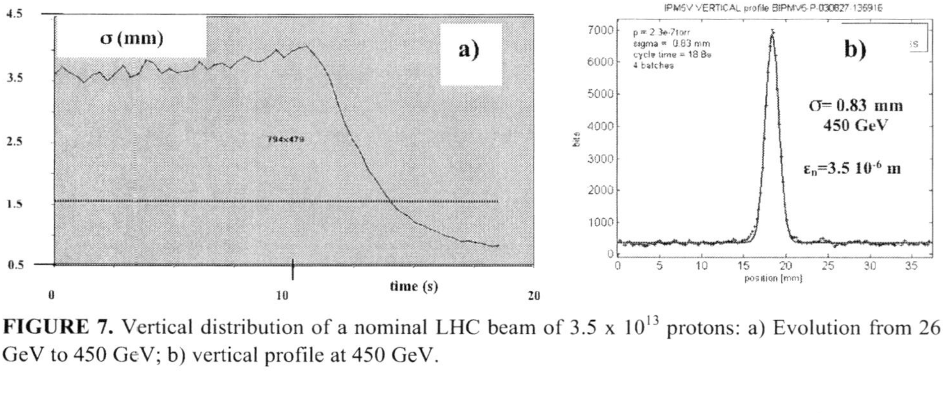

FIGURE 7. Vertical distribution of a nominal LHC beam of 3.5 x 10^{13} protons: a) Evolution from 26 GeV to 450 GeV; b) vertical profile at 450 GeV.

Figure 7 a) gives an example of the evolution of a nominal LHC beam vertical profile rms value throughout a complete SPS acceleration cycle with the corresponding profile obtained at the end, at 450 GeV.[Fig. 7 b)] During the 10.8 seconds flat bottom at 26 GeV, needed for the injection of 4 batches of 72 bunches each, the rms value is constant and starts to slightly increase between 8 and 10.8 seconds, when the third batch is injected. The injection of the fourth batch occurs at t = 10.8 s and then the acceleration from 26 GeV to 450 GeV starts. The rms value shrinks down to 0.83 mm while the energy is progressively increased to 450 GeV.

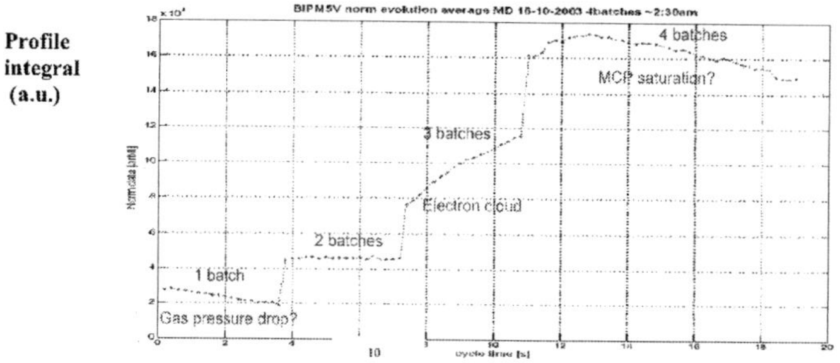

FIGURE 8. Profile integral evolution of a nominal LHC beam of 3.5 x 10^{13} protons.

Figure 8 shows the evolution of the beam profile integral along the SPS cycle. The steady signal increase observed after injection of the third batch is the signature of an electron cloud effect mechanism, reduced but still showing up within the detector, in spite of the NEG coating applied on the electrodes.

During acceleration from 26 GeV to 450 GeV, between 10.8 s and 19 s, the profile integral decreases, whereas, without beam loss, it is expected to remain constant. This may indicate a saturation effect within the monitor, possibly at the level of the MCP.

The monitor was also operated with beams having a normalized emittance value lower than the nominal LHC beam normalized emittance of 3.5 x 10^{-6} m. Comparisons with the wire scanners were made and the results are presented in Fig. 9. The agreement is good at 26 GeV but when the beam rms value decreases during acceleration to values lower than 0.7 mm the gas monitor data become more and more pessimistic with discrepancies increasing for the lowest emittances. The contribution of the monitor resolution limit to the measured rms value is of the order of 0.35 mm.

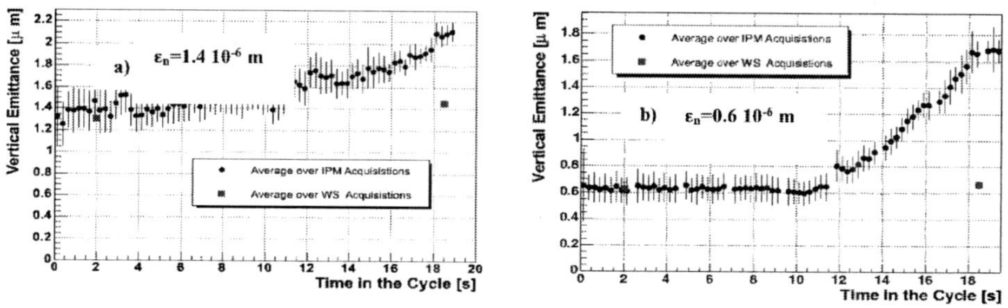

FIGURE 9. Profile integral evolution of low emittance beams: a) ε_n = 1.4 x 10-6 m; b) ε_n = 0.6 x 10-6 m; • Gas monitor data, ■ comparative wire scanner data.

MODIFICATIONS AND PERSPECTIVES FOR 2004

Based on the 2003 data, some modifications have been implemented for 2004.

After some feasibility tests, the input face of the new MCP has also been coated with NEG to further reduce the electron cloud effect within the detector.

Several steps have been implemented to improve the resolution of the monitor:

- A few components will be replaced: a single new MCP will replace the previous chevron stacked MCP amplification system for better S/N ratio and resolution. A new $Y_3Al_5O_{12}$:Ce (P46) phosphor coating, replacing the previous CdS:In, will be used at the entrance face of the prism, less demanding from the MCP for the same signal amplitude. A reflective aluminum layer has been deposited onto the prism back face extracting the light to the outside to ensure total internal reflection and hence, a more homogeneous signal distribution.
- For high precision measurements, the detector will be coupled to a new Low Light Level camera with low noise and higher resolution, which is expected to be insensitive to magnetic fields. The optics placed in front of the camera have also been reconsidered for an improved aperture. A splitter has been incorporated to add a signal path used by the fast turn-by-turn system equipped with a 32-channel photomultiplier tube.
- On the cathode grid, the number of wires has been reduced by a factor of 3 with a view to reduce their contribution to tail distributions, via secondary electron emission, and to RF field generation.
- A non-reflective treatment was applied at the surface of relevant detector supporting parts to suppress parasitic reflections.

New RF absorbers have been installed close to the HV supplies for better noise filtering, and the possibility to vary the MCP gain during the beam acceleration cycle, in order to stay below the saturation level, is now foreseen in the HV control program.

Hence, a performance improvement is anticipated. With upgrades of the control software, the monitor is expected to soon reach an operational stage for the LHC era.

ACKNOWLEDGMENTS

M. Jimenez and his section helped us when monitor conditioning was necessary, TS/MME and PH/TA1 dealt with the coatings, and D. Cornuet and his team did the magnet measurements. G. Arduini helped with the monitor insertion in the SPS, A. Beuret addressed the problem of the magnet power supplies, A. Guerrero of the control software, and B. Dehning and F. Roncarolo provided the wire scanner data. We appreciated the help of many other colleagues who cannot all be acknowledged.

REFERENCES

1. G. Ferioli et al., "Beam Studies made with the SPS Ionization Profile Monitor," DIPAC2003, Mainz, Germany.
2. J. M. Jimenez et al., "Electron Cloud with LHC Type Beams in the SPS," LHC Project Report 632.
3. B. Dehning et al., "Comparative Transverse Distribution Measurements between the New SPS Rest Gas Ionization Monitor and the Wire Scanner Monitors," contribution to EPAC, Lucerne , Switzerland, July 2004.

Reliability of Beam Loss Monitors System for the Large Hadron Collider

G. Guaglio*+, B. Dehning*, and C. Santoni+

*CERN, CH-1211 Genève 23 Switzerland
+UNIVERSITE BLAISE PASCAL, 34 avenue Carnot BP 185, 63006 Clermont-Ferrand France

Abstract. The employment of superconducting magnets in high energy colliders opens challenging failure scenarios and brings new criticalities for the whole system protection. For the LHC beam loss protection system, the failure rate and the availability requirements have been evaluated using the Safety Integrity Level (SIL) approach. A downtime cost evaluation is used as input for the SIL approach. The most critical systems, which contribute to the final SIL value, are the dump system, the interlock system, the beam loss monitors system and the energy monitor system. The Beam Loss Monitors System (BLMS) is critical for short and intense particle losses, while at medium and higher loss time it is assisted by other systems, such as the quench protection system and the cryogenic system. For BLMS, hardware and software have been evaluated in detail. The reliability input figures have been collected using historical data from the SPS, using temperature and radiation damage experimental data as well as using standard databases. All the data have been processed by reliability software (Isograph). The analysis ranges from the components data to the system configuration.

INTRODUCTION

The Large Hadron Collider (LHC) is the next CERN particle accelerator that will try to penetrate further into the matter structure, accelerating protons up to 7 TeV. The innovative characteristic is the wide-scale use of superconducting magnets to reach fields closed to 9 Tesla so as to bend the high-energy particle beam. The superconducting technology involves different challenges, mainly addressed to generate and maintain the magnets in the superconductive state. One of the critical applications is the Machine Protection System, which intends to avoid machine damages caused by the heating following a beam loss.

In this work we will introduce the general LHC approach to prevent these events with the utilization of different systems. Later we will focus on the Beam Loss Monitors System, analyzing the design aspects and the dependability figures.

SIL APPROACH

In this paragraph we will discuss the possibility to use the Safety Integrity Levels (SIL) approach for machine protection purposes. In fact, strictly speaking, the standard IEC 61508 [1] is used as guideline for personal safety and not for machine protection.

CP732, *Beam Instrumentation Workshop 2004: Eleventh Workshop*,
edited by T. Shea and R. C. Sibley III

TABLE 1. Category table used for LHC consequences definition.

Category	Injury to personnel		Damage to equipment	
	Criteria	**N. fatalities (indicative)**	**CHF Loss**	**Downtime**
Catastrophic	Events capable of resulting in multiple fatalities	≥ 1	$> 5*10^7$	> 6 months
Major	Events capable of resulting in a fatality	0.1 (or 1 over 10 accidents)	$10^6 - 5*10^7$	20 days to 6 months
Severe	Events which may lead to serious, but not fatal, injury	0.01 (or 1 over 100 accidents)	$10^5 - 10^6$	3 to 20 days
Minor	Events which may lead to minor injuries	0.001 (or 1 over 1000 accidents)	$0 - 10^5$	< 3 days

In any case, it gives general figures and hints that could wisely be used to protect the machine rather than people.

The aim of the SIL is to suggest failure rate figures to face a risk occurrence. In the following we will define the risk concept and we describe the procedure used to extract the reliability figures for the system.

RISKS FOR LHC BEAM LOSSES

Generally speaking, we can define the risk associated to an event as a product of the probability that this event will happen multiplied by a general cost of that consequence, which is "how often we pay something."

The SIL standard simply gives a way to estimate and to fix that probability to minimize the risk to pay a cost.

In fact, the probability P is strongly connected to the system designed to minimize the risk of the outcome: its range runs from the event frequency F, when there is no protection system, down to a level defined by the system functionality.

In the following we will extract from the SIL standard the figures for a safe operation (the system works when it has to work) and for an efficient one (the system does not work when it does not have to work).

The two main failures that can occur are Magnet Destruction and False Dump. We will have a Magnet Destruction (MaDe) if there is a dangerous loss and the beam is not extracted. This eventuality could cause a downtime of around 30 days to substitute the 250 kCHF superconductive magnet with a spare one. Currently, only 16 spare main quadrupole magnets are foreseen.

TABLE 2. Frequency table used for LHC risk definition.

Category	Description	Frequency (per year)
Frequent	Events which are very likely to occur	> 1
Probable	Events that are likely to occur	10^{-1} - 1
Occasional	Events which are possible and expected to occur	$10^{-2} - 10^{-1}$
Remote	Events which are possible but not expected to occur	$10^{-3} - 10^{-2}$
Improbable	Events which are unlikely to occur	$10^{-4} - 10^{-3}$
Negligible	Events which are extremely unlikely to occur	$< 10^{-4}$

TABLE 3. Failure rate (SIL) and risk table used for LHC risk evaluation. See text for details.

Event Likelihood	Consequence							
	Catastrophic		Major		Severe		Minor	
Frequent	SIL 4	I	SIL 3	I	SIL 3	I	SIL 2	II
Probable	SIL 3	I	SIL 3	I	SIL 3	II	SIL 2	III
Occasional	SIL 3	I	SIL 3	II	SIL 2	III	SIL 1	III
Remote	SIL 3	II	SIL 2	II	SIL 2	III	SIL 1	IV
Improbable	SIL 3	II	SIL 2	III	SIL 1	IV	SIL 1	IV
Negligible	SIL 2	III	SIL 1	IV	SIL 1	IV	SIL 1	IV

A False Dump (FaDu) generation occurs when the system generates a false alarm, followed by a dump, even if there were no dangers for the superconductive magnets. This generates on average almost 3 hours of downtime to return to the previous beam status.

Inspired by the IEC 61508-5 annex B [1], LHC expert judgment created the consequences table, the frequency table and the risk table (Tables 1-3).

For our scenarios, we will have a major consequence for the MaDe occurrence, a minor (or less) for the FaDu, as shown in Table 1.

The second step is to estimate the frequency of the event.

Due to the fact that there are no historical data on machines that are comparable for technology, size and luminosity, we have to estimate the order of magnitude of the event frequency. We can guess, on LEP experience, that we can expect probably more than 10 dangerous losses per year. What's more, due to the fact that every dump comports around 3 hours of downtime to come back to the original conditions, we hope to have less than 1000 dumps per year. Consequently, we try to face an event that we guess to have a frequency of 100/y.

To avoid the Magnet Destruction, we will develop a system that can generate some False Dumps per year, caused by its internal failures. As already said, every dump brings around 3 hours of no operational beam, consequently we will try to keep this number on the order of 30 per year (1 per week), so as not to seriously decrease the LHC luminosity time.

The frequency of both events is "Frequent," as shown in Table 2.

Now we have to decide which approach we want to follow. In the standard IEC 61508-5 Annex A and B, two possibilities are present: Functional Approach (FA) and Malfunction Approach (MA).

FA is based on the definition of the SIL levels and it requires that the system has to work with that failure rate to be considered safe; later it verifies that the risk is acceptable. MA defines first the acceptable risk and then tries to reduce it to *as low as reasonably practicable*.

In both philosophies, we have to define what the maximum tolerable risks for our events are. For MaDe we can say that, due to the facts that LHC will work for 20 years and that there will be 16 spares magnets, we tolerate a maximum of 0.8 MaDe per year. For FaDu, we have already said that we would prefer to stay around 30 FaDu per year so as not to deteriorate operational efficiency.

FA and MA give us the failure probability per hour, then we have to multiply that number by the operational hours per year to have the failure probability per year and finally, with the proper factors, we will have our failures per year. This should be less than the tolerated value. In other words

$$Failure_{MaDe}/h \cdot 4000\ h/y \cdot 100 \leq Tolerated_{MaDe}/y \quad (1)$$

$$Failure_{FaDu}/h \cdot 4000\ h/y \leq Tolerated_{FaDu}/y \quad (2)$$

TABLE 4. Probability of a dangerous failure per hour.

SIL	Probability of a dangerous failure per hour
4	10-9 < Pr < 10-8
3	10-8 < Pr < 10-7
2	10-7 < Pr < 10-6
1	10-6 < Pr < 10-5

TABLE 5. Risk table with the definitions of the risk extrapolated from Table 3.

RISK	DEFINITIONS
I	Intolerable
II	Tolerable if reduction is impracticable or cost is disproportionate
III	Tolerable if cost exceeds improvement
IV	Negligible

In Eq. (1), we multiply the probability to have a MaDe by the expected number of annual dangerous losses, because we will "throw the dice" 100 times in the year.

For simplicity, we will develop the FaDu concept in the FA and MA.

In the FA we have, from Table 3, that a frequent-minor consequence has to reach SIL2 level: that means, from Table 4, a failure probability less than 10^{-6}/h. So we have to design the system to have a failure rate, for the FaDu, which is less than this value. Substituting in Eq. (2) the higher limits, we calculate 4 10^{-3} failures per year: probably we are too conservative if we compare it with the tolerated 30 failures per year. Table 3 is also useful to estimate what the risk is (Roman number and Table 5) if we don't reach the suggested SIL level with our system. Suppose we design a system with a failure probability of 10^{-4} for FaDu. With Eq. 2 we calculate an error Frequency of 4 10^{-1} failures per year that, in Table 2, is an "occasional" event. With Table 3 we read that we have to face a III risk, which means that we are allowed to not improve the system only if the changing "costs" more than the improving.

In the MA we start assigning the tolerated FaDu per year and calculating the Failure figures: with 30 FaDu/y we have 7.5 10^{-3} failures per hour. Note that entering in Table 3 with the 30 tolerated FaDu, a frequent-minor event, we have a II risk. That means (Table 5) we have to reduce it if it is possible and the costs are not disproportionate; as it could be expected, due to the fact that 30 is our maximum tolerated value.

It has to be noted that Table 3 is extracted by the suggestion of [1], but it could and should be adapted to the different situations with expert judgment, as well as Tables 1 and 2.

Following the same procedure for the MaDe, we obtain Table 6. From this table we can see that there is a good agreement between the two approaches in the MaDe predictions, whereas for FaDu they are deeply different, probably because there is an overestimation of the gravity of the event (3 days of downtime is much greater than 3 hours) that brings the FA to be too conservative.

TABLE 6. FA and MA results for LHC machine protection. In **bold**, the input parameters

		Failure/h	Total/y
MaDe	FA	**10^{-7}**	0.04
	MA	$2\ 10^{-6}$	**0.8**
FaDu	FA	**10^{-6}**	$4\ 10^{-3}$
	MA	$7.5\ 10^{-3}$	**30**

LHC SYSTEM: MAIN ACTORS

In the LHC the safe philosophy will be: whenever there will be a dangerous proton loss, we extract the beam from the machine. The first line safety systems are: the Beam Loss Monitors System (BLMS), which detects the dangerous loss and inhibits a beam permit through the Beam Interlock System (BIS), so that the LHC Beam Dump System (LBDS) can extract the beams from the machine in a safe way. This extracts the beam in function of the beam energy signal given by the Beam Energy Meter (BEM). These 4 main systems are assisted by other second line systems that additionally protect the machine but with slower time constants: the first line system has to act for 100μs intense losses as well for 100s low losses, the second line systems react after 10 ms of time, so they cannot help for fast losses. Currently there are also ideas to extend the Beam Position Monitor and the lifetime system to protect the machine also for the fast losses, but they are still not well defined.

Nevertheless, several systems can generate dumps but not exclusively for the machine protection aim. From the operational systems to the safety ones we can have almost 20 systems that can request for a dump, as show in Fig 1.

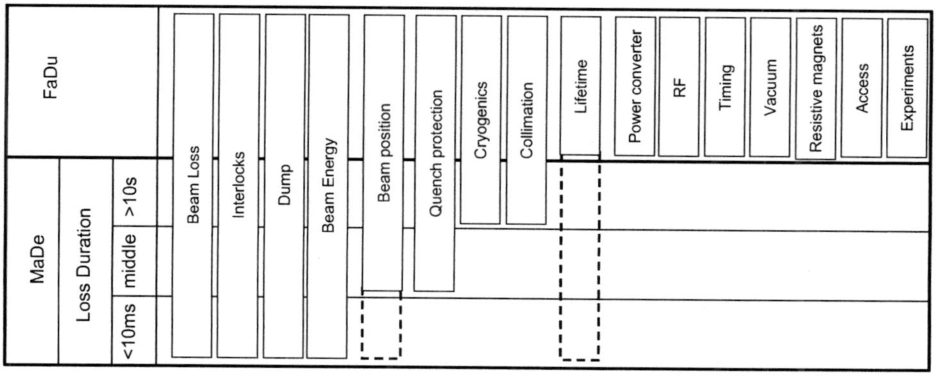

FIGURE 1. Main LHC systems connected to the Beam Interlock System and that can generate a dump, and classification of the protective systems.

If we assume that the previously calculated failure rate has to be equally shared between the different systems, we should guarantee a failure rate, with MA, less than 6 10^{-8} for MaDe and less than 3.8 10^{-4} for FaDu. The equal distribution of the failure rate could be mainly unrealistic for the FaDu event, due to the fact that some of these systems, like Access, historically have really high reliability.

BLMS FOR MAGNET PROTECTION

As reported in [2] and calculated in [3], the Beam Loss Monitors System has to protect the superconducting quadrupoles against losses of different duration and intensity. The quadrupole locations have been chosen because they are expected to be more loss sensible due to the larger beam dimensions and the limits in physical aperture.

As shown in Fig. 2, there is also a strong dependence with the energy and with the loss duration that brings to a dynamics of 9 orders of magnitude. Note that Fig.2 has been calculated for the dipole magnet. Better definition of the levels for the quadrupoles is still ongoing.

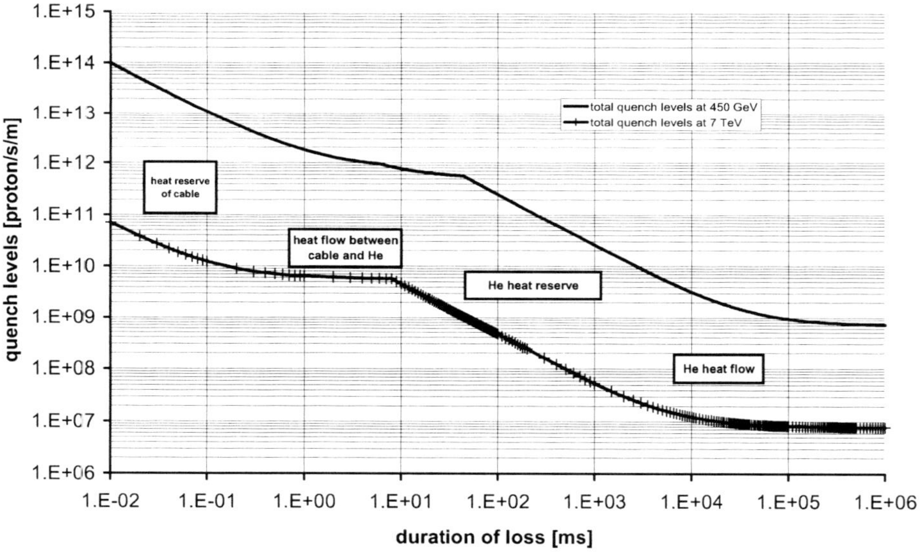

FIGURE 2. Calculated quench level for LHC dipole.

The BLMS is mainly constituted by Ionization Chambers (ICs) around the quadrupole magnets in the LHC tunnel. There will be 6 ICs per quadrupole, in different locations, to cover all of the quadrupole, as calculated in [4]. These chambers send a current, which ranges from 1 pA to 1 mA, to a Current-to-Frequency Converter that digitizes the current into pulses. These pulses are then counted by the digital part of the front-end electronics. The digital part, hosted in a FPGA, multiplexes 8 different channels and several status bits; it doubles the signal and sends it to the surface through 2 optical lines. At the surface the signals are checked and compared, to avoid transmission error, de-multiplexed and then compared with the threshold levels corresponding with the current beam energy. The measured signal, that arrives every 40 μs, is then averaged to compare it with the other threshold levels [5].

TABLE 7. Summary of the calculated failure rate.

<table>
<tr><th rowspan="2">Element</th><th colspan="3">Failure rate λ [10⁻⁸ 1/h]</th><th rowspan="2">Inspection interval [h]</th><th rowspan="2">Notes</th></tr>
<tr><th>Single</th><th>Not redundant</th><th>Redundant</th></tr>
<tr><td>IC+cable+terminations</td><td>2.5</td><td colspan="2" rowspan="3">24</td><td>20</td><td>Experience SPS</td></tr>
<tr><td>Integrator</td><td>2.0</td><td rowspan="8">Continuous (40 μs)</td><td rowspan="4">Dose and fluence tested</td></tr>
<tr><td>Switch</td><td>8.7</td></tr>
<tr><td>FPGA TX*</td><td>200</td><td rowspan="6">840</td><td rowspan="6">0.014</td></tr>
<tr><td>Laser</td><td>510</td></tr>
<tr><td>2 Optical connectors</td><td>20</td><td></td></tr>
<tr><td>Optical fibre</td><td>20</td><td></td></tr>
<tr><td>Photodiode</td><td>3.2</td><td></td></tr>
<tr><td>FPGA RX*</td><td>70</td><td></td></tr>
</table>

RELIABILITY DATA COLLECTION

The failure rates of the different components are calculated using the Military Handbook 217F [6]. The general inputs are: expected ambient temperatures of 10°C into the tunnel and 30° at the surface, fixed ground environment (a factor of 2 in the failure rate) for the tunnel, benign ground for the surface (factor 0.5); the average time to substitute a failed unit is 1 hour. Then all the usual component failure rates are evaluated with the military standard. For unusual components we have evaluated the failure rates using historical data from similar components. For example, for the Ionization Chambers we have 140 LHC like chambers installed in SPS, which have been operational for 30 years without changing in the chamber sensitivity. It is a general procedure that the upper $\alpha\cdot 100\%$ confidence level of failure rate for a test of t_{tot} hours over N component and F fails after t_i hours is given by

$$cumulative\_time = \sum_{i=1}^{F} t_i + (N - F)\cdot t_{tot} \tag{3}$$

$$\lambda = \frac{\chi^2(1-\alpha, 2\cdot F)}{2\cdot F\cdot cumulative\_time} \tag{4}$$

Assuming F = 0 in Eq. (3) and F = 1, i.e. 1 failure just after the 30th year, for Eq. (4), we have for IC a failure rate of 2.5 10^{-8}/h. Table 7 summarizes some significant figures.

From the table we can see that the lasers are the weakest components; this is the reason why we have decided to double the optical line, with the improvements reported in the column "Redundant". The FPGA figures, here, are overestimated with the MIL standard; experimental data will substitute for them. The power supply is not considered here, because there are actions ongoing for their final layout definition.

In the previous table are also reported the inspection intervals; this could be continuous, (that means an on line measurement of the functionalities), every operational dump, (roughly every 20 hours), or every year (IC gas is checked during every shut down with a radioactive source). A frequent inspection interval decreases the probability of finding the system not ready when required and so decreases the Magnet Destruction probability.

ANALYSIS

The entire system has been studied with commercial software (Isograph). A fault tree analysis has been performed, with particular attention to the unavailability of the system (the probability to find the system not ready to act), for the MaDe, and to the failure rate of the system for the FaDu generation. The unavailability of a single BLMS channel, without power supply, is 4.9 10^{-7}/h, 55% given by failure of the IC, mainly for the reason that it is the least checked component. The single channel failure rate, on the other hand, is around 2.4 10^{-7}/h, 70% given by the switch systems into the CFC. Considering that we have 3200 channels in the system, we have a failure rate of 7.7 10^{-4}. The FaDu number is quite close to the maximum accepted by the MA, after the LHC systems apportionment. In this way BLMS generates, in 4000 operational hours per year, 3 false dumps. We are trying to reduce this number, which is in any case not so worrying, by improving the switch electronics. For the MaDe further considerations are required. The previously given figure is the unavailability (U_1) of a single channel. For the system, we have to consider the probability that the losses could be seen by just 1 or more channels. It is in fact common experience that a single dangerous loss could affect more than one location around the ring and so more than one channel that could detect the loss. So, if we define N_i to be the number of losses per year that occur in i locations and U_i to be the probability to have i unavailable channels, the system unavailability per year (U_s) is:

$$U_s = \sum U_i \cdot N_i \approx \sum U_1^{\,i} \cdot N_i \approx U_1 \cdot N_1 . \tag{5}$$

In fact, the probability that 2 channels (so distant to avoid common failure causes) fail at the same time is U_1 squared, and so on for more channels. In the last step of Eq. (5) we neglected the term higher than U_1, due to the fact that U_1 is $\ll 1$. So the question now is: how many losses, of the 100 initially foreseen, are affecting just one channel? If we suppose that $N_1 = 100$, we will lose in 4000 hours 0.2 magnets, the maximum MaDe per year. If $N = N_1 + N_2 = 5 + 95$, we will decrease U_s to the FA requirements. To estimate what the loss distribution along the rings is and their correlation, a beam dynamics simulation project is required and it has already been launched.

CONCLUSIONS

The IEC 61508 standard has been used as a guideline to estimate the failure rate of the Beam Loss Monitors System. Either for the main function failure, the Magnet Destruction, or the system induced failure, False Dump, our current design is on the border of the tolerated risk; further analysis is required to better estimate the LHC loss distribution and correlation. Improved electronics are also in the process of being developed.

REFERENCES

1. *IEC 61508 International Standard*, First edition, 1998-12.
2. J. B. Jeanneret and H. Burkhardt, "Measurement of the Beam Losses in the LHC Rings," *LHC Project Document*, LHC-BLM-ES-0001.00, 2003.
3. J. B. Jeanneret, D. Leroy, L. Oberli, and T. Trenkler, "Quench Levels and Transient Beam Losses in LHC Magnets," *LHC Project report 44*, 1996.
4. G. A.. Arauzo and B. Dehning, "Configuration of the Beam Loss Monitors for the LHC arcs," *LHC Project note 238*, 2000.
5. B. Dehning, G. Ferioli, W. Friesenbichler, E. Gschwendtner and J. Koopman, "LHC Beam Loss Monitors System Design," in *Beam Instrumentation Workshop-2002*, edited by G. A. Smith and T. Russo, AIP Conference Proceedings 648, New York: American Institute of Physics, 2002, pp. 229-236.
6. Military Handbook, "Reliability Prediction of Electronic Equipment," *MIL-HDBK-217F*, 1991.

Development of a Beam Loss Detection System for the CLIC Test Facility 3

T. Lefèvre*, M. Velasco*, M. Wood*, H. H. Braun†, R. Corsini†, M. Gasior†, and F. Tecker†

*Northwestern University, Evanston, IL, USA
†CERN, Geneva, Switzerland

Abstract. The CLIC test facility 3 (CTF3) provides a 3.5-A, 1.6-μs electron beam pulse of 150 MeV at the end of the linac. The average beam power is 4 kW. Beam losses will be monitored all along the linac in order to keep the radiation level as low as possible. The heavy beam loading of the linac can lead to time transients of beam position, size, and energy along the pulse. To compensate for these transients effectively, the beam loss monitor (BLM) technology must have a time response faster than a few nanoseconds. Preliminary tests have been performed in 2003 on the already existing part of the accelerator with the aim of studying the requirements for the system to be built in the future. The experimental data are compared to the results of Geant3 simulations. Based on these results, a complete beam loss detection system is currently designed for the observation of the beam transient loss and its minimization.

INTRODUCTION

A new test facility, named CTF3 [1], is under construction at CERN with the aim of demonstrating the feasibility of the Compact LInear Collider (CLIC) [2], as a high luminosity, 3-TeV center of mass energy e^+-e^- Collider. The CTF3 linac is composed of 18 consecutive fully loaded 3-GHz cavities which will accelerate a 3.5-A, 1.6-μs-long electron beam pulse up to a final energy of 150 MeV. Housed in the LEP injector tunnel, the linac is scheduled for completion by the end of 2004. The installation started in January 2003. The injector, including the 3-GHz bunching system and two accelerating cavities, has already delivered 20-MeV electrons in July 2003. Two extra cavities were installed during the rest of the summer and were commissioned toward the end of the year.

Northwestern University is currently developing the beam loss detection system for the CTF3 linac. Thermal calculations have shown that losses are dangerous only if they exceed 10% of the total beam current. This means that the machine protection system can be based on wall current monitors [3]. Beam losses are measured and minimized in order to keep the radiation level and the activation as low as possible all along the linac. It is, however, important to remind that for the CLIC Drive beam linac beam losses would have to be measured with an accuracy which is not achievable with wall current monitors. And the development of beam loss monitor on CTF3 must be pursued with the aim of acquiring the required technology relevant for CLIC. Moreover, even on CTF3, the control of beam losses will become much more important in the future, in particular for the CLIC Experimental areas (CLEX) [4]. Quantitative beam loss measurements will

CP732, *Beam Instrumentation Workshop 2004: Eleventh Workshop*,
edited by T. Shea and R. C. Sibley III

be necessary to ensure the good operation of the machine. As a first step, simulations of beam losses and the corresponding e^+/e^- showers have been performed using Geant3 [5] in order to define the requirements for the BLM system to be built.

In this paper we present the results of a preliminary test performed in November 2003 on the already existing part of the linac. First the layout of the beam line is described, showing the main components of the accelerator. The technical characteristics of the detector chosen for this test are then described. Finally measurements are compared to Geant3 predictions and the perspectives for the final design of the system are presented.

EXPERIMENTAL SETUP ON CTF3

The machine layout is shown in Fig. 1. The beam line is composed of three main parts: the injector, the magnetic chicane, and a first accelerating module. A 1.6-μs-long high current beam is emitted from a thermionic DC gun. The electrons then enter a 3-GHz bunching and accelerating section, which brings their energy up to 20 MeV. The heavy beam loading in the accelerating cavities induces strong time transients and generates energy dispersion. The beam head experiences a higher accelerating field than the rest of the beam. The energy gain per accelerating cavity for the leading edge of the beam pulse can be as high as twice the nominal value (7 MeV). On top of that, any variation of the beam current along the pulse duration will induce a corresponding energy variation. A cleaning chicane consisting of a set of four dipole magnets and a tungsten slit-collimator is used to remove the undesired part of the beam energy spectrum. Dipole magnets create a dispersive region, where the horizontal position of the particles is correlated to their energy. By displacing the slit and varying its width, a well-defined part of the beam energy spectrum can be selected. The beam is then sent to the linac modules. Every module is composed of quadrupoles, which ensure the transverse focusing, then two 3-GHz accelerating structures and one monitor (BPM) measuring the beam position and current [6]. The beam misalignment is corrected using vertical and horizontal steerers installed in each module just upstream of the first cavity. In November 2003, only the first module was installed.

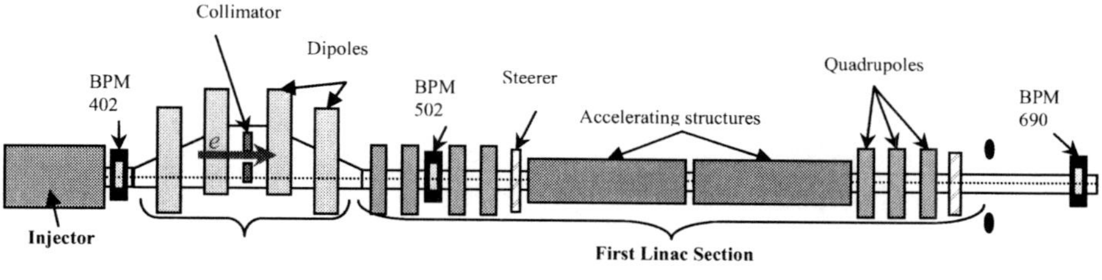

FIGURE 1. The CTF3 beam layout.

The design transverse optic [1] has been calculated so that the beam is focused between the two accelerating cavities. Its size re-expands then as electrons propagate to the next set of quadrupoles. For this reason, there is a high probability for beam losses to be concentrated near the quadrupole regions where the beam is large.

We have installed two beam loss monitors downstream of the second set of quadrupoles, as shown in Figure 1. As we will show later, the beam transient from the first part of the linac is lost at this location if it is not stopped in the cleaning chicane. The detectors are Aluminum Cathode Electron Multipliers (ACEMs) from Hamamatsu [7]. They are only sensitive to charged particles, generating secondary electrons on a thin aluminum layer (100 nm). Several amplification stages bring up the current of these secondary electrons up to a level strong enough to be measured directly. The gain of this amplification is controlled by adjusting the bias voltage on the different stages. ACEMs have a fast time response of the order of few nanoseconds, which allows the observation of the time evolution of the beam losses with good resolution. This is a very important feature in order to identify the time slice of the beam when losses are concentrated. A gain calibration curve made at CERN using a radioactive ^{137}Cs source is shown in Fig. 2.

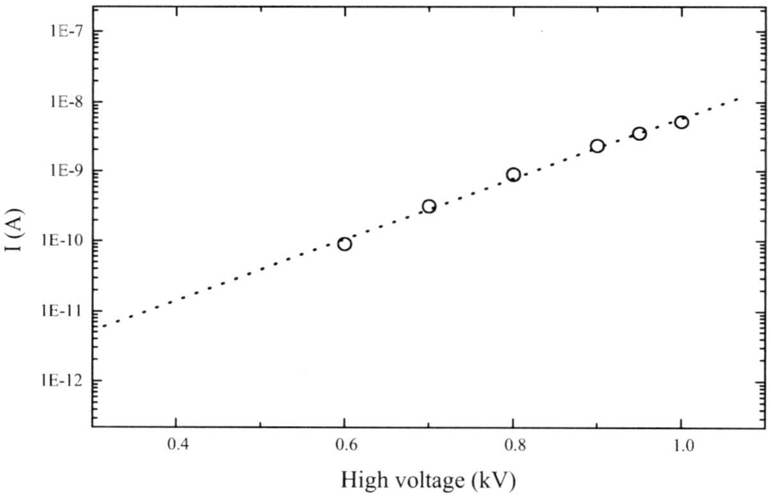

FIGURE 2. Calibration curve of an ACEM using a radioactive source.

The ^{137}Cs source is a β emitter with energies up to 514 keV (94%) and 1.176 MeV (6%). Using an intense source with an activity of 330 Mq, the output current of the ACEM is directly measured as a function of the high voltage bias. This detector is suitable for the detection of beam current from 100 nA to 100 mA.

During the test on CTF3, the ACEM's output signals are directly acquired on a 1-GHz bandwidth oscilloscope. The gain calibration made using the ^{137}Cs source has been verified with beam losses in the linac. A value equivalent to the slope of the fitted curve of Fig. 2 is found for operating voltages from 300 V to 500 V. The results are summarized in Table 1. The values, indicated in mA, correspond to the incident electrons current passing through the detector to measure a signal of 1 V amplitude on the oscilloscope.

Table 1. ACEM calibration.

High Voltage (V)	500	400	300
Incident beam current (mA)	26.5	75.7	177

SUPPRESSING THE BEAM TRANSIENT IN THE CHICANE

A first set of measurements has been performed in order to observe the initial beam transient suppression in the cleaning chicane. This is done by varying the width of the slit and looking at the beam current transmission through the chicane. The beam current signals seen by the BPM502, located just after the chicane, are shown in Fig. 3.

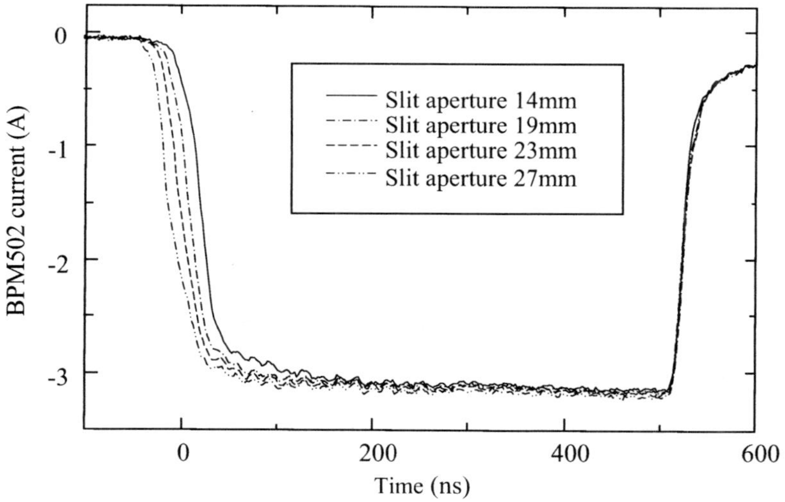

FIGURE 3. Time evolution of the beam current on BPM502 for different slit widths.

When the slit is being closed, the first part of the beam is stopped in the collimator, and in consequence the signal on the BPM502 is shortened. This illustrates the time energy correlation due to transient effects. At the same moment, beam loss measurements are done as shown in Fig. 4. The ACEM signal has a maximum when the slit is completely opened and it is being reduced as the slit is closed. The correlation between the signals from the BPM502 and the ACEM is very good. When the slit is opened, the full beam enters the accelerating section. The steady-state part of the beam and the transient are then accelerated further up to 35 MeV and 80 MeV respectively. Since the transverse optic is matched to 35-MeV electrons, the beam transient is not propagated correctly and lost near the quadrupole region where the ACEM sits.

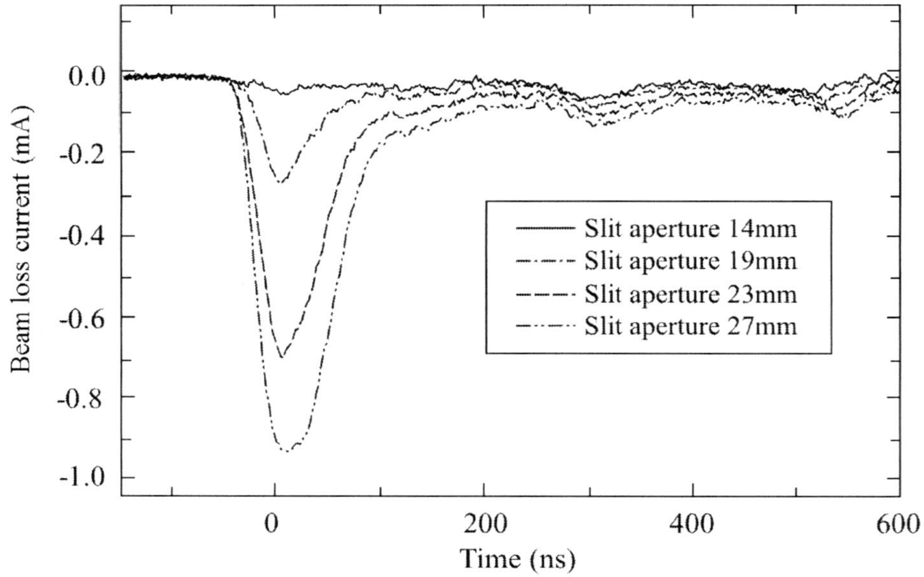

FIGURE 4. Time evolution of the beam loss measured by the ACEM for different slit width.

LOCALIZING THE BEAM LOSS USING BPMS AND BLMS

As a second step, the BLM measurements are compared to the expected shower simulations from Geant3. It is shown [5] that in order to obtain quantitative measurements of the beam losses, a large number of detectors must be installed all along the accelerator. To measure the beam loss intensity and position with good accuracy, detectors should to be installed every about 50 cm. In our test, only a few monitors had been installed, and the BLM measurements alone cannot be used to determine both the intensity and the position of the beam losses. In order to benchmark Geant3 predictions, the intensity of the beam loss is measured using two consecutive BPMs. The lost beam current is calculated from subtracting the signal of the BPM690 from the signal of the BPM502. Once the beam loss intensity is known, the signal from the ACEM is used to determine the longitudinal position of the losses.

In the actual test, by changing the steering conditions we displace the beam position and thus vary the position of the beam loss along the linac. We compare the beam current lost evaluated using the BPMs in Fig. 5a with the beam losses measured by the ACEM in Fig. 5b. The two curves correspond to two different steering conditions.

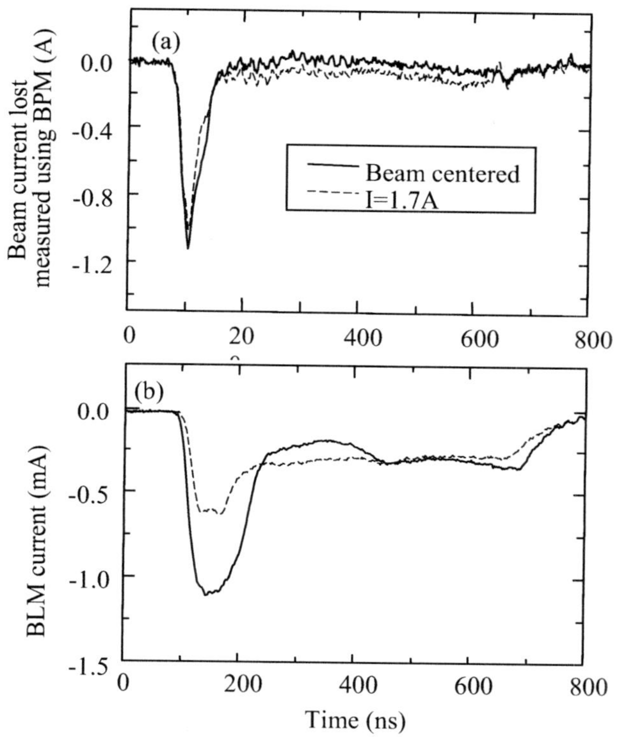

FIGURE 5. (a) Lost beam current as a function of time. (b) Beam loss current measured by the ACEM as a function of time.

To interpret these measurements, some simulations have been performed in order to evaluate the transverse distribution of the e^+/e^- showers in the realistic environment of the machine assuming. Since the shower depends on the beam energy, two extreme cases have been considered assuming 35-MeV and 80-MeV electrons. Each simulation assumes a punctual beam loss occurring at different locations: center of the 1st quadrupole, center of the 2nd quadrupole, center of the 3rd quadrupole, and two positions closer to the ACEM where the electrons are lost on the beam pipe. The results are expressed in terms of the shower efficiency, which represents the e^+/e^- shower current measured by the ACEM normalized to the beam loss current. A summary of the shower efficiencies obtained for the two different beam energies is given in Table 2.

Table 2. e^+/e^- Shower efficiency from Geant3 simulations.

Beam loss position	1st Quad	2nd Quad	3rd Quad	90cm	30cm
Shower efficiency (%) at 35 MeV	5.25 10^{-4}	1.4 10^{-3}	9.3 10^{-3}	2.7 10^{-2}	0.2
Shower efficiency (%) at 80 MeV	4.6 10^{-3}	8.9 10^{-3}	4.5 10^{-2}	9.9 10^{-2}	0.28

The shower efficiencies can be calculated by normalizing the curve in Fig. 5b with the curve in Fig. 5a. The measured efficiencies are reported as a function of the current in a steerer for the transient in Fig. 6a and for the steady-state part of the beam in Fig. 6b.

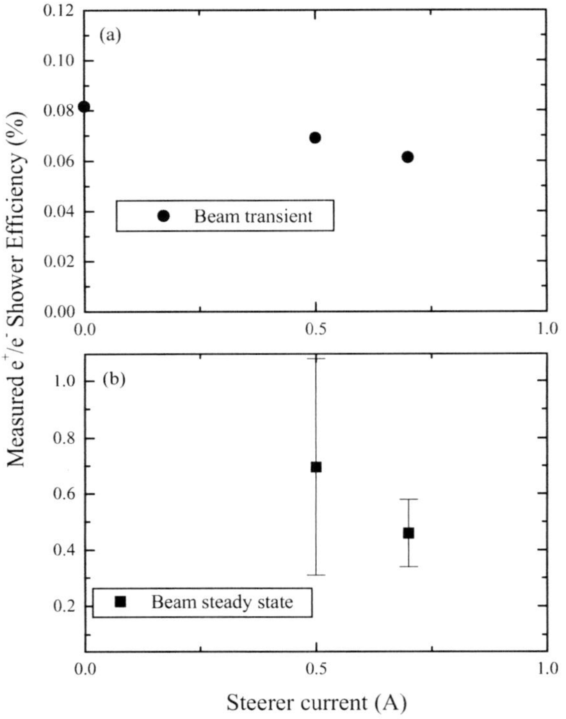

FIGURE 6. Shower efficiency measurement under different steering conditions. (a) for the beam transient. (b) for the "steady-state" part of the beam.

When the beam is centered, the measured efficiency for the beam transient is compatible with a loss located downstream of the 3rd quadrupole. This value decreases as the beam is steered; the position of the loss is displaced upstream of the detector in the direction of the 3rd quadrupole. The big error bars present in the case of the beam core come from the noise of the beam current signal shown in Fig. 5a. This illustrates the fact that the beam current losses cannot be measured from two consecutive BPMs with accuracy better than the 1% level just for one beam passage.

CONCLUSION

A beam loss monitoring system is under development for the CTF3 linac, and a preliminary test was done on the machine. Due to the heavy beam loading in the accelerating structures, beam losses were concentrated at the beam leading edge. During the test the beam current lost in the first part (~50 ns) of the pulse can be as high as 100%. However, the initial beam transients can be easily suppressed in the momentum cleaning chicane by stopping the highest energy electrons.

Beam loss measurements were in good agreement with the Geant3 simulations. The e^+/e^- showers measured by our 40-mm-diameter detector were as high as 1 mA for 35-MeV electrons. This corresponds to shower efficiencies of the order of 10^{-4}. During the test, the beam losses were concentrated at a well-defined position of the linac section, near the quadrupoles region, where the beam size is larger.

During the test, some of the principle features of the system to be built have been tested. To distinguish beam losses from the transient or from the beam core, a 10- to 20-ns time response is required. In order to get the transverse (X-Y) position of the loss, each BLM unit is going to be segmented into 4 individual detectors.

To get a complete mapping of the beam loss, and to provide quantitative measurements, the system becomes very complex in the sense that a large amount of detectors must be installed. The risk of damaging a machine component from beam losses is relatively small and corresponds to an extreme situation that can be detected by other means like the wall current monitors or the beam position and current monitors. Our system will be used as a tool for machine operation and the optimization of the beam transient compensation.

ACKNOWLEDGMENTS

This work was supported by the Illinois Consortium for Accelerator Research, agreement number 228-1001.

REFERENCES

1. G. Geschonke et al., *CTF3 note 047* (2002).
2. G. Guignard et al., *CERN-2000-008* (2000) and *CERN-2003-007* (2003).
3. Odier, *CTF3 note 055*, 2003.
4. H. H. Braun et al., *CLIC note 364*, 1998.
5. M. Wood, *CTF3 note 064*, 2004.
6. M. Gasior, *CTF3 note 056*, 2003.
7. Radiation monitor model R3807: http://usa.hamamatsu.com/default.htm.

YO! - A Time-of-Arrival Receiver for Removal of Femtosecond Helicity-Correlated Beam Effects*

J. Musson†, T. Allison†, A. Freyberger†, J. Kuhn¶, and B. Quinn¶

†*TJNAF, Newport News, VA 23606, USA*
¶*Carnegie-Mellon University, Pittsburgh, PA 15213, USA*

Abstract. The G0 parity violation experiment at Jefferson Lab is based on time-of-flight measurements, and is sensitive to timing effects between the two electron helicity states of the beam. Photon counters triggered by time-of-arrival at the target mandate that timing must be independent of delays associated with different orbits taken by the two helicity states. In addition, the standard 499 MHz beam structure is altered such that 1 of every 16 microbunches are filled, resulting in an arrival frequency of 31.1875 (31) MHz, and an average current of 40 μA. Helicity correction involves identifying and tracking the 31 MHz sub-harmonic, applying a fast/fine phase correction, and finally producing a clean 31 MHz trigger and a 499 MHz clock train. These signals are phase-matched to the beam arrival at the target on the order of femtoseconds. The 10 kHz output bandwidth is sufficiently greater than the 30 Hz helicity flip settling time (500 μs). This permits the system to correct each helicity bin for any orbit-induced timing inequalities. A sampling phase detection scheme is used in order to eliminate the unavoidable $2\pi/n$ phase shifts associated with frequency dividers. Conventional receiver architecture and DSP techniques are combined for maximum sensitivity, bandwidth, and flexibility. Results of bench tests, commissioning and production data will be presented.

INTRODUCTION

G0 is an experiment at Jefferson Lab designed to measure parity-violating asymmetries in elastic electron-nucleon scattering as a function of momentum transferred to the nucleon. This will provide better understanding of proton charge and magnetic structure by describing contributions from up, down and strange quarks [1].

At the heart of the experiment is JLAB's ability to deliver highly polarized (~75%) electrons, and to alternately "flip" the helicity from +z to -z at a 30 Hz rate using a pseudo-random code. It is possible that small helicity-correlated offsets associated with the launching time of the electron bunches could occur, on order of femtoseconds. This effect would produce a perceived asymmetry in the calculated time-of-flight of the particles from the target to the focal plane detectors.

---

*This work was supported by the U.S. Department of Energy under Contract Number DE-AC05-84-ER40150.

CP732, *Beam Instrumentation Workshop 2004: Eleventh Workshop*,
edited by T. Shea and R. C. Sibley III

The YO! Box produces two beam-derived signals, CLK and Y_0, which are synchronized to the arrival of the electron bunches in experimental Hall C, rather than to the laser-pulse associated with the time of launch in the Injector. In addition to tracking any time-of-flight variations, it is mandatory that the receiver be unaffected by variations in beam intensity and position. Also, RF sampling was chosen as an alternative to prescaling in an effort to remove the inevitable $2\pi/n$ phase ambiguity associated with frequency dividing. This ambiguity would create arbitrary cycle slips, catastrophically confusing the triggering of the photon detectors. Tight phase control and monitoring are also required to prevent cycle slips from within the control system.

The G0 beam is a modified 499 MHz CEBAF bunch structure, whereby only every 16th bucket of 2 ps duration contains electrons, but with 16 times the intensity. This results in a 31 MHz sub-harmonic whose lower-limit level is nominally -95 dBm after redistributing the pulse-power energy and combining with the response from a standard stripline BPM designed for 1497 MHz operation. Although the S/N is too low for high precision timing, a filtered signal is easily derived for use as a sampling reference. Also available is energy from a resonant 1497 MHz BCM cavity, which is much greater (-20 dBm) and affords good S/N ratio for fine/fast tracking.

Timing is derived by phase-locking a stable 31.1875 MHz VCXO to the -95 dBm 31.1875 MHz sub-harmonic. The clean VCXO signal output is used to stroboscopically sample the 1497 MHz beam-derived signal with a Sampling Phase Detector (SPD), establishing a high resolution (ps) zero crossing with 10 kHz bandwidth. This phase correction is applied back to the 31 MHz VCXO signal, defining the time-of-arrival. Finally, the corrected VCXO signal is again used to sample a Master Oscillator-derived (MO) 499 MHz signal, applying the phase correction and aligning the 499 signal to create a CLK output. Figure 1 is an overview of the system.

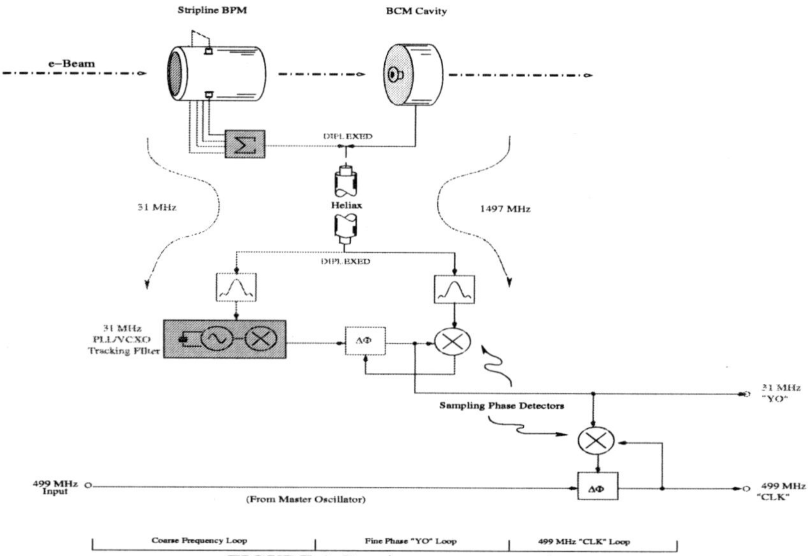

FIGURE 1. Receiver system overview.

RECEIVER ARCHITECTURE

A hybrid analog/DSP architecture was selected primarily to piggyback other development efforts at JLAB. Although the baseband FPGA-based DSP processor board was specifically designed for the YO! Receiver, it also resulted in a development tool for similar projects, and provided an ideal platform from which to develop the control algorithm. Due to the low noise figure requirement, combined with moderate output bandwidth, conventional RF front-ends were designed, but also included the use of Sampling Phase Detectors (SPDs) for direct-conversion to baseband. Figure 2 is the block diagram of the entire receiver system.

In addition to bench testing sub-assemblies, extensive modeling using SystemView (Matlab) was performed in order to verify SPD operation, generate filter coefficients, and also check initial PID parameters for the control algorithm.

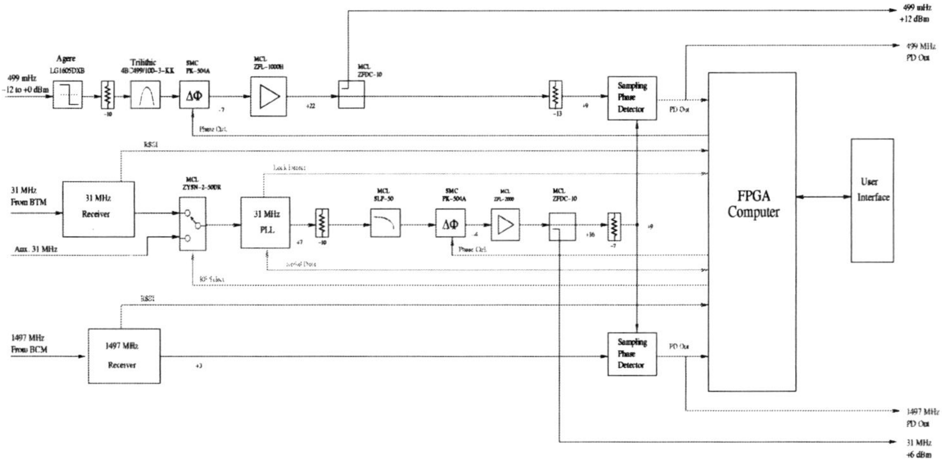

FIGURE 2. Receiver system block diagram.

SAMPLING PHASE DETECTOR (SPD)

The fundamental advantage of the SPD, exploited by the YO! Receiver, is the ability to compare the phases of harmonically related signals, as opposed to conventional phase detectors which require identical frequency inputs. The concept of stroboscopically sampling a fast signal with an integral sub-harmonic has a rich and fascinating history [2]. References from as early as 1904 appear in literature [3], and were the basis for advancing vacuum tube technology to the point of observing microwave frequencies, resulting in affordable microwave instrumentation [4].

Although modern passive SPDs incorporating step-recovery diodes were initially investigated, they were found to be too lossy and fussy, and only moderate success was achievable in trial runs [5]. Ultimately, the selected SPD was based on a circuit which first appeared in a General Electric Transistor Manual in 1964 [6]. NIST-based equivalent-time sampling circuits using comparators also provided promising results, but not at the intended operating microwave frequency [7,8]. A circuit using a PECL flip-flop and integrator was found which operates on the principle of balancing the Q and /Q output durations by precisely holding the input at the logic decision threshold [9]. Feedback is used to balance the bias point at a metastable region, making the system highly sensitive to input perturbations. Input disturbances appear as pulse-width modulation on the Q-outputs, which are low-pass filtered and integrated, appearing as a voltage after the integrator. Since PECL operates well into the microwave region, its speed was adequate for 1497 MHz operation.

Since the sampling is synchronous, it possesses inherent noise immunity (much like a lock-in). In addition, the circuit is lossless, preserving the overall system noise figure [9].

A serious drawback to the single SPD is 2-quadrant operation, resulting in ambiguity between quadrants I/III and II/IV. This could either be resolved by dithering, or by using two SPDs in a quadrature configuration.

DSP BASEBAND PROCESSOR

Figure 3 illustrates the general functionality of the DSP Board. It consists of 4 analog-to-digital converter (ADC) stages, 4 digital-to-analog converter (DAC) stages, an FPGA, and multiple status and control inputs/outputs. The ADCs and DACs have a resolution of 16-bits and are updated at a rate of 150 kHz. Each ADC is preceded by an op-amp stage used for anti-alias filtering, scaling, and offsetting. The DAC outputs are also buffered with an op-amp stage used for scaling and filtering update-related noise.

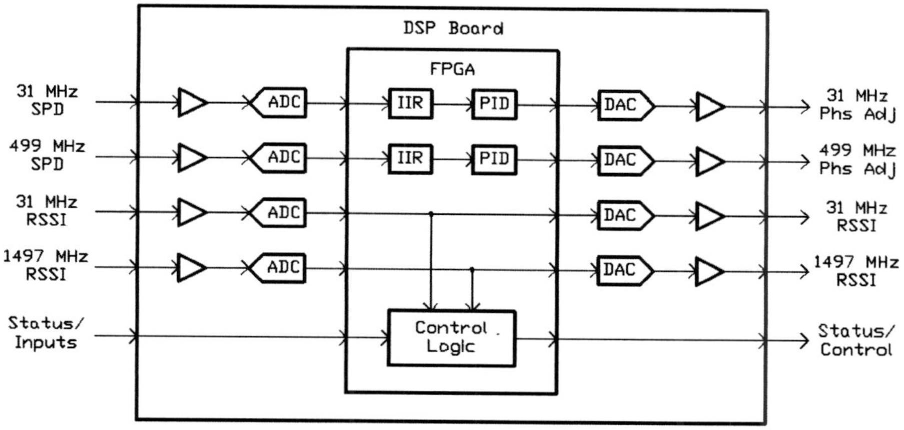

FIGURE 3. FPGA-based DSP processor.

The digitized Sampling Phase Detector (SPD) signals are sent to the FPGA where they are first passed through an IIR filter then a PID algorithm. The IIR filters have a 3 dB roll off of 500 Hz and are implemented using Eq. (1). The filter constants (a_1, b_0, and b_1) where derived using simulation software and scaled appropriately for this application. The variable n represents a discrete sample time, $x(n)$ is the signal input, and $y(n)$ is the filter output.

$$y(n) = a_1 * y(n-1) + b_0 * x(n) + b_1 * x(n-1) \tag{1}$$

The PID algorithm is accomplished with Eq. (2) and (3). The expression $e(n)$ is the filtered SPD error signal input, $S(n)$ is the integral term, and $u(n)$ is the PID output. The PID constants (K_P, K_I, and K_D) can be selected with jumpers, updated using an optical encoder knob, and displayed via status LEDs. Once the control loop has been tuned, the associated PID constants are hard coded into the FPGA.

$$u(n) = K_P * e(n) + K_I * S(n) + K_D * [e(n) - e(n-1)] \tag{2}$$

$$S(n) = \frac{S(n-1) + [e(n) + e(n-1)]}{2} \tag{3}$$

The control logic monitors the Received Signal Strength Indicators (RSSI) and the 31 MHz Phase Adjust in order to intelligently switch the 31 MHz input signal and establish lock. The RSSIs are used to determine if the signal levels are high enough to maintain lock. When the RSSIs become too low, the DSP Board freezes the IIR-PID-31 MHz Phase Adjust output chain and switches the chassis to an external 31 MHz reference. When the RSSIs return to an acceptable level, the chassis is switched back to the 31 MHz input. The IIR-PID-31 MHz Phase Adjust chain remains frozen for 50 ms to permit the 31 MHz input signal to propagate through the chassis and into the DSP board. This allows for seamless switching back and forth between the input signal and reference. If the chassis loses lock (the 31 MHz Phase Adjust rails) the DSP Board automatically resets the IIR-PID-31 MHz Phase Adjust chain in an effort to re-establish lock. The Phase Adjust outputs are limited to discourage multiple lock points and to ensure that the same lock point is found each time. The control logic also provides various status outputs to indicate the state of the chassis.

TESTS

Along with component and sub-assembly testing, system performance tests were required as a condition of customer certification. Of these, static bench tests were performed which measured such parameters as phase detector scaling, residual offset voltages, sensitivities to varying beam currents and positions, and slow signal tracking. Dynamic measurements included open and closed-loop responses, output BW, loop damping/settling time, and tangential sensitivity to 31 and 1497 MHz. Finally, beam-based tests were performed to provide the in-situ calibration data needed for accurate conversion from voltage data to engineering units.

Attempts to fully characterize the dynamic behavior of the YO! Receiver were performed using an HP 89410 Dynamic Signal Analyzer. The following stimulus-response tests were performed by modulating a pre-characterized voltage-variable phase shifter to induce a known phase delay, verifying frequency response of the PID in closed-loop, as well as to determine tangential sensitivity at a 30 Hz modulation frequency (ultimately reported in fs). Figure 4 shows the measured results of the final 499 MHz phase loop, confirming the well-behaved 10 kHz loop BW (100 us response time) for the entire system. Figure 5 demonstrates 10 dB S/N for a measurement of a 1 ps equivalent delay. Subsequent tests using lower phase noise RF sources produced tangential sensitivities at the 10 fs level. All tests were performed with nominal 40 uA simulated beam: P_{31} = -80 dBm, P_{1497} = -20 dBm, S/N >= 20 dB.

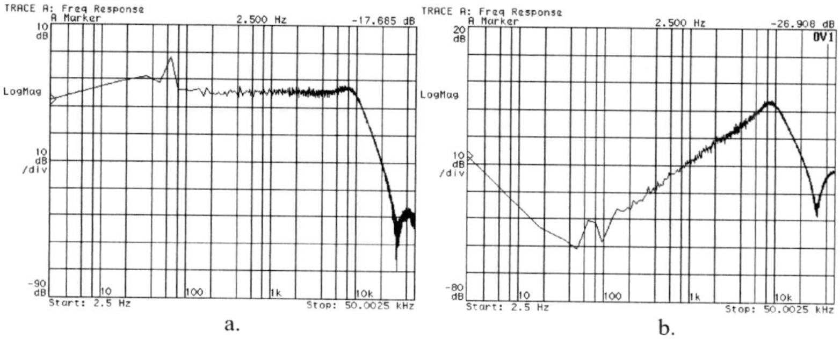

FIGURE 4. Dynamic stimulus-response test plots of 499 MHz phase-loop control (a) and error signals (b), performed using HP89410 Dynamic Signal Analyzer.

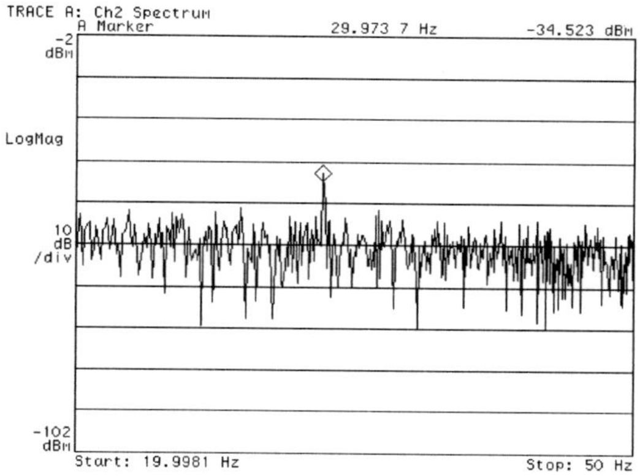

FIGURE 5. Measurement of 1ps equivalent phase delay at 30 Hz modulation, demonstrating > 10 dB S/N.

After installation in the experimental hall, test plans were executed, which verified the sensitivity as well as the proper scaling of the time-of-flight measurements. Since the relationship between magnet current and resulting path length within the accelerator is reasonably well known, the easiest method was to wiggle a dogleg magnet in the accelerator arc, intentionally modulating the path length by +/- 2 ps. Data was gathered using a PC104-type single-board computer serving EPICS, and producing real-time statistics and FFTs. Figure 6 displays the measured delay (in ps) vs. dogleg magnet flux (g-cm).

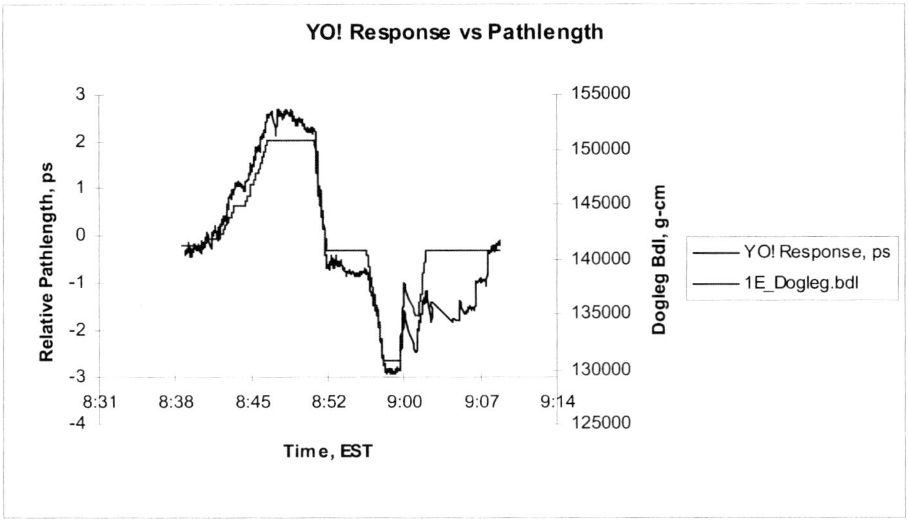

FIGURE 6. YO! Receiver output response of 499 MHz phase control signal vs. dogleg magnet setting.

Although the G0 experiment was in-progress at the time of this writing, off-line data was analyzed at the 50% mark to verify proper operation. Of most interest were the immunities to current intensity fluctuations, as well as position. Figure 7 demonstrates both effects, resulting in certification of the receiver system by the G0 Collaboration [10].

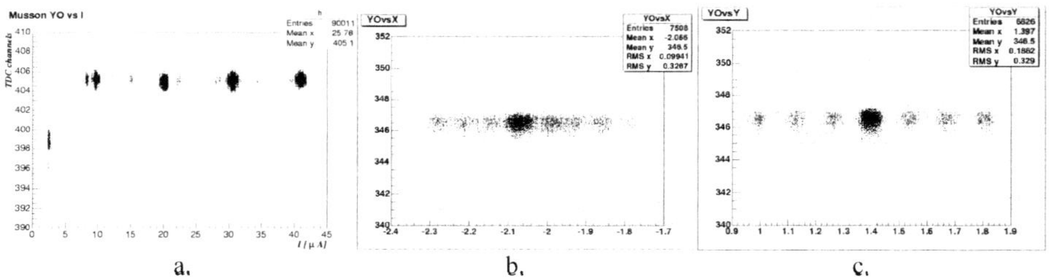

a. b. c.

FIGURE 7. Demonstration of immunities to beam intensity (a) and beam position for X (b) and Y (c). TDC bins each represent 35 ps.

OTHER APPLICATIONS

As is usually the case for new systems, measured data often describes an effect previously unmeasureable by existing hardware. The ability to measure pathlength at the end-station with high speed and resolution provides a feedback mechanism for energy locks, helicity control at the target, as well as corrections associated with seasonal machine growth. Finally, a multi-channel, multi-frequency vector voltmeter spinoff project is being produced for the purpose of Injector RF Phase monitoring, whereby an MO-derived 1497 MHz signal is used as a reference to quadrature SPDs, providing ps-level resolution with 1 Hz update rates.

ACKNOWLEDGEMENTS

The authors would like to express their gratitude to the Jefferson Lab EES Group and CMU. Also to the Experimenters, with special thanks to Paul King for his assistance in YO! operations and data gathering, and Mark Pitt for his generous input on early design phases.

REFERENCES

1. G0 Collaboration homepage, http://www.npl.uiuc.edu/exp/G0/G0Main.html.
2. M. Kahrs, "50 Years of RF and Microwave Sampling," IEEE Trans. On Microwave Theory and Techniques, Vol. 51, No. 6, June 2003, pp. 1787-1805.
3. E. Hospitalier, "The Slow Registration of Rapid Phenomena by Stroboscopic Methods," *The Electrical Engineer,* Jan. 1904, pp. 40-44.
4. M. Kahrs, "Applications of RF and Microwave Sampling to Instrumentation and Measurement," Proc. Of The 2003 International Microwave Symposium, Philadelphia, PA, 2003, pp. 1503 – 1506.
5. "Sampling Phase Detector Application Note," Metelics Corp. http://www.metelics.com/sampling_phase_detector.htm.
6. J. F. Cleary, *GE Transistor Manual, 7th Ed. (Circuits, Applications, Characteristics, Theory)*, General Electric Co., Syracuse, NY, 1964, p. 372.
7. T. F. Souders, H. K. Schoenwetter, and P. S. Hetrick, "Characterization of a Sampling Voltage Tracker for Measuring Fast, Repetitive Signals," IEEE Trans. On Instrumentation and Measurement, vol. IM-36, No. 4, December 1987.
8. O. B. Laug, M. Souders, and D. R. Flach, "A Custom Integrated Circuit Comparator for High-Performance Sampling Applications," IEEE Trans. On Instrumentation and Measurement, vol. 41, No. 6, December 1992.
9. http://www3.sympatico.ca/add.automation (AD Automation, Inc).
10. G0 Certification Document, "Results from the G0 Engineering Runs (Answers to the Certification Questions)," Submitted to JLAB March 5, 2004.

BPMs with Precise Alignment for TTF2

D. Noelle, G. Priebe, M. Wendt, and M. Werner

Deutsches Elektronen Synchrotron DESY, *Notkestr. 85, D-22603 Hamburg, Germany*

Abstract. Design and technology of the new, standardized BPM-system for the warm sections of the TESLA Test Facility phase II (TTF2) are presented. Stripline- and button-BPM pickups are read-out with an upgraded version of the AM/PM BPM-electronics of TTF1. The Stripline-BPMs are fixed inside the quadrupole magnets. A stretched wire measurement was used to calibrate the electrical axis of the BPM wrt. to the magnetic axis of the quadrupole.

INTRODUCTION

The control of the beam orbit is essential for the operation of linear accelerators for future linear colliders (LC), as well as for free electron laser (FEL) drive linacs. Beam position monitors (BPM) with high resolution and single-bunch, single-pass measurement capability are mandatory. The transport of the beam, by preserving its low emittance, requires a precise measurement of the beam orbit with respect to the magnetic axis of the quadrupoles. As a beam based alignment procedure is not always applicable (common quadrupole power supplies), or sometimes may not give satisfactory results (shot-to shot beam jitter), a stretched wire alignment measurement for quadrupole and BPM pickup can be used as an alternative or add-on.

THE TESLA TEST FACILITY PHASE II

In phase II of the TESLA Test Facility (TTF2), currently under commissioning at DESY, the beam accelerated up to 1 GeV/c by the use of 5 (6) TESLA cryo-modules (Fig. 1.). The task of the TTF2 facility is twofold:

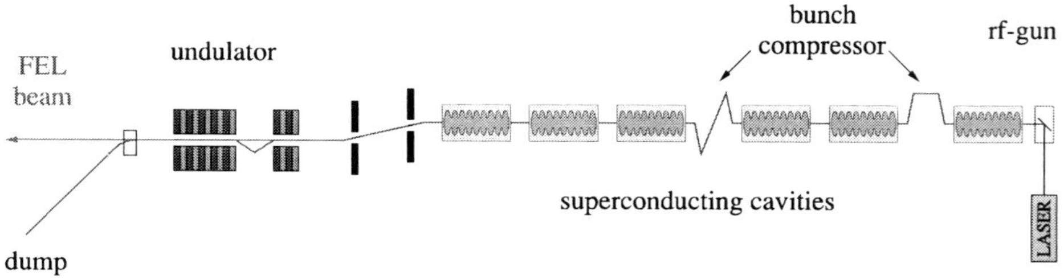

FIGURE 1. Schematic overview of TTF2 (bypass beamline not shown).

CP732, *Beam Instrumentation Workshop 2004: Eleventh Workshop,*
edited by T. Shea and R. C. Sibley III

- Linac test facility for beam test of high-gradient superconducting L-Band cavities and subsystems in terms of linear collider development.
- Drive linac for a 30 m long VUV (wavelength 6 nm) SASE-FEL user facility [1].

TABLE 1. Parameters of the TTF2 electron beam.

max. beam energy	=	800 MV (1 GeV)
max. rep. rate f_{rep}	=	10 Hz
macro pulse length t_{pulse}	=	800 μs
max. # of bunches per pulse	=	7200
bunch spacing Δt_b	=	110 ns or 1 μs
N_e per bunch	=	0.5...4 nC (typical 1 nC)
bunch length σ_z	$\approx$	50 μm
norm. emittance ε_{norm}	<	2 mm mrad
rms energy spread	=	0.1 %

Table 1 gives an overview of the TTF2 electron beam parameters. In order to establish standard components to improve operational and maintenance aspects we limit to the following BPM types:

- $\approx$ 30 *button* BPMs located in the injector, bunch-compressors and in and between the FEL-undulator sections.
- $\approx$ 30 *stripline* BPMs in all other "warm" parts of the machine.
- 4 *resonant* and one *re-entrant cavity* BPMs in the accelerating modules.

For a compact design the stripline BPMs are mounted inside the corresponding quadrupole magnets. With a proper mounting this allows to align and to fix the BPM and the magnetic axis of the devices with respect to each other. Due to the absence of the synchrotron radiation power, this is much easier for LINAC drive devices compared to storage rings.

Standardized read-out electronics, based on the *AM/PM normalization* principle, are used for all button- and stripline BPMs. Although this analog signal processing seems to be today somewhat "conservative," it matches best to the TTF data acquisition system.

THE STRIPLINE BEAM POSITION MONITOR

Principle

The impulse response of a stripline electrode of length l, terminated at both ends in its characteristic impedance Z_0, is given to:

$$z(t) = \frac{Z_0}{2}\left[\delta(t) - \delta(t - \frac{2l}{c_0})\right] \tag{1}$$

The characteristic impedance Z_0 of the "semi"-coaxial tube-electrodes of the TTF2 stripline BPMs are matched to the usual 50 Ω impedance of rf and microwave equipment.

The *Fourier* transform of (1) is the transfer impedance:

$$Z(\omega) = jZ_0 e^{-j\frac{\omega l}{c_0}} \sin\left(\frac{\omega l}{c_0}\right) \tag{2}$$

The magnitude of (2) has a maximum at frequencies where the length is an odd multiple of the quarter-wavelength:

$$f_{\text{center}} = \frac{c_0}{4l}(2n-1) \tag{3}$$

Usually the stripline pickup operates at the first "lobe" (n=1). Its 3dB-bandwidth exceeds an octave:

$$f_{\text{lo}} = \frac{1}{2} f_{\text{center}} \qquad\qquad f_{\text{hi}} = 3 f_{\text{lo}}$$

The physical length of the stripline electrodes in TTF2 is fixed to $l = 20$ cm, which leads to $f_{\text{center}} = 375$ MHz ($n = 1$).

The signal voltage delivered by a BPM electrode (stripline, button) is:

$$V_{\text{electrode}}(\omega, x, y) = s(x, y) Z(\omega) I_{\text{beam}}(\omega) \tag{4}$$

which is proportional to the current $I_{\text{beam}}(\omega)$ of the passing beam and to the beam-to-electrode distance (x,y) (transverse beam displacement) or coupling, which is described by a sensitivity function $s(x,y)$. Because of the short TTF2 electron bunches the frequency spectra of $I_{\text{beam}}(\omega)$ is of no concern; for the BPM system the bunches behave like *Dirac* impulse excitation signals. $s(x,y) = \Phi(x,y)$ is evaluated by analyzing the 2-dimensional cross-section geometry of the stripline BPM, applying *Laplace*'s equation $\Delta\Phi = 0$ with help of a numerical electromagnetic solver, e.g., *MAFIA*, *EM Studio*, etc. For a centered beam the coupling to a TTF2 stripline electrode is $s(0,0) = k \approx 0.065$.

The TTF2 Stripline BPMs

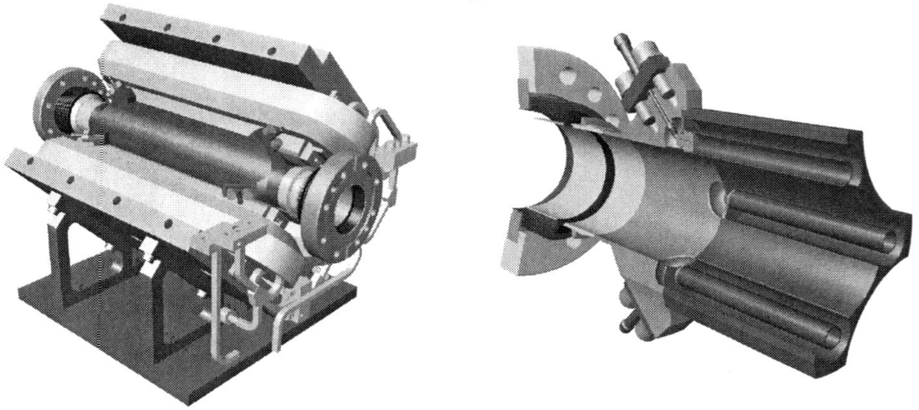

FIGURE 2. TTF2 stripline BPM (*type B*: 44 mm beam-pipe diameter, 8 mm electrode diameter), inside quadrupole installation and sectional view.

In order to save space and for alignment purposes the cross-section of the TTF2 stripline monitors are matched to the poles of the corresponding quadropole magnets (Fig. 2). The gold-plated stainless-steel tube electrodes provide high mechanical stiffness. They are fixed with ceramic spacers to the monitor body, which is shaped by an EDM process. A micro-bellow keeps the electric contact between the spherical shaped electrode-end and the pin of the SMA UHV feedthrough. Because of the high gradient superconducting cavities at TTF2 dust or particle emission has to be avoided. Thus the stripline BPM production had to fulfill cleanroom class 100 conditions.

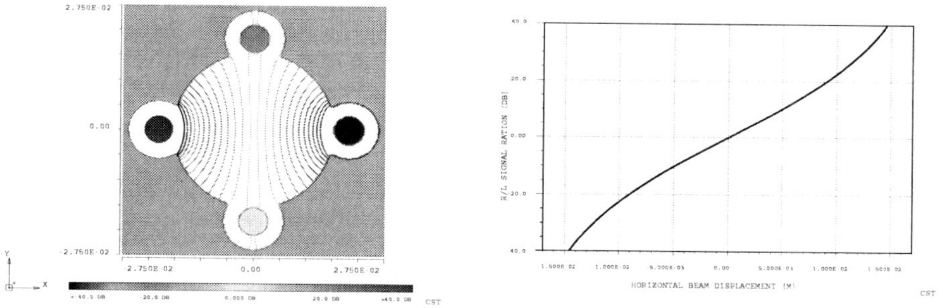

FIGURE 3. Equipotentials of the horizontal stripline electrodes (logarithmic scaled for $s(x,y)$ and $s(x,0)$).

The 4 electrodes are arranged orthogonally, along horizontal and vertical planes. One channel of the read-out electronics processes the signals of two opposite electrodes, e.g.:

$$\text{normalized hor. beam displacement} = \frac{V_{\text{right-electrode}}(x,y)}{V_{\text{left-electrode}}(x,y)} \tag{5}$$

with $V_{\text{right-electrode}}(x,y) = V_{\text{left-electrode}}(-x,y)$. This *normalization* procedure results in an intensity independent beam position signal, which also reduces the nonlinearities of $s(x,y)$ of a single electrode. Fig. 3 shows this normalized equipotential pattern for the horizontal electrodes. The sensitivity around the center is ≈ 2 dB/mm.

Currently 30 stripline BPMs of two different apertures (34 mm and 44 mm beam pipe diameter) are installed in TTF2. Further stripline BPMs are foreseen for SASE-FEL seeding upgrade. Advantages of stripline BPMs, compared to the more simple, lower cost button BPMs, are:

- High signal levels, already at moderate frequencies (here: 375 MHz), simplify the read-out electronics (no downconversion) and results in a good S/N-ratio. Button BPMs have a high-pass behavior and forces for complicated high frequency read-out electronics.
- A resistive 50 Ω source impedance (button BPMs: capacitive source impedance) minimizes reflection effects between BPM pickup and read-out electronics and therefore resolves single bunches in a better way.

PRECISION ALIGNMENT OF STRIPLINE BPM AND QUADRUPOLE MAGNET

The stripline BPM is fixed tightly at both ends with the quadrupole magnet, so they form a rigid mechanical unit. A stretched wire setup is used to measure both:

- The magnetic center of the quadrupole magnet.
- The electric center of the stripline BPM.

and calibrate them without using an external reference. This idea [2] was already adapted for a few stripline BPM-quadrupole units, used at the S-Band Test Facility linac [3], and is now applied under cleanroom conditions for larger quantities in the frame of TTF2 beam instrumentation.

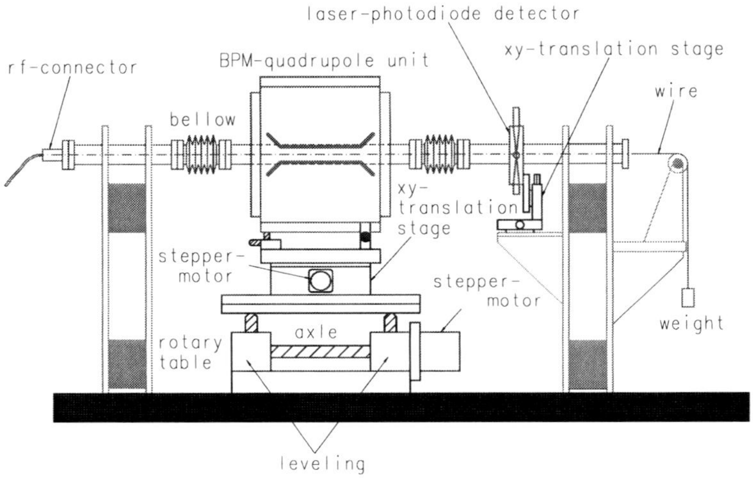

FIGURE 4. Schematic view of the stretched wire alignment setup.

Fig. 4 shows a schematic view of the stretched wire test bench with its basic elements. An aluminium plate carries the BPM-quadrupole unit on a stepper-motor movable xy-translation stage. Two rigid pillars fix the beam-pipe sections, which adapt to the movable BPM pipe in the center via bellows. A 130 μm diameter, copper-beryllium wire is stretched in the center of the beam pipe and forms a coaxial transmission line. Pulse- resp. rf-signals are feed from the upstream N-connector for calibration of quadrupole or BPM axis. A laser-photodiode detection system is located on the downstream pillar for calibration of the magnetic center. The distance between this detector and the end of the wire has to be longer than the half of the magnet length, in order to separate original and reflected signals. Other parts of the test bench are a defined weight to stretch the wire and mechanical elements to compensate a tilt between BPM-quadrupole and wire. The detector can also be moved by a manual xy-translation stage.

The calibration procedure is done in two steps:

Calibration of the Magnetic Center of the Quadrupole

A method to measure the effect of a magnetic field on a stretched wire, excited with a strong, but short pulse of charge Q is described in [4], [5]. The *Lorenz* force accelerates the part of the wire, on which the magnetic field B acts, in transverse direction. This displacement of the wire

$$x(z_0,t) = \frac{Q}{2\mu c} \int_{z_0}^{z_0-ct} B(\tilde{z})\, d\tilde{z} \tag{6}$$

moves with the wave velocity $c = \sqrt{T/\mu}$ towards upstream and downstream fixpoints of the wire and can be detected at a location z_0 behind the magnet (μ is the weight per unit length of the wire, T is the tensile force to which the wire is stretched).

Before starting the calibration procedure the magnet was cycled using a bipolar power supply. Then the magnet was powered with 25% of its nominal current ($\approx$ 100 A). A pulse of 400 V, 20 A and 10 μs length was feed into the stretched wire. The magnet was then moved in a way that the signal from the wire gets zero. This reading of course defines the magnetic axis of the quadrupole.
The movement of the the BPM-quadrupole unit was done by computer controlled stepper motors with a resolution of 5 μm (hor.) resp. 0.6 μm (vert.) Being close to the magnetic center also the tilt between wire and magnetic axis was minimized.

From this *reference position* – set micrometer gauges and step counter to zero – we started the second step of the procedure:

Calibration of the Electrical Center of the Stripline BPM

A Δ-signal of two opposite electrodes of the stripline BPM was produced by wiring well calibrated, phase-stable semi-rigid coaxial cables and a *M/A-COM H-9* 180^0 broadband hybrid. With a network analyzer a frequency-domain $|S_{21}(f)|$ measurement was setup between stretched wire input and Δ-signal output. We didn't care about the reflections in the not terminated stretched wire beam-pipe, as the measurement was performed for a single frequency (zero-span mode). By moving the BPM-quadrupole unit with the stepper motors and appropriate settings of the network-analyzer ($f_{\text{center}} = 180$ MHz, $f_{\text{RBW}} = 100$ Hz, $t_{\text{sweep}} = 5$ s) it was simple to minimize $|S_{21}|$ down to the -100 dBm noise level. The signal minimum could be identified clearly within a single step in the horizontal (step size: 5 μm) and 2..3 steps in the vertical plane (step size: 0.6 μm).

When both planes show a minimum of the Δ-signal transfer, the stretched wire was in the electrical center of the stripline BPM. The *xy-offset* between magnetic center of the quadrupole and electrical center of the stripline BPM was evaluated by counting the driven steps, cross-checking with the micrometer gauge readings.

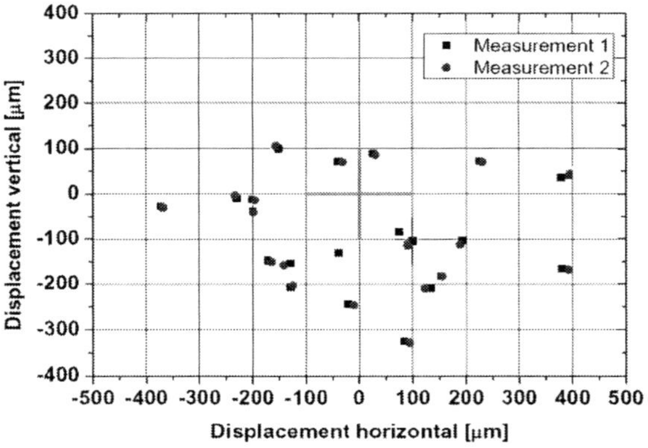

FIGURE 5. Results of the stretched wire alignment procedure.

Results

23 BPM-quadrupole units were calibrated with the stretched wire alignment setup. Each measurement was performed twice, the setup was de-adjusted and partially demounted between individual measurements. The result are shown in Fig. 5. As expected 200...300 μm offsets typical for mechanical construction show up. While the resolution to identify the BPM center is in the range 1...2 μm, the identification of the magnetic center is limited to 10...20 μm, dominating the the resolution of the complete setup. This is due to several facts, like stray fields, wire diameter, laser focussing, mechanical vibration, etc. and, in the horizontal plane, the rather large step size.

BPM READ-OUT ELECTRONICS

A set of standardized read-out electronics has been developed, which applies the AM/PM-principle for normalization to the signals of two opposite BPM electrodes (5). Let us simplify the electrode signals to to stationary sine-wave voltage functions v_A and v_B:

$$v_A(t) = \hat{a}\, e^{j\omega t} \qquad v_B(t) = \hat{b}\, e^{j\omega t}$$

of the same frequency and in phase, but with different amplitudes $\hat{a}$ and $\hat{b}$ due to the beam displacement. At the outputs C and D of the hybrid-with-delay circuit the signals:

$$v_C(t) = \mathscr{A} \arctan\left[\frac{\hat{a}\sin(\omega t) + \hat{b}\cos(\omega t)}{\hat{a}\cos(\omega t) - \hat{b}\sin(\omega t)}\right] \qquad v_D(t) = \mathscr{A} \arctan\left[\frac{\hat{a}\sin(\omega t) - \hat{b}\cos(\omega t)}{\hat{a}\cos(\omega t) + \hat{b}\sin(\omega t)}\right]$$

have the same amplitude $\mathscr{A} = \sqrt{\frac{\hat{a}^2+\hat{b}^2}{2}}$, but the amplitude-ratio of A and B is converted into a phase-difference:

$$\Psi_{C-D} = 2\,\text{arccot}\left(\frac{\hat{a}}{\hat{b}}\right) \tag{7}$$

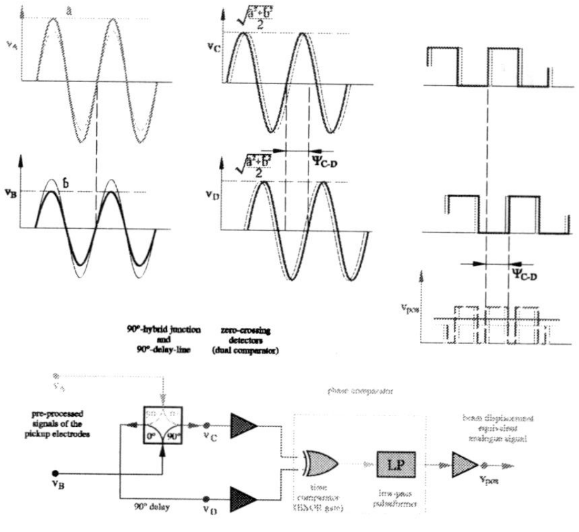

FIGURE 6. AM/PM-principle for normalization of the BPM electrode signals.

The phase (time) detection and further signal processing results in beam intensity independent position signal (Fig. 6).

The TTF2 BPM read-out electronics is realized on a VXI mainboard housing a set of 9 rf-modules. Button BPMs are processed with a broadband “monopulse” technique, which was already successfully operated on the FEL-undulator button BPMs during the TTF1 run [6]. By changing the input low-pass towards “ringing” band-pass filters and a few other modifications the same board is capable for processing the stripline BPM signals. In this case the measurement (integration) time is longer: $\approx$ 80 ns, instead of 12 ns for the monopulse technique.

The expected single-bunch resolution for a complete setup – BPM-pickup and read-out electronics – is in the range 10...30 μm.

REFERENCES

1. B. Faatz, et. al., “The SASE FEL at the TESLA Test Facility as User Facility,” Proc. of the 24rd FEL Conf., Argonne (2002), USA.
2. G. E. Fischer, et. al., “Finding the Magnetic Center of a Quadrupole to High Resolution,” SLAC-TN-89-01, 1989, USA.
3. F. Brinker, et. al., “Precision Alignment of BPM’s with Quadrupole Magnets,” Proc. of the 18rd Int. Linac Conf. LINAC 96, Geneva (1996), Switzerland, pp. 502-504.
4. R. W. Warren, “Limitations on the Use of the Pulsed-Wire Field Measuring Technique,” NIM A272 (1988), pp. 257-263.
5. A. Geisler, et. al., “The Pulsed-Wire Method,” DELTA Int. Rep. 94-4 (1994), University of Dortmund, Germany.
6. M. Wendt, “BPM Read-Out Electronics based on the Broadband AM/PM Normalization Schema,” Proc. of the DIPAC 2001, Grenoble (2001), France, pp. 63-65.

Fast Orbit Feedback and Beam Stability at the Swiss Light Source

V. Schlott, M. Böge, B. Keil, P. Pollet, and T. Schilcher

Paul Scherrer Institute, Villigen/PSI, Switzerland

Abstract. A global, fast orbit feedback (FOFB) based on the digital beam position monitor (DBPM) system has been in user operation at the Swiss Light Source (SLS) since November 2003. The SVD-based correction scheme acts at a sampling rate of 4 kHz using position information from all 72 DBPM stations and applying corrections with all 72 horizontal and 72 vertical corrector magnets. As a result, the FOFB successfully damps orbit distortions, which are mainly caused by ground and girder vibrations as well as the 3-Hz booster crosstalk. It also allows fast and independent ID gap changes, which are completely transparent to all SLS users. With top-up as a regular operation mode at SLS, global beam stability on a μm-level has been achieved from days to milliseconds.

INTRODUCTION

The Swiss Light Source (SLS) is a third generation synchrotron radiation facility, which has been in operation since the middle of 2001. Successful user operation requires the reproduction and stabilization of a previously established reference orbit ("golden orbit") well below the resolving power of the experimental stations. Photon energy resolution of crystal monochromators in the order of 10^{-4} to 10^{-5} translate into requirements for angular electron beam stability of < 1 μrad corresponding to photon beam motion of < 10 μm at the first optical elements of the beamlines. Photon beam intensity modulations of < 1% on the samples require stabilization of the electron beam position within $1/10^{th}$ of its size at the radiation source points. For the low β insertion device (ID) straights of the SLS, this corresponds to tolerable vertical electron beam motions of < 1 μm.

During the first two years of SLS operation, the above mentioned stability requirements were achieved by a central high level application, the so called slow orbit feedback (SOFB), which automatically performed global orbit corrections at an update rate of ~ 0.5 Hz [1]. However, the growing number of beamlines performing fast and independent ID gap-scans as well as the increasing sensitivity of the experiments to orbit distortions caused by ground vibrations and environmental noise, required an increase of orbit correction bandwidth. This has been realized by commissioning the SLS fast orbit feedback (FOFB) during the year 2003, which was designed to correct orbit perturbations in a frequency range of up to 100 Hz to a μm level.

CP732, *Beam Instrumentation Workshop 2004: Eleventh Workshop*,
edited by T. Shea and R. C. Sibley III

FOFB ARCHITECTURE

As a result of the localized structure of the SVD-inverted corrector/BPM response matrix, where only the diagonal and their adjacent coefficients have non-zero values [2], the global SLS FOFB has been realized in a de-centralized architecture. The FOFB is therefore an integral part of the SLS digital BPM (DBPM) system [3], which is distributed over the 12 sectors of the SLS storage ring. Each DBPM consists of an RF front end and a quad digital receiver. The horizontal and vertical position readings of all six BPMs per sector are processed on a single DSP board. For orbit display in the control room and in case of centralized orbit manipulation via a high level SW application (like the SOFB), all 12 DSP boards are connected to the control system via the VME-bus. In case of the FOFB, the feedback algorithms are performed in parallel on the 12 (local) DSP boards and data are directly exchanged between adjacent sectors via fast fiber optic links. This structure allows the calculation of the required corrector magnet kicks per sector based on 18 beam position readings (three adjacent sectors) at a rate of 4 kHz. The resulting corrector kicks are fed into one PID controller per corrector. The FOFB is initialized and monitored by a central PC-based beam dynamics server, which takes the number of available BPMs and correctors into account. Any off-energy trajectories in the horizontal plane are corrected by the central RF frequency. Since the resulting frequency corrections are carried out by a slow high-level application on the central beam dynamics server, dispersion orbits must not be corrected by the FOFB and have to be subtracted before each orbit correction.

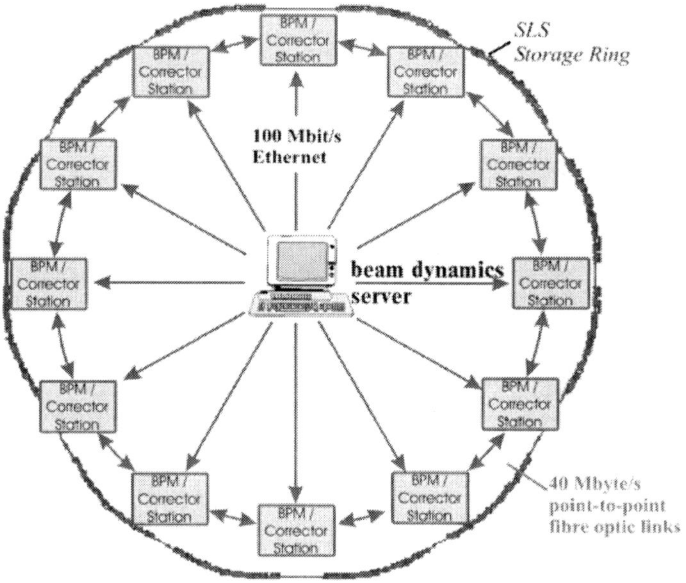

FIGURE 1. De-centralized architecture of the SLS fast-orbit feedback. The FOFB is an integral part of the SLS DBPM system and therefore integrated in the twelve BPM/corrector stations (sectors) of the SLS storage ring. A dedicated fiber optic network provides communication between adjacent sectors. The FOFB is initialized and monitored by a central beam dynamics server.

FOFB OPERATIONAL EXPERIENCE AND PERFORMANCE

The FOFB has been operational during SLS user shifts since November 2003. The overall reliability of hardware and software has been the main focus during the first phase of operation. Therefore, the FOFB has been operated at moderate PID loop gains and DBPM filter settings providing a regulation bandwidth of about 60 Hz. In order to avoid beam loss due to malfunctioning FOFB components, a low level security package has been implemented monitoring FOFB parameters and sub-systems performance. When exceeding pre-defined deadbands, the FOFB is automatically stopped and a corresponding alarm is displayed to the operator, who can thus decide on calling in an expert or continuing with FOFB operation. Operational experience of the FOFB shows up to now an overall availability of > 95%. Although most of the automatic interruptions of FOFB have been caused by malfunctioning of DBPM electronics and corrector magnet power supplies, a growing number of FOFB interruptions are caused by users, who are performing beamline commissioning during regular user shifts.

The actual bandwidth of the FOFB as it is presently operated at SLS is 100 Hz in both transverse planes. Although the overall loop delay, which has been measured to be in the order of 1.6 ms, could not be reduced by a DBPM firmware upgrade, the synchronization of all DBPM electronics allowed the use of a more stringent set of PID parameters, which finally leads to the operation of the FOFB close to its design parameters [2]. Figure 2 shows the measured FOFB closed loop transfer function for both transverse planes. Orbit perturbations up to 100 Hz are effectively damped, while noise sources between 100 – 300 Hz are moderately excited.

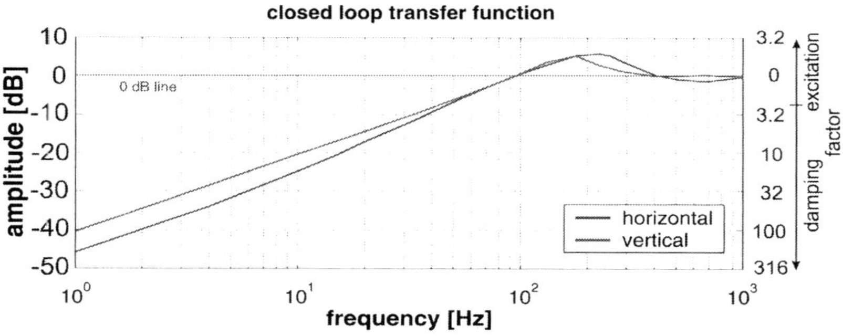

FIGURE 2. Measurement of FOFB closed loop transfer functions in both transverse planes. Effective damping of orbit perturbations has been achieved close to the FOFB design goal of 100 Hz.

Similar to other light sources, the main orbit perturbations at SLS in the frequency range up to 100 Hz are dominated by the following features:

- booster respectively injector operation (3 Hz),
- vibrations at the girder eigenmodes (15 – 30 Hz),
- 50 Hz induced by power supply noise and/or asynchronous pumps.

Snapshots of the horizontal and vertical power spectral densities and integrated spectral densities (see inlets) with and without FOFB measured at the tune BPM outside the feedback loop are shown in Fig. 3. It can clearly be seen, that orbit perturbations caused by booster operation, which is in the same tunnel as the storage ring and which is continuously running during "top-up" operation of the SLS, as well as girder vibrations are significantly damped. The 50 Hz peaks are still reduced by a factor of 3 - 5, while noise contributions above 100 Hz are moderately increased.

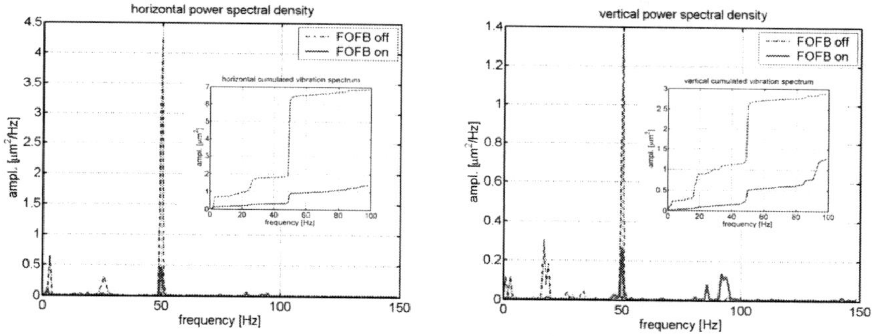

FIGURE 3. Snapshots of the horizontal and vertical power spectral densities and integrated spectral densities (inlets) measured at the tune BPM outside the FOFB loop, with FOFB off and on. Orbit perturbations within the FOFB bandwidth (100 Hz) are significantly damped.

Table 1 summarizes the improvements of beam stability at SLS with the FOFB running compared to the situation without feedback. The values still contain the noise contribution of the DBPM system, which has been measured to be < 0.13 μm within the FOFB bandwidth. Orbit distortions caused by ID gap changes, which would emphasize the effectiveness of the FOFB even more, are not included. The temporal RMS values are integrated over the active range of the FOFB (up to 100 Hz) and up to 150 Hz, where significant noise contributions could still be observed. The temporal RMS values at a location s in the storage ring are obtained from the table by multiplication with $\sqrt{\beta(s)}$. At the location of the tune BPM with $\beta_y = 18$ m (see Fig. 3) this translates into vertical temporal RMS values of $\sigma_y = 1.15$ μm, while at the source points of the SLS low gap insertion devices ($\beta_y = 0.9$ m) the integrated vertical beam motion up to 100 Hz yields to $\sigma_y = 0.25$ μm.

	horizontal		vertical	
FOFB	off	on	off	on
1- 100 Hz	**0.83 μm · $\sqrt{\beta_x}$**	**0.38 μm · $\sqrt{\beta_x}$**	**0.40 μm · $\sqrt{\beta_y}$**	**0.27 μm · $\sqrt{\beta_y}$**
100-150 Hz	0.08 μm · $\sqrt{\beta_x}$	0.17 μm · $\sqrt{\beta_x}$	0.06 μm · $\sqrt{\beta_y}$	0.11 μm · $\sqrt{\beta_y}$
1-150 Hz	0.83 μm · $\sqrt{\beta_x}$	0.41 μm · $\sqrt{\beta_x}$	0.41 μm · $\sqrt{\beta_y}$	0.29 μm · $\sqrt{\beta_y}$

TABLE 1. Integrated beam position temporal RMS values with FOFB off and on, without moving of ID gaps. Values at a location s in the storage ring can be obtained by multiplication with $\sqrt{\beta(s)}$.

Apart from the improved integrated beam stability up to 100 Hz, the FOFB allows autonomous and independent changes of ID gaps as well as beamline optimization by the users. Tests with rapidly moving ID gaps and correctors have shown that the resulting orbit kicks are invisible to all other users, when the FOFB is active. Figure 4 shows the improved long-term stability at the SLS with the FOFB running, compared to the situation, when orbit corrections were performed by the SOFB. Each data point represents the spatial RMS values of all 72 BPMs in the feedback loop with respect to a pre-defined "golden orbit". The BPM readings have been averaged over a period of 0.32 ms, which is a quite relevant bandwidth for most of the present SLS users. While RF frequency and ID gap changes have still resulted in orbit disturbances in case of the SOFB (see left side of Fig. 4), the FOFB completely rules out any orbit transients. Likewise, the remaining RMS noise drops significantly to about 0.71 μm horizontally and 0.06 μm vertically (@ ≈ 3 Hz bandwidth) due to the higher sampling and correction rate (4 kS/s) of the FOFB. Since beam position information in the SLS FOFB architecture is only distributed locally and dispersion fits in each sector are based on a total of 18 beam position readings at a rate of 4 kHz, the low level part of the feedback corrects off-energy trajectories only on a micron level. Resulting slow drifts of the dispersion orbit and horizontal mean corrector kicks are avoided by a high-level beam dynamics application, which adds suitable offsets to horizontal corrector currents and RF frequency every 20 – 30 minutes. As a result, a blurring of the global horizontal spatial RMS distribution can be observed (see inlet in the lower right graph of Fig. 4), which however does not affect the stability of the source points in the ID straight sections since they are dispersion free.

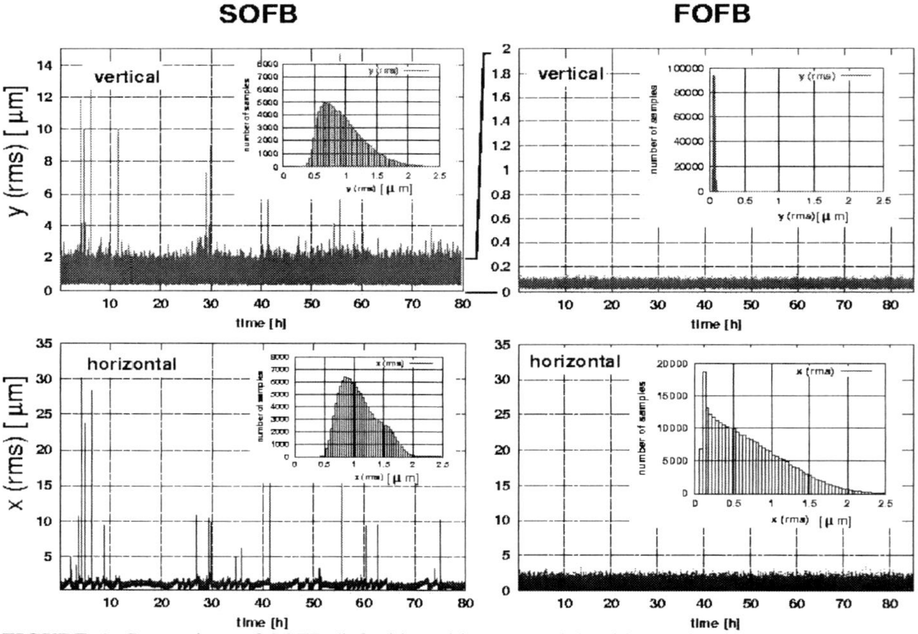

FIGURE 4. Comparison of SOFB (left side) with FOFB (right side) performance. Graphs show the horizontal and vertical spatial RMS orbits of all 72 in-loop BPMs with respect to a pre-defined reference orbit ("golden orbit") over a period of three days. Orbit disturbances due to RF frequency and ID gap changes, which are still visible on the left side (SOFB) are completely ruled out by the FOFB. The mean value of the spatial RMS orbit with FOFB is $x\,(RMS) \approx$ 0.71 μm horizontally and $y\,(RMS) \approx$ 0.06 μm vertically.

Photon BPM (PBPM) readings have been taken as external references to demonstrate the efficiency of electron beam stabilization for the beamlines. Although the PBPM signals at SLS are presently still subject to position offsets whenever the ID gaps are changed, Fig. 5 shows horizontal and vertical photon beam positions on the X06S protein crystallography (PX) beamline over a measurement period of 19 h with constant ID gap. The data points are not (yet) synchronized to the DBPM readings and have been averaged over a period of 1 s.

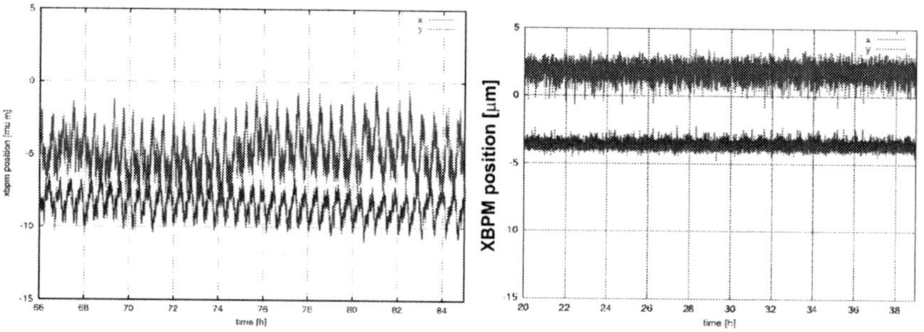

FIGURE 5. Left side: horizontal and vertical PBPM readings at the X06 protein crystallography (PX) beamline indicating photon beam stability of ± 1.5 μm peak to peak with the FOFB active. The remaining systematic oscillation of the photon beam with an ~ 45 min period has been stabilized to sub-μm RMS level at the first optical element of the beamline, implementing a slow, high-level feedback on the PBPM readings (right side).

The left side of Fig. 5 shows the situation with the FOFB active. It stabilizes the photon beam at the location of the PBPMs about 10 m from the source point to ± 1.5 μm peak to peak. A systematic oscillation of an ~ 45 min period is left on the photon beam, which was originally suspected to be a temperature effect in the four channel DBPM electronics. However, measurements of the DBPM rack temperatures did not show any correlation with the observed systematic drifts. Just lately, the time constant of this residual oscillation could be correlated to the injection clock cycle, which constantly proceeds over the buckets to be filled in the storage ring during "top-up" operation. A corresponding bunch pattern dependency of the DBPM electronics has been demonstrated to be the reason for the remaining orbit oscillations and a bunch pattern feedback in the storage ring is presently under development to eliminate this effect. Presently, a slow, high-level feedback application on the PBPM readings has been implemented resulting in sub-μm RMS beam stability at the first optical elements of the SLS beamlines (see PBPM readings on the right side of Fig. 5).

MONITORING OF MEDIUM TERM STABILITY

While the FOFB has proven to stabilize the electron beam orbit at the SLS on a sub-μm level in a frequency range from ~ 0.1 to 100 Hz, medium term stability ranging from minutes to days represents a similar important feature for the experimentalists. Due to impedance optimization, the SLS vacuum system has no bellows over a length of 19 m throughout the arcs of the TBA lattice. Thus, during its design phase finite element analyses (FEA) have been performed, showing that the vacuum chamber bends significantly under thermal loads and thermal transients. As a result of these studies, it has been decided to support the vacuum chamber at the positions of the BPMs using transversely stiff supports and allowing motion in the beam direction at the same time. While a 1 mm clearance prevents any touching of the vacuum chamber and the storage ring magnets, the transverse mechanical positions of all BPM blocks in the storage ring are monitored by the so called POMS (position monitoring system) [3]. Two linear dial gauges, which are rigidly attached to the adjacent quadrupole magnets, are housing a set of optical encoders providing 0.5 μm position resolution in respect to the magnetic axis of the machine, which has been previously determined by the method of beam based alignment. As an example of the POMS performance, Fig. 6 shows the measurement of the horizontal position of the 02-SE BPM block. It can clearly be seen that the mechanical position of the BPM is changing whenever thermal transients – in this case due to beam loss and decaying beam – are present. In case of "top-up" operation providing thermal equilibrium in the storage ring no mechanical movements of BPM blocks within the resolution of the system are visible. Thus, monitoring of the BPM block positions with the POMS system allows the conclusion that even the medium term stability of the SLS under "top-up" conditions is in the μm-level.

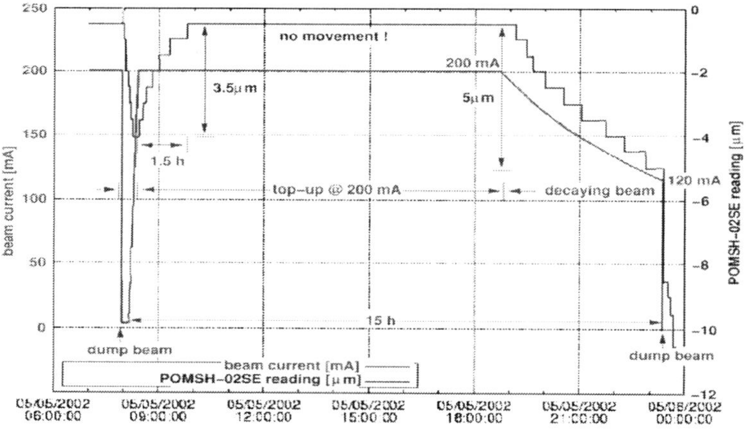

FIGURE 6. Readout of horizontal POMS sensor attached to BPM-02SE showing that the mechanical position of the BPM block is unchanged during "top-up" operation of the SLS storage ring. Only thermal transients caused by beam losses and/or decaying beam force a movement of the BPM block.

CONCLUSIONS AND OUTLOOK

The global FOFB with its distributed hardware structure has been in regular user operation since November 2003 and has lead to sub-μm electron beam stability at SLS in a frequency range from ~ 0.1 – 100 Hz. Stability-wise, it has exceeded its original design parameters and satisfies the present stability requirements of the SLS users. The regular "top-up" operation of the SLS presently at 330 mA (± 0.5 mA) provides the most stable and reproducible conditions for the DBPM electronics and leaves the storage ring in a thermal equilibrium. Thus the medium term stability (from days to minutes) of the SLS, which is constantly monitored by the POMS system, resides on a μm-level as well. A residual systematic beam oscillation of ± 1.5 μm at the location of the photon BPMs has been correlated to the SLS injection cycle clock and has been shown to be a filling pattern dependence of the DBPM electronics. A slow, high-level feedback application has been implemented at the protein crystallography and material sciences beamlines to correct for these oscillations, stabilizing the photon beam at the location of the first optical elements to a sub-μm level.

The commissioning of the "fs pulse slicing" beamline as well as the construction of additional bending magnet beamlines during the next year will break the symmetry of the present FOFB architecture due to additional RF and photon BPMs, which are needed to provide the required beam stability for these experiments. Therefore, an extension of the FOFB architecture is already foreseen allowing the integration of fully synchronized position readings from additional sensors on the DSP bus of the FOFB system.

ACKNOWLEDGMENTS

The achievement of beam stability on a μm-level has shown to be only possible, if all people working on the FOFB subsystems like beam dynamics, instrumentation, control system as well as electrical and mechanical engineering are open for new concepts and continuous discussion pursuing the common goal of beam stability. We would like to take the opportunity to thank all these colleagues at SLS as well as the colleagues from other light sources for many fruitful discussions and their impact on our strategies on orbit stabilization.

REFERENCES

1. M. Boege, J. Chrin, P. Pollet, T. Schilcher, and V. Schlott, *Orbit Control at the SLS Storage Ring*, Proceedings EPAC 2002, Paris, France, pp. 2067-2069. See also: http://accelconf.web.cern.ch/AccelConf/e02/PAPERS/THPRI030.pdf.
2. M. Boege, M. Dehler, T. Schilcher, V. Schlott, and R. Ursic, Fast Closed *Orbit Control at the SLS Storage Ring*, Proceedings PAC 1999, New York, USA, pp. 1129-1131. See also: http://accelconf.web.cern.ch/AccelConf/p99/PAPERS/TUA26.PDF.
3. V. Schlott, M. Dach, M. Dehler, R. Kramert, P. Pollet, T. Schilcher, M. Ferianis, R. DeMonte, A. Kosicek, and R. Ursic, *Performance of the Digital BPM System for the Swiss Light Source*, Proceedings DIPAC 2001, Grenoble, France. See also: http://www.esrf.fr/conferences/DIPAC/Proceedings/stampedpdfs/CT-03.pdf.

DIAMOND Storage Ring Optical and X-Ray Diagnostics

C. A. Thomas and G. Rehm

Diamond Light Source Ltd

Abstract. This paper gives an overview of the optical and X-ray diagnostics for the 3-GeV stored electron beam of the DIAMOND storage ring. The specification and the requirement for the diagnostics are given in detail. The diagnostics are performed using the properties of the synchrotron radiation from two bending magnets in the visible and in the X-ray range. Four measurements will provide information on the beam longitudinal distribution, the relative energy spread, the transverse size, and emittance of the beam.

INTRODUCTION

The DIAMOND synchrotron light source is the largest scientific facility to be built in the UK for nearly thirty years. As a third generation synchrotron light source, the characteristics of the electron beam stored in the ring match a number of drastic requirements. The brightness of the electron bunches must be high to ensure a high photon flux. This implies that the electron bunch length and the transverse size have to be very small, as do the emittance and the energy spread. Moreover, excellent stability in terms of position and dimension has to be achieved so that the combination of the brightness and stability will guarantee the success of the synchrotron radiation experiments carried on there. Indeed synchrotron radiation is used to analyse a large assortment of different materials like crystalline and non-crystalline solids, liquids and gases, and all kinds of complex materials. To carry out these analyses several techniques are used, in particular X-ray crystallography, to look at the structure of chemical compounds and composite materials; X-ray spectroscopy, to detect chemical elements and their state and to follow chemical reactions; microscopy, which allows nanometer structures to be seen, and which can be combined with other methods like spectroscopy for example (see [1] and [2] for more information). A high brightness and highly stable beam ensures the full capabilities for these techniques using the synchrotron radiation.
Diagnostic systems in DIAMOND will be essential for the rapid commissioning and successful operation of the new facility. In particular, optical diagnostics will be based on the synchrotron radiation properties to measure the 6 dimensional phase space characteristics of the stored electron beam in stable condition. Observation and the study of instabilities occurring in the beam will be also provided by the synchrotron radiation properties. The measurements we want to perform are:

1. From the measure of the electron beam size in horizontal and vertical plane, and in two locations, one can deduce at the same time the emittance and the energy spread

CP732, *Beam Instrumentation Workshop 2004: Eleventh Workshop*, edited by T. Shea and R. C. Sibley III

of the beam, and the vertical coupling emittance.

2. Analysis of CCD images in the visible will provide information on the stability of the source; variations of centroid position are related to the transverse instabilities.
3. The bunch length measurement is obtained from the second order moment of the bunch longitudinal distribution, provided by streak camera images. Longitudinal instabilities can be observed and studied as well with the streak camera.
4. Purity of buckets in the single bunch mode will be measured with a multi-channel analyser. Any parasitic bunches will add parasitic light for the users.

From these measurements, one can retrieve the horizontal and vertical emittance (beam size × beam divergence), as well as the longitudinal emittance (bunch length × energy spread). As already mentioned above, the brightness, which is inversely proportional to the product of horizontal, vertical and longitudinal emittance, is the main achievement of the third generation synchrotron light sources. Diagnostics will be performed in the visible and in the X-ray range, extracted from two different bending magnets (Dipole 1 and 2). The two extraction lines are placed tangentially at 0.859 degrees from the entrance of the bending magnet, on the arc made by the electron path. Dipole 1 and 2 lines go through the extraction ports ID 1 and BM 1, and the value of the dispersion function is $\eta = 0.9$ cm and $\eta = 3.3$ cm respectively. The measurement of the beam transverse size, with a pinhole X-ray camera in the two location, gives a measure of the emittance and of the energy spread.
The visible light will be extracted from BM 1, for the visible diagnostics, which will be done using a streak camera, a CCD camera, and a multi-channel analyser. The quality of the extracted visible beam is important for the quality of the measurement described above, and also allows for the possibility of other new and/or complementary measurements.

DIAMOND ELECTRON BEAM CHARACTERISTICS

At the source points of the Dipole 1 and 2 lines, the electron beam characteristics are determined by the values of the lattice function. In the transverse plane the Twiss parameters, the dispersive function, the emittance and the energy spread define the transverse size of the beam. In the longitudinal plane the natural electron bunch length is given by the momentum compaction factor, the energy spread and the synchrotron frequency [3]. Table 1 gives the values of the Twiss parameters and of the dispersive function at the source point in the two locations.
The nominal emittance in the horizontal plane is $\varepsilon = 2.7 \times 10^{-9}$ m.rad. In the vertical plane, the coupling emittance, assumed to be 1 % of the horizontal emittance, is then $\varepsilon_{\perp} = 2.7 \times 10^{-11}$ m.rad.
Using the values given in table 1, and assuming a natural relative energy spread $\Delta\sigma_{\varepsilon,0} = 0.00096$, the sizes of the electron beam at the point sources are of the order of 50 μm in the horizontal plane, and 25 μm in the vertical plane.
The natural bunch duration is proportional to the natural relative energy spread, $\sigma_{\tau,0} = \frac{\alpha_c}{\Omega_s}\sigma_{\varepsilon,0}$ (given in ps), where $\alpha_c = 1.7 \times 10^{-4}$ and $\Omega_s = 18.3$ kHz account for the

TABLE 1. Twiss parameters (α, β, $\gamma = \frac{1+\alpha^2}{\beta}$) and dispersive functions calculated for Dipole 1 and 2 in the horizontal (x) and vertical(y) planes.

	β_x	α_x	γ_x	β_y	α_y	γ_y	η	η'
Dipole 1	0.963	1.089	2.2703	21.711	1.8071	0.1964	0.009	-0.019
Dipole 2	0.552	0.528	2.3162	23.8661	2.0397	0.2162	0.033	-0.086

compaction momentum factor and the synchrotron (radial) frequency respectively. The natural bunch (r.m.s) duration is then $\sigma_{\tau,0} = 9.3$ ps.
These are the values that our set of optical diagnostic will have to measure accurately. In the following sections we describe the measurement in the visible and X-ray domain to characterize the 6 dimensional phase space.

MEASUREMENT IN THE LONGITUDINAL PLANE

The intensity of the synchrotron radiation is proportional to the number of radiating electrons. As a result, in the time domain, the structure of the synchrotron radiation is the image of the time structure of the electron beam. In the following subsections, we describe the techniques employed to measure the electron bunch distribution, from which we retrieve the bunch length, and the electron beam time structure, which shows the eventual parasitic bunches.

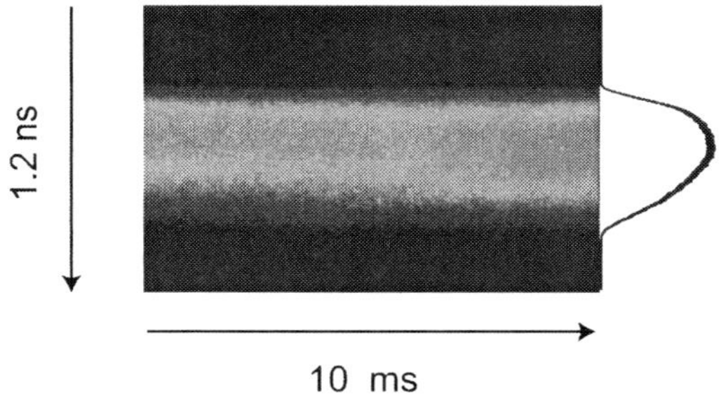

FIGURE 1. Image of the bunch longitudinal distribution obtained with a Hamamatsu steak camera at Super ACO with the total current $I = 41$ mA. The moment method, applied on the distribution shown on the right side of the image, gives the bunch length $\sigma_\tau = 140 \pm 1$ (ps).

Electron Bunch Distribution

The synchrotron radiation has the longitudinal profile of electron beam. Because the r.m.s bunch duration is of the order of 10 ps, very few devices could give accurately the profile of the electron bunch. One of them is the double sweep streak camera, which has already been used successfully by many different groups on storage rings [4]. The camera uses the following principle: the radiation illuminates a photocathode, which delivers electrons proportionally to the radiation intensity. The electrons go through deflection electrode plates which have a time varying voltage triggered to the main RF cavity signal. The deflected electrons hit a phosphore screen, forming an image, which contains time structure information. Figure 1 shows such an image that has been taken at Super ACO with the Hamamatsu C5680 streak camera. A cut of the image, taken at a given beam current, along the fast sweep (vertical) gives the longitudinal bunch distribution, and the time evolution of the distribution can be followed along the slow time scale (horizontal). The bunch duration is given by the second order moment of the distribution. In our case, a fast sweep at 250 MHz, which is half the RF cavity frequency is important because is allows the streak camera to acquire consecutive bunches. The slow sweep should cover the range from twenty nanoseconds to several hundred milliseconds. This will allow characterization of all the longitudinal instabilities of the beam. For the acquisition, a firewire CCD camera at the back of the phosphor screen is preferred, so that the images could be read directly and analysed at the same time with a software such as Matlab.

Electron Beam Train Measurement

For some of the applications using the synchrotron light, the time structure of the photon beam is essential. Therefore, it is necessary to deliver an electron beam with a perfect time structure, having the homogeneous electron bunches in the buckets they have to be. Therefore, a diagnostic of the quality of the electron bunches train is necessary. In order to achieve such a measurement, a collimating - attenuating system coupled to a new generation of PCI card multi-channel analyser is currently being designed. The statistics of the photons arrival time in the multi-channel analyser are recorded, and the resulting histogram gives the electron beam time-structure. The new feature of this system is to be able to continuously acquire several photons per electron revolution in the ring.

Resolution of the Measurement

The quality of the electron bunch train is described by the quality of the empty buckets, and the homogeneity of the beam current. The diagnostic measurement has to render account of both. Therefore, the number of points in the histogram should be high enough to reveal the average time structure of the electron bunches and the empty buckets, and the dynamical resolution should be high enough to show the trailing edge of the bunches and the possible residual electrons that could be stored in supposedly empty buckets, as well as the parasitic bunches adjacent to the main bunches. In other words, the multi-channel analyser should have a large number of channels per bucket and a large number of bits per channel, and the signal to noise ratio of the system should be very high.
More quantitatively, 40 to 50 points per bucket of 2 ns, makes 50 to 40 ps per channel, and a dynamical resolution of 10^6, to see residual bunches, imposes 20 bits per channel. The revolution period of the electrons in the ring is 1.872 μm, so that the histogram for the ring contains at least 2^{15} channels. The system under development matches all

these requirements, with 10^{16} channels of 32 ps, dynamical resolution limited by the dark current, and 5 MHz acquisition time, which translates to 8 photons per turn for DIAMOND.

MEASUREMENT OF THE RELATIVE ENERGY SPREAD AND EMITTANCE

The basic scheme to measure the relative energy spread and the emittance of the beam is shown in Fig. 2. The X-ray beam is taken from a bending magnet. A thick cooled Cu block should absorb most of the Hard X-ray beam, so that only a fraction of the radiated power passing the Al plate (max. $\sim$ 5 - 10 W) is hitting the pinhole The radiation above 15 keV (< 0.1 nm) is selected with a water-cooled aluminum plate ($\sim$ 3 mm). Therefore no cooling should be needed for the pinhole. The pinhole is located as close as possible to the source point (4 m). It is made from tungsten blocks 5 mm thick, as shown in Fig. 3.

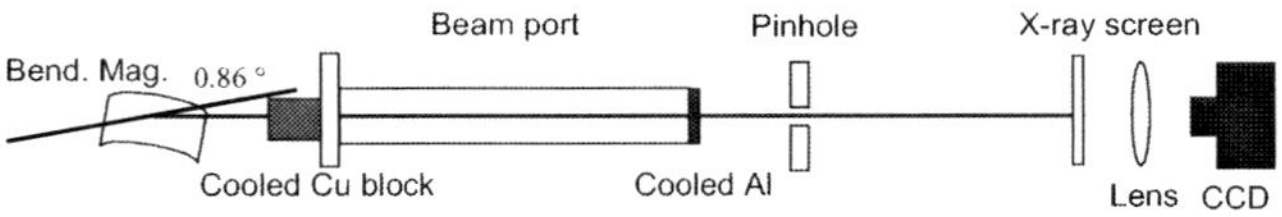

FIGURE 2. Scheme of the pinhole camera system for measuring the electron beam transverse profile. Electron beam size, emittance, and energy spread can be calculated from the image of the beam profile.

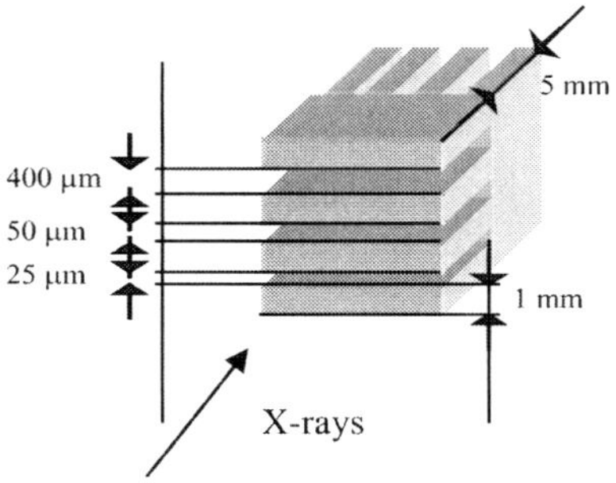

FIGURE 3. Scheme of the three calibrated X-ray pinholes.

Size of the Image at the Fluorescent Screen

In this subsection, we present the theoretical model we use to calculate the size of the image at the fluorescent screen. We assume first that betatron oscillation in the vertical and horizontal plane are not coupled. Let us define <f> the statistical average over all the electrons of the physical quantity f. In each plane, the relative energy spread, σ_ε, and the emittance, ε, of the electron beam are related to the following quantities [3]:

$$\begin{aligned}\langle y_e^2\rangle &= \beta_i\varepsilon+(\eta_i\sigma_\varepsilon)^2 &(1)\\ \langle y_e'^2\rangle &= \gamma_i\varepsilon+(\eta_i{}'\sigma_\varepsilon)^2 &(2)\\ \langle y_e y_e'\rangle &= -\alpha_i\varepsilon+\eta_i\eta_i{}'\sigma_\varepsilon^2\ , &(3)\end{aligned}$$

where the index i refers to the horizontal (x) or vertical (y) plane, γ_i is defined as $\gamma_i=\dfrac{1+\alpha_i^2}{\beta_i}$, and α_i, β_i, η_i and $\eta_i{}'$ are the lattice parameters defined by Courant and Snyder [5]. The r.m.s beam size is then given by $\langle y_e^2\rangle$, and its divergence is $\langle y_e'^2\rangle$.
The size of the image on the screen is given by:

$$\Sigma^2=S^2+S^2_{diffraction}+S^2_{pinhole}+S^2_{screen}\ , \tag{4}$$

where S is the r.m.s size of the image of the photon beam located at the distance D from the pinhole, $S_{diffraction}$ is the diffraction spot size, $S_{pinhole}$ is the geometrical contribution due to the finite size of the pinhole and S_{screen} is the spatial resolution of the screen. The size of image of the photon beam is given by [6]:

$$S^2=D^2\frac{E}{B+2Ad+Gd^2}\ . \tag{5}$$

Where d is the distance between the source and the pinhole, E is the emittance of the photon beam, and A, B, G are the generalised Twiss parameters of the photon beam. We consider the source of the photon beam as the sum of all individual electrons. Therefore, the Twiss parameters of the photon beam are given by:

$$\begin{aligned}BE=\langle y_e^2\rangle+\langle y_p^2\rangle &\approx \beta_x\varepsilon+(\eta\,\sigma_\varepsilon)^2 &(6)\\ GE=\langle y_e{}'^2\rangle+\langle y_p{}'^2\rangle &\approx \gamma_x\varepsilon+(\eta'\sigma_\varepsilon)^2+\langle y_p{}'^2\rangle &(7)\\ AE=-\langle y_e y_e{}'\rangle-\langle y_p y_p{}'\rangle &\approx \alpha_x\varepsilon-\eta\,\eta'\sigma_\varepsilon^2 &(8)\end{aligned}$$

The quantities y_p and y_p represent the position and the angle of any photon emitted by the electrons. We note that the quantity $<y_p y_p'>$ is zero due to the symmetry of the emission process. As a further remark, the diffraction law imposes the condition:

$$\langle y_p^2\rangle\langle y_p{}'^2\rangle\le(\frac{\lambda}{4\pi})^2\ , \tag{9}$$

which is used together with $<y_p'^2>\approx\dfrac{1}{\gamma^2}$ (γ being the Lorentz factor linked to the energy of the electrons) and $\lambda\approx 0.1$ nm to evaluate $<y_p^2>$ and conclude that it is absolutely negligible compare to $<y_e{}^2>$.
In addition, for very large G E, i.e. a large divergence of the photon beam compared to its size, the size of the beam is given by

$$S^*=D\frac{\sigma_i}{d}, \tag{10}$$

with $i=x,y$

Procedure for the Evaluation of the Relative Energy Spread and Emittance

Optical System Imaging the Screen

The image of the electron beam provided by the fluorescent screen can be acquired with an off-the-shelves objective-lens mounted on a firewire CCD camera. The optimum magnification m^* given by the result of preliminary tests with WinLens software (from LINOS) is between 1 and 2. This system gives a spot size of the order of 5 μm. The size of the image on the fluorescent screen, Σ, can be directly measured by the analysis of the CCD image. A fit of the two-dimensional Gaussian image gives the horizontal and vertical r.m.s sizes in pixels, which can be directly scaled in micrometers at the fluorescent screen.

Analysis of the Results

Having the size of the photon beam at two different locations with two different dispersive function values and using expression 5, leads to the solution of a non-linear system of two equations with two unknowns, which are the emittance and the energy spread. A least square method is used to solve this system, using the horizontal sizes measured in ID 1, and in BM 1.

Precision of the Measurement

The minimum spot size of the image is:

$$\Sigma^{*\,2} = S^2_{diffraction} + S^2_{pinhole} + S^2_{screen} + S^2_{CCD}\,, \tag{11}$$

In our case, estimation with a thin screen and a 25 μm^2 pinhole gives $\Sigma^* = 25\ \mu$m.
The total error on the measure of the energy spread and emittance is given by sum of the uncertainties on the knowledge of the Twiss parameters for the estimation of S and the error on the fit of the Gaussian curve. The typical error due to the uncertainty on the Twiss parameters has been computed in a Matlab file: the typical error on the Twiss parameters, given an energy spread and emittance, generates an uncertainty on the size on the screens in ID1 and BM 1. From this, the codes generates a distribution set of sizes. This set is used to generate, with the least square analysis, a set of values for the emittance and relative energy spread, using the nominal Twiss parameters values. The mean values of these two distribution sets are compared with the initial given value, and the typical relative error is deduced from the standard deviations of the distributions. Table 2 summarizes the results.

TABLE 2. Estimation of the errors vs. the uncertainties on the Twiss parameters (error); The Matlab program gives the relative difference between the initial values and the computed values of the relative energy spread and emittance ($\Delta\sigma_{\varepsilon,0}$ and $\Delta\varepsilon_0$), and the relative uncertainties on the final result, $\Delta\sigma_\varepsilon$ and $\Delta\varepsilon$.

error	$\Delta\sigma_{\varepsilon,0}$	$\Delta\varepsilon_0$	$\Delta\sigma_\varepsilon$	$\Delta\varepsilon$
0.01	0.002	0.001	0.035	0.034
0.05	0.003	0.013	0.164	0.181
0.1	0.067	0.060	0.387	0.358

CONCLUDING REMARKS

A set of optical measurements to diagnose the quality of the DIAMOND electron beam has been designed. The small dimensions of the electron beam imposes drastic requirements to provide the transverse size, the emittance, the energy spread, the bunch distribution of the electron beam and the bunch train structure with sufficient accuracy. The design presented here shows that these requirement can be satisfied, so that reliable measurements are expected.

ACKNOWLEDGMENTS

The authors would like to thank Pascal Elleaume, Graham Naylor and Kees Scheidt from ESRF for their support and the enlightening discussions that contributed to the achievement of this design.

REFERENCES

1. European Synchrotron Facility Highlights 2003 (2003), http://www.esrf.fr/UsersAndScience/Publications/Highlights/2003.
2. Research with Diamond Light Source (2004), http://www.diamond.ac.uk/Activity/ACTIVITY=Research.
3. M. Sands, *The physics of electron storage ring, an introduction*, vol. 121, SLAC report, 1970.
4. K. Scheidt, "Review of streak cameras for accelerators : features, applications and results," in *Proc. EPAC 2000, Vienna*, 2000, p. 182.
5. E. D. Courant and H. S. Snyder, *Annals of Physics*, **3**, 1 (1958).
6. P. Elleaume, C. Fortgang, C. Penel, and E. Tarazona, *J. Synchrotron Rad.*, **2**, 209 (1995).

Fermilab Recycler Ring BPM Upgrade Based on Digital Receiver Technology

R. Webber, J. Crisp, P. Prieto, D. Voy, C. Briegel, C. McClure, R. West, S. Pordes, and M. Mengel

Fermi National Accelerator Laboratory, Batavia, IL 60501-0500, USA

Abstract. Electronics for the 237 BPMs in the Fermilab Recycler Ring have been upgraded from a log-amplifier based system to a commercially produced digitizer-digital down converter based system. The hardware consists of a pre-amplifier connected to a split-plate BPM, an analog differential receiver-filter module and an 8-channel 80-MHz digital down converter VME board. The system produces position and intensity with a dynamic range of 30 dB and a resolution of ±10 microns. The position measurements are made on 2.5-MHz bunched beam and barrier buckets of the un-bunched beam. The digital receiver system operates in one of six different signal processing modes that include 2.5-MHz average, 2.5-MHz bunch-by-bunch, 2.5-MHz narrow band, un-bunched average, un-bunched head/tail and 89-kHz narrow band. Receiver data is acquired on any of up to sixteen clock events related to Recycler beam transfers and other machine activities. Data from the digital receiver board are transferred to the front-end CPU for position and intensity computation on an on-demand basis through the VME bus. Data buffers are maintained for each of the acquisition events and support flash, closed orbit and turn-by-turn measurements. A calibration system provides evaluation of the BPM signal path and application programs.

SYSTEM OVERVIEW

1. Beam position split-plate capacitive pickups in the beamline vacuum.
2. Signal preamplifiers in the beam enclosure near the beamline.
3. Analog signal receiver (transition modules) and digital signal processing electronics in the service buildings.
4. Timing signal generators.
5. VME based "front-end" data acquisition and control computers and support hardware.
6. Integrated position and intensity calibration system.

There are 211 BPM locations (104 horizontal, 107 vertical) served by this system in the Recycler Ring and 26 locations in the associated beam lines.

CP732, *Beam Instrumentation Workshop 2004: Eleventh Workshop*,
edited by T. Shea and R. C. Sibley III

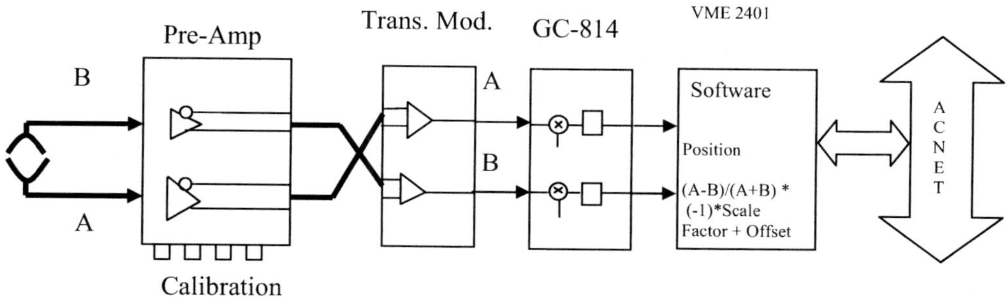

FIGURE 1. BPM system.

Beam position is determined from the relative amplitude from opposing electrodes in a BPM pick-up. In this design, the individual A and B signals follow parallel, but independent, signal processing paths from the pick-up through digitization and hardware digital processing. Signal differencing and normalization to obtain beam position information is done in software in the VME CPU. The accuracy of the BPM system is limited by how well the A and B channel pairs are matched, calibrated, and controlled in gain, offset, and frequency response. The BPM system is required to report the A + B signal sum as well as beam position.

The BPM signals are a function of pick-up design, beam intensity, the temporal structure of the beam current, and the beam position. BPM system design is thus influenced by each of these parameters. The Recycler functional specification identifies two beam structures (2.5 MHz, and un-bunched) to be measured, the range of beam intensities for each, and the required accuracy for each. This sets the required scale for determination of signal magnitude dynamic range, bandwidth, and A/B channel matching. This BPM system uses existing tunnel-to-service building twisted pair cabling with runs ranging from 150 ft to 1300 ft. Signal attenuation over this range of cable lengths varies by about a factor of six, impacting the overall system dynamic range requirements

Signal digitization and hardware digital processing is performed in EchoTek GC814 VME digital down-converter/digital receiver (DDC) modules. The EchoTek module requires different set-ups for processing signals from the different beam structures and for different measurement modes, e.g. single bunch, head vs. tail, etc.

The Timing Signal Generator interfaces to the Recycler Beam Sync Clock and provides programmable triggers to the EchoTek modules. This is an Industry Pack (IP) style module that resides on a carrier board on each VME CPU module.

The VME front-end computer receives digital A and B signal data from the EchoTek modules, performs final beam position and intensity (sum) signal processing, and controls the numerous aspects of timing and the EchoTek module. It also provides calibration system set-up, and interfaces to ACNET.

BEAMS IN THE RECYCLER

The BPM system is required to measure proton beams and anti-proton beams. The Recycler has up to three partitions of beam at a time - cold "de-bunched" beam, warm "de-bunched" beam, and injected or to be extracted beam with 2.5-MHz structure.

The 2.5-MHz structure is made up of four bunches in successive 2.5-MHz RF buckets (396-ns spacing). Bunch lengths range from a Gaussian sigma of 25 ns to 50 ns. The intensity range for bunched beam is 2.0e10 to 30e10 particles in four bunches. The required absolute position relative to the center of the BPM is ±1.0 mm. The relative difference between two measurements on subsequent pulses with stable beam is ±0.40 mm.

A partition of de-bunched beam is contained between RF barriers created by a wideband RF system. The width of a typical barrier is 906 ns; it can be as small as 679 ns and as large as 1132 ns. The de-bunched beam has no intentional modulation structure over the length of the bunch. The dynamic range is 20e10 to 400e10 particles for de-bunched beam.

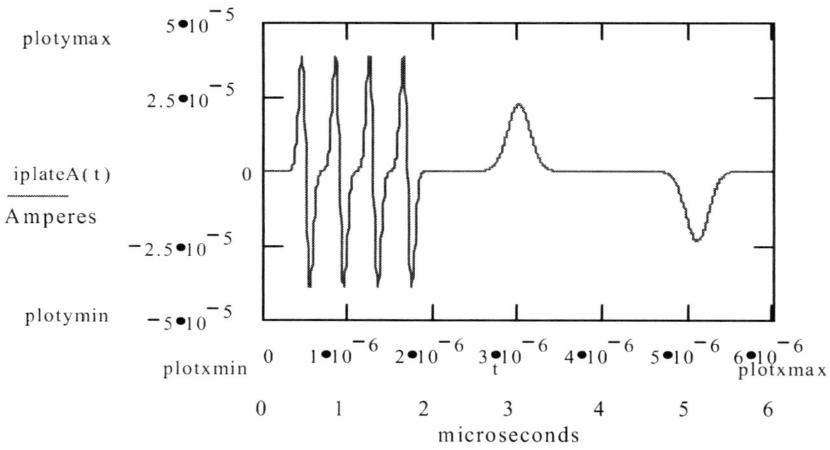

FIGURE 2. Electrode current for 2E10 total in four 2.5 MHz $\sigma = 50$ ns bunches and 20E10 in 2.094 μsec bunch with 566 ns edges.

HARDWARE SIGNAL PROCESSING MODES

The Recycler BPM system allows selection of one of six signal-processing modes in the EchoTek digital receiver boards. The modes are:

1. **2.5-MHz Ensemble.** This mode results in measurement of the magnitude of the 2.5-MHz signal integrated over an interval corresponding to about 1.6 microseconds (four 2.5-MHz periods), the full extent of the 4-bunch beam structure.

2. **2.5-MHz Bunch-by-Bunch.** This mode results in measurement of the magnitude of the 2.5-MHz signal integrated over an interval corresponding to about 0.4 microseconds (one 2.5-MHz period). The measurement duration is about 1.6 microseconds (four 2.5-MHz periods) allowing measurement of each of the four 2.5-MHz bunches.
3. **2.5-MHz Narrow Band.** This mode results in measurement of the magnitude of the 2.5-MHz signal integrated over an interval corresponding to about 700 microseconds (about 64 TBD Recycler turns). This mode may be used to measure the closed orbit of all beam in the Recycler Ring that generates 2.5-MHz signal frequency components.
4. **Un-bunched Ensemble.** This mode results in measurement of the magnitude of the signal from beam confined in a barrier bucket otherwise lacking intentional modulation. The signal is effectively integrated over an interval corresponding to the full barrier bucket length. The actual EchoTek setup is determined algorithmically based on the barrier bucket length as determined from MDAT, the accelerator data distribution system. This mode is used to measure the "average" position of the barrier bucket structure beam.
5. **Un-bunched Head or Tail.** This mode results in measurement of the magnitude of the signal from beam in the leading or trailing edges of a barrier bucket. The signal is effectively integrated over an interval corresponding to the typical rising or falling time of the barrier bucket beam structure. This mode measures the position of the head and/or the tail of the barrier bucket structure beam.
6. **89-kHz Narrow Band.** This mode results in measurement of the magnitude of the 89-kHz signal integrated over an interval corresponding to about 700 microseconds (about 64 Recycler turns). This mode is used to measure the closed orbit of all beam in the Recycler Ring that generates 89-kHz signal frequency components.

PREAMP DESIGN

Each preamp is a two-channel circuit required to:

1. Receive via coaxial cables and into well-controlled impedances two beam signals (A and B) from the respective BPM electrodes.
2. Provide suitable gain and impedance matching to drive the beam signals on twisted-pair cables to the service buildings.
3. Provide suitable connections and necessary DC isolation from the electronics to apply ion-clearing voltages up to 500 volts to the BPM electrodes.
4. Provide connections, control, and signal conditioning for BPM calibration signals for application to the normal beam signal paths at the preamp inputs.
5. Sustain without damage the peak input signal levels presented by specified beams (e.g. 53 MHz) outside the normal functional range for beam position measurement.

The preamp is designed to present a resistive load impedance of 1.5 kΩ to beam signals from the pick-up electrodes. This resistance coupled with the parallel electrode and coaxial cable capacitance provides a "high-pass" transfer function from beam current to preamp input voltage with a corner frequency of about 1.5 MHz. The preamplifier is designed for a mid-band voltage gain of about 3.5 with a bandwidth of approximately 23 MHz. The final amplifier stage provides a differential output to drive a 100-Ω twisted-pair cable in a balanced fashion. Series resistors provide a back termination for the cable. The preamp is capable of driving a differential voltage up to 3.7 V peak-to-peak on the cable downstream of the back-terminating resistors.

RG58 cables with SMA connectors connect the preamp to the BPM electrodes. The calibration signal waveform is connected to and daisy chained through the preamp on RG58 cables with BNC connectors. A calibration control signal is connected to and daisy chained through the preamp on RG58 cables with BNC connectors. Ion clearing voltages connect to and daisy chain through the preamp on RG58 cables with SHV connectors. A cable with four twisted-pair wires connects the preamp in the tunnel and the Transition Module crate in a service building to carry the preamp A and B output signals from the tunnel and to provide DC power to the preamp.

TRANSITION MODULE

The transition module input is configured using a differential receiver op-amp (AD8130) converting the differential input to a single-ended signal. The main function of the transition module is to compensate for cable attenuation, provide an anti-aliasing filter, and normalize the input voltage to use the full dynamic range of the A/D converter.

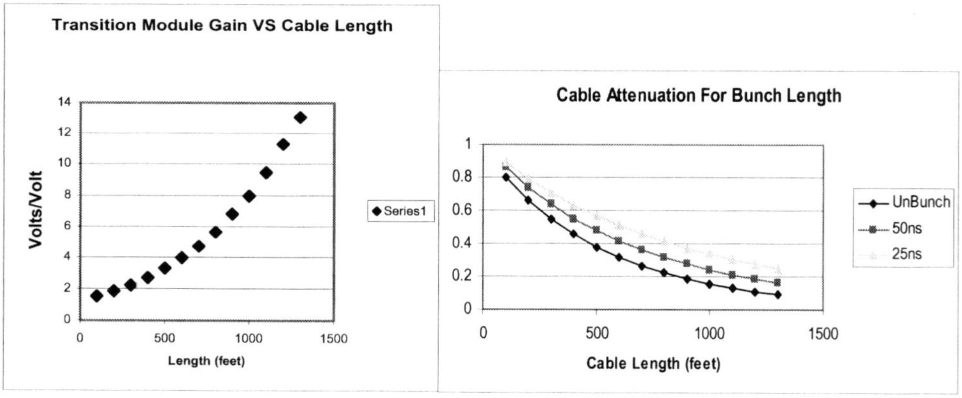

FIGURE 3. Left: range of gains vs cable lengths needed to normalize the pre-amp signal to 1.1 Vpp at the input of the DDC. Right: attenuation due to cable length on bunches of different sigma_t.

Each of the 6 houses in the ring has 10-12 eight-channel transition modules, their gains were selected based on the cable lengths attenuation.

- 150 to 400 ft, G = 2.0
- 400 to 600 ft, G = 2.5
- 601 to 900 ft, G = 4.0
- 900 to 1300 ft, G = 6.0

SYSTEM FREQUENCY RESPONSE MEASUREMENTS

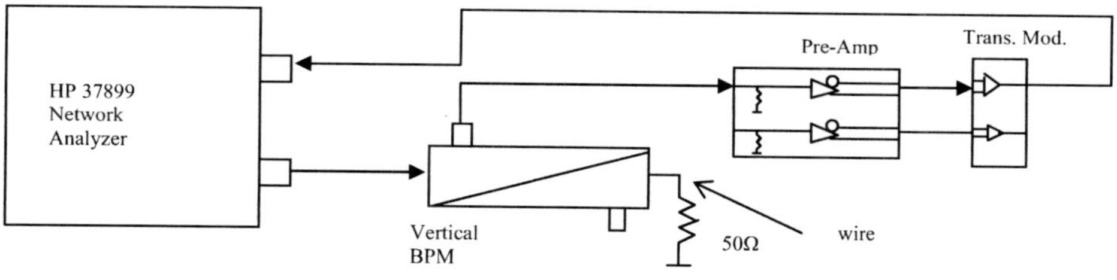

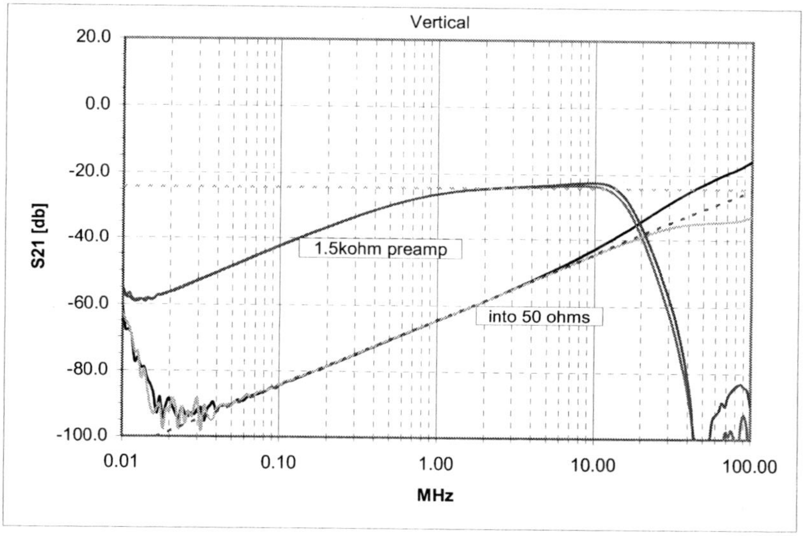

FIGURE 4. Transfer function of a vertical split plate BPM with the transition module as well as the BPM by itself. The network analyzer drives the wire in the BPM terminated in 50 Ω. The first measurement made was with the BPM plate directly terminated in 50 Ω, shown as the lower trace as the high pass response. In the second measurement the BPM plate was terminated in 1.5 kΩ at the input of the pre-amp. The measurement was made at the output of the transition module. The upper trace shows the frequency response of the BPM and the corner frequency shifted down in frequency toward 1.5 MHz.

SIGNAL DIGITIZATION AND PROCESSING ON ECHOTEK BOARD (DDC)

EchoTek GC814 cards have eight independent channels each structured as a set of A/D converter, digital down-converter and FIFO memory elements. Each channel digitizes and filters one of the A and B beam signals from the transition module at 80 MHz asynchronously.

Each digital down converter (DDC) device has four channels in parallel of a NCO, CIC filter and two user-programmable LPF. Each channel can produce an I and Q pair as a 16-bit word and stored in the FIFO as one 32-bit longword. The data are transferred to the front-end processor where position and intensity are computed by processing the I and Q pair.

Our system uses two of the four DDC device channels, one channel generates the I term, and the second channel the Q term. This reduces decimation in the CIC filter, increases computation speed, and allows for the design of a fairly wide bandwidth filter. It also allowed us to meet the more narrow bandwidth requirements where high decimation rates were needed in the case of turn averaging.

The DDC programmable filters were setup in asymmetric mode. The first filter stage (CFIR) was coded with an 11 tap Gaussian filter with the purpose of reducing data jitter introduced by asynchronous sampling. The second filter (PFIR) was setup as a boxcar averaging filter whose length was determined by the clock rate at the input of the 64 tap filter, the number of taps used, and the period of the data being sampled.

Data transfers between the Echotek board and the front-end processor (MVME 2401) are done under DMA control (DMA32 or DMA64) or direct bus transfers depending on the type of measurement performed. Data transfer rate measurement provide a figure-of-merit determining the number of data points produced at the output of different types of filters as well as set the maximum number of transfers achievable in the most stringent case where data is obtained every turn (turn-by-turn) as opposed to single shot data acquisition mode.

CALIBRATION SIGNAL DISTRIBUTION

The calibration signal from the Calibration Signal Driver module is attenuated by about 14dB (a factor of five in voltage) down the 1300 ft RG58 calibration signal cables. The signal is locally adjusted to a common level in each preamp by a resistive attenuator network and split equally into A and B channels. Remotely controlled JFET switches connect/disconnect the A and B calibration signal channels to the respective preamp inputs. The A channel contains a JFET controlled attenuator to allow reduction of that signal by a factor of two relative to the B channel. Circuitry in the preamp decodes the –5 to +5 volt state control signal to set the JFET switches as required. The three calibration circuit states, OFF, ON w/A=B, ON w/ $A = \frac{1}{2}B$, are thus supported.

The calibration signal source impedance presented to each preamp input is approximately 2KΩ. This is suitable to verify the voltage gain of the preamp and the transition module. However, the impedance is low enough to load the 1.5KΩ preamp input and thus provide only a rather insensitive test of the preamp input circuit, i.e. the pick-up and signal cable capacitance and the preamp input impedance.

To meet the requirements specification, the relative A and B signal amplitudes within a preamp must be controlled to approximately 0.5%.

The calibration application lets the operator do a thorough set of calibration tests, and generates graphs and reports that let him/her quickly spot problem BPM's. It does this by taking advantage of the Closed Orbit mode of the BPM Data Acquisition hardware, and the signal generating hardware. The results are also logged in a database, allowing historical review of any particular BPM's calibration results. To perform a calibration test set, at the operator's request, we do 6 separate iterations of turning on the signal generation, and then do a Closed Orbit 16-turn reading of that signal. Each iteration gives us an average position, and RMS error for each BPM. The six iterations consist of three voltage levels with a balanced (1:1) and unbalanced signal (2:1). The operator is then presented with a choice of graphs to examine, or the option of generating a "bad BPM" list. Two graph types provide useful diagnostic information, the un-scaled position graphs, and the A/B ratio graph. In the former, one expects to see the un-scaled position come back with the balanced (1:1) signal reading dead center, and the unbalanced (2:1) signal reading just over 0.3. Also one expects the RMS error to decrease as the voltage goes up. In the A/B ratio graph, we examine the A/B ratio for the balanced(1:1) signal at all three voltages; this shows if we have a different amplification gain between the two channels, which would lead to an intensity-dependent position error. The operator also decides whether to log this run in a database, or save the raw data to a spreadsheet. In this fashion, the application lets the operator do a set of calibration tests, and generates graphs and reports that quickly spot problem BPMs, and usually the type of problem with a glance at the graphs.

POSITION ERROR DUE TO COMPONENT NOISE

The Position error has contributions due to noise in the circuits in the pre-amp, the transition module, and the noise of the A/D converter. The noise of the pre-amp and the transition module can be expressed as power noise density calculated using the following expression:

$$V_{Calc\ \ RMS} = \left[\left(7.1\frac{nV}{\sqrt{Hz}}\right)^2 + \left(\frac{1}{2}\right)^2\left(42\frac{nV}{\sqrt{Hz}}\right)^2\right]^{\frac{1}{2}}. \tag{1}$$

The voltage noise density is multiplied by the bandwidth at the final stage to obtain the noise factor that can be added in quadrature with the noise factor of the A/D to be applied in the computation of position noise.

A/D CONVERTER NOISE

SNR for a sinusoidal input is the ratio of the RMS signal at the output to the RMS noise excluding harmonics and DC term. If harmonic and DC terms are included, one obtains the term SINAD.

$$ENOB = N_{effective} = \frac{SINAD - 20\log\frac{\sqrt{3}}{\sqrt{2}}}{20\log 2}. \tag{2}$$

The effective number of bits, ENOB, reflects the accuracy the A/D will have when digitizing a signal relative to the inherent noise.

Over-sampling can compensate for loss of bits from the ideal. The increased resolution (R) is given by the expression:

$$R_{oversampling} = ENOB + \frac{1}{2}\log_2\left(\frac{f_{sample}}{2f_{signal}}\right) . \tag{3}$$

The A/D RMS voltage noise can be computed from the ratio of the full input RMS voltage to the ratio of the RMS noise or the ENOB.

$$V_{A/D \quad noise} = \frac{V_{peak-to-peak}}{2\sqrt{2}} 10^{-\frac{SINAD}{20}} ; \tag{4}$$

$$V_{A/D \quad noise} = \frac{V_{peak-to-peak}}{2\sqrt{2}} 2^{-\ ENOB} . \tag{5}$$

The total noise can be calculated from the expression

$$A_{noise} = \sqrt{V^2_{PreAmp+Trans.} + V^2_{A/D \quad noise}} . \tag{6}$$

The system dynamic range, $Range_{DYN} = 20\log\left(\frac{V_{fscale}}{A_{noise}}\right)$. (7)

The position error, $P_{noise} = \frac{A_{noise} - B_{noise}}{A_{signal} + B_{signal}} * BPM_{scalefactor}$. (8)

Making the approximation A = B for beam on-center and since noise is un-correlated, $\sqrt{2} * A_{noise}$ can be substituted in the numerator.

$$P_{noise} = \frac{4 * A_{noise}}{V_{signal}} BPM_{scalefactor} * \frac{1}{\sqrt{\frac{\# \quad A/D \quad samples}{\# \quad DDC\,output \quad samples}}} . \tag{9}$$

The factor of 4 is the multiplication of $\sqrt{2}$ with 2 $\sqrt{2}$ factor, which is required to convert peak-to-peak voltage at the input of the A/D to RMS.

NUMERICAL CALCULATION

$$V_{PreAmp+Trans \quad rms} = V_{Calc \quad rms} * \sqrt{2BW} \text{ , and } 0Hz \le f \le 12MHz . \tag{10}$$

The $\sqrt{2}$ factor accounts for noise present beyond the cutoff frequency and it follows the filter attenuation.

$$V_{PreAmp+Trans.\ rms} = 22.17 \ \frac{nV_{RMS}}{\sqrt{Hz}} \ 4.899e3\sqrt{Hz} \ = 108.6\mu V_{RMS} \ . \quad (11)$$

Using Eq. (4) to compute the A/D noise SINAD = 75 dB from AD6645 data sheet,

$$V_{A/D\ noise} = 69.16\mu V_{rms} \ , \quad (12)$$

then the total RMS noise is given by

$$A_{noise} = \sqrt{(108.6\mu V)^2 + (69.16\mu V)^2} = 128.8\mu V_{rms} \ . \quad (13)$$

The largest BPM scale factor in the Recycler is 62 mm per Volt, the number of A/D samples =128 and the number of DDC output samples are 4 after averaging and decimation.

$$P_{noise\ \ rms} = \frac{(4)\ \ (128.75\mu V)}{1.1V} * 62 \ \ \frac{mm}{V} * \frac{1}{\sqrt{\frac{128}{4}}} = 5.1\mu m \ . \quad (14)$$

The position error must be scaled by the difference between the maximum allowed signal and the signal intensity at which the measurements were taken, given that at lower signal levels fractional noise level increases.

$$P_{error} corrected = 5.1\mu m * \frac{70e10}{20e10} = 17.9\mu m_{rms} \ . \quad (15)$$

RECYCLER RING BPM MEASUREMENTS

Turn-by-Turn BPM data shown below were obtained during injection. All horizontal and vertical BPMs in the ring Data for turns 20 up to turn 50 were fitted to a sine wave, shown as dots on the plot. The sine wave was extended to earlier and later turns to show the uniformity of measurements before and after the selected turns.

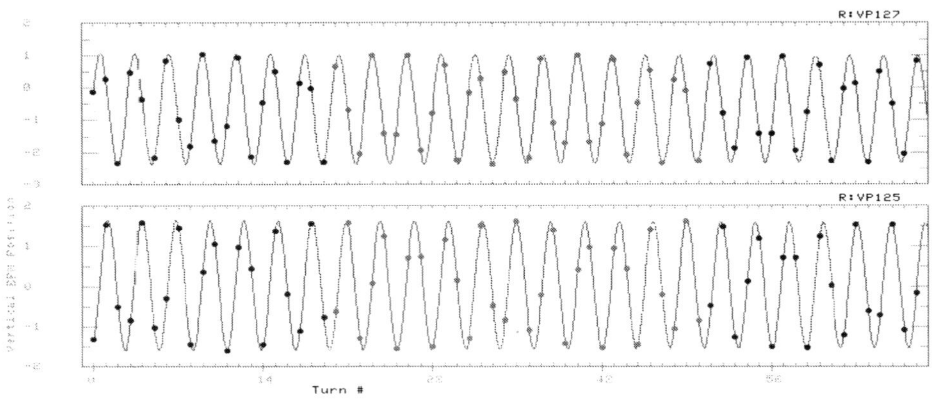

FIGURE 5. Turn-by-turn BPM data obtained during injection.

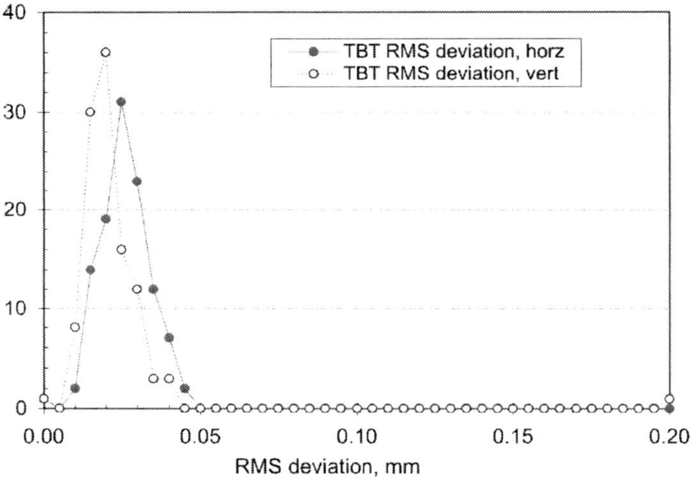

FIGURE 6. Histogram of RMS position deviation for vertical and horizontal BPMs.

The plot above shows a histogram of the deviation of the measured BPM data (turns 20 to 50) from the fitted sine wave.

ACKNOWLEDGEMENTS

We want to thank Dr. Ming-Jen Yang for providing this Recycler BPM data.

Posters

11th Beam Instrumentation Workshop
May 3–6, 2004

CTF3 Injector Profile Monitor

C. Bal,* E. Bravin,* S. Burger*, and T. Lefèvre†

*CERN, Geneva, Switzerland
†Northwestern University, Evanston, IL, USA

Abstract. The electron gun of the CLIC Test Facility 3 (CTF3) produces 15.6-μs-long pulses with an energy of 140 keV and a current that can be as high as 9 A. For the nominal beam operation, a 5.4-A beam current is generated and sent downstream into the bunching system and the rest of the accelerator. The corresponding beam charge will induce a thermal load that most of the materials, considered as radiators, would not withstand. With this problem in mind, we have built a beam imaging system, installed just after the gun and equipped with two screens. The first one is a phosphorescent screen which yields a high signal and can be for low beam currents. The second screen, a thin graphite foil, is used as a forward OTR radiator and can stand the full beam intensity. Moreover, the time resolution of OTR is very good, in the femtosecond range. This allows for observation of the evolution in time of the beam size during the pulse by using a gated camera. We present in this paper the first results obtained using this system.

INTRODUCTION

The injector of the CTF3 facility in the nominal phase will produce intense beams [1], which impose severe constraints on the design of beam profile monitors. Very few materials can be used to produce screens capable of standing the full beam induced thermal load [2]. For this reason the use of a thermal-resistant material like graphite has been considered as an Optical Transition Radiator (OTR). Even if they are not commonly used for non-relativistic particles, enough photons will be emitted to be measured using an intensified camera. A preliminary test [3] was already made in 2002 on an 80-keV electron beam produced at the CERN photocathode R&D laboratory. The characteristic of backward OTR generated from an aluminium screen was measured and found to be in good agreement with our expectations. However, an additional difficulty arises when using materials like graphite which have a low reflectivity coefficient. The backward OTR is then strongly suppressed. To overcome this limitation, the screen is tilted in order to observe the forward OTR light emitted at 90°, which according to our calculations will produce even more photons than in the backward direction.

Even if the system is designed to measure the full beam charge, it is always helpful during the commissioning period to have the possibility to image lower intensity beams. For this purpose, scintillating screens, which yield a considerable amount of photons, are classically used [4]. However, one must be careful because phosphorescent screens are subject to saturation, limiting the accuracy of the

CP732, *Beam Instrumentation Workshop 2004: Eleventh Workshop,*
edited by T. Shea and R. C. Sibley III

measurement. Moreover their use must be limited to the observation of small charges, since above a given charge density the risk of damaging the screen is high.

The beam imaging system we designed for the CTF3 injector is described below, presenting both the mechanical assembly with the two screens and the details of the optical system. The performance and limitations obtained with the phosphorescent screen are reported. Measurements obtained with the OTR screen are shown as well, emphasizing the linearity of the light emission, which provides precise quantitative measurements. OTR light has a very fast time response, and measurements relative to the temporal evolution of the beam size are presented.

OVERVIEW OF THE SYSTEM

Installation of the new CLIC Test Facility 3 started in January 2003. By the end of the year, the injector and the first part of the accelerator had already been installed and commissioned [5]. The layout of the CTF3 injector is presented in Fig. 1.

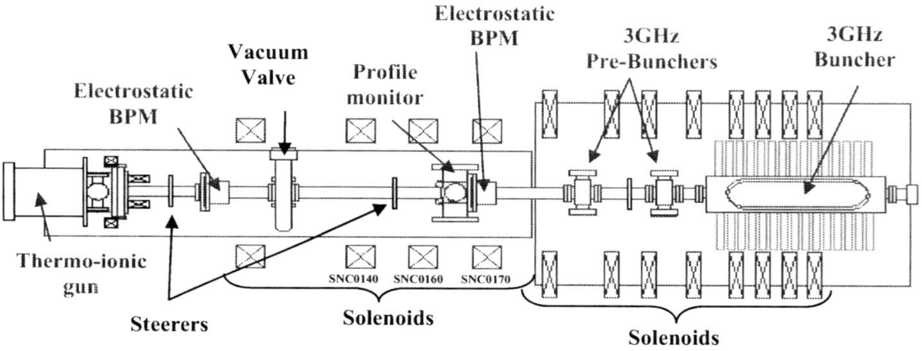

FIGURE 1. The CTF 3 injector beam layout.

A 140-keV electron beam is produced by a thermo-ionic gun. The beam current and the pulse length can be adjusted from 100 mA to 9 A and from 300 ns to 1.6 μs by changing the voltage and the duration of the gun pulser. Using a set of solenoids and two steering magnets, the beam is transported to the 3-GHz bunching system, itself composed of two standing wave cavities (prebunchers) and a traveling wave structure (buncher). The beam current and positions are monitored using two electrostatic pickups [6].

Due to very strong space constraints, the beam imaging system is attached to the second beam position monitor. The details of the beam profile monitor are given in Fig. 2. This device is composed of a long cylinder which provides 3 positions for two screens and one to allow the beam to pass through. The vacuum chamber has an extension tube in order to keep the pneumatic insertion mechanism out of the solenoid magnetic field. The first screen is a 5-μm-thick P47 phosphor (Y_2SiO_5 : CeTb) deposit on a 10-μm-thick aluminum foil. It emits visible photons in the wavelength range from

370 nm to 480 nm with a decay time of 100 ns. The total number of photons emitted per electron in a 2π solid angle is high, of the order of 400. This light yield makes the P47 phosphor screen a perfect choice for the observation of small beam charge. The second screen is a 5-μm-thick carbon screen used as an OTR target, dedicated to the imaging of the full beam charge. The normal of the screen is tilted with respect to the beam trajectory by 30° so that the forward radiation emitted at the back plane of the screen is observed. Only 10^{-4} photons per electron are emitted in the visible range from 400 to 600 nm [3], and an intensified camera must be foreseen to allow its detection. The optical system is composed of a set of achromat lenses. A first lens, placed very close to the vacuum tank view port, is used to optimize the light collection. A semitransparent mirror splits the optical line in two distinct paths sending 50% of the light onto a normal CCD camera dedicated to the observation of the phosphor screen and the remaining 50% onto a gated and intensified camera bought from Proxitronic [7]. In the right side of Fig. 2, one can see two images representing the screens as seen by the corresponding camera. These pictures are used for calibration using references on the screen support. The active area of the screen has a diameter of 40 mm, and the optical calibration gives a spatial resolution of 200 μm per pixel.

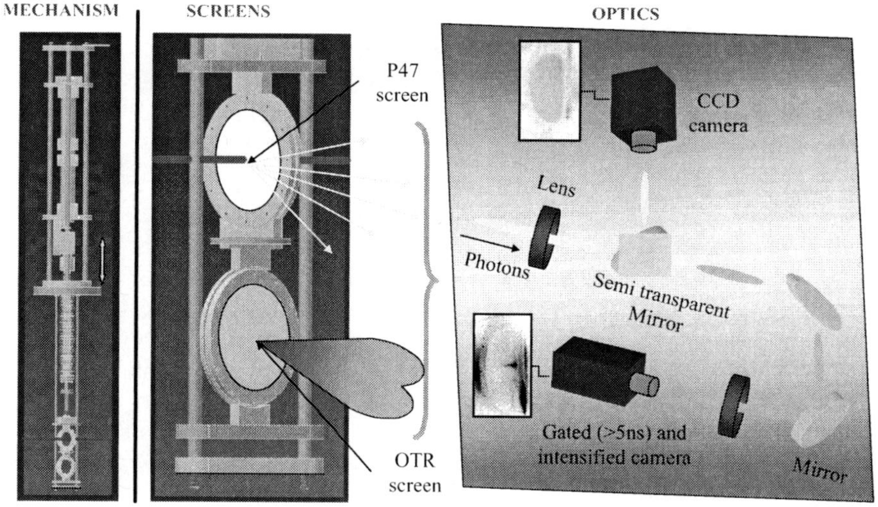

FIGURE 2. The CTF 3 injector beam profile monitor.

Several tests have been performed using the two screens. The most representative results are reported in the following paragraphs.

MEASUREMENTS WITH THE P47 PHOSPHOR SCREEN

High Sensitivity Measurement of the Gun Dark Current

If the gun pulser is off, only the dark current emitted from the gun triode grid can hit the screen. Two pictures of this dark current are presented in Fig. 3. The shape of the intermediate grid electrode is imaged on the screen, and the magnification of this image can be adjusted by varying the current in one of the solenoids. By measuring the total light intensity on each image and taking into account the screen light yield, the solid angle of our optical line (1/4000 of the total number of photons emitted), and the camera sensitivity, the dark current is estimated to be in the range of 10 to 100 μA.

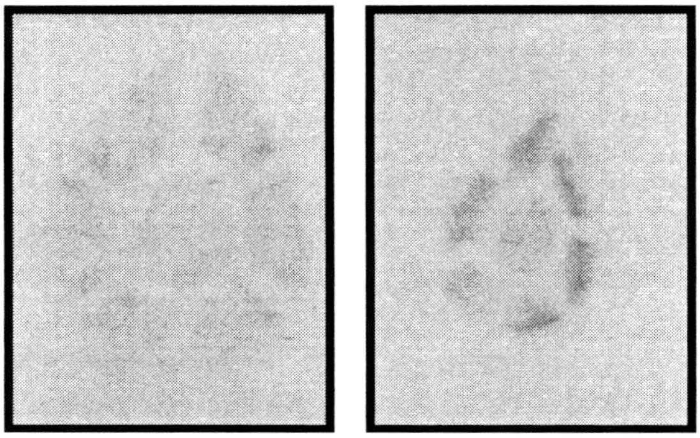

FIGURE 3. Images of the dark current flowing from the gun triode grid for two different focusing conditions.

Sign of Damage on the Screen

After some running time, a clear sign of damage is observed on the screen, as depicted in Fig. 4. The P47 deposit has been removed at least partially over a zone exposed to the beam. The supplier [7] indicates that after a beam irradiation corresponding to a total charge of 5 C/cm^2, the sensitivity of the screen would decrease by a factor 2.

Taken with a reduced beam current of 0.5 A and a 400-ns pulse length, two different beam images are shown in Fig. 5. The beam position is moved in and out of the damaged zone of the screen using a steering magnet so that the light yield can be checked. A reduction of the screen efficiency, equivalent to 27% of the nominal light yield of the screen, is measured. According to our estimations, the total charge seen by the screen should have been lower than 0.1 C/cm^2. This would indicate that this damage is the consequence of a bad manipulation sending onto the screen a beam with a much too high charge, which basically heats up and removes the phosphor deposit.

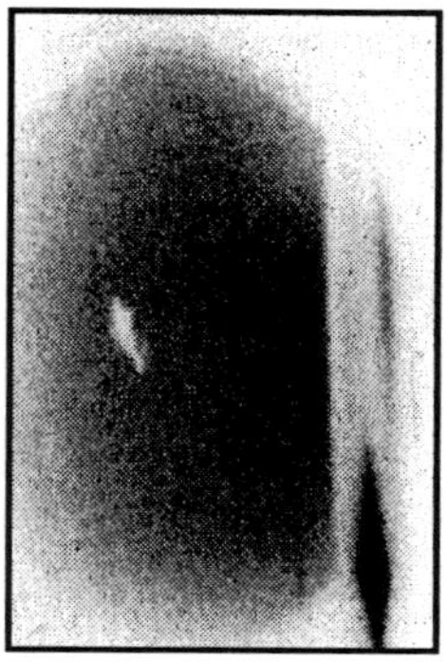

FIGURE 4. Images of the phosphor screen after high beam current irradiation.

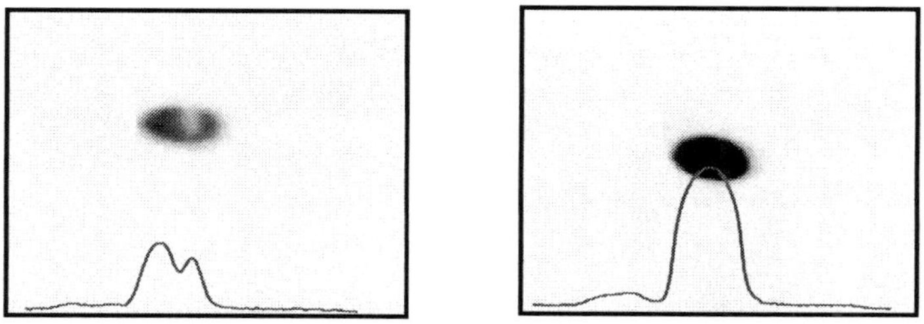

FIGURE 5. The two images are acquired with the same beam current and under identical focusing conditions. The beam has just been displaced in the down direction by few millimeters using a steerer.

MEASUREMENT WITH THE GRAPHITE OTR SCREEN

Precise Beam Size Measurements

For nominal operating conditions, an electron beam of 5.4 A current and 1.6-μs pulse duration is observed using the graphite OTR screen. Beam images obtained using the intensified camera and a 500-ns time gate are shown in Fig. 6. The pictures are taken under different focusing conditions by adjusting the current in a solenoid.

The corresponding vertical and horizontal R.M.S beam sizes are calculated and shown in Fig. 7. The error bars correspond to the fluctuations observed on 10 consecutive measurements. Beam sizes from 3 mm to 0.5 mm are measured here with good accuracy. The precision is nevertheless not very good when the beam size is big. The number of photons is limited, and the signal to noise ratio is in this case reduced.

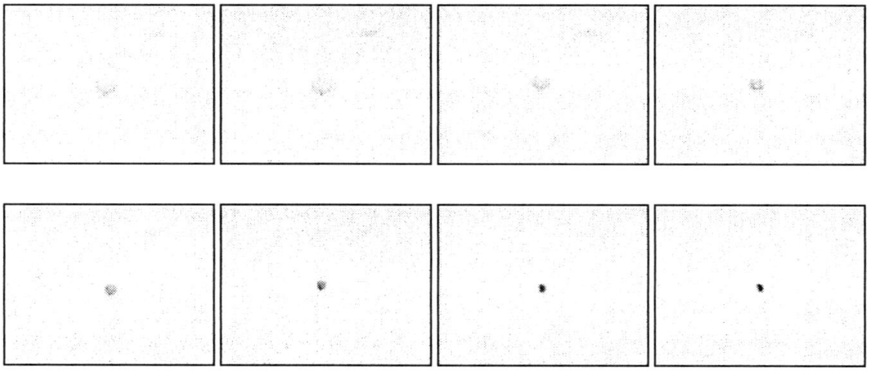

FIGURE 6. Set of beam images obtained for different focusing conditions.

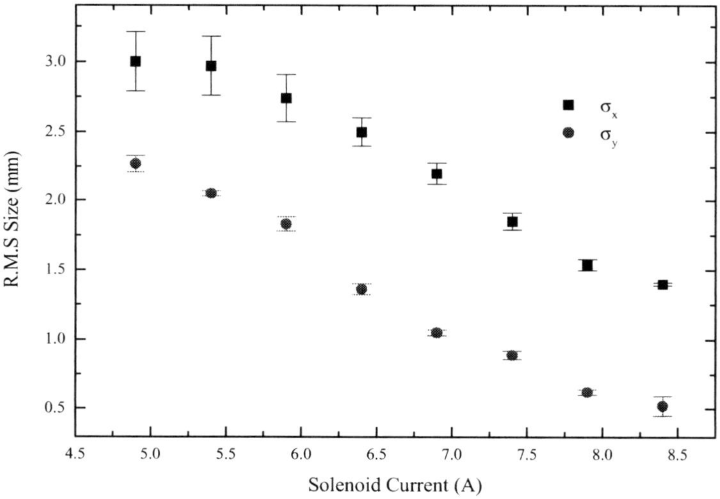

FIGURE 7. Evolution of the R.M.S. transverse beam size under different focusing conditions.

Temporal Evolution of the Beam Size within the Pulse Duration

A second set of measurements was taken using a different camera setting with a shorter gate duration. By adjusting the camera trigger, we can check the temporal evolution of the transverse beam size within the overall pulse duration. The results are given in Fig. 8, where 8 images of the beam acquired at different timings are presented. The R.M.S. beam transverse sizes, σ_x and σ_y, are computed and reported in Fig. 9.

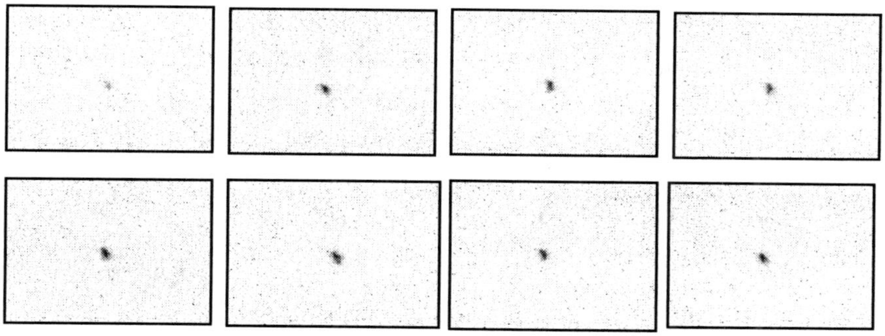

FIGURE 8. Set of 100-ns gate time images taken at different timing positions in the 1.56 μs of the pulse duration.

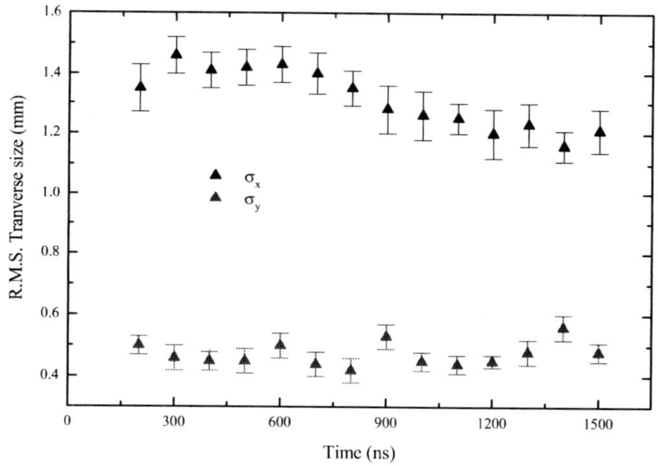

FIGURE 9. Temporal evolution of the transverse R.M.S. beam size within the pulse duration.

A reduction of the transverse size, σ_x, from 1.4 mm to 1.2 mm is measured between the beam head and the beam tail. The position of the beam centroid is also moving by some few hundred microns. Due to the lack of beam time, very few tests have been done in order to investigate the reasons of this observation. However, two possible explanations can be mentioned. The first one would be due to a small reduction of the beam energy at the end of the pulse due to a capacitive drop of the gun high voltage platform. This would be correlated to a smaller beam size onto the screen when the beam is focused by the solenoid magnetic field. A second explanation of this focusing would come from the presence of positive ions released by the graphite screen when heated up by the beam. This effect is well known when using high charge density beams and has already been observed several times [8]. In our case, some simulations of this effect have already been performed and have indicated that a 10 to 20% beam size reduction can be expected. This result is within the range of what has been observed experimentally.

CONCLUSION

A beam imaging system has been developed and commissioned for the 140-keV high charge beam produced in the CTF3 injector.

The P47 phosphor screen has shown its good efficiency to measure the 10 to 100 μA from the triode dark current. However, a sign of damage is already observed and illustrates the risk of using this type of screen on a high charge beam.

To our knowledge it is the first time that a beam imaging system for non-relativistic particles is designed to operate using the forward OTR emission from a graphite foil. The results obtained so far with this screen are very encouraging, giving the possibility to obtain precise beam size measurements, which deliver helpful information to characterize the injector performance. Moreover, giving access to the time evolution of the transverse beam sizes, these OTR measurements provide a powerful way of studying the beam stability all along the pulse duration. However, few tests have been performed so far, and to have conclusive results, more measurements will be done in the future.

ACKNOWLEDGMENTS

The authors would like to thank the CTF3 team and Northwestern University for their support. We are especially grateful to C. Vermare from CEA for help on the simulations of the ion hose instability.

This work was partially supported by the Illinois Consortium for Accelerator Research, agreement number 228-1001.

REFERENCES

1. G. Geschonke et al., *CTF3 note* 047, 2002.
2. E. Bravin, *CTF3 note* 019, 2001.
3. C. Bal, E. Bravin, E. Chevallay, T. Lefevre. and G. Suberlucq, Proceeding of the 6th European Workshop on Beam Diagnostics and Instrumentation for Particle Accelerators, Mainz, 2003, pp. 92. *CLIC note* 574-577.
4. M. E. Globus and B. V. Grinev, *Inorganic Phosphorescent Substances: Traditional and New Materials*, Kharkov: Akta, 2000.
5. R. Corsini et al., to be presented at the European Conference on Particle Accelerator, Lucerne, 2004.
6. L. Söby, Private communication, 2004.
7. http://www.proxitronic.de/.
8. E. Bravin, T. Lefevre, and C. Vermare, Proceeding of the Particle Accelerator Conference, Portland, 2003, pp. 2464. *CLIC note* 570-573.

Automated System Calibration and Verification of the Position Measurements for the Los Alamos Isotope Production Facility and the Switchyard Kicker Facilities*

D. Barr, J. D. Gilpatrick, D. Martinez, and R. B. Shurter

LANSCE-1 BDIT, Los Alamos National Laboratory, Los Alamos, NM 87545

Abstract. The Los Alamos Neutron Science Center (LANSCE) facility at Los Alamos National Laboratory has constructed both an Isotope Production Facility (IPF) and a Switchyard Kicker (XDK) as additions to the H+ and H- accelerator. These additions contain eleven Beam Position Monitors (BPMs) that measure the beam's position throughout the transport. The analog electronics within each processing module determines the beam position using the log-ratio technique. For system reliability, calibrations compensate for various temperature drifts and other imperfections in the processing electronics components. Additionally, verifications are periodically implemented by a PC running a National Instruments LabVIEW virtual instrument (VI) to verify continued system and cable integrity. The VI communicates with the processor cards via a PCI/MXI-3 VXI-crate communication module. Previously, accelerator operators performed BPM system calibrations typically once per day while beam was explicitly turned off. One of this new measurement system's unique achievements is its automated calibration and verification capability. Taking advantage of the pulsed nature of the LANSCE-facility beams, the integrated electronics hardware and VI perform calibration and verification operations between beam pulses without interrupting production beam delivery. The design, construction, and performance results of the automated calibration and verification portion of this position measurement system will be the topic of this paper.

INTRODUCTION

The LANSCE facility has constructed an Isotope Production Facility (IPF) to provide radioisotopes for diagnosis and treatment of various human medical conditions. This spur beamline starts at the 100-MeV transition region of the accelerator and transports H+ beam to a target area where samples may be irradiated and safely handled. The new beam line contains eight Beam Position Monitors or BPMs that are used to diagnose the beam's position throughout the transport and verify the beam's placement on the target/sample region during either a 10-kHz raster or static operation.

At the end of the 800-MeV LANSCE accelerator, a switchyard directs the beam to either Line D or to Line X. Line D contains such facilities as the Lujan Neutron Scattering Center and Line X contains such facilities as the Proton Radiography and

*Work supported by the U.S. DOE Office of Nuclear Energy, Science and Technology and the Office of Isotopes for Medicine and Science.

CP732, *Beam Instrumentation Workshop 2004: Eleventh Workshop,*
edited by T. Shea and R. C. Sibley III

Ultra Cold Neutron Projects. Prior to the installation of this upgraded switchyard kicker beam line, a lengthy process was required to reconfigure the switchyard beam line in order to transport the beam to either Lines D or X. With the completion of the Switchyard Kicker (XDK), LANSCE can now simultaneously provide H- beams to both beam lines by rapidly switching between them using fast kickers located in the new XDK. Three BPMs, located in the new XDK, provide beam position information to the facility operators.

SYSTEM CONFIGURATION

The beamline device that actually senses the beam position is commonly called a beam position monitor (BPM) and is pictured in Fig. 1. The BPMs used for XDK and IPF are micro-stripline image-current devices designed to sense the 201.25-MHz bunched beam that transports through the device. Figure 2 shows an XDK BPM and its associated beam line components in schematic form [1].

FIGURE 1. The 4" and 6" Isotope Production Facility BPMs. The 4" Switchyard Kicker Upgrade BPMs are very similar.

In addition to the physical BPM (also referred to as a probe which consists of four electrodes), the hardware system includes the cable plant. This consists of five coaxial cables, a cable strain relief plate that has a 4-way RF splitter/divider, and a set of four 20-dB attenuators. The five cables transport the four image-current signals from the BPM to the processing electronics and also transport the verification signal from the processing electronics to the BPM. The schematic in Fig. 2 shows a simplified block diagram of the cable plant that performs several functions. In order to maintain proper BPM operation, it terminates the downstream end of the BPM electrodes with the characteristic impedance of 50 Ω while providing greater than 70 dB of isolation between BPM electrodes. Additionally, it provides a signal injection path for the BPM and cable plant verification test. Since the components in this "loop" are linear, a single mid-dynamic-range power measurement for each cable/electrode loop path is sufficient to determine the health of any of the serial components within the loop. If a cable is inadvertently crimped or a BPM electrode is injured to the point of losing its 50-ohm characteristics, the total loop attenuation will change. With the help of a software process, this attenuation measurement is performed so that facility operators can have a quantitative method of detecting BPM and cable health. This is referred to as "single power-point system verification" or simply "single-point verification."

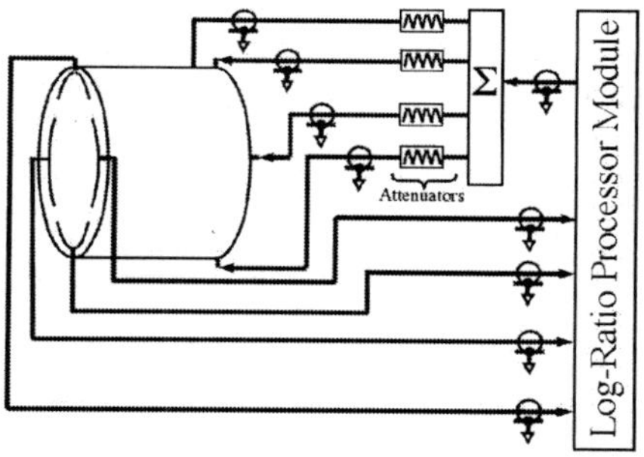

FIGURE 2. A simplified schematic of the BPM cable plant including verification test hardware.

The last component of the hardware system is the electronics processor or "Log-Ratio Processor Module" as shown in the schematic in Fig. 2. This device processes the four signals from the BPM using a calibrated log-ratio technique. The processor also sends signals to itself via a VXI daughter card calibration board to perform a full dynamic range system calibration.

The system software is comprised of two main components. The first is an IBM-clone personal computer (PC) running a LabVIEW virtual instrument (VI). The PC connects to the VXI electronics hardware system via a PCI-to-VXI MXI-3 device. This device maps portions of the VXI backplane onto the PC's PCI address space, thus enabling the LabVIEW VI to communicate directly with each VXI electronics module. The VI instructs each electronics processor how and when to collect data. This timing aspect of data collection is assisted by a pair of externally generated and controlled TTL timing signals supplied by the main accelerator control system. The VI also performs many other tasks including both the on-demand and automated calibration and verification procedures, which will be covered in more detail in a later section of this paper.

The second component of the software system is the EPICS and LANSCE facility interface that receives the process variable data served by the LabVIEW VI. This is accomplished via an Ethernet-based local area network (LAN) and displays these data in a fashion that the LANSCE facility operators can effectively use. The EPICS interface is also responsible for the delivery of the two triggers mentioned above. One is required to sample the data at a particular time within the LANSCE beam operation macropulse, and the second to acquire automated calibration and verification data.

STANDARD CALIBRATION PROCEDURES

The predecessor of the current BPM LabVIEW VI was designed to monitor the BPMs at the Low-Energy Demonstration Accelerator (LEDA) at LANL. It monitored the 15 BPMs spread throughout the LEDA beamline and reported the values to the EPICS control system. Due to varying conditions it was necessary to calibrate each BPM electronics module at least once a shift (about 8 hours). This required an operator to stop beam operations and perform a somewhat lengthy series of BPM tests. This typically took one operator the best of about 15 minutes. This was viewed as a minor annoyance all the way up to a full-blown hassle depending on schedules, conditions, and personnel.

The basic goal of the calibration is simple. The ideal response of a logarithmic amplifier can be stated as

$$V_{out} = K \cdot \log\left(\frac{v_{in}}{v_{ref}}\right), \tag{1}$$

where v_{in} is the amplitude of the input signal, v_{ref} is a reference voltage, and K is an amplitude constant [2]. Both v_{ref} and K are dependent on the particular log amp chosen. Since the log amp is non-ideal, there are small perturbations in its transfer function and distortion effects throughout its dynamic range. By using digital error-correction the non-ideal performance in the log amps and other system non-linearity can be taken out of the overall transfer function of the system. The system calibration provides a method to determine the error correction values. These values affect the position output by modifying a set of lookup tables referred to as RAMLUTs (Random Access Memory Look-Up Tables). The analog front end (AFE) of the electronics module processes the raw beam signal through various stages including the logarithmic amplifier ending up at a 14-bit A/D converter. The least significant twelve bits are used to address a 4096-location RAMLUT. See [3] for more details. The 12-bit digitized lobe signals from the log amp are used as an address for the 4096-point deep RAMLUT. The 12-bit wide output word is used as the new corrected lobe value and fed to a local DSP for acquisition and storage until requested by the LabVIEW VI.

The system calibration is achieved by placing known values into the AFE inputs for each BPM electrode and measuring the generated output signal at the DSP. The RAMLUTs are bypassed for this measurement. This calibration output data is compared to the theoretical values that would be obtained at the DSP with the known input power levels. These theoretical values take into account the individual characteristics of the attenuators used in the calibration process. They are known a priori by measurements made in the laboratory using a precision network analyzer.

Since the response of the system transfer function depends so much on the dynamic range of the log amp, the calibration must be done over a large range of power levels. This is accomplished by means of a bank of attenuators on the calibration board. A fixed amount of RF power is generated on the calibrator board and then attenuated in 1-db steps over a range of 90 steps. With the proper amount of input power, it is possible to calibrate over the full dynamic range of the log amplifiers. See [4] for details on the log amp's dynamic range. Once this data is collected, a linear interpolation routine is used to generate the remaining values in the 4096-deep RAMLUT. This effectively fills in the gaps between the measured 1-dB steps. All that is left is to load the new values into the RAMLUTs and give them control of the input data stream. This is referred to as a "full dynamic range calibration."

A test routine was written to check how well the RAMLUTs worked once loaded. This test calibration routine or "full dynamic range calibration test" is performed immediately after a successful calibration and works as follows. A range of signal powers is placed into AFE inputs for each BPM electrode by means of the attenuators on the calibrator board as with the calibration routine. The values for these signals are measured at the DSPs. In the test routine case, the RAMLUTs are used to process the data. This measured data is subtracted from the theoretical values that would be obtained at the DSP with the known input power levels. These values are plotted versus attenuation, thus obtaining an estimate of system performance over the full dynamic range of the log amp.

ON-DEMAND VERSUS AUTOMATED OPERATIONAL MODES

As mentioned above, the normal procedure for initiating a calibration operation (either single-point verification, full dynamic range calibration, or full dynamic range calibration test) is at the request of an accelerator system operator. This is only done when there is no accelerator beam present. It can be initiated from either the PC running LabVIEW or from an operation's EPICS terminal. This form of user calibration request is referred to as "on-demand mode."

The system presented in this paper addresses the problems encountered with the on-demand mode as it can operate in an additional "automated mode." This mode takes advantage of the pulsed nature of the beam. The LANSCE beam uses at most a few hundred microseconds out of every 8.3 milliseconds of beam time. This leaves the BPM LabVIEW VI plenty of time to perform various calibration procedures between beam pulses so as not to affect data acquisition operations. It is only required that the system be able to acquire the data from the next beam pulse. This usable time space between pulses is referred to as a "beam gap" or just "gap" for short. The length of the gap is assumed to be the worst-case condition. At the maximum operational beam repetition rate of 120 Hz this allows at least 8.3 milliseconds between pulses.

The engineering problem presented by these requirements would be solved using a dedicated real-time processor. It is quite a bit tougher using LabVIEW running on a PC. Note that National Instruments also has a real-time version of LabVIEW (titled LabVIEW RT), which is not being used due to its need of additional expensive hardware modules. By utilizing a hardware TTL input located on the PC-VXI interface card (VXI slot 0), the LabVIEW VI can determine when the beam pulse is finished and it is safe to switch to calibration data acquisition and processing. Once this input is received, calibration processing can begin and continue until just before the next beam pulse. At this point the system returns to beam acquisition mode, obtains the next pulse, waits for the beam gap indicator pulse and the cycle starts all over again. Note that it is up to the operators to change the timing of this pulse if the length of the beam changes. During actual periods of automated calibration, the BPM system has to guarantee the main control system a display update of at least 5 Hz. The beam gap indicator timing pulse is obtained from the main accelerator timing system and is altered to deal with changes in beam pulse length. This timing pulse is referred to as the AutoCal (automated full dynamic range calibration)/AutoCalTest (automated full dynamic range calibration test)/AutoVerification (automated single-point verification) trigger and is associated with Delay_2 as discussed in the next section.

Timing Considerations

It is very important to understand the timing issues related to the BPM systems. The following figure does a good job of showing the relationship between the required BPM timing signals and the beam signal itself.

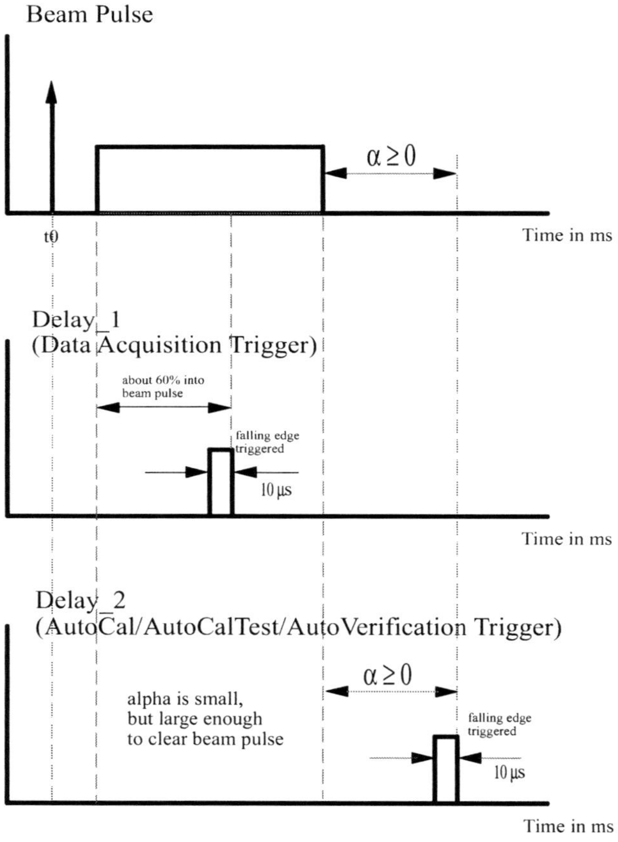

FIGURE 3. Timing diagram required for automated calibration.

The main beam pulse is triggered by the t_0 pulse from the master timing system (or some fixed delay after t_0) where t_0 is the main accelerator trigger.

The basic premise of Delay_1 and Delay_2 is simple. Delay_1 is set so that the falling edge occurs approximately 60% into the beam pulse. [Note that both Delay_1 and Delay_2 feed falling-edge triggered TTL inputs.] This is the time at which data acquisition takes place on the BPM VXI boards. The signal is fed directly into the front panel of the BPM electronics module. The module is programmed to acquire one data point for each BPM electrode (Top, Bottom, Left, and Right).

Delay_2 represents the time at which the beam pulse is guaranteed to be over and hence acceptable to commence automated calibration and verification functions. Delay_2 is chosen based on the beam pulse width and α. This is set to some small value after the end of the beam pulse that gives the system enough time to collect its automated data, but makes sure that the AutoCal/AutoCalTest/AutoVerification trigger samples after the beam is gone. Around 10 to 20 μs seems to work well in actual accelerator operations. If this trigger overlaps the beam pulse, the autoverification routine will give erroneous values. If it is delayed too long, the automated routine will not have enough time to complete execution before the next beam pulse.

The LANSCE Master Timing System controls the values for Delay_1, Delay_2, and their associated widths. Only the delays can be changed. The widths are hard coded to 10 μs.

Automated System Operation in Practice

Automated Calibration consists of a repeating cycle. The cycle starts at Time = 0 and executes each function (Single-point Verify -- actually a single-point in the dynamic range of the calibrator, but averaged over many beam gaps, Dynamic Calibration, and Dynamic Calibration Test) at the programmed time in the cycle. The routine contains a maximum cycle timeout. This lets the operator set a maximum cycle time, which is used if the main cycle ever gets out of sync or has problems. The cycle will start over after this timeout expires, regardless of the current operation. The default automated cycle is set to repeat on about a one-hour period.

Figure 4 shows plots of the deterministic or systematic error in dB for each lobe of a typical IPF BPM over the full range of the calibration attenuators. The plots are fairly close to zero and stay that way until around 60 dB of attenuation. The results are similar for all four lobes and are very consistent with similar plots generated by properly functioning and calibrated BPMs. Figure 5 shows data from the same BPM in the form of the averages (which reveals the deterministic error) and standard deviations (which reveals the random error) of the horizontal and vertical errors (given in mm) plotted against beam current in mA.

The new additions to the BPM systems have worked extremely well in actual beam operations. The automated routines run without operator assistance and produce results of a similar quality to those of the on-demand routines.

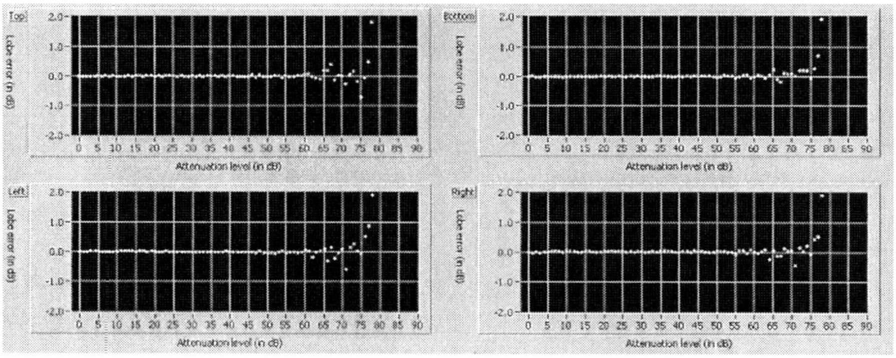

FIGURE 4. Test calibration response in dB of attenuation versus deterministic output error for the four electrodes of a typical IPF BPM system.

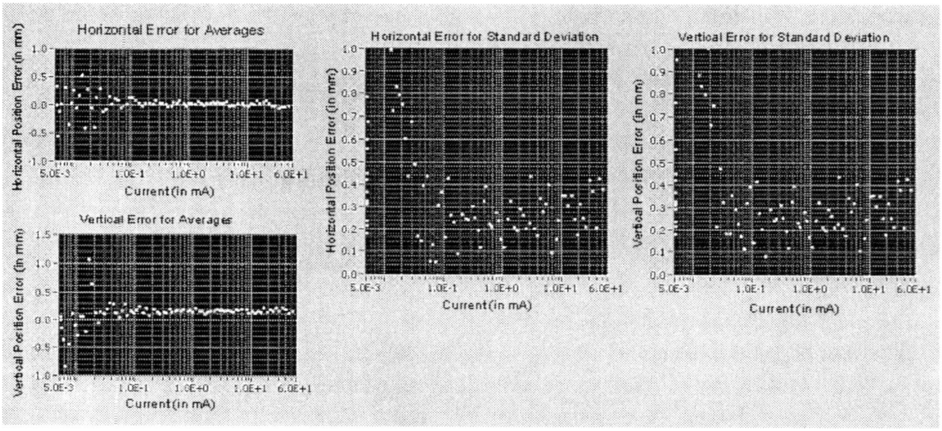

FIGURE 5. Averages and standard deviations of the horizontal and vertical errors (given in mm) plotted against beam current in mA. The left two plots (errors in the averages) show the deterministic error while the two plots on the right (errors in the standard deviations) show the random errors.

REFERENCES

1. J. D. Gilpatrick et al., "Beam Position Measurements at Los Alamos: Isotope Production Facility and Switchyard Kicker Upgrade," Proceedings of the 2003 Particle Accelerator Conference, Portland, Oregon.
2. C. R. Rose et al., "Description and Operation of the LEDA Beam-Position Intensity Measurement Module," Proceedings of the 1997 Particle Accelerator Conference, Vancouver, B.C., Canada.
3. D. Barr et al., "LEDA Beam Diagnostics Instrumentation: Beam Position Monitors," Proceedings of the 2000 Beam Instrumentation Workshop, Boston, Mass.
4. R. B. Shurter et al., "BPM Analog Front-End Electronics Based on the AD8307 Log Amplifier," Proceedings of the 2000 Beam Instrumentation Workshop, Boston, Mass.

Experience with the SNS* LabVIEW/EPICS-Based Diagnostics Instruments

Willem Blokland and Saeed Assadi

Spallation Neutron Source, Oak Ridge National Laboratory, P.O. Box 2008, Oak Ridge, TN 37831, USA

Abstract. The SNS Diagnostics Group uses rack-mounted PCs running LabVIEW and EPICS to implement its instruments as network attached devices. Many of these instruments, such as wire scanners and beam position monitors, come from the partner labs. The final integration and testing is done at SNS. We have now gone through several commissioning runs with success. This paper describes the integration, the commissioning, and the physics performance of the devices.

INTRODUCTION

The Diagnostics Group predominantly uses rack-mounted PCs running LabVIEW to acquire and process the data and EPICS IOC to communicate with the SNS control system. The data-acquisition boards are either off the shelf or custom made. A rack with PCs is shown in Fig/ 1. The PCs are configured as network attached devices (NADs) see [1]. The NAD concept prescribes that each pickup or sensor to have its own resources, such as timing, data acquisition, and processing. This means that a single failure of the NADs hardware only affects one particular channel and not multiple channels.

FIGURE 1. BPM systems mounted in rack.

*SNS is managed by UT-Battelle, LLC, under contract DE-AC05-00OR22725 for the U.S. Department of Energy.

CP732, *Beam Instrumentation Workshop 2004: Eleventh Workshop*, edited by T. Shea and R. C. Sibley III

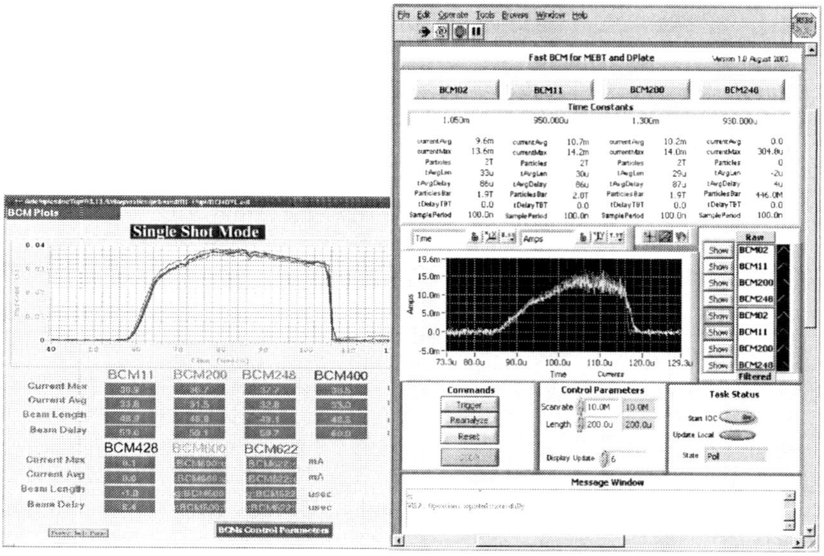

FIGURE 2. The Fast BCM screen in the control room is on the left, and the LabVIEW application with Shared Memory IOC on the instrument is in the middle and on the right.

An example of a NAD program is shown Fig. 2. On the right is the front panel of the Fast Beam Current Monitor LabVIEW program. The program acquires the data from a 1 GS/s Acqiris digitizer, then analyzes the data and writes the results to the Shared Memory. The IOC, shown in the black window, accesses the Shared Memory and communicates the results to the rest of the control system. This data are displayed on the console screen shown on the left.

Table 1 shows the commissioned diagnostic instruments. The table shows the name of the instrument, the number installed in the MEBT, the number installed in the DTL, the type of EPICS software (IOC or channel access only), the hardware (commercial or custom), and comments. Not listed in the table is the custom-made timing card that will be installed, once ready, in all NADs with data acquisition. Currently we are using timing signals from a VME timing decoder.

Several instruments have been completely designed and implemented at SNS. The need for these instruments came after the original design plans. Many of these instruments are located in the Diagnostics Box (Dbox) in the MEBT. The Dbox replaces the anti-chopper, which is no longer needed. Another instrument built at SNS is the laserwire. The laserwire replaces the wire scanners in the superconducting linac (SCL). In total we have dealt with 39 instruments of 18 different types.

TABLE 1. Commissioned diagnostics instruments.

Instrument	MEBT	DTL	Type	Hardware	Comments
Beam Position	6	5	IOC	custom	1 Hor & 1 Ver plane
Wire Scanner	5	3	CA	Commercial with custom pre-amps	3 wires per fork
Beam Stop	1		IOC	Commercial	Dbox
Beam Aperture	1		IOC	Commercial	
Faraday Cup MEBT	1		IOC	Commercial	
Arc Detector	1		IOC	Commercial	2-4 channels/node
Neutron Detector	1		CA	Berkeley	loan from Berkeley
Source Current Logger	1		IOC	Commercial	2-4 channels/node
Fast Beam Current	1	1	IOC	Commercial	1-4 channels/node
Beam Current Prototype	1		IOC	Custom	2 channels
Laserwire Prototype	1		--	Commercial	No EPICS IOC yet
Faraday Cup DTL		3	IOC	Commercial	
Halo Scraper/Beam Stop		1	IOC	Commercial	
Fast Beam Loss		1	IOC	Commercial	2 channels
Beam Loss Prototype		1	IOC (VME)	Commercial	5 channels
Bunch Shape Monitor		1			
Emittance		2	IOC (VME)	Commercial	32 wires per system
Physics Test Node	1		IOC	Commercial	Test Java apps

INTEGRATION

The SNS team is the primary team responsible for the integration of the instruments into the accelerator complex. This paper defines the integration to include the receiving and installation of the electronics and computer, software restructuring, PC management, and off-line testing but excludes beam line device installation, cable pulls, rack infrastructure installation, and cable termination (see Table 2).

TABLE 2. Integration activities.

Step	Work	Tools	Documentation
Installation	Receiving	Forklift, etc	Traveler
	Installing	Hands, power tools, etc	Diagrams
Software	Structuring program	SNS Template, manuals, examples, clone tool	Auto documentation generator
	Add analysis/acquisition	Libraries	
PC Management	Create hard disk image	XP Embedded toolkit	XPe files
	PC management	Altiris, Timbuktu, Oracle, Virus scanners	Oracle database entries
Test	Test electronics and code	Simulation code, LabVIEW Client, signal generator	Hand-off document
	Calibration	(built-in) signal generator	

The software integration includes restructuring the program to match the SNS template if this has not already been done. This template includes the Shared Memory interface and has tools to automatically generate the files the IOC needs to initialize and instantiate the PVs. Other tools generate program documentation and can rename the whole program hierarchy to create a new project from an example or different project.

Special attention is given to the management of the NADs. Due to the large number of NADs (more than 300), tools are necessary to efficiently maintain our instrument. We can choose from a variety of management tools as our NADs are PC-based. We chose to use Altiris, which is also used for desktop PC management at SNS. We use XP Embedded as the operating system, as it supports removal of system modules, thereby reducing OS footprint and virus susceptibility. The Altiris server can remotely reformat the hard drives and install the OS and the program software on multiple similar systems in about 15 minutes, see Ref. [2].

Many of the instruments include testing software. For example, the wire scanner comes with software to put the electronics in a special mode to test the operation of the electronics. The beam current monitor has a simulation to test the overall operation of the program analysis and log without needing actual beam. Table 3 summarizes the integration activities per device.

To further simplify the integration of the various instruments, SNS and Brookhaven National Laboratory (BNL) share a common rack mount PC for all their purchases.

TABLE 3. Integration activities per instrument.

Instrument	Activities
Beam Position	LANL provided tested electronics and software package. SNS restructured program to template structure. Both labs commissioned.
Wire scanner	BNL provided actuator. BNL and SNS cooperated on fork improvements. LANL provided electronics and software. SNS added analysis routines. SNS and LANL commissioned.
Beam Stop	All activities by SNS.
Beam Aperture	All activities by SNS.
Faraday Cup MEBT	All activities by SNS.
Arc Detector	All activities by SNS.
Neutron Detector	LBNL provided turnkey system. SNS installed system.
Source Current Logger	All activities by SNS.
Fast Beam Current	BNL provided toroid and filter program, all other activities by SNS.
Beam Current Prototype	BNL provided toroid, electronics, and software program. SNS and BNL installed and tested.
Laserwire MEBT	All activities by SNS (initial laserwire studies by BNL).
Faraday Cup DTL	LANL provided the hardware and demo software. SNS wrote NAD software, console screens, and commissioned the device.
Halo Scraper/Beam Stop	LANL provided the hardware and demo software. SNS wrote NAD software, console screens, and commissioned the device.
Fast Beam Loss	All activities by SNS (detectors delivered by BNL).
Beam Loss Prototype	SNS and BNL wrote VME software, console screens, and commissioned.
Bunch Shape Monitor	INR delivered instrument and commissioned. SNS added IOC interface.
Emittance	SNS wrote VME software, console screens, and commissioned.
Physics Test Node	SNS wrote NAD software.

COMMISSIONING

Commissioners and operators expect not only fully functional instruments but also fully commissioned diagnostics prior to beam startup. This idea is a novel idea but unrealistic. A number of steps are taken to prepare new instruments without requiring beam. As stated in the integration section above, simulation of the beam signal within the NAD thoroughly debugs the logic and analysis routines of the program. Automatic pulse finding routines allow the timing to be approximately right, rather than exactly right. This has helped us measure the very first beam pulse through the first drift-tube linac tank.

In general, most diagnostic systems came up quickly and successfully last year and this year. The LabVIEW and Shared Memory IOC based systems have shown excellent reliability with no crashes during operation. The problems we ran into during commissioning include the following:

1. Loss of all wires in one wire-scanner of DTL-2; one wire lost in a MEBT wire scanner. The other eight wire scanners were sufficient to complete the commissioning without compromising physics needs.
2. One block of the radiation shield interfered with a Faraday Cup actuator. The shielding block was moved away from the actuator.
3. The remaining NADs with the ActiveX/Shared Memory Interface had update problems and required a reboot. The update problems have been fixed, and the ActiveX will be replaced with the IOC version as planned.
4. Noise on signal cables. Various improvements have been made, but we expect this to be an ongoing problem with gradual improvements through better grounding, better shielding of RF and high-power components, rerouting of signal or power cables, and use of analog and digital filters.

Fine tuning of the programs, e.g., adjustments of the analysis and modifications requests, can be handled efficiently because all NADs run the full development version of LabVIEW. The data flow within the NAD is instantly visualized by opening the program's subroutines. The data are graphed as they are acquired and processed. Issues such as noise can be quickly investigated and documented by adding spectrum analysis and data-logging to the NADs during commissioning.

The primary remaining issues are the EMI noise environment, as mentioned above, and better GUI interface of the NAD systems to allow complex instrument run without expert involvement.

PHYSICS PERFORMANCE

This section discusses performance of the diagnostic instruments as it relates to the physics requirements. The data are obtained using the EPICS data-archiver, Java data-archiver, and the EDM screens, from the instruments directly, and from the LabVIEW-based Channel Access loggers. All instruments are designed to meet the physics requirements listed in Table 4. The table shows the requirements and the analyzed results to this date, but further analysis of the data is under way. The piecewise commissioning allows us to implement a few systems at a time and test their performance with beam on. We have the flexibility to improve and install the next-generation devices with minimal cost impact at the next commissioning milestone. For example, the phase measurements of the BPM have improved from many degrees to smaller than one degree.

TABLE 4. Physics requirements and measurements.

Device	Location	Range	Accuracy Required	Experimental Results
BPM position	MEBT	+/-0.5*apert	+/-0.5 mm	+/-0.2 mm
	DTL	+/-0.5*apert	+/-0.125 mm	+/-0.125 mm
BPM phase	MEBT	+/-180 deg	+/-2 deg	+/-0.1 Deg
	DTL	+/-180 deg	+/-2 deg	+/-0.2 Deg
BLM (35 kHz)	Ion	30-2.5e5 rem/h		2.40%
Neutron detector	DTL	30-2.5e5 rem/h		3%
Faraday Cup	MEBT	50 mA	0.3%	0.25%
	DTL	50 mA	0.3%	0.25% (@20 mA)
Wire scanner	MEBT	+/-15 mm	10%	0.15 mm
	DTL		10% (+/- 0.4 mm)	0.63 - 50. μA 0.13 mm jitter
Harp	DTL	+/- 10 mm		0.08 mm
Laser wire	MEBT, DTL	+/- 15 mm		Measure out to 3 sigma
BCM	MEBT-to-HEBT	15 mA - 52 mA	+/- 1%	1%
	Ring-RTBT	15 mA-100 A	+/- 1%	--

Figure 3 shows the zero-centered histograms of two MEBT and two DTL BCMs with the average value and standard deviation. The standard deviation shows a minimum jitter (jitter includes noise and real beam variation) of about 0.13 mA. For a 15-mA beam this is less than 1%.

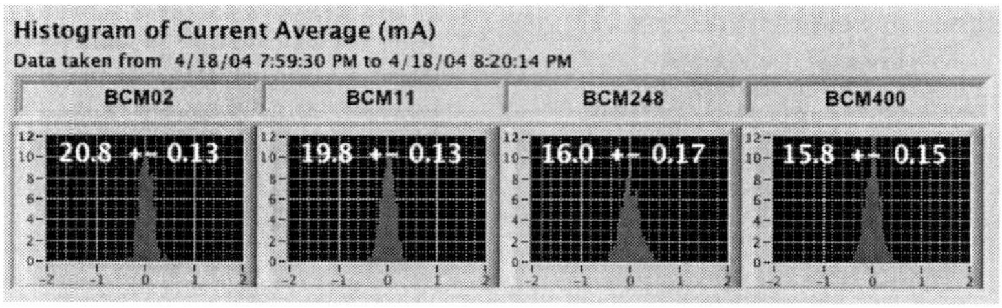

FIGURE 3. Histogram of four BCMs.

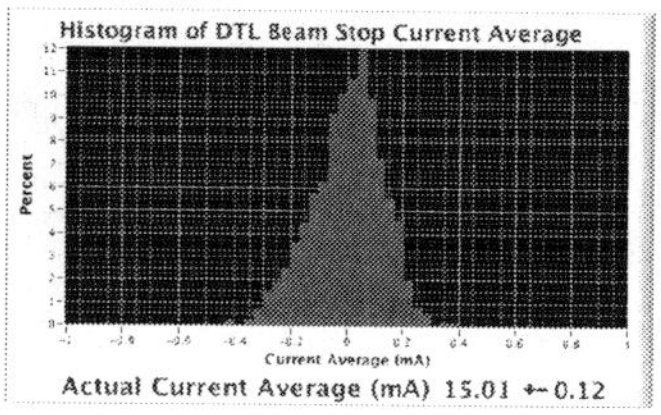

FIGURE 4. Histogram of the data from the beam stop at end of DTL3.

Figure 4 shows the histogram of the measured beam current at the beam stop. In general this signal shows slightly less noise than the BCMs and has a total jitter of less than 1%. Figure 5 shows the phase histograms of two MEBT and two DTL BPMs along the diagonal of the matrix and the phase difference histograms at the other elements of the matrix. It shows that the phase jitter due to the electronics is at most 0.10 degrees. The phase difference between BPM203 and BPM209 shows a double peak due to cavity RF field variations. We presume that these variations are due to the coarse control of the cooling water for the cavity.

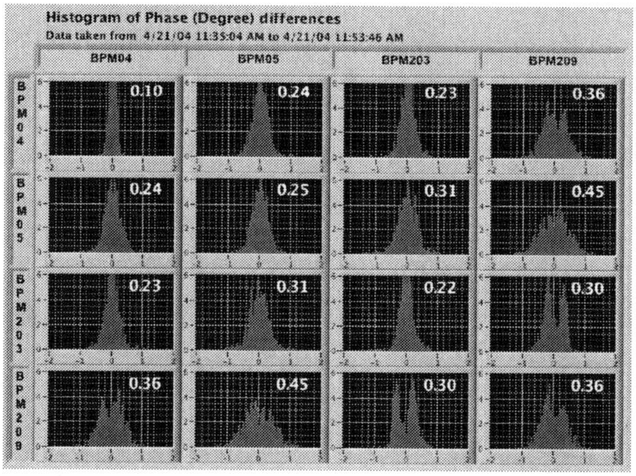

FIGURE 5. Histogram of the phase differences between BPMs.

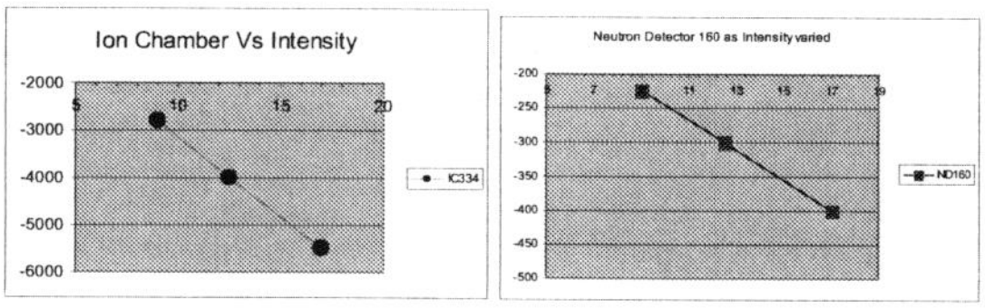

FIGURE 6. Beam loss signal of ion chamber (left) and neutron detector (right) versus beam intensity.

Figure 6 shows initial data from one of the ion chambers and neutron detectors. It confirms a linear relation of their signal intensities to the beam intensity.

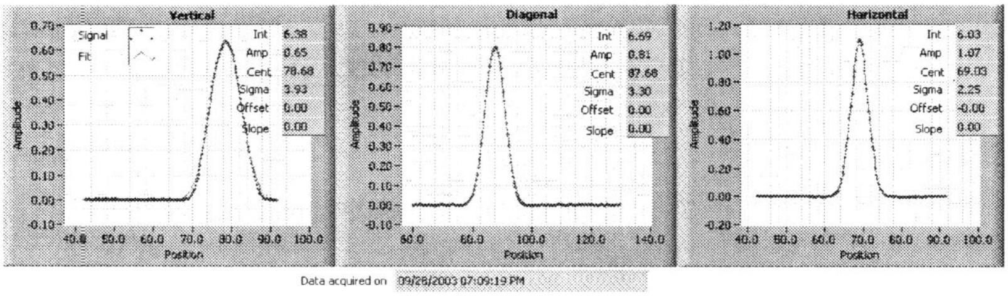

FIGURE 7. Example of the wire scanner profile points (blue) and the Gaussian fit (red).

The wire scanners produce very nice profiles. We found the measured widths to correspond to the model within 10% and a variation in the beam width measurements with no changes to the settings of the accelerator of less than 2% (see Fig. 7).

SUMMARY

We have successfully deployed 19 different instruments. Data analysis so far shows that all are within the physics requirements. All NADs, with the exception of the wire scanners, have been converted into the SNS template with the Shared Memory IOC. All NADs are being equipped with the XP-embedded operating system and remote management to ease deployment and maintenance.

FUTURE

To fully implement the NAD concept, all instruments are going to be equipped with their own timing card. We plan to have timing cards running for the next run. Further analysis of the data is under way to further specify the performance of the instruments. The next run will see more instruments of the type used for the last run but also a new MEBT emittance scanner.

ACKNOWLEDGMENTS

The authors wish to thank the many people at all the involved labs for their tremendous effort to get the SNS diagnostics to this stage.

REFERENCES

1. W. Blokland, T. Shea, and M. Stettler, "Network Attached Devices at SNS," Proceedings of DIPAC 2003, Mainz, Germany, May 2003, pp. 146-148.
2. D. J. Murphy and D. J. Purcell, "Large-Scale PC Management and Configuration for SNS Diagnostics," this proceedings, BIW04, Knoxville, USA, May 3-6 2004.

Solid-State Video Camera for the Accelerator Environment

R. L. V. Brown, B. H. Roster, and C. K. Yee

*Stanford Linear Accelerator Center, Menlo Park, CA 94309**

Abstract. Solid-State video cameras employing CMOS technology have been developed and tested for several years in the SLAC accelerator, notably the PEPII (BaBar) injection lines. They have proven much more robust than their CCD counterparts in radiation areas. Repair is simple, inexpensive, and generates very little radioactive waste.

INTRODUCTION

Ever since SLAC was built, vacuum tube video cameras have been used to aid the machine physicists and machine operators in their endeavors. When CCD cameras were introduced, tube cameras quickly faded from the marketplace and SLAC was left with a problem.

SOLID-STATE CAMERAS

CCDs are notoriously susceptible to accelerator radiation. They soon develop "ON" pixels (either steady or blinking) which, as far as we can tell, remain forever. Blinking pixels are especially difficult to mask using background subtraction.

CMOS cameras made a brief appearance in the early 90s but, like tube cameras, they couldn't compete with the sensitivity, resolution, and compactness of CCDs.

Judging whether a video camera is good or bad is somewhat more subtle and subjective than judging, say, a fuse. In reality, a camera in the housing rarely "dies," but instead the picture degrades as the sensor is irradiated. Eventually the image becomes too useless or too annoying to watch. This pivotal point (when a good camera goes bad) is dependent upon too many factors to be truly objective. Thus the evidence presented is more anecdotal than quantitative.

Cameras degrade in different ways: Ultricons (the vacuum tubes in most cameras at SLAC) develop "streaks" which eventually wash out the desired image. CCDs may be the most annoying as they develop damaged pixels (either saturated "on" or blinking). CMOS cameras tend to fade uniformly until there is insufficient contrast to be useful. All damage described is permanent.

One point which is important to note is that suppliers, with the best of intentions and in the interest of abiding by some standard, sometimes rate their cameras in terms of MTBF/RAD. The important thing to remember when reviewing these numbers is

*Work supported by the U.S. Department of Energy under contract DE-AC03-76SF00515.

CP732, *Beam Instrumentation Workshop 2004: Eleventh Workshop*,
edited by T. Shea and R. C. Sibley III

that they are usually based upon exposure to Cobalt-60, which of course is a very different type of radiation than a particle accelerator has to offer.

We measure radiation dosage using liquid-filled gamma dosimeters, in pairs, with overlapping ranges. We found that exposure to 5200 Rads in the Cobalt-60 well had no lasting effects upon a CCD camera. In fact, not one "ON" pixel was created. When an identical camera with identical dosimeters was placed near a 1.2-GeV electron storage ring for a short time, we measured only 2200 Rads when it was removed. This camera was severely damaged, with a large number of pixels on or blinking. We then tried a third camera in the Plutonium-Beryllium source well for 24 hours and observed three or four "ON" pixels when it was removed.

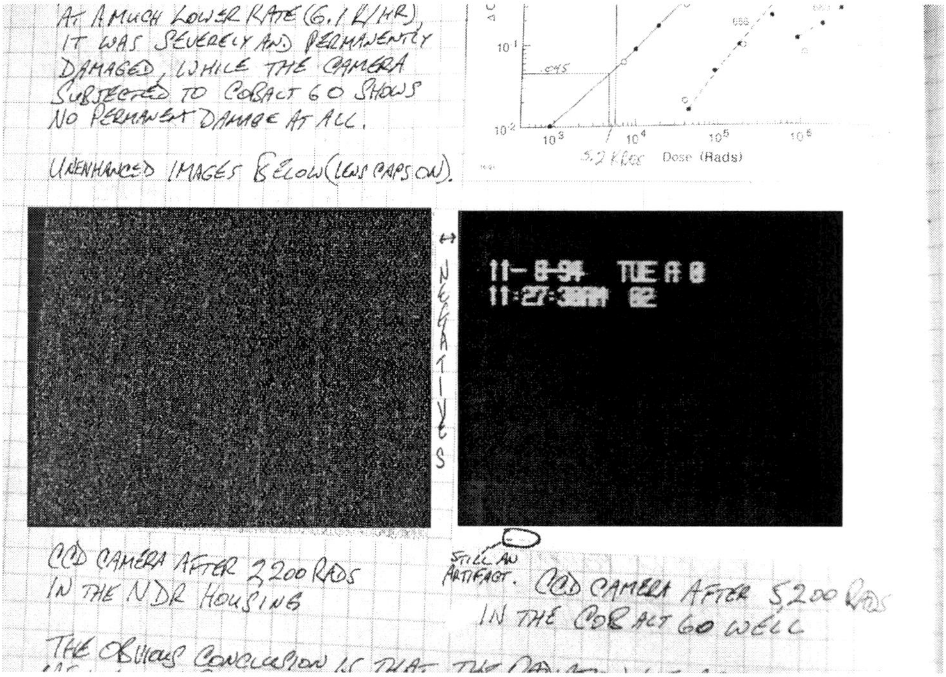

The above experiments do not prove that neutrons are responsible for damage to cameras in the accelerator. However it is interesting to note that, in a 1990 lecture, Sherra E. Kerns commented that "MOS devices are 'immune:' to neutron damage." [1]

The CMOS camera we developed was designed to live a relatively moderate but useful lifetime in low-to-medium radiation areas. In high radiation areas we actually pipe the photon image to the surface and watch from there. [2] This is an expensive and complex solution.

SLAC is approximately synced to the power grid. Strange things happen when observing a 1-Hz beam with a crystal-controlled camera running at 59.9 Hz. The beam seems to fade in and out, sometimes prompting the operator to search for problems in the accelerator. It is only an electro-optical illusion, however, and one solution is to line-lock the camera to prevent the beat-frequency effect. We chose to pick off the ripple voltage from the AC adapter, amplify it with a 120-Hz active filter, apply it to a comparator against its own average, divide the 120 Hz by 4, trigger a variable time delay (controllable either locally or remotely by DAC), and then create a 30-Hz TTL pulse which line-locks the camera. This is all done in the interface box.

The cameras we built have a Delrin™ housing (DuPont de Nemours and Company), a CS mounting (C mounting with a 5-mm spacer), and a single S-Video cord for power and signals. Oversized walls of Delrin permit replacing existing cameras. They are tripod mountable and can support relatively heavy lenses. The chips are wholesaled by Omnivision (OVT.COM) and products using these chips are retailed by www.smartvisionproducts.com.

We repair our cameras in-house since they may have been activated in the accelerator and cannot be sent off-site. With Ultricon cameras the cost of repair was nearly $1000. The last Ultricon tube sold by SLAC Stores cost over $650. In addition the sync and video chips were always replaced at every opportunity. Add in labor for a moderately difficult repair job, then labor for calibration and alignment, and it might seem more sensible to buy a new camera. However, a new camera meant about a Kilogram of mixed waste added to our collection. (970 gms). We chose to repair them ourselves, generating about 50 gms of waste.

Repairing the CMOS cameras generates about 7 gms of waste, and the entire camera weighs about 150 gms. This allows us to mount the camera on a mechanically anchored lens rather than the other way around. Repair is well under $100 per B&W.

The NTSC chips we use are true-interlaced and are also available in Color, PAL, VGA, Scientific Square Pixel, etc., with a wide variety of packaging choices. There are other vendors available; a search engine will be useful to help locate a CMOS device which meets your specific needs.

SUMMARY

CMOS Cameras outlast CCD Cameras when installed near the SLAC beam lines.

ACKNOWLEDGMENTS

These people and more were instrumental in making this camera possible: Mer Baldoza, Carolyn Burton, Earl Hamner, Frank Hoang, Young Kim, Robert Simon, Amalia Russell, Rudy Toledo, and Joseph Yu.

REFERENCES

1. S. E. Kerns, "Radiation Effects on Electronics," Vanderbilt University and University Consortium for Research on Electronics in Space, 1990.
2. F.-J. Decker, "Improved Controls and Optics for the Linac Off-Axis Profile Monitors," SLAC Internal Memo, July 9, 1990.

Beam Diagnostics for the BNL Energy Recovery Linac Test Facility*

Peter Cameron, Ilan Ben-Zvi, Michael Blaskiewicz, Michael Brennan, Roger Connolly, William Dawson, Chris Degen, Al DellaPenna, David Gassner, Martin Kesselman, Jorg Kewish, Vladimir Litvinenko, Joseph Mead, Brian Oerter, Tom Russo, Kurt Vetter, and Vitaly Yakimenko

Brookhaven National Laboratory, Upton, NY 11973, USA

Abstract. An Energy Recovery Linac (ERL) test facility is presently under construction at BNL. The goals of this test facility are first to demonstrate stable intense CW electron beam with parameters typical for the RHIC e-cooling project (and potentially for eRHIC), second to test novel elements of the ERL (high current CW photo-cathode, superconducting RF cavity with HOM dampers, and feedback systems), and finally to test lattice dependence of stability criteria. Planned diagnostics include position monitors, loss monitors, transverse profile monitors (both optical and wires), scrapers/halo monitors, a high resolution differential current monitor, phase monitors, an energy spread monitor, and a fast transverse monitor (for beam break-up studies and the energy feedback system). We discuss diagnostics challenges that are unique to this project, and present preliminary system specifications. In addition, we include a brief discussion of the timing system.

INTRODUCTION

The motivation for the development of the Energy Recovery Linac at BNL is to have high-quality, high-current electron beams (with acceptable RF power requirements) for high-energy bunched-beam electron cooling of ions [1], and possibly for an electron-ion collider [2]. By high quality, we mean beams with minimal emittance, and whose profiles are sharper than Gaussian. The ERL approach proposes to accomplish this not by relying upon equilibrium with radiation to establish the emittance (as in the case of storage rings) but rather by preserving the low emittance (typical of a linac) of a carefully designed source during acceleration for a single pass through the interaction region, then recovering the beam energy with a second pass through the acceleration cavities. This method permits significant reduction in beam emittance compared to a storage ring, and also permits high current without requiring unreasonable RF power. Figure 1 shows a possible layout of the test facility (others are under consideration). Expected parameters are shown in Table 1.

---

*Work performed under the auspices of the U.S. Department of Energy.

CP732, *Beam Instrumentation Workshop 2004: Eleventh Workshop*,
edited by T. Shea and R. C. Sibley III

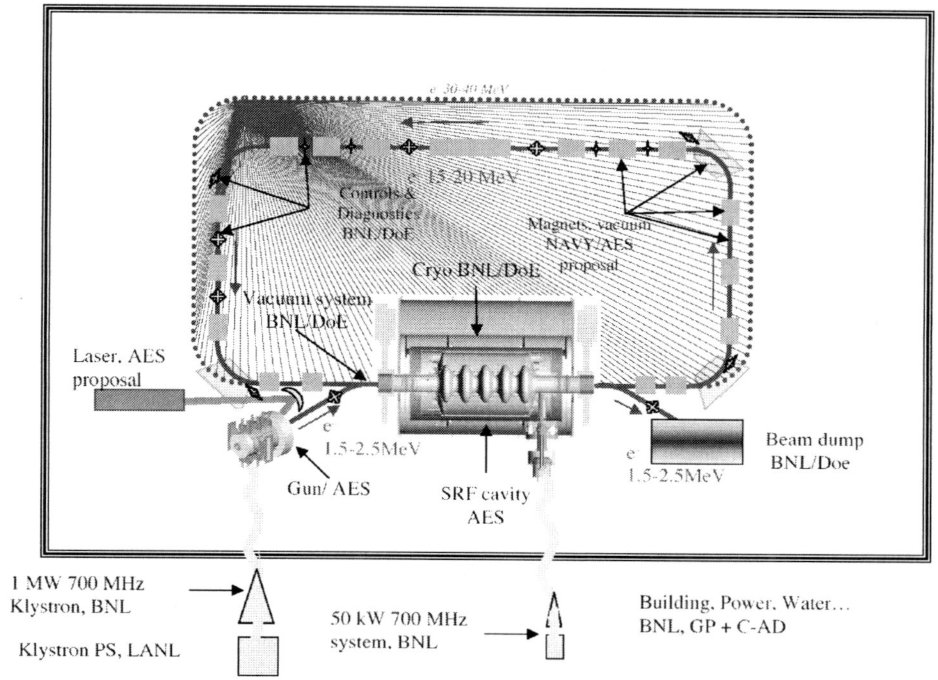

FIGURE 1. A Possible Layout of the Test Facility.

ERL-SPECIFIC REQUIREMENTS

The ERL imposes diagnostics requirements [3, 4] beyond those normally present in linacs and storage rings. One such requirement is to measure with high resolution the difference in currents between the accelerated and decelerated beams. In principle this could be accomplished by measuring the power requirement of the SRF cavity, but that presumes that the low- and high-energy beams are accurately anti-phased, and proper calibration of such a measurement is not completely straightforward. A simple and elegant method is to utilize two toroids placed immediately after the line to the beam dump, one in the dump line and one in the accelerated beam line, and to link those toroids with a figure eight winding. The output of one toroid is used to drive a nulling current through the figure eight, and the output of the second toroid is then the differential current measurement. This overcomes the dynamic range problem of measuring a small current difference in the presence of a large current signal.

As implied in the above discussion of current monitoring, a second requirement is to have the bunches properly phased through the SRF cavity in both the acceleration and deceleration passes, to minimize momentum spread in the accelerated beam (crucial for electron cooling) and to maximize energy recovery from the decelerated beam. Phasing for the acceleration pass can be accomplished by measuring beam position and/or profile in dispersive regions. Phasing for the deceleration pass will be accomplished by adjusting path length with a small chicane, and can be monitored in a variety of ways. In a method similar to that used for the accelerated beam, beam position and/or profile can be measured in the dump beamline. A second possibility is to directly monitor the SRF cavity drive power, a minimum in the required power indicating good phasing. A third method is direct measurement of phase of either or both beams. Our intent is to I/Q demodulate the outputs of all BPMs. This will provide accurate phase measurement with single beams, but would impose excessive demands on time resolution if both beams were monitored by a single pickup. BPM phase measurements will require a means of accurately correcting time delays relative to the SRF. There is the possibility to construct a longitudinal pickup based on the same principle as the SRF cavity, where a null measurement is accomplished by the anti-phasing of the two beams. However, it is not yet clear that there is a need for such a pickup, that it would provide any advantage over simply monitoring the SRF cavity drive power.

TABLE 1. Machine parameters

Parameter [units]	**high chg**	**low emit**
injection energy [MeV]	5	5
beam energy [MeV]	15-20	15-20
rms bunch length [ps]	~20	~20
RF frequency [MHz]	704	704
revolution freq [MHz]	9.4	9.4
bunching freq [MHz]	9.4/ 28.2	704
charges/bunch	~1e1 1	4e9
beam current [mA]	~150/ 450	500
rms energy spread	10e-3	10e-4
εx, εy [mm-mrad]	30	5
beampipe dia [cm]	5	5
current recovery [%]	99.95	99.95

A third ERL-specific requirement is to have the accelerated and decelerated beams on a common center and in addition to have that common center centered in the SRF cavity, to minimize excitation of higher-order transverse modes and thereby raise the threshold for the beam breakup (BBU) instability. This implies position monitors immediately before and after the SRF cavity. In low duty cycle operation the time separation (~100 ns) of the low and high energy pass through the pickup would impose a not-unreasonable lower limit on position monitor electronics bandwidth, so that independent position measurement and correction of both beams is feasible. Centering of the aligned beams on the SRF cavity might then be accomplished either by reliance on survey data, or as beam current goes up by measuring higher-order mode power in the SRF cavity. However, this measurement could not be accomplished when most crucial, during normal full-current operations. Alignment of the two beams relative to each other during normal operations might be accomplished via the TM120 and 210 modes of a rectangular resonant cavity pickup. Such a pickup might be operated in the vicinity of the third harmonic of the SRF cavity frequency to minimize impedance seen by the beam. Positioning such pickups both immediately before and after the SRF cavity would permit proper transverse alignment through the cavity, and the TM111 mode might also provide a high quality longitudinal phase measurement. A method [3] which is similar in principle but not yet demonstrated in practice, would center the two beams relative to each other by looking at a conventional BPM output at the RF fundamental and second harmonic and tuning to minimize the fundamental and maximize the second harmonic.

Bunch length measurement for the BNL ERL is perhaps not so critical as in ERLs designed to produce ultra-short bunches. Ultra-short bunches are not a requirement for either e-cooling or eRHIC. Bunch length can be measured most directly with a streak camera. Less direct measurement can be accomplished by comparing transverse profiles in dispersive and non-dispersive regions, perhaps in combination with variation of the RF phase.

Finally, it is crucial to monitor and localize the development of halo in the ERL. Useful information on halo monitoring may be found in the proceedings of a recent workshop [5]. The measurement requirements in the ERL will require careful implementation of the best methods described in that workshop.

ARCHITECTURE AND SPECIFICATIONS

One responsibility of the Diagnostics Group for the Brookhaven ERL will be to provide an economic, reliable and uniform interface to the RHIC timing and control systems. We intend to utilize well-developed expertise to accomplish this with a straightforward extension of the approach taken in the RHIC BPM system [6], and later refined in SNS Diagnostics [7, 8]. In both cases the interface to timing is encoder logic embedded in a gate array in each data acquisition module. In RHIC the data acquisition is controlled by a DSP in each custom module, where calibration and initial processing occurs before data is transmitted via firewire to VME and the control system. Additional processing and display are accomplished at the application level. In the SNS, data acquisition is controlled by LabVIEW running in Windows, and communication to the control system is via Ethernet. The timing decoder gate array board also provides the interface to the PCI bus, serves as a motherboard for the analog electronics and digitizers, and permits the possibility of fast pre-processing before delivery of data to LabVIEW. The intent is to employ a similar architecture wherever possible for the ERL diagnostics systems, a plan that may also be compatible with future upgrade plans for the RHIC BPM system.

A second responsibility of the Diagnostics Group will be the instruments themselves. Design, implementation, and utilization of the instruments is always a collaboration between Diagnostics and Accelerator Physics. In the case of several of the systems (position monitors, loss monitors, wire scanners, scrapers, current monitors, etc.) most of the needed expertise already resides within the Diagnostics Group, and involvement of Accelerator Physics will probably not extend beyond specification of system requirements. However, for those systems specific to electron machines (synchrotron light monitors, transition radiation monitors, streak cameras, etc.) and more specifically to the ERL (phase monitor, energy spread monitor, high resolution differential current monitor, etc.), the experience within the Diagnostics Group is somewhat limited, and there will be greater reliance on Accelerator Physics in design, implementation, and operation.

As suggested by Table 1, the ERL will operate in either of two modes. In the first (high bunch charge) mode, the 9.4/28.2 MHz bunch frequency is appropriate for electron cooling of 120/360 bunches in RHIC (although other bunching frequencies are under consideration). In the second, every bucket of the 703.75 MHz RF will be filled for low emittance studies. The additional dynamic range required by these two modes adds only minimal complication or expense. However, some forethought is required as a result of the sparse spectrum with the 703.75 MHz bunching frequency. This makes it more difficult to work away from the RF frequency, and imposes more stringent demands on RF shielding. Our intent is to provide flexibility where possible (for instance, by using programmable synthesizers to generate local oscillator frequencies) to permit avoiding the RF fundamental when operating with the 9.4/28.2 MHz bunching frequency. Processing at 14.1 MHz looks particularly attractive, as this will permit the same LO frequency for either e-cooling or high current mode, and the resulting 56.4 MHz clock frequency for BPM I/Q demodulation is comfortable for the intended digitizer.

A preliminary list of proposed Diagnostics, together with their preliminary Accelerator Physics specifications, is shown in Table 2. In addition to the two modes (high charge/low emittance) mentioned above, these diagnostics must also meet the needs of both low duty-cycle commissioning and high current operations. Refinement of specifications to include this consideration is in progress.

CONCLUSION

We have discussed ERL-specific diagnostics requirements, presented a proposed data acquisition platform and timing/controls interface, and made a first attempt to define diagnostics devices, quantities, locations, and specifications. More detailed system-by-system designs will be presented in a series of forthcoming tech notes.

TABLE 2. Diagnostic devices and AP Specifications

Device	Qty	Range	Accuracy	Resolution	Comments
Position/Phase					
BPM (button)	12	1/2 pipe rad	500μ	1μ (av)/100μ	Dual plane
BBU/Energy Feedback	1				Sample scope
Beam Transfer Function	1				Include BTF kicker
Energy Spread	2		10^{-4}	10^{-5}	Dispersive BPMs
Phase	8	+/- 180 deg	+/- 2 deg	0.2 deg	BPMs w/ I/Q
Loss					
BLM (PMT)	10	1-1000 rem/h	30%	0.5 rem/h	20μsec and 1sec
BLM (cable ion chamber)	10	1-1000 rem/h	30%	0.5 rem/h	20μsec and 1sec
Current					
Current	12		5%	1%	BPM sum signal
Differential	1		10^{-4}	10^{-5}	2 toroids w/ null
Profile					
Flags	4?		0.2σ	0.1σ	Phosphor + TR 3 inj line, 1 dump
Wire Scanner - profile	4?	Full aperture	0.2σ		SEM mode, 1 inj & dump, 2 ring
Wire Scanner - halo	3?			10^{-6}	BLM mode
Scraper	2?		0.2σ		SEM + BLM
Synch Light	8?		0.2σ		Every bend mag
Compton	2?		0.2σ		Inj line
Streak Camera	1?		0.2σ		Dual sweep

REFERENCES

1. Ilan Ben-Zvi et al., "R&D Towards Cooling of the RHIC Collider," PAC 2003, Portland.
2. M. Farkhondeh and V. Ptitsyn, "eRHIC Zeroth Order Design Report," BNL CAD AP note 142, March 2004. (available at http://www.agsrhichome.bnl.gov/AP/ap_notes/cad_ap_index.html)
3. G. A. Krafft and J-C Denard, "Diagnostics for Recirculating and Energy Recovered Linacs," BIW 2002, Brookhaven.
4. P. Piot et al., "Performance of the Electron Beam Diagnostics at Jefferson Lab's High Power Free Electron Laser," PAC 1999, New York.
5. P. Cameron and K. Wittenberg, "Halo Diagnostics Summary," Halo 2003, Montauk, and references therein.
6. T. J. Shea and R.L. Witkover, "RHIC Instrumentation," BIW 1998, Stanford.
7. W. Dawson et al., "BPM System for the SNS Ring and Transfer Lines," BIW 2002, Brookhaven.
8. K. Vetter et al., "RF Beam Position Monitor for the SNS Ring," these proceedings.

Femtosecond Resolution Mid-Level RF Phase Jitter and Drift Measurement Using Passive Components

Daniel H. Cheever, G. Townsend Zwart, and W.S. Graves

MIT Bates Linear Accelerator Center, 49 Manning Ave, Middleton, MA 01949-1526

Abstract. The Bates Linear Accelerator uses six 5-MW 2.856-GHz RF transmitters. Phase drifts between the transmitters' 12 klystrons of greater than 0.5 degrees, or any amplitude drifts, compromise beam delivery and are a common occurrence. Instrumenting the phase shifts of these klystrons for machine diagnostics has received high priority. The hardware at the Bates facility that has been developed over the past seven years has proven very flexible. It is employed as a slow monitoring system of 14 phase and amplitude channels with VME hardware under EPICS, and as a fast jitter measurement system. We have recently measured phase jitters between the master oscillator reference line and klystron outputs, RF cavity outputs, drive laser detectors, waveguide loads and beam position monitors at two facilities. The resolution and repeatability of the medium level passive mixer implementation are better than 0.2 degrees for measurements of 5 to 60 seconds. This represents an R.M.S. jitter resolution of 20 fs. Data are presented for the above implementations and measurements. We are investigating the use of harmonic cavities and frequency up-converters for further improvements in jitter measurement resolution.

INTRODUCTION

For many facilities, control and instrumentation of RF fields with a resolution of ~1 degree provide sufficient accuracy for reliable accelerator operation. However, a new class of accelerators and RF or laser timing systems is emerging with extremely demanding phase detection specifications. These systems, including seeded free electron lasers and ultra-stable laser-based RF sources, will require timing measurement errors to better than 50 fs and control of RF phases to significantly better than 0.1 degrees at GHz frequencies.

In systems where abundant RF power taps are available for diagnostics and controls systems it was found that the use of simple passive double balanced mixers for the measurement of phase drifts and jitter can meet this performance level. This paper documents the efforts of the Bates Linear Accelerator Center in the measurement of phase drifts at pulsed S-band and higher RF frequencies. While no novel use of RF components is claimed, this work is a reference for other engineers tasked with the monitoring and control of RF systems.

CP732, *Beam Instrumentation Workshop 2004: Eleventh Workshop*,
edited by T. Shea and R. C. Sibley III

IMPLEMENTATIONS

Three phase detection systems have been developed: a 14-channel phase and amplitude monitoring system with dedicated hardware at the transmitters with digitization and display under EPICS measuring output klystron level deviations and phase drifts against the reference line; a single-channel, portable, self contained phase and amplitude system with analog IF output; and a wideband (2 – 18 GHz) phase detector prototype.

EPICS Based Phase and Amplitude Monitoring of the MIT Bates Accelerator RF

The MIT Bates accelerator is a pulsed RF structure operating at 2.856 GHz. Twelve CPI klystrons are situated along the length of the 200 m LINAC. Each klystron is capable of delivering 5 MW of RF power for 30 μs at a repetition rate of 600 Hz. Reliable operation of the machine requires that the relative phase of the 12 klystrons must be maintained to better than 1°. Similarly, the relative amplitude stability of each klystron must be maintained to better than 10^{-3}. In addition two other RF devices are instrumented, the RF chopper and pre-buncher.

To better meet these requirements Bates has constructed an RF phase and amplitude monitoring system (PAM). This system samples a small fraction of the RF immediately downstream of each of the twelve klystrons. The complete system includes several main components: RF phase and amplitude detectors, analog signal conditioning, VME based digitizing and EPICS software for monitoring and calibration. The system has been in service for six years. The phase detection analog IF portions performance, in situ at the transmitter with the RF port fed a copy of the LO port usual master oscillator signal, measures a 50 uV R.M.S. deviation[1] over one minute, or 0.013 degrees or ~12 fs.

FIGURE 1. RF chassis for two klystrons.

[1] All data presented as an "R.M.S. deviation" is measured as follows: single-ended, DC coupled, 50-Ω internally terminated oscilloscope input (Lecroy LT355). 360 consecutive 6-Hz externally triggered IF pulses build the standard deviation of the **mean** of 100 nS worth of the middle of the ~1-uS RF pulse width.

Shown in Fig. 1 is one of the RF chassis installed in the RF vault. The input is via N connectors and semi-rigid coax cables from the waveguide directional couplers. Off-the-shelf RF components (listed in Appendix I) are used. The Reference line tap is attenuated to ~ +21 dBM and split to the two mixers' LO ports. The RF inputs are individually attenuated to normalize to +18 dBM and 100-MHz wide band-pass filters[2] are utilized. 360-degree electronic phase shifters are implemented and under software control null the mixer output and calibrate the digitized data to the corresponding phase shifts in absolute degrees. The output of the mixers and pin diode detectors is impedance matched, fanned out, and routed to custom gated integrators and a custom video matrix switch. The integrators integrate and hold a gated portion of the pulsed signals and are reset immediately before the next gate pulse. This s/h signal is routed to the VME digitizers. Further details of this system can be found in a paper titled "The Bates Phase and Amplitude Monitoring System," submitted to ICALEPS 2001[3].

Over the past years the PAM system has proved enormously useful for catching step discontinuities in the unregulated high voltage applied to the klystrons; for monitoring and correcting phase drift of one or more klystrons; and in establishing a clear correlation between a slow 3° phase drift on all klystrons and the primary water cooling system temperature. This PAM information is presented to the operations group and accelerator scientists through a set of EPICS tools, MEDM strip charts, bar charts and archived data. This information has enabled the operations group to more easily set and restore the phasing of the machine. It allows them to more effectively troubleshoot malfunctioning components of the RF system.

A Portable S-Band Phase and Amplitude Detector

As mentioned in the introduction, new classes of accelerators under investigation will require very tight phase drift and jitter performance. Some effort was spent in investigating if a corresponding class of RF diagnostic tools existed, but no commercial equipment was found with specifications in the sub-picosecond range. The Bates PAM system, as described in the previous section, was shown to measure sub-0.1-degree phase jitter between the reference line and some output klystrons. A portable system was built to explore the limits of performance. As shown below in Figs. 2 & 3, this unit houses much of the same RF components as the RF vault PAM system. In addition, a 1-db/step rotary S-band attenuator is included in-line with the RF port to readily normalize differing RF levels. The manual 360-degree S-band phase shifter allows mixer calibration and nulling[4].

This portable system was used for jitter analysis both at the Bates facility and at the Argonne National Laboratories' Advanced Photon Source. By no means is the presentation of this data to be a comparison of the RF systems, simply a set of data from two 2.856-GHz RF systems.

---

[2] There are no fast phase shifts outside this range.

[3] The paper "EPICS Based Phase and Amplitude Monitoring of the MIT Bates Accelerator RF" is also available from the **arxiv** server as "WEAP012."

[4] The manual phase shifter eliminates the possible introduction of any noise due to the electronic phase shifter itself or from the bias signal.

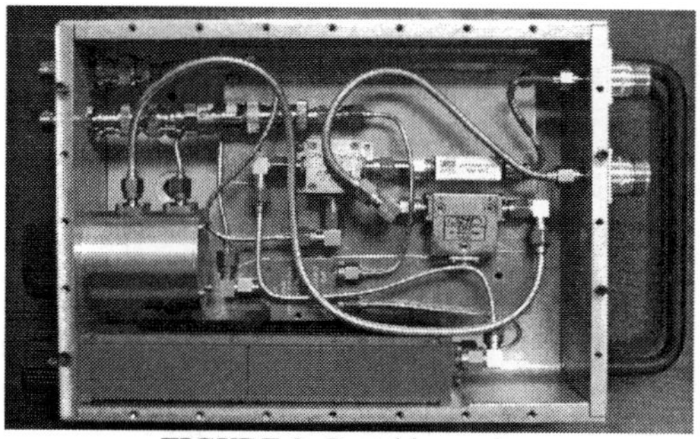

FIGURE 2. Portable version.

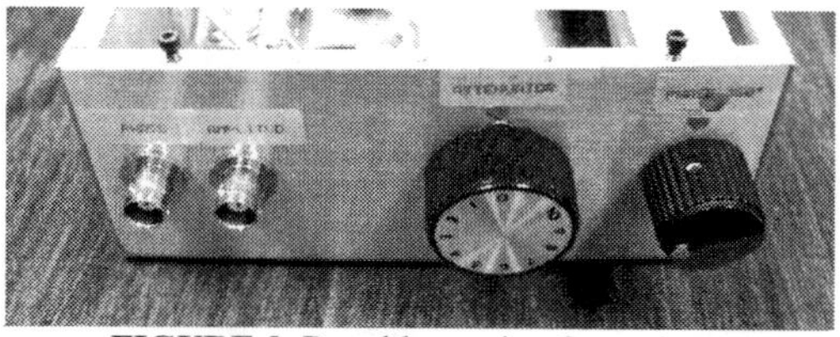

FIGURE 3. Portable version front view.

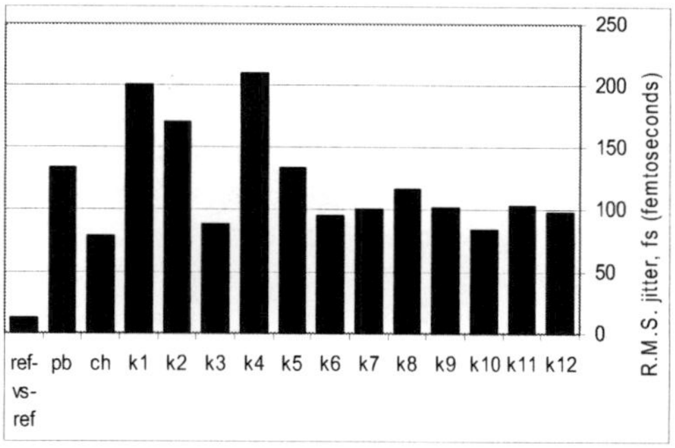

FIGURE 4. Jitter results of the Bates RF system.

Figure 4 is a summary of the results of the distribution of 1 minute of 6-Hz RF pulses from the Bates accelerator RF system. The data point furthest to the left is the limit of the phase detector resolution, 12 fs. This is calculated by multiplying the sigma spread of the 360 IF voltage means (50 uV[6]) with the slope of the mixer (determined via shifting the phase 360+ degrees). Most of the RF systems performance is sub-100-fs R.M.S. of pulse-to-pulse jitter, or sub-0.10-degrees R.M.S.

Figure 5 is a summary of measurement results at the ANL APS RF system. In the APS case the available reference line power was ~10 dbM less than at Bates, accounting for the difference in the "ref-vs-ref" data point versus the Bates similar

[6] The oscilloscope "noise floor" (RF off) in this mode measures 19 uV. The vertical amplifier is set to 5 mV/div, or, 10 degrees of 2.856 GHz, full-scale.

data point. Although discussion is outside of the scope of this paper, note the reduced phase jitter of the SLED output versus its input for sections L4 & L5. Many other jitters were measured at the APS, including the jitters of the 119-MHz to 2.856-GHz drive laser PLL, chicane loads, and BPM outputs. APS personnel indicated that they were interested in continuous phase jitter diagnostics using this simple passive mixer approach.

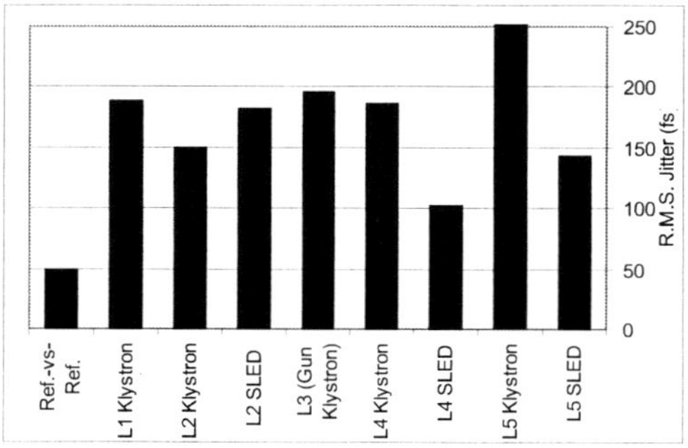

FIGURE 5. Jitter results of the ANL APS RF system.

Wideband (2 - 18 GHz) Phase Detector Prototype

Overdriving of the mixer input ports will create junction heating correlated compression resulting in phase calibration errors and drifts, and perhaps reduced service life. On the other hand, raising the RF level on either the RF or LO ports did not increase the IF noise level. It follows therefore that maximum s/n is realized with high power levels. The mixer used, a Mini-Circuits ZEM-4300MH, is specified for +13- dBm LO and "up to +9-dBm RF". When tested with +18 dBm on both ports no thermal drift was measured, but the output was quadrupled. Note below in Fig. 6, +18-dBm RF is where this mid-level mixers output compresses when tested over a 360+ degree phase shift.

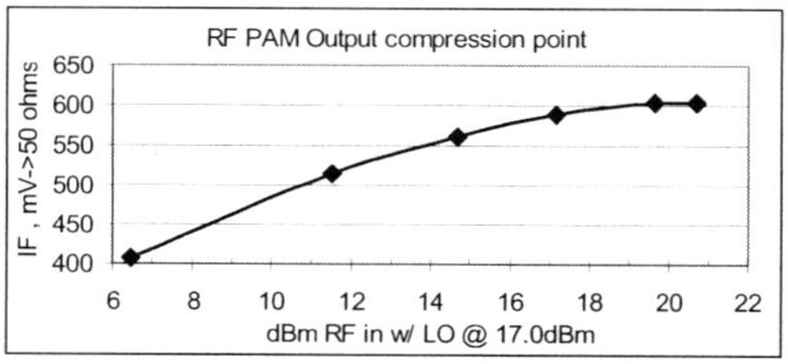

FIGURE 6. ZEM4300 Mixer output compression.

To further improve upon the mixer resolution, raise the slope by the use of higher frequencies. This is a realistic implementation because frequency multipliers, dividers, and high-order non-linear cavities are available. The configuration we envision is a dedicated phase jitter/timing diagnostic reference line at a multiple of the linac frequency and band-pass filtered non-linear cavities designed with strong high order components. We explored this via using an M/A-COM "MY85" 2-18 GHz double-balanced mixer. Table 1 shows the results for 1f, 2f, and 4f of 2.856 GHz. The column "mV/180 °" refers to the output swing for a full phase shift via a broadband trombone type phase shifter. Again, the data is the R.M.S. spread of the mean of 360, 6-Hz IF pulses with the mixer nulled.

TABLE 1. High-order results

GHz	mV/180°	mV R.M.S. jitter	R.M.S. fs jitter
2.85	678	0.045	7.4
5.71	665	0.043	3.6
11.4	532.4	0.041	2.1

These results clearly show promise for further extension of the timing jitter resolution via higher frequencies. An area we have yet to explore is pluses and minuses of passive harmonic multiplication over frequency division; passive or electronic.

Work supported by the U.S. Department of Energy under cooperative agreement DE-FC02-94ER40818.

APPENDIX I

Installed PAM RF Chassis Components

Qty.	Manufacturer	Model No.	Description
(3)	Mini-circuits	ZAPDQ-4	Splitter
(2)	Hewlett Packard	8473D	Detector diode
(2)	Mini-circuits	ZEM-4300MH	Mixer
(2)	Lorch	VP-360-2800	Phase shifter
(2)	Lorch	5BP7-2850/100	B.P. filter
(3)	S.M.T.	SM22052	Isolator

APPENDIX II

Test Equipment Listing

Hewlett Packard 8350B Sweep Oscillator, 0.01 – 26 GHz
Hewlett Packard 83595C RF plug-in, +23 dBM output
Hewlett Packard 8510C Network Analyzer
Narda 3752 Dial Coax Phase Shifter
Lecroy LT344 oscilloscope with JTA & WAVA opts.
Hewlett Packard 3457A Multi-meter
Advantest R3162 Spectrum Analyzer, 9 KHz - 8.1 GHz

Profile, Current, and Halo Monitors of the PROSCAN Beam Lines

Rudolf Dölling

Paul Scherrer Institute (PSI), CH-5232 Villigen, Switzerland

Abstract. PROSCAN, an extended medical facility using proton beams for the treatment of deep-seated tumors and eye melanoma, is under construction at PSI. Ionization chambers and secondary emission monitors will be used as current monitors and in a multi-strip configuration as profile monitors at the PROSCAN beam lines. A thin and a thick version of these detectors are in preparation as well as a 4-segment ionization chamber to detect the beam halo. Electromagnetic and microphonic noise from the signal and high-voltage cables, saturation due to recombination, and the evaluation of the profiles are discussed, as well as measures to detect failures of the detectors during operation.

INTRODUCTION

In the PROSCAN facility (Fig. 1) a 250-MeV proton beam of 1 to 500 nA will be extracted from the COMET cyclotron. After passing through a degrader, where the energy can be adjusted in the range from 230 to 70 MeV, it can be delivered, at a maximum current of 10 nA, into one of four areas: two gantries, an eye treatment room and a material irradiation area. Fast changes of beam energy are foreseen for the spot-scanning treatment of deep-seated tumors in the new Gantry 2. Several diagnostics will be used to monitor the beam parameters in the various modes of operation (Fig. 1). The first components will be taken into operation this year.

THIN PROFILE MONITORS

In front of the degrader and at the "control point" in front of the gantries, profile and current monitors are inserted permanently in the beam. The currents and their ratios are rapidly monitored as a safety measure. In addition, thin retractable monitors are available at the exit of the cyclotron. All these monitors must be very thin to prevent excessive scattering of the beam.

Planar ionization chambers (ICs) and multi-strip ionization chambers (MSICs) filled with ambient air are used to obtain sufficient signal from the small beam currents (0.1 to 500 nA). These are formed by a stack of alternating high-voltage (HV) and measurement planes made from titanium foils of 6-μm thickness (Fig. 2). "1 broader + 30 regular" + 1 broader-strip patterns with 1-mm pitch are used for the measurement of (the projections of) the vertical and horizontal beam profiles. The two total-current measurements are used by the machine control system and by the patient

CP732, *Beam Instrumentation Workshop 2004: Eleventh Workshop*,
edited by T. Shea and R. C. Sibley III

safety system. Additional grounded foils are added at both sides of the stack to limit the active region and thereby decrease the maximum HV current load.

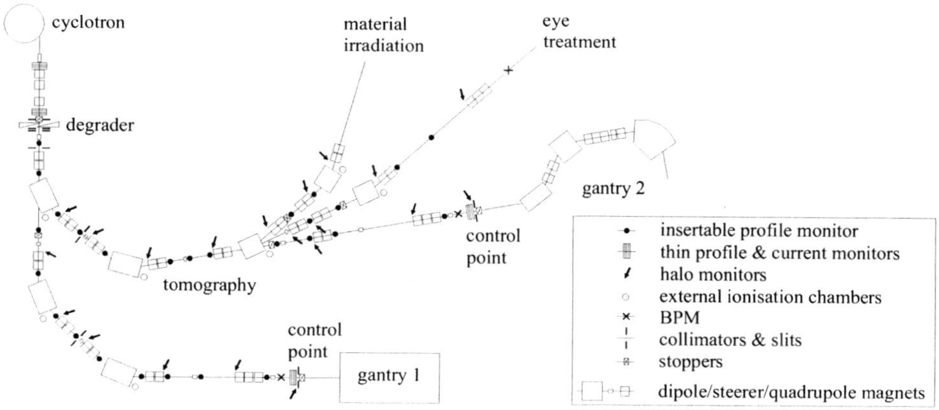

FIGURE 1. Overview of beam lines.

The pre-tensioned foils (full or with the etched strip pattern) are mounted on the supporting frames made from thick-film plated ceramic board. This is done by soldering, which requires a thin sputtered multi-layer metal coating of the outer parts of the foil. The IC/MSIC measurement head is separated from the vacuum by a box with thin (25 or 50 μm) titanium windows, which are clamped between flanges.

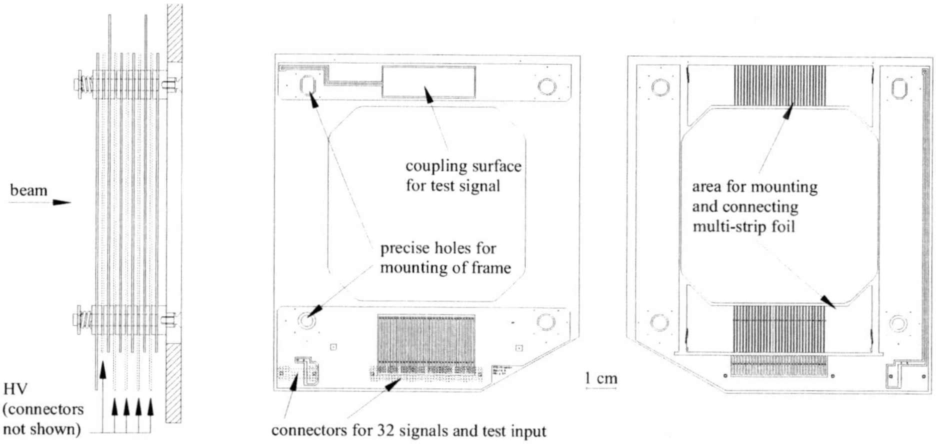

FIGURE 2. Left: Stack of frames with foils, forming 2 MSICs and 2 ICs (vacuum enclosure and windows not shown). Middle and right: Supporting frame for multi-strip foil (front and back side).

At low beam currents recombination effects are negligible and the ratio "signal current/beam current" varies from 48 to 113 for proton energies of 250 to 70 MeV.

With a gap of only 2 mm separating the planes and a bias of the HV-electrodes of +1.5 kV, the high electric field counteracts recombination and reduces the charge collection time t_{coll} = (plane separation)^2 * (high voltage)$^{-1}$ * (k_{air})$^{-1}$ to approximately 16 μs (with the mobility of air k_{air} ~$1.7 \cdot 10^{-4}$ m^2 s^{-1} V^{-1}). Nevertheless, with smaller beam diameters, recombination effects are dominant at higher beam currents (Fig. 3).

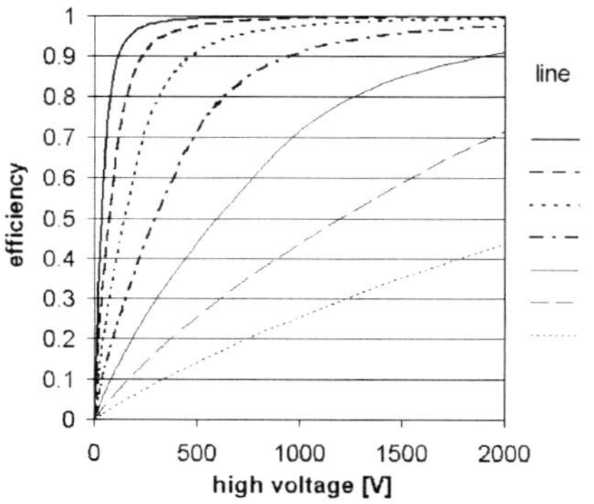

$\frac{(\text{beam current}[nA])(\text{gap}[mm])^4}{(\text{beam diameter}[mm])^2}$ @beam 250 MeV	beam diameter [mm] @beam 250 MeV, 512 nA and gap 2 mm or @beam 250 MeV, 32 nA and gap 4 mm or @beam 70 MeV, 13.6 nA and gap 4 mm
2	64
8	32
32	16
128	8
512	4
2048	2
8192	1

(round proton beam with homogeneous current density assumed)

FIGURE 3. Saturation of air-filled ionization chambers due to recombination (calculated according to [1]). The efficiency is a function of (plane separation)^4 * (beam current density) * (high voltage)$^{-2}$. Hence the curves can be scaled to other values of beam current, beam diameter, plane separation or high voltage than given in the right column of the legend.

In order to get signals not compromised by recombination, the same devices are placed in vacuum and used as multi-strip secondary emission monitors (MSSEMs) and current monitors (SEMs) at the locations near the cyclotron and in front of the degrader. Nevertheless, at low beam current the signal current, a few percent of the former, is too low for the electronics used. The stability of the secondary emission coefficient of titanium against ageing was one major reason for its choice [2].

INSERTABLE THICK PROFILE MONITORS

Insertable MSIC monitors, successively introduced into the beam by compressed-air actuators, yield the input information for the calculation of a beam envelope with the "Transport" code [3]. Since these monitors are not thin, only a measurement at one location at a time is possible. (As in the case of the thin monitors, the electronics used allow for the measurement of the temporal development of a beam profile.)

Metalized ceramic 0.63-mm thick boards separated by 4-mm wide air gaps provide the strip patterns and HV-electrodes. With one exception, an HV of +0.6 kV is sufficient to suppress the effect of recombination on the measured currents to below 10% at the expected beam current densities and beam energies.

The strip pattern will be adjusted in the directions transversal to the beam to the requested accuracy of 0.1 mm to the reference given by the flange of the vacuum box.

The pitch of the metalized strips varies from 0.5 to 1 and 2 mm (plus one broader strip at each side). Strips of 2 x 68 are fed out by flexible-printed-circuit cables.

With two exchangeable printed circuit boards placed in an electrically shielded box at the top of the profile-monitor feed-through, the signals are routed to the 2 x 16 channels of the electronics. With this arrangement, a strip pitch of 0.5, 1, 1.5, 2, 3 or 4 mm can be chosen for a "1 broader + 14 regular + 1 broader"-strip arrangement in each plane (Fig. 4). This variability allows for the adaptation of the strip pitch to the expected range of beam profile width. This is required due to the limited number of only 16 channels per plane that is foreseen for most of the monitors.

At three successive locations, a higher resolution is foreseen for beam tomography. A 1 "broader + 30 regular + 1 broader"-strip arrangement in each plane will be realized by doubling the number of cables and electronic modules. At these locations, the available strip pitch is 0.5, 1 or 2 mm.

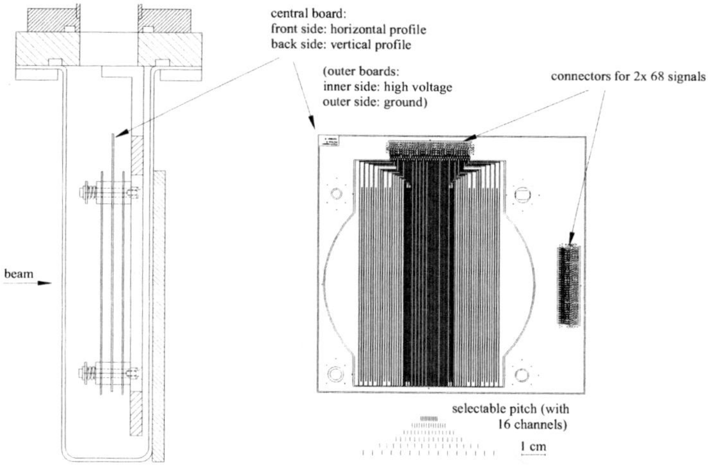

FIGURE 4. Insertable profile monitor.

PROFILE EVALUATION

The beam profile can be reconstructed from the measured currents to the strips by assuming a constant line-current density over the width of a strip. Nevertheless, in order to get a realistic representation of a broad beam profile in the region of the outer broader strips, it is required to combine the current readings from the outer broader strips with the information from the next inner strips (Fig. 5, [4]). This is essential for a correct determination of beam center and beam width, which are evaluated from the first and second moments of the measured distribution. Without this technique, the width of broad beam profiles is usually overestimated.

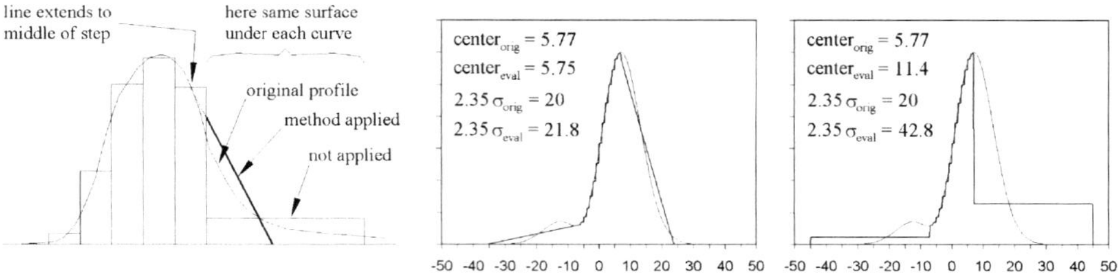

FIGURE 5. Treatment of broad beam profiles. Left: method, middle: applied to a fictitious profile, right: not applied. (Strip configuration: 1 strip from -45 to -7 units, 14 strips 1 unit wide from -7 to 7 units, 1 strip from 7 to 45 units. A precise current measurement without noise is assumed.)

For very narrow beams with a width comparable to or smaller than the strip pitch, the width is also usually overestimated. For beams with a width down to the strip pitch s_{pitch} this can be largely corrected by an empirical correction:

$$\sigma_{corr} = \sqrt{\sigma_{uncorr}^2 - 0.16\, s_{pitch}^2}\ . \tag{1}$$

Simulations [5] indicate that by a profile measurement using "1 + 14 + 1" strips beam position and width can be measured accurately when the FWHM beam width is in the range 1x to 10x strip pitch and the profile is of conventional shape. A pattern with varying strip pitch can also be chosen to further enlarge the range of beam profile width accessible with a fixed configuration.

HALO MONITORS

"Halo monitors" are placed around modified bellows adjacent to the quadrupole doublets and triplets (Fig. 6). These 4-segment ionization chambers, which protrude circumferentially 5 mm into the 90-mm diameter beam pipe, give enough signal to detect traversing beam current fractions of below 1 pA. This provides a much more sensitive loss control than the external ionization chambers located near the dipole magnets and should also give an online control of the stability of the beam settings.

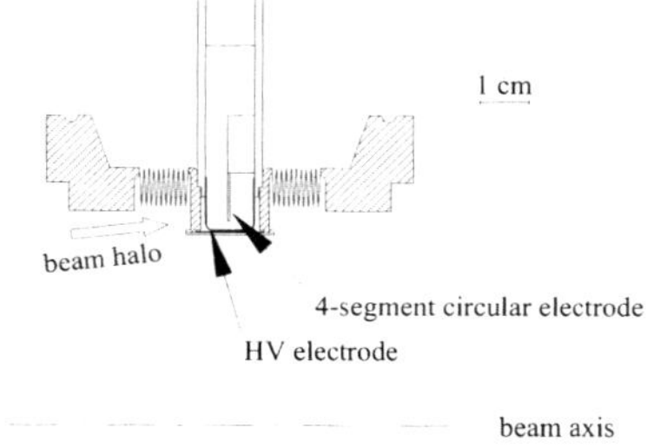

FIGURE 6. Halo monitor.

ELECTRONICS AND GROUNDING

With beam currents of 0.1 nA to 500 nA, the signal currents from MSICs, ICs, MSSEMs, SEMs, Faraday-cups and halo monitors are of the order of 0.1 pA to 25 μA per single electrode. All currents are measured with multi-channel logarithmic-amplifier modules with a current range of 20 pA to >200 μA. These allow single channel readout as well as the measurement of up to 4096 profiles with a minimum time step of 1 ms. Algorithms for data evaluation and the generation of interlocks can be implemented. A trigger input allows the observation of time-dependent machine behavior by simultaneous operation of several modules [6]. The electronics, including the HV-modules for the detectors, are located some 40 meters away outside of the concrete shielding to prevent radiation damage and to allow for service without opening the bunker.

The very low signal levels require the elimination of ground loops. Therefore, the internal shield and "ground" of the analogue amplifier electronics are only grounded via the shield of the measurement cable, which in turn is grounded at the diagnostic head at the beam line (Fig. 7). The ground of the digital electronics is connected to the (CAMAC- or VME-) crate ground. The ground transition for the signals is provided by a differential amplifier. Originally we intended to use this type of grounding for the HV-modules as well. However, locally grounded modules are now foreseen due to better availability.

CABLES, SHIELDING, NOISE, AND FILTERS

In order to measure signal currents below 10 nA, it is imperative to use cables with a low production of microphonic noise and to prevent ground loops. Several cables have been tested for their susceptibility to microphonic noise and differences of several orders of magnitude were found. Ground loops are adequately prevented by the single-point grounding scheme (see above) for the low frequencies of concern (<<300 kHz corresponding to $l_{cable} << \lambda/30$). A good electrical shield is also of importance. The effectiveness of the cable shield against external AC electrical fields was also tested and large differences were found corresponding to the tightness of the shield.

A good standard in both aspects is the coaxial cable G03130HT, which is used for all single signals. It is a declared "low noise" cable using a semiconductor layer between the inner isolator and the dense copper braid. As signal cable for the MSIC and MSSEM, a 40-wire twisted-pair cable (Huber & Suhner No. 12 566 226) is used. A plastic coated aluminum tape and a copper braid (enclosing the whole bundle of 40 wires) give a compact shield. Thirty-two wires carry the signals. The return path is provided by the cable shield and the residual wires (which as well could have been left floating). Hence no advantage is taken from the twisted-pair property. This cable performs as well as the G03130HT. A similar cable with fewer wires (Huber & Suhner No. 12 568 329) is foreseen for the halo monitors. It should be noted that an individual shielding of each wire would give no advantage because the cross talk between the wires is already negligible due to the very low voltage burden of the electronics used.

Additional electrical shielding is provided by enclosing measurement- and HV-cables from each diagnostic head together in a second isolated shield (copper braid), which is grounded at many locations along the way. (With the HV-supplies grounded, it is disputable whether or not to include the HV-cable in this shield.) No attempts are made to magnetically shield the signal cables, but some efforts were made to separate the diagnostic cables from magnet, AC and stepping motor cables. Nevertheless, this was possible only to a limited degree due to boundary conditions from the infrastructure and concrete shielding. Furthermore, the diagnostic cables are placed on their own support, separate from water pipes, in order to reduce microphonic noise.

Grounding of both ends of the HV-supply cable causes a ground loop, and noise is coupled to the HV-planes and from there capacitively to the signal electrodes. Hence, a filter for noise suppression is required (Fig 7., filter 1). The filter parameters are chosen to fulfill several requirements: a noise input voltage of 0.1 V to the cable (which is considered to be the worst case) should yield a noise signal below 1 pA, at least at frequencies above 50 Hz. With full signal, the high voltage should not drop below 90% of the zero-current value in order to preserve some stability against recombination. The slowing of the detector speed due to the "soft" high voltage must be moderate; i.e., 90% of the final value should be reached after approximately 10 μs (the charge collection time is not included in this number).

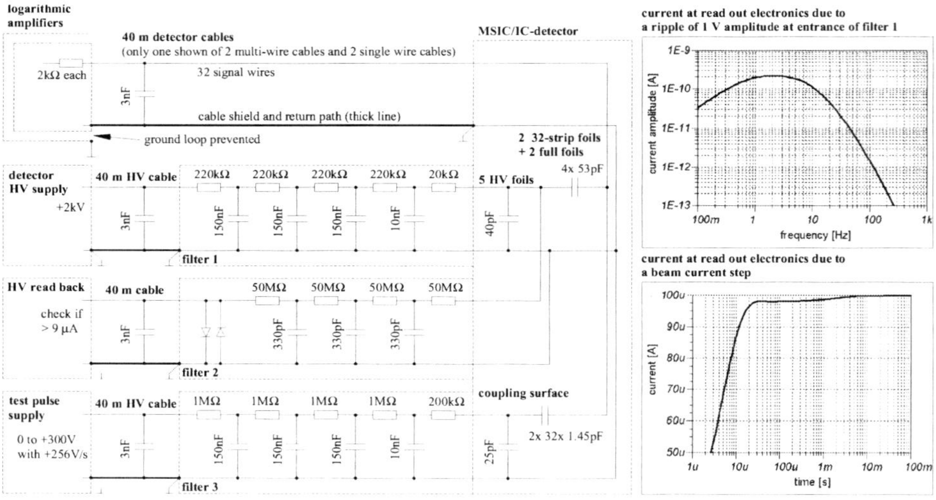

FIGURE 7. Filtering of HV input, HV read-back output (only one shown) and test-pulse input of the thin profile (MSIC) and current (IC) detector head. Detector and cable capacitances are roughly estimated. Also shown are cable shields and grounding. (With the given setup, noise coupling to the HV read-back output and the test-pulse input is less critical than to the HV input.)

DETECTOR FAILURE SURVEILLANCE

In order to get the correct current reading from each electrode, it is required to have: intact foils in their correct position, high voltage at the HV-foils, signal path uninterrupted, signal path and signal foil isolated (i.e. not short circuited to ground or HV or other signal path), air present in ICs respectively vacuum in SEMs, and intact current measurement electronics.

Possibilities for failures are:

- Break of foil or solder joint (spontaneous or due to a break of a vacuum window) resulting in a misplacement or a loss of isolation or an interruption of the path to all or part of the signal.
- Bad connectors or break of PCB or wires resulting in interruption of the signal path.
- Humidity, radiation damage, sparking or sputter effects leading to a loss of isolation in the detector or in cables or connectors, resulting in an erroneous current reading or a loss of high voltage.
- Local disintegration of foils due to radiation damage or sparking or spark induced burning of foil (in air) which results - if the holes in the foil are larger than the gap between the planes - in too low current readings ("blind spot")
- Local change of the secondary emission coefficient of SEM planes ("blind spot").

With all detectors, the mechanical integrity of the foils, the presence of the high voltage, the integrity of signal path and isolation and the readout electronics can be checked from time to time without beam: The high voltage is switched off and on with defined slopes, thereby capacitively inducing a signal to the measurement electrodes. This response can be compared to previously recorded data. The signal should be chosen small enough to detect even small leakage currents. Nevertheless, this method is hardly sensitive enough to detect e.g. a hole of a few mm diameter in a foil of 70-mm diameter or a break of the foil far from the readout side.

The strip-foils can be checked in addition by capacitively coupling a test pulse (or AC voltage) only to the end of the strips opposite to the readout side (Figs. 2, 7). The setup from Fig. 7 results in a pulse of 210-pA height and 1.7-s FWHM.

At the safety relevant current monitors, the presence of the high voltage at all electrodes is checked online. The high voltage is provided to one electrode and via a chain of connectors to the other HV-electrodes. From the last HV-electrode, the potential is fed back to the readout electronics, which is part of the patient-safety system. (An additional readout is provided to the machine control system.) This feedback is filtered so that any retroaction of the readout electronics to the high voltage at the HV-electrode is negligible.

As a further control, the total currents (with beam) of the individual measurement planes can be compared. This is the only available means to detect small holes in the foils. (It is assumed that a larger fraction of the beam passes through the hole and that holes in different planes are not formed at the same time.) At the same time the readout electronics are checked against each other. This check should be performed regularly.

In the same way a change of the secondary emission coefficient can be detected by comparing the output of SEMs and ICs.

REFERENCES

1. G. Mie, "Der elektrische Strom in ionisierter Luft in einem ebenen Kondensator," *Ann. Phys. (Leipzig)* **13,** 857-889 (1904). See also: H. Attix, "Boag's Treatment of Mie's Theory," in *Introduction to Radiological Physics and Radiation Dosimetry*, Wiley, 1986, pp. 334-336.
2. G. Ferioli and R. Jung, "Evolution of the Secondary Emission Efficiencies of Various Materials Measured in the CERN SPS Secondary Beam Lines," DIPAC97, 1997, pp. 168-170.
3. http://people.web.psi.ch/rohrer_u/trans.htm.
4. R. Dölling, "Methode der Auswertung von MSIC-Messungen," PROSCAN-document P24/DR84-401 (2004).
5. R. Dölling, "Auswertung von MSIC-Messungen und Einfluss der Streifenzahl auf die Genauigkeit von Profilschwerpunkt und –breite," PROSCAN-document P24/DR84-305 (2003).
6. P.-A. Duperrex et al., these proceedings.
7. R. Dölling, "Messkabel für kleine Ströme bei der Strahlführungsdiagnostik PROSCAN," PROSCAN-document P24/DR84-403 (2004).

Noise Reduction and Correction in the IPNS Linac ESEM

J. C. Dooling, F. R. Brumwell, L. Donley,
G. E. McMichael, and V. F. Stipp

Argonne National Laboratory, 9700 South Cass Avenue, Argonne, IL 60439, USA

Abstract. The Energy Spread and Energy Monitor (ESEM) is an on-line, non-intrusive diagnostic used to characterize the output beam from the 200-MHz, 50-MeV IPNS linac. The energy spread is determined from a 3-size, longitudinal emittance measurement; whereas the energy is derived from time of flight (TOF) analysis. Signals are detected on 50-ohm, stripline beam position monitors (BPMs) terminated in their characteristic impedance. Each BPM is constructed with four striplines: top, bottom, left and right. The ESEM signals are taken from the bottom stripline in four separate BPM locations in the 50-MeV transport line between the linac and the synchrotron. Deterministic linac noise is sampled before and after the 70-microsecond macropulse. The noise phasor is vectorially subtracted from the beam signal. Noise subtraction is required at several frequencies, especially the fundamental and fifth harmonics (200 MHz and 1 GHz). It is also necessary to correct for attenuation and dispersion in the co-axial signal cables. Presently, the analysis assumes a single particle distribution to determine energy and energy spread. Work is on-going to allow for more realistic longitudinal distributions to be included in the analysis.

INTRODUCTION

Stripline beam position monitors (BPMs) are useful diagnostics for measuring important beam parameters such as beam centroid [1], transverse emittance (transverse beam size) [2,3], energy, and energy spread (longitudinal emittance) [4]. In the 50-MeV transport line between the Intense Pulsed Neutron Source (IPNS) linac and the rapid cycling synchrotron (RCS), BPMs provide both longitudinal beam characteristics as well as transverse centroid information.

The energy spread and energy monitor (ESEM) detects beam bunches using four stripline BPMs as shown in Fig. 1. Since it was first described, the ESEM has undergone several modifications to improve its reliability; the two most significant changes are discussed here. First, the sampling period was extended to provide greater frequency resolution in the FFT spectra. Increasing the frequency resolution reduces the time resolution, but the change was necessary to remove a specific noise component discussed in greater detail below. Second, data acquisition hardware was added in the form of a deep-memory oscilloscope. The new hardware permits the entire macropulse to be sampled at sufficient rate to allow its microbunch structure to be discerned.

CP732, *Beam Instrumentation Workshop 2004: Eleventh Workshop*,
edited by T. Shea and R. C. Sibley III

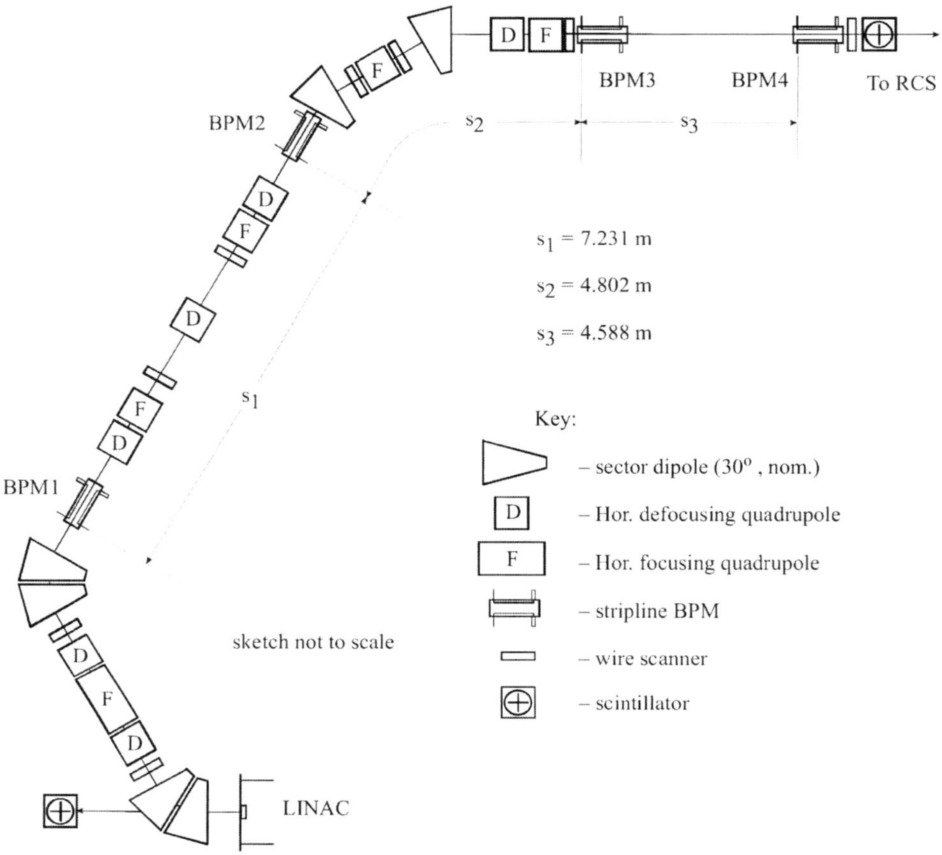

FIGURE 1. 50-MeV transport line with location of the ESEM stripline BPMs.

With regard to frequency resolution, for the BPM task, a local oscillator (LO) tone at 205 MHz is mixed with the micro-bunch signal from the striplines (primarily 200.07 MHz) to yield a difference waveform of approximately 5 MHz. The signalfrom the stripline is split between the digitizing oscilloscope performing the ESEM measurement and the BPM electronics chassis. Initially, interference from the 205- MHz LO tone was not considered important because the ESEM signal was split upstream of the BPM chassis. Sampling of the ESEM signal was done at either 5 GS/s or 10 GS/s. Once obtained, data files were then divided into 100-ns windows for FFT analysis yielding a frequency resolution, Δf of 10 MHz. Anomalous results in energy and energy spread measurements led us to look more closely at the noise contributions from both 200-MHz linac rf pulse and the LO. Increasing the sample window to 400 ns provided sufficient resolution (Δf = 2.5 MHz) to differentiate between the two noise sources, as shown in Fig. 2. The spectra presented in the figure come from the same raw BPM time data and indicate a substantial noise contribution from both linac rf and the LO [Fig. 2b)]. When reconstructing the temporal profile of the BPM pulse, only harmonics of the linac frequency are included in the inverse FFT; therefore, using higher frequency resolution resolved the problem with LO noise, but not the contribution from linac rf. Figure 2b also shows substantial noise components at the second harmonic frequencies (400 and 410 MHz); additional higher noise harmonics are present in the data as well, as discussed below.

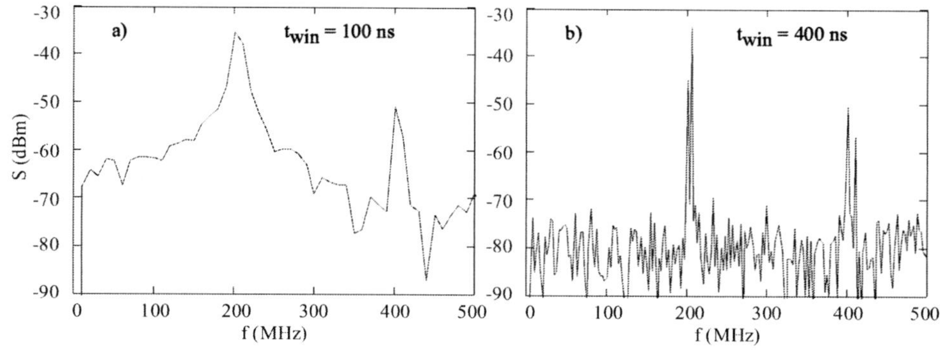

FIGURE 2. Resolving Linac and Local Oscillator noise just before the beam pulse on BPM 1.

New oscilloscope hardware allowed the entire macropulse waveform from four separately-positioned BPM striplines along the 50-MeV transport beamline to be obtained simultaneously; however, we were limited to a sample rate of 5 GS/s. With the possible exception of BPM 1, essentially all of the microbunch temporal profile information is found in the first 10 harmonics (≤ 2 GHz); therefore a 5-GS/s sampling rate, without bandlimiting, is considered acceptable.

ANALYSIS

The rf signal detected on a stripline is a vector sum of the beam image current and the noise current, represented as voltage phasors seen at the digitizing oscilloscope, as shown in Fig. 3. Expressing the signal, noise, and measured voltage phasors as

$$\mathbf{V_s} = V_s e^{j\theta_s} e^{j\omega t}, \quad \mathbf{V_n} = V_n e^{j\theta_n} e^{j\omega t}, \quad \text{and } \mathbf{V_m} = V_m e^{j\theta_m} e^{j\omega t}, \tag{1}$$

where $j = \sqrt{-1}$ and ω is the relevant harmonic frequency, the measured signal phasor is the real part of V_m. The noise vector is known from ESEM data before and after the beam. The FFT provides initial and final amplitude and phase in 400-ns time increments. A linear fit is made of the noise amplitude and phase data at a given frequency; the fit is then used to correct the measured time data. To obtain the real part of the beam data, $V_{beam}(t) = \mathrm{Re}\{\mathbf{V_s}\}$, the real part of the noise is subtracted from the measured signal

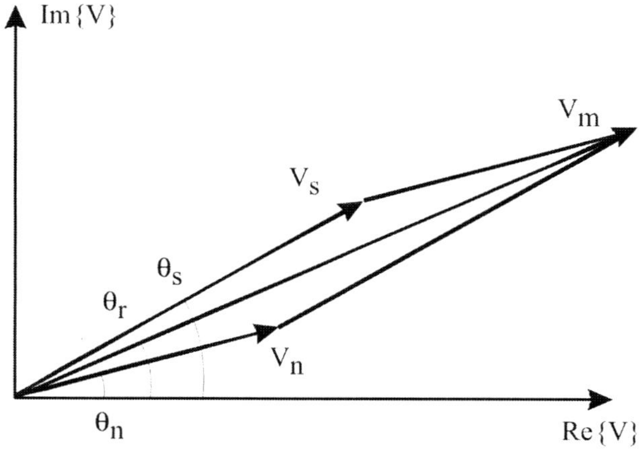

FIGURE 3. Phasor diagram of signal, noise, and resultant vectors measured with the BPM striplines.

$$\begin{aligned} V_s(t) &= V_{BPM}(t) - \mathrm{Re}\{V_n(t)\} \\ &= V_{BPM}(t) - A_n(t)\cos(\omega t - \theta_n(t)) \end{aligned}, \tag{2}$$

where $A_n(t)$ and $\theta_n(t)$ are determined from a linear fitting of the data. Specifically, An(t) and $\theta_n(t)$ are expressed as

$$A_n(t) = m_A t + b_A \text{ , and} \tag{3a}$$

$$\theta_n(t) = m_\theta t + b_\theta \text{ .} \tag{3b}$$

Generalizing Eq. (2) to cover all beam harmonics where noise may be present, the noise correction may be expressed as

$$V_s(t) = V_m(t) - \sum_{i=1}^{N} A_{n_i} \cos(i\omega t - \theta_{n_i}(t)) \text{ .} \tag{4}$$

The phase coefficient, $m_\theta = \Delta\theta/\Delta t$, represents a constant frequency. In the present case, it should be dominated by the difference between the actual linac frequency and the frequency bin where the signal resides, as specified by the FFT. Given the 2.5-MHz resolution determined by the sample window (400 ns), the FFT places essentially all the linac fundamental harmonic signal in bin 80 (200 MHz).

EXPERIMENTAL RESULTS—PROCESSED DATA

The linac main tube (Burle 7835) supplies 5 MW of rf power to the Alvarez structure during the 70- to 80-μs macropulse. The signal on the stripline closest to the linac, BPM1, is particularly susceptible to interference from the high-power rf pulse. An example of amplitude and phase of the fundamental frequency on the bottom stripline of BPM1 is presented in Fig. 4; note the apparent reduction in amplitude near the start of the macropulse. As mentioned above, the linear phase shift is the result of the constant frequency difference between 200 MHz and 200.07 MHz.

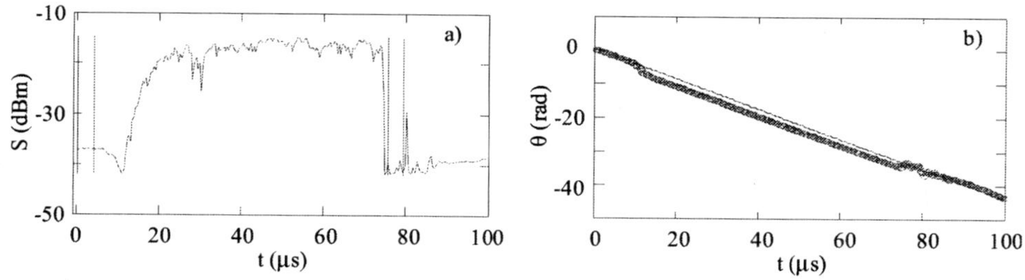

FIGURE 4. a) Amplitude and b) phase of the 200-MHz macropulse component on BPM 1, bottom stripline ($\Delta f = 2.5$ MHz). The pre- and post-pulse sampling windows are shown in a).

In Fig. 5, the 200-MHz component has been subtracted as prescribed in Eq. (2).

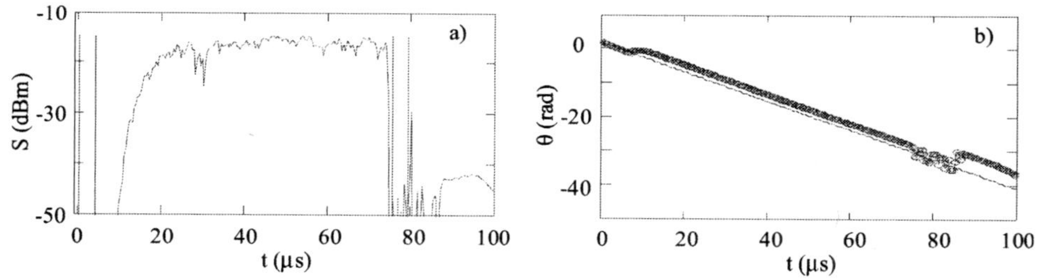

FIGURE 5. a) Amplitude and b) phase of the same data presented in Fig. 3 with the noise component removed. The pre- and post-pulse sampling windows are again shown in a).

The apparent dip of early signal intensity in Fig. 4a) appears to originate from destructive interference between the noise and signal phasors. As mentioned earlier, other noise harmonics are present on the stripline data. The linac main tube has a resonance near 1 GHz that is detectable on the striplines, especially BPM1. In Fig. 6, two successive noise reduction operations are carried out on the data showing improvement in both cases. This suggests that more than one noise component is present within the 1-GHz band. Note the increased phase shift for the 1-GHz data set (#3), nominally 5 times the 200-MHz case.

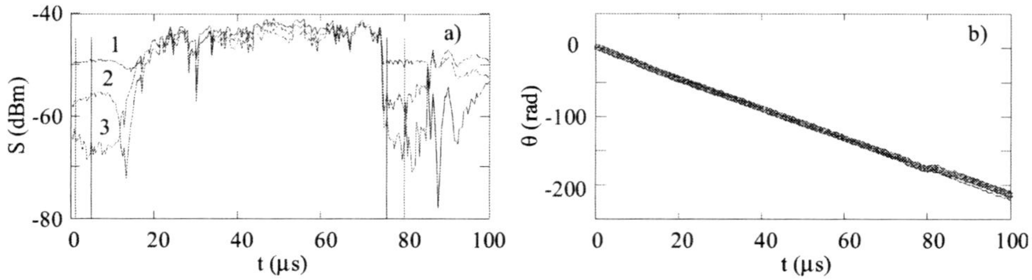

FIGURE 6. a) Amplitude and b) phase of the 1-GHz macropulse component on BPM 1, bottom stripline ($\Delta f = 2.5$ MHz). In a), the noise reduction operation is repeated with the result shown in curve 3.

During the macropulse, the linac field strength varies by less than 2 percent; therefore we expect that the linac noise component also remains relatively constant over this period. Energy and energy spread results are compared for pre- and post-noise removal in Figs. 7 and 8. In both sets of graphs, noise removal is done at 200 MHz and 1 GHz, only. Little change in the energy is noted between the two cases; however, energy spread does show a substantial reduction in value relative to that calculated without noise removal. PARMILA studies suggest the lower value of $\Delta E \sim 0.4$ MeV to be the more accurate result for our linac.

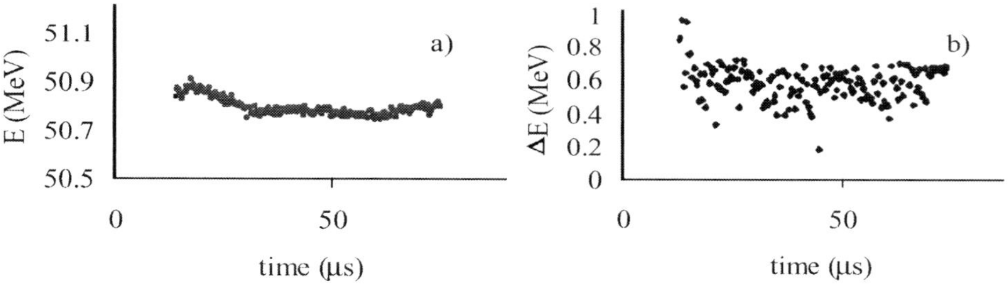

FIGURE 7. Energy and energy spread from preceding data without noise removal.

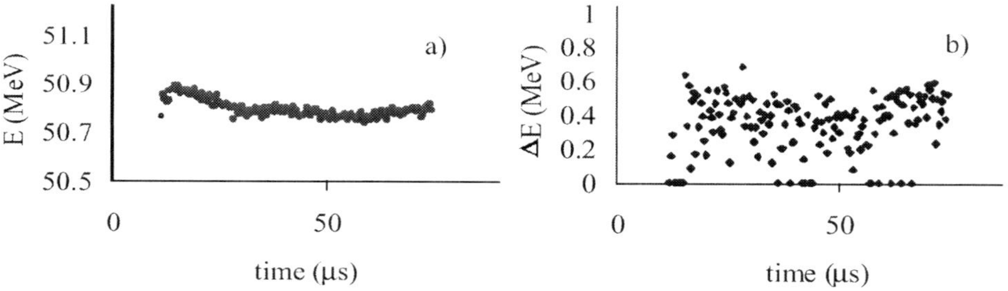

FIGURE 8. Energy and energy spread from preceding data with noise removal.

ATTENUATION AND DISPERSION CONSIDERATIONS

ESEM signal attenuation is accounted for using published data for the type of coaxial cable employed (mainly RG-213). The signals from the four BPM striplines travel over differing cable lengths, though each BPM unit is provided with the same length of cable for all four of its striplines. After the ESEM data is corrected for linac noise in the time domain by Eq. (4), it is converted to the frequency domain where it is then corrected for insertion loss and attenuation. Figure 9 gives the cable attenuation as a function of frequency for each of the ESEM stripline channels. The spectrum from BPM1 stripline, before and after attenuation correction, is presented in Fig. 10.

Restoring the amplitude by itself does not necessarily restore the phase shift introduced by attenuation. Presently, when calculating the width of the microbunch, used to determine the energy spread, rectangular pulse train (RPT) phasing is assumed. The RPT phase relationship between harmonics is expressed as [5]:

$$\phi_n = -n\omega_o\left(t_o + \frac{\tau_o}{2}\right), \tag{5}$$

where $t_o = T/2$ and τ_o = FWHM pulsewidth. RPT phasing tends to minimize the pulsewidth for a given harmonic content. For the energy calculation, FFT phasing is retained after attenuation correction. A comparison between RPT and FFT phasing is presented in Fig. 11 for the same harmonic content.

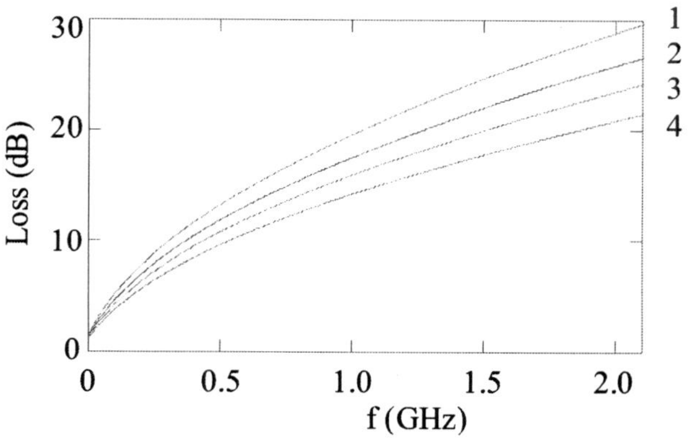

FIGURE 9. Attenuation correction for the BPM ESEM channels.

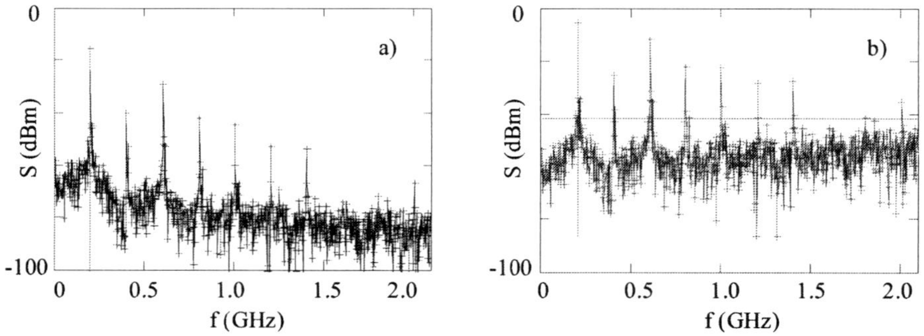

FIGURE 10. BPM1 spectra at 40 μs, a) uncorrected and b) corrected for cable attenuation. Also in b) a –42-dBm threshold is indicated.

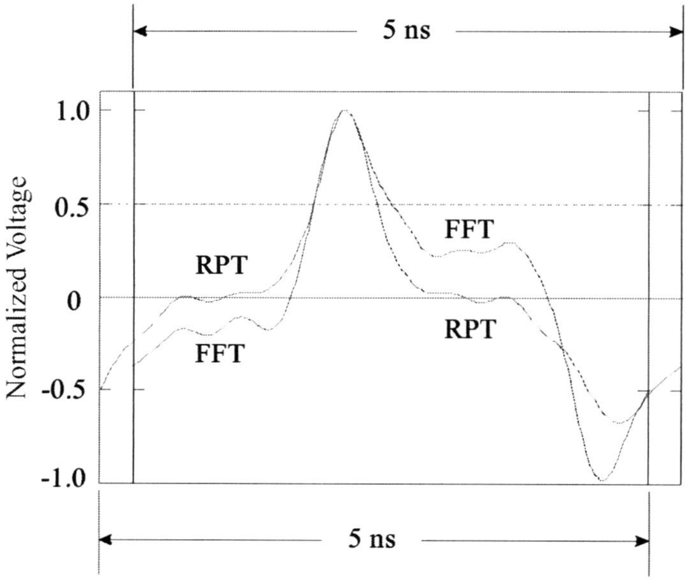

FIGURE 11. Comparison of a reconstructed microbunch using RPT and FFT phasing (same data).

RPT phasing works well if the linac output energy distribution is single-valued; i.e., it can be described by a single profile (e.g., gaussian); however, PARMILA modeling [6] and coarse energy analysis data suggest this may not always be the case. Both scintillator and wire scanner data in the high dispersion region of the 50-MeV transport line (between BPM 3 and 4) indicate a multi-valued distribution in the linac output energy. One goal of this work is to determine if it is possible for the ESEM to resolve such a multi-valued distribution. This may be difficult with the present hardware since PARMILA suggests the centroids of the two main lobes are separated by less than 400 mrad or approximately 280 ps.

ACKNOWLEDGMENTS

This work is made possible by the IPNS Accelerator Group and Division staff.

REFERENCES

1. R. E. Shafer, *IEEE Trans. Nucl. Sci.* **32**(5), 1933 (1985).
2. R. H. Miller et al., *Proc. 12th International Conf. on High-Energy Accelerators*, Fermilab, 1983.
3. A. Jansson, *Phys. Rev. STAB*, **5** 072303 (2002).
4. J. C. Dooling et al., *Proc. 20th LINAC Conference*, SLAC-R-561, Monterey, CA, August 2000, p 193.
5. W. H. Hayt and J. E. Kemmerly, *Engineering Circuit Analysis*, McGraw-Hill, New York, 1978. p. 631.
6. J. C. Dooling et al., *Proc. PAC 2001*, Chicago, June 18-22, 2001, p. 1653.

LabVIEW-Based Data Acquisition, Control, and Analysis Programs for BESSY as Versatile Tools for Optimization and Machine Controls*

O. Dressler, J. Feikes, P. Kuske, and J. Kuszynski

BESSY II, Albert-Einstein-Strasse 15, 12489 Berlin, Germany

Abstract. Complex machines like synchrotron light sources or newly proposed Free Electron Lasers (FEL) posses a variety of coupled parameters that need complex optimization procedures to achieve best possible working conditions. A programming tool like LabVIEW, with its emphasis on easy data acquisition and its very high flexibility, is used extensively to simultaneously access diverse measurement instruments like Scopes, Spectrum Analyzers, and Waveform Generators and combine them into measurement routines which can access all process variables available under EPICS.

INTRODUCTION

The Bessy II control system is based on HP-UX consoles, VME controllers with VxWorks, a relational ORACLE database, and EPICS. EPICS is a set of software tools and applications developed collaboratively and used worldwide to create distributed real-time control systems for large scientific instruments. Whereas EPICS is used to control the machine during normal operation, we also need tools for optimization and machine studies. The tools have to fulfil these requirements:

- easy user interface
- modular structure
- expandable
- automatic execution
- universal data interface for easy analysis and documentation
- running on PC under Win2000/WinXP with access to EPICS process variables (PVs)

LABVIEW AND EPICS

LabVIEW, a graphical development environment used for test and measurement applications in BESSY during the construction phase, has many features for programming, analyzing, and documentation. The concept of "virtual instruments"

*Work supported by Bundesministerium fuer Forschung und Entwicklung and by the Land Berlin.

CP732, *Beam Instrumentation Workshop 2004: Eleventh Workshop*,
edited by T. Shea and R. C. Sibley III

helps to build self-explanatory GUIs (graphical user interfaces). Only a direct interface to the EPICS world is missing. With the build-in ActivX functionality and the ActivX channel access (CA) interface for EPICS developed by K.-U. Kasemir*, we have easy access to communication with all available PVs. The following functions are provided for the channel access client:

ActivX methods: get/put callback
ActivX properties: connected, name, type, units, value

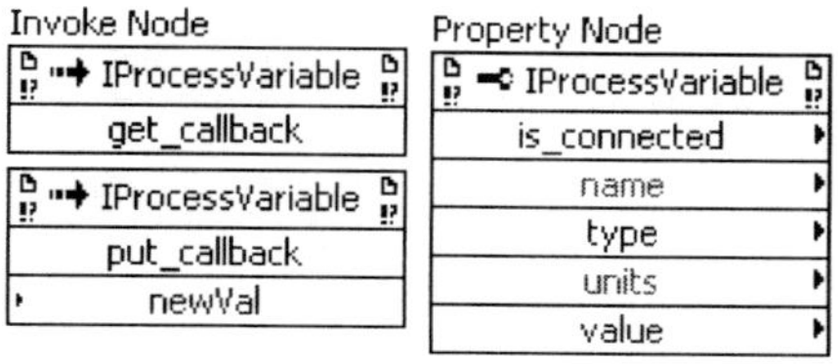

FIGURE 1. ActivX functions for the LabVIEW channel access client.

PRINCIPLE PROGRAM STRUCTURE

All programs are modularly constructed with the parts initialization, scan loop(s), error handling, and re-initialization (Fig. 2). The initialization file is an easy-to-edit ASCI script with the names of PVs to read and the scan parameters (e.g., scan area, step width) of PVs to write. An integration of additional measurement instruments is simplified by LabVIEW that includes lots of I/O drivers (GPIB, Ethernet, USB, serial, and CAN). Other preparing tasks are building scan parameter arrays, saving initial state and open EPICS connections. Most scans would interfere with normal user operation so we have to use special machine (night) shifts. Therefore, the programs must be optimized for automatic execution through intelligent error handling, e.g., self-controlled new injection after beam loss. Another important aspect is the optimization of scan cycle time because of the limited available time.

Especially the optimal trigger points of instruments and settling times of changes, e.g., lifetime, have to be found. Data are stored as ASCI tables for easy further processing or binary for large amounts of data. At the end of scans or cancellation by the program or user interrupt all changed machine parameters should have their initial state.

LAUNCHER FOR PROGRAM MANAGEMENT

For the increasing number of programs we need an easy access. We have developed a launcher which is permanently displayed on the desktop (Fig. 3). Program groups are represented through buttons. An ASCI file includes the directory names of these groups. A new program must only be copied in the adequate directory.

---

*E-mail: kasemir@lanl.gov.

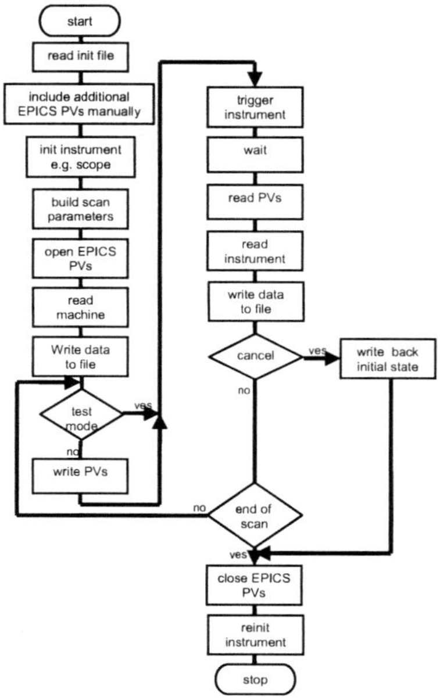

FIGURE 2. Program structure.

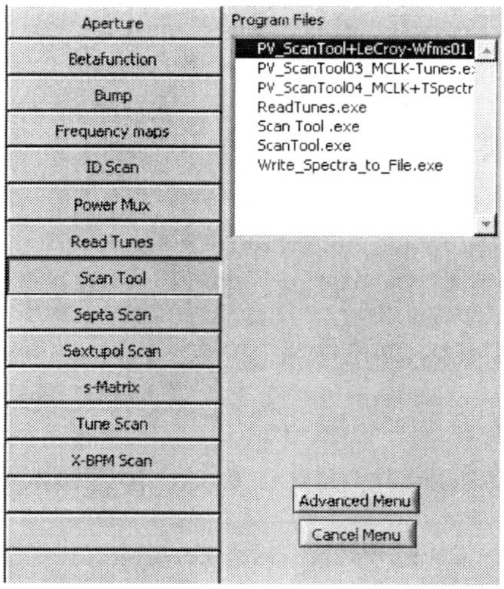

FIGURE 3. Launcher panel with selected program group "Scan Tool." Buttons without text are not used yet.

EXAMPLES

Tune Scan

BESSY II as a 3rd-generation light source has some insertion devices (IDs) in user operation which have an impact on machine performance when they are in operation. An important issue is the reduction of beam lifetime due to the closing of the undulators to minimal gaps. The typical lifetime at 250 mA stored current is about 8 h with all gaps opened. Lifetime reductions can reach 50% when the most harmful IDs are closed. Systematic tune scans around the working points have been performed by acquiring lifetime, beam loss monitor rates and source sizes at the pin hole monitors [1]. As an example in Fig. 4, the product current * lifetime (normalized to a constant current) is painted in a grey code vs the horizontal and vertical frequency of the tune [2]. With the beam lifetime measurement taking 10 s and the scan matrix of 41*41 points, the scan time is about 4 h.

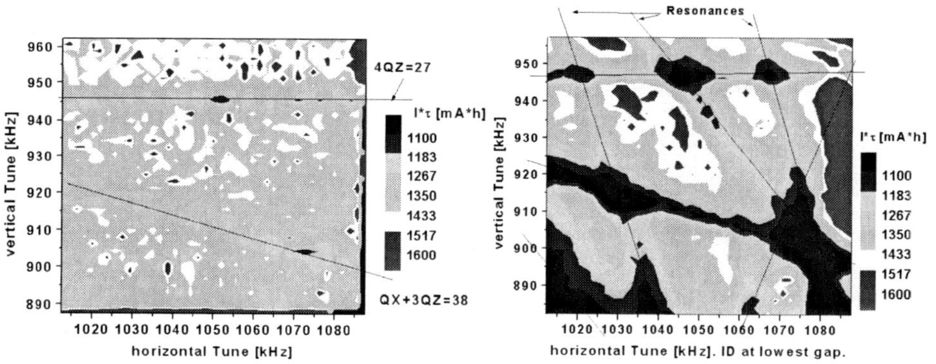

FIGURE 4. Tune scans with ID UE56D3 open (left) and closed to 15.7 mm (right). The tune shift range is +/- 37.5 kHz from start point fx = 1049kHz, fz = 925kHz.

With minimal gap (15.7 mm) of UE56D3, the QX+3QZ = 38 resonance appears with a lifetime degradation of up to 50% relative to its value at open gap. New higher order resonances are visible. The area of tunes with good lifetime values decrease significantly.

The program scans the area from the working point (centre) to the left bottom and then as spiral line back to the centre (Fig. 5). So the quadrupol magnets see current changes analogue to the cycling procedure and the hysteresis effects will be minimized.

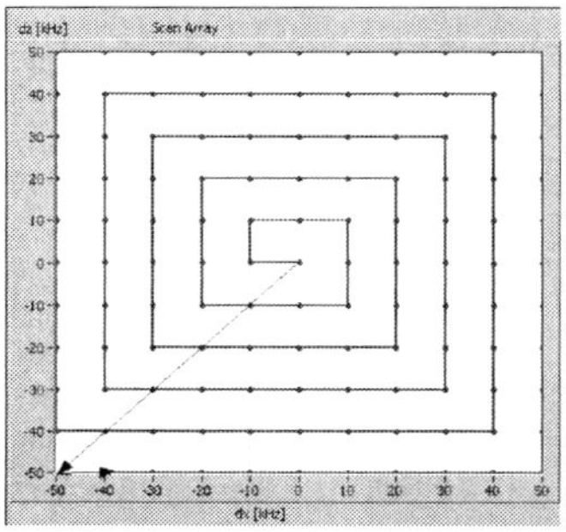

FIGURE 5. Example of an 11*11 area scan at +/- 50 kHz transversal and vertical range.

Scan for Adjusting of the Harmonic Sextupoles

The BESSY II lattice has two sextupole families at positions with non-zero dispersion to allow for chromaticity correction. The sextupole magnets of each family are located at positions of low or high horizontal betafunction. The dispersion is comparatively low, so these two families have to be strongly excited, leading to a drastic break down of dynamic aperture. In order to counteract this effect, there are two additional sextupole families, the harmonic sextupoles, present at positions with zero dispersion. They do not change the chromaticity but restore the dynamic aperture. Their optimum settings could not be predicted correctly from theory but were found by scanning all four harmonic sextupole circuits against each other. Adjusting all harmonic sextupoles to their best values (obtained after 2 iterations of scans) the BESSY lifetime increased by 30% [3]. In Fig. 6 an example of the influence of the sextupole current variation to the product beam current* lifetime (I*τ) (normalized to the reference current) is shown.

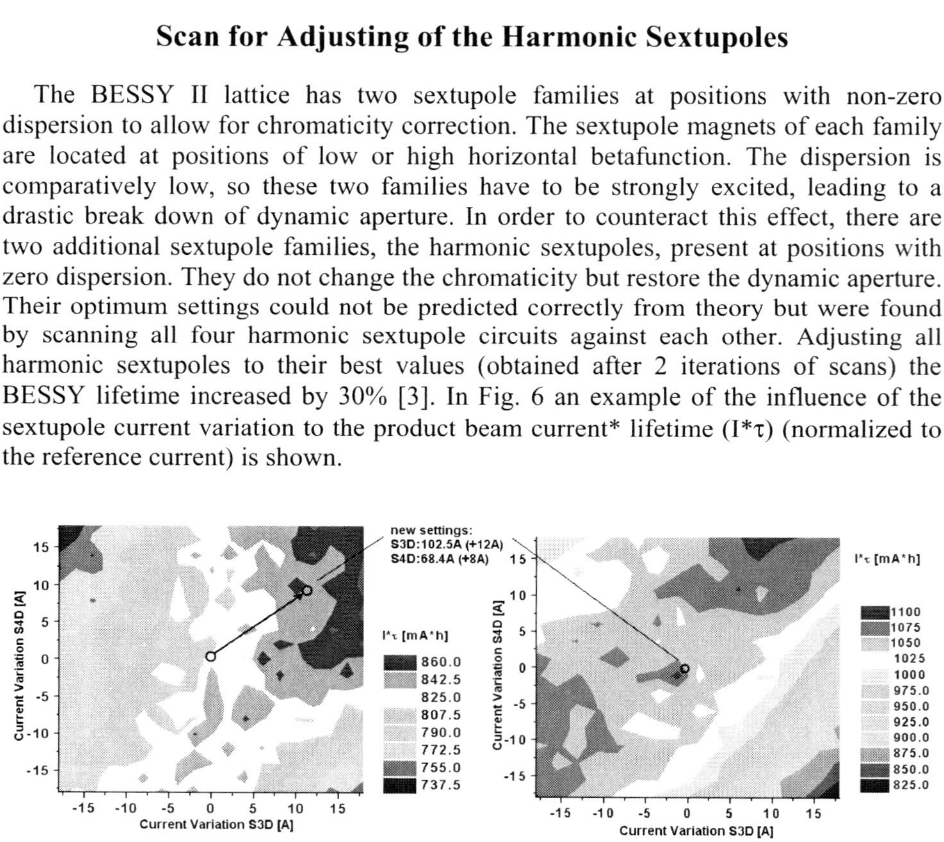

FIGURE 6. Current* lifetime (I*τ) after first (left) and second (right) iteration of the S3D vs S4D scan.

Sextupole Scan for "Low Alpha" Mode Optimization

The "low alpha" optics is a dedicated storage ring mode, where stored electron bunches are compressed, leading to emission of intense THz radiation. The bunch length and shape have been manipulated by tuning the storage ring optics into the "low alpha" mode, where alpha is the momentum compaction factor describing the orbit length variation with beam energy [4]. Most machine settings are much more critical than in the standard BESSY user optics. With the sextupole scans, settings were found which increase the beam lifetime significantly without degradation of infrared radiation and source size on pin hole monitors. These settings were successfully used during the "low alpha" user run in Nov. 2003 at BESSY II.

CONCLUSIONS

For the BESSY II synchrotron light source, a collection of programs for active and passive machine scans was developed. By the intuitive graphical development environment short programming times are achieved. The possibilities of access to the machine control system are expanded by an easy access to external instruments. A wide range of interfaces is supported. Better machine settings with the result of, e.g., significant higher lifetimes, could be found with systematic scans around the working points. The LabVIEW-based programs work without negative influence on the machine stability.

ACKNOWLEDGMENTS

The authors thanks the colleagues of the BESSY machine group, who have helped them to implement the measurement PC in the control system net, and all operators of machine night shifts for their engagement.

REFERENCES

1. J. Feikes and G. Wuestefeld, "Lifetime Reduction Due to Insertion Devices at BESSY II," Proceedings of the 2003 Particle Accelerator Conference, Portland, pp. 845-847.
2. J. Feikes, "Lebensdauer, BLM-Verlustrate, und Quellgroesse in Abhaengigkeit des Arbeitspunktes von BESSY II und Einfluss des UE56D3," internal note, Nov. 14, 2001.
3. J. Feikes, et al., "Beam Lifetime Optimization of BESSII By Systematically Adjusting the Harmonic Sextupoles," Proceedings of EPAC 2002, Paris, France, pp. 775-777.
4. M. Abo-Bakr et al., "Brilliant, Coherent Far-Infrared (THz) Synchrotron Radiation," in *Physical Review Letters,* Vol. 90, No. 9, March 2003.

Latest Diagnostic Electronics Development for the PROSCAN Proton Accelerator

P.A. Duperrex, U. Frei, G. Gamma, U. Müller, and L. Rezzonico

Paul Scherrer Institute, CH - 5232 Villigen, Switzerland

Abstract. New VME-based diagnostic electronics are being developed for PROSCAN, a proton accelerator for medical application presently under construction at PSI. One new development is a VME-based multi-channel logarithmic amplifier for converting current to voltage (LogIV). The LogIV boards are used for measuring current from the multiple wire (harp) profile monitors. The LogIV calibration method, current dependant bandwidth and temperature stability are presented. Another development is a BPM front end, based on the newest digital receiver techniques. Features of this new system are the remote control of the preamplifier stage and the continuous monitoring of each individual signal overall gain. Characteristics of the developed prototype are given.

THE PROSCAN PROJECT

PROSCAN is a proton accelerator for medical application presently under construction at PSI. The new superconducting 250-MeV cyclotron is delivered by ACCEL Instruments GmbH, and the first beam line elements are now being assembled at the PSI site. The PROSCAN project is an interdisciplinary project of several laboratories and divisions within PSI [1] and will give improved scanning technologies for radiotherapy and the treatments of additional medical conditions (e.g., movable tumors). The degrader and the beam lines have been designed and built by PSI. For the beam line diagnostics, new VME-based electronics [2] are being developed and their basic features are presented here.

DEVELOPMENT OF VME-BASED ELECTRONICS

The VME-based electronics under development will eventually replace the existing CAMAC-based boards used for the initial phase of the PROSCAN project. For these electronics, the chosen approach was to take advantage of the general structure of a VME system [2]. By clearly defining the tasks of the main VME module (generic VPC-Board) and the transition module, the development of these modules can be performed in parallel. The transition module electronics will be dedicated to the specific needs of a given diagnostic such as the beam position monitors (BPMs) or current measurements (LogIV32) for the profile monitor.

The generic "VPC-Board" is a VME-64x live insertion compliant carrier board with two slots for PMC mezzanine board plugins [3]. All user pins on the backplane

CP732, *Beam Instrumentation Workshop 2004: Eleventh Workshop*,
edited by T. Shea and R. C. Sibley III

VME connectors P0 and P2 can be redirected by the user to any destination on the VME, to the PMC mezzanines using the 66-MHz PCI bus or to an external location using an optical Gigabit link that is based on the RocketIO™ protocol. Embedded PowerPC™, fast memory, and a Sharc™ DSP allow the user to acquire and process data in real time. PMC slots allow the use of a wide range of commercially available boards for specific tasks (Ethernet, ADC/DAC, Digital I/O modules, etc.).

LOGARITHMIC AMPLIFIER FOR SMALL CURRENTS

Basic Principle

Logarithmic amplifiers (see Fig. 1) use the logarithmic relationship between emitter voltage V_E and the collector current I_c of a transistor [4].

$$-V_E = \frac{kT}{q} \ln\left(\frac{I_c}{I_o}\right), \tag{1}$$

where k is the Bolzmann's constant (1.38 x 10^{-23} J/K), T the temperature in K, q the electronic charge (1.6x10^{-19} C), and I_o a reference current depending on the exact amplifier configuration. Differentiating (1) gives the small signal feedback resistance

$$r_E = \frac{\partial V_E}{\partial I_c} = \frac{kT}{qI_c} \approx \frac{1}{40I_c} . \tag{2}$$

To ensure stable closed-loop operation at high operating current, a capacitor is connected between the op amp output terminal and the summing point. The capacitor however places a severe restriction on the bandwidth and output slewing rate at the lower levels of operating current because of large r_E values for small currents. A resistor R_E must also be connected in series to reduce the load of the op amp at high operating current (at 1 mA the feedback resistance r_E is ~25Ω).

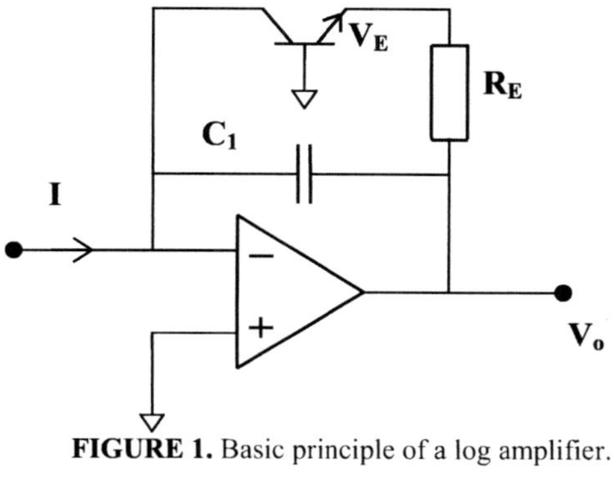

FIGURE 1. Basic principle of a log amplifier.

Existing CAMAC Boards

Previously, LogIV electronics for the PSI proton accelerator have been based on the CAMAC format [5]. They will still be used during the initial phase of operation for PROSCAN [6].

The CAMAC LogIV board has 16 LogIV channels, and current from 20 pA to 200 μA can be measured. This sensitive circuit is mounted in a metallic enclosure for shielding from external electromagnetic noise. To avoid ground loop problems, the electrical ground of the CAMAC board and that of the LogIV amplifier stage (being at the machine ground) are separated. Readout and further processing of the simultaneously sampled signals are performed every millisecond by the DASH (Data Acquisition module with Hitachi SH2 microcontroller) back-end electronics. The 10-bit sampled raw data are also averaged over a 20-ms period, allowing both an improved resolution (2 bits or 12d-B improvement) and also filtering of 50-Hz mains noise that is relatively large at low (<1 nA) signal current.

The calibration procedures have been automated with LabVIEW. The response for each channel is measured at different currents using a calibrated Keithley220 current source. Offset and gain correction are then calculated and automatically stored in an EEPROM of the CAMAC board. Figure 2 shows the typical performance achieved using this procedure.

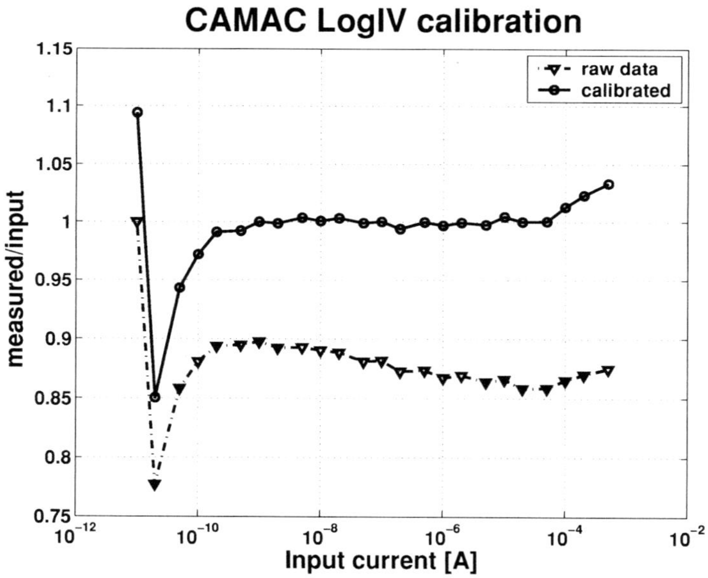

FIGURE 2. Current measurement before (broken line) and after calibration (thick line). Precision better than ±2% is expected between 100 pA and 100 μA.

New VME-Based Boards

The new VME-based LogIV electronics is made of two parts, the General Acquisition Transition Interface (transition board) and a module containing the LogIV electronics for 32 channels (see Fig. 3).

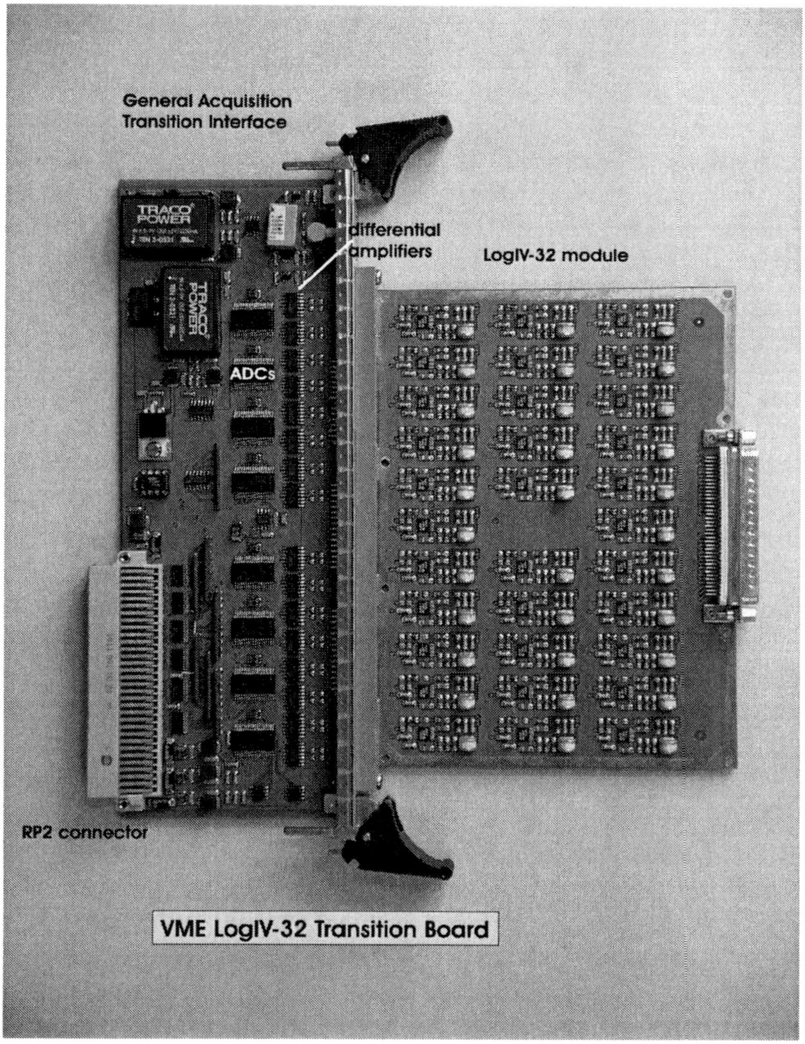

FIGURE 3. VME LogIV-32 board consisting of a LogIV-32 module and a General Acquisition Transition Interface.

The LogIV-32 module electronics uses commercially available logarithmic amplifiers (AD8304 from Analog Devices) allowing current from 20 pA to 500 μA to be measured. A metallic enclosure shields the electronics from external electromagnetic noise. The electrical grounds of the transition board and the enclosure electronics are separated to avoid ground loop problems. The 32 channels are synchronously digitized using Maxim MAX125 14-bit ADCs and are then further processed on the generic board.

The performance of this system is expected to be the same as the original CAMAC-based version. Every channel will be separately calibrated with the corresponding calibration parameters (offset and gain correction) and stored in an EEPROM of the transition board. The precision is expected to be better than ±2% for current between 100 pA and 100 μA.

Temperature Stability and Dynamic Response

Temperature sensitivity has been checked for both CAMAC and VME solutions using a climatisation chamber. The results indicate a variation less than ±5% from 20 to 50°C. Most of the variations occur between 20 and 35°C.

The response time of the logarithmic amplifier is dependent on the measured input current as explained in the basic principle section. To measure this dependence, either a small sine or a square wave modulation was superimposed to a given input current. For the sine modulation technique, the -3 dB frequency was measured; whereas, for the square wave modulation, the rise and fall time have been measured and the corresponding cut-off frequency calculated. Both methods deliver similar results (Fig. 4). However, the square wave modulation is faster and the measurements are more easily averaged. The measured results show that the new VME version is typically five times faster than the former CAMAC one.

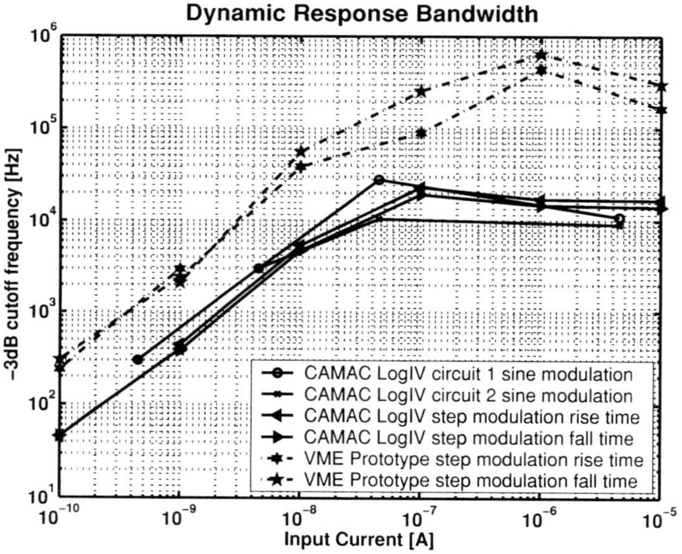

FIGURE 4. Small signal bandwidth: comparison of results obtained by two different methods and for two different types of LogIV circuit.

DIGITAL RECEIVER ELECTRONICS FOR BPM

The new circuit in Fig. 5 will be used for the 72-MHz BPM front end on PROSCAN. In 2005 the same circuit will also replace the obsolete electronics on the proton accelerator.

The circuit is based on the digital receiver techniques. The RF beam signals will be directly digitized without the usual analog mixers. Larger dynamic range, true 90 degree I/Q demodulation, increased temperature, and long time stability are among the advantages of the digital receiver over the analog receiver. Furthermore, the flexibility of the digital process techniques makes the system easily reconfigurable.

Features of this new system are the remote gain control of the preamplifier stage and the continuous monitoring of each individual signal overall gain (amplifier gains and various attenuations). The remote gain control maximizes the dynamic range, expected to be 104 dB. The overall gain of each signal is measured using a pilot signal fed at the pre-amplification stage. This will improve the precision and quality of the measurement by taking into account the variations of factors affecting the channel overall gain.

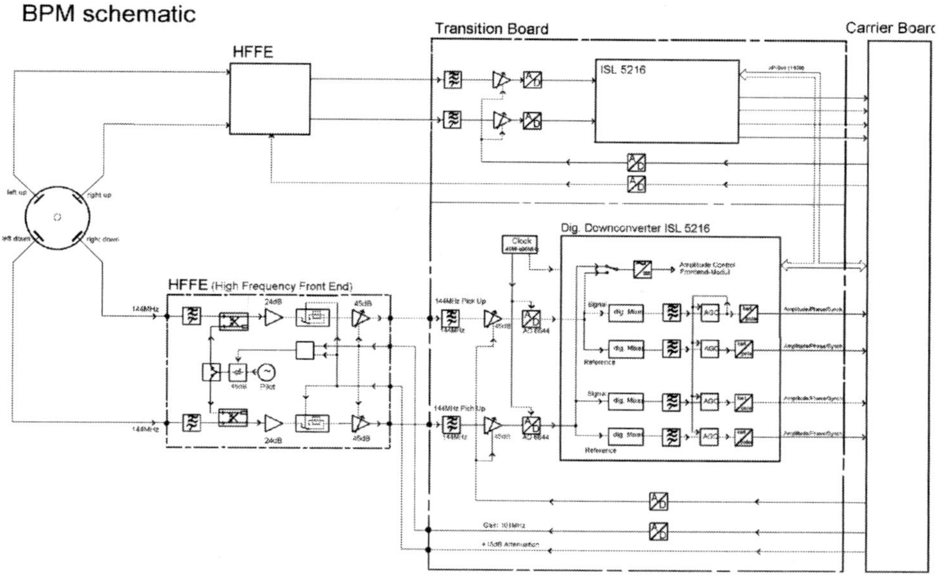

FIGURE 5. Schematics showing the electronics for the new BPM system.

In the High Frequency Front End (HFFE) circuit, the signals from the BPM pick-up coils are first filtered around the RF second harmonic, i.e., 144 MHz. This frequency is used because the direct RF field contribution is expected to be smaller than the one at the fundamental ($f_0 = 72$ MHz) or at the third harmonic ($3 \cdot f_0$). The beam and pilot signals are then pre-amplified to minimize the noise effects (typical cable lengths between HFFEs and the VME electronics are 75 to 100 meters). The pilot signal is set at a slightly different frequency from the RF signal and is separately filtered and processed after the gain stages.

The first block of the VME transition board is a band-pass filter centered on the RF first harmonic, i.e., 144 MHz. The signal is then amplified to use the full dynamic range of the fast 14-bit AD6644 ADCs. The ISL5216 digital down converter (DDC) translates the pilot and beam signals into the baseband.

Digitally filtered base-band signals are then further processed on the carrier board using the Xilinx Virtex2 Pro FPGA. In particular, the beam signal amplitudes are normalized using the pilot signals, adaptive filtering is applied, and the beam position (horizontal and vertical component) is computed. A built-in-test (BIT) procedure will also be implemented to localize and identify problems or failures.

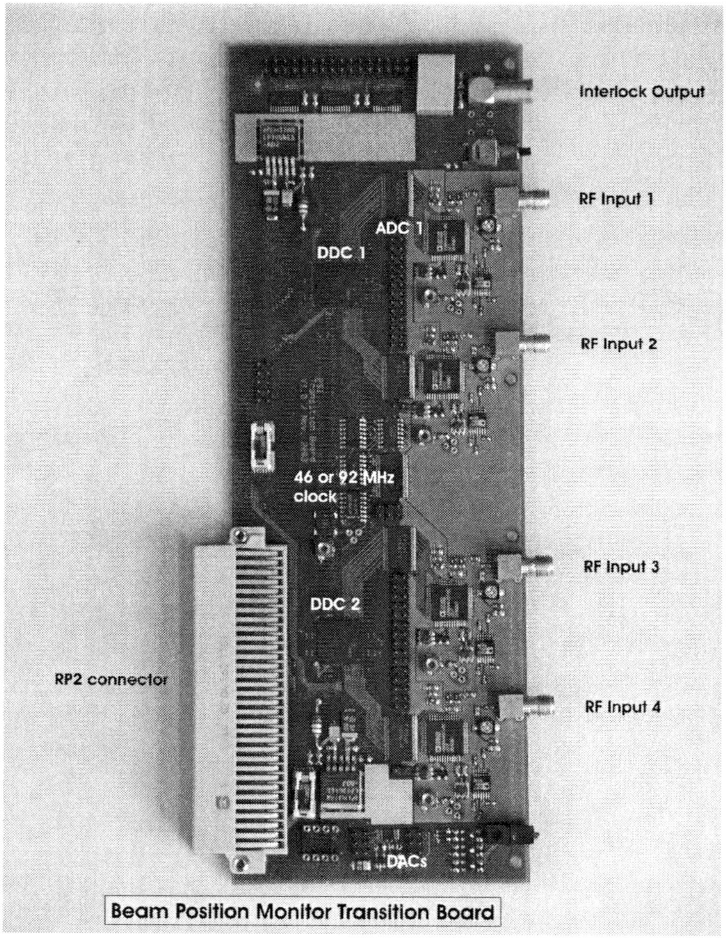

FIGURE 6. Transition board prototype for the new BPM electronics.

A task of the FPGA firmware is to set the gain of the electronics amplification stages. The procedure has been defined so that the average signal level uses the full dynamic range of the digitizer. At the same time, it optimizes the various gains to minimize the total noise figure of the receiver. The total noise figure referenced to the input port can be computed as follows:

$$F_{total} = F_1 + \frac{F_2 - 1}{G_1} + \frac{F_3 - 1}{G_1 . G_2} + \ldots , \tag{3}$$

where F_x is the noise factors for each of the stages and G_x is the gains of the stages. For similar F_x values at the different stages, it is advantageous to have larger gains in the initial stage of the receiver. Spurious effects near amplifier saturation must however be avoided.

The expected dynamic range of the input RF signal level is from -120 dBm to -16 dBm, and the sampling rate for position measurement is 10 kHz. Figure 6 shows a transition board prototype presently under test.

ACKNOWLEDGMENTS

The authors thank P. Pollet, B. Keil, and R. Kramert for the discussions concerning the VME generic PMC carrier board and their support. The authors are also grateful to C. Gough for manuscript reading.

REFERENCES

1. M. Jermann et al., "The PROSCAN Project: a Progress Report," in *PSI Scientific and Technical Report 2003 Vol.6 Large Research Facilities,* PSI, 2003, pp. 90-91.
2. Wade D. Peterson, *The VME Handbook 4th edition*, VMEbus International Trade Association, 1997.
3. P. Pollet et al., "Introducing the VME Generic PMC Carrier Board: A Modular Approach for Beam Instrumentation at PSI," in *PSI Scientific and Technical Report 2003 Vol. 6 Large Research Facilities,* PSI, 2003, pp. 38-39.
4. G. B. Clayton, *Operational Amplifiers 2nd Edition*, Butterworth & Co Publishers Ltd, 1983, pp. 163-222.
5. R. Dölling, "Profile Measurement of Scanning Proton Beam for LiSoR Using Carbon Fibre Harps," in *Beam Instrumentation Workshop-2002*, edited by G. A. Smith and T. Russo., AIP Conference Proceedings 648, New York: American Institute of Physics, 2002, pp. 361-367.
6. R. Dölling, "Profile, Current, and Halo-Monitors of the PROSCAN Beam Lines," this proceedings.

Improving FFT Frequency Measurement Resolution by Parabolic and Gaussian Spectrum Interpolation

M. Gasior and J. L. Gonzalez

CERN, CH-1211, Geneva 23, Switzerland

Abstract. Discrete spectra can be used to measure frequencies of sinusoidal signal components. Such a measurement consists of digitizing a compound signal, performing windowing of the signal samples and computing their discrete magnitude spectrum, usually by means of the Fast Fourier Transform algorithm. Frequencies of individual components can be evaluated from their locations in the discrete spectrum with a resolution depending on the number of samples. However, the frequency of a sinusoidal component can be determined with improved resolution by fitting an interpolating parabola through the three largest consecutive spectrum bins corresponding to the component. The abscissa of its maximum constitutes a better frequency approximation. Such a method has been used for tune measurement systems in circular accelerators. This paper describes the efficiency of the method, depending on the windowing function applied to the signal samples. A typical interpolation gain is one order of magnitude. Better results are obtained with Gaussian interpolation, offering frequency resolution improvement by more than two orders of magnitude when used with windows having fast sidelobe decay. An improvement beyond three orders of magnitude is possible with steep Gaussian windows. These results are confirmed by laboratory measurements. Both methods assume the measured frequency to be constant during acquisition and the spectral peak corresponding to the measured component to constitute a local maximum in a given band of the input signal discrete spectrum.

FFT FREQUENCY MEASUREMENT

Assume that a band-limited compound signal $s(t)$ has been uniformly sampled with frequency f_s and contains a sinusoidal component $s_{in}(t)$, whose frequency f_{in} is to be measured. The Discrete Fourier Transform (DFT) magnitude spectrum of the signal sample sequence $s[n] = s(nT_s)$, usually computed using the Fast Fourier Transform (FFT) algorithm, is given by

$$S[k] = \left| \sum_{n=0}^{N-1} s[n] \exp\left(-\mathrm{j} \frac{2\pi n k}{N} \right) \right| , \qquad (1)$$

where $T_s = f_s^{-1}$ is the sampling period and N is the total number of samples. The discrete spectrum (1) is calculated at frequencies that are integer multiples of $\Delta_f = f_s/N$. If $S[k]$ has an observable local maximum corresponding to $s_{in}(t)$ at the spectrum bin k_m, f_{in} can be approximated as

CP732, *Beam Instrumentation Workshop 2004: Eleventh Workshop*,
edited by T. Shea and R. C. Sibley III

$$f_{in} \cong k_m \Delta_f = k_m \frac{f_s}{N} = \frac{k_m}{N T_s} = \frac{k_m}{L} , \tag{2}$$

where $L = NT_s$ is the sampling duration of the input signal. The largest approximation error $\varepsilon = \max(|k_m\Delta_f - f_{in}|)$ occurs for a frequency located exactly between two bins. This error can be considered as the resolution of the FFT frequency measurement and

$$\varepsilon = \frac{1}{2}\Delta_f = \frac{f_s}{2N} = \frac{1}{2NT_s} = \frac{1}{2L} . \tag{3}$$

The resolution can be increased considerably by discrete spectrum interpolation, which has been used in tune measurement systems [1]. The method principle is sketched in Fig. 1. When the frequency component f_{in} is located exactly on a local maximum of the discrete magnitude spectrum at bin k_m, f_{in} can be calculated according to (2) with no error. This is the ideal case shown in Fig. 1(a). When f_{in} increases, the amplitude of bin $k_m - 1$ gets smaller and of bin $k_m + 1$ bigger, as presented in Fig. 1(b). The spectrum value $S[k_m]$ remains the biggest, until f_{in} is equidistant between bins k_m and $k_m + 1$, see Fig. 1(c). In this case, determining f_{in} with (2) leads to the largest error (3). When f_{in} is increased further, as shown in Fig. 1(d), bin $k_m + 1$ becomes the biggest. For f_{in} smaller than $k_m \Delta_f$ the analysis is similar, resulting in symmetrical cases.

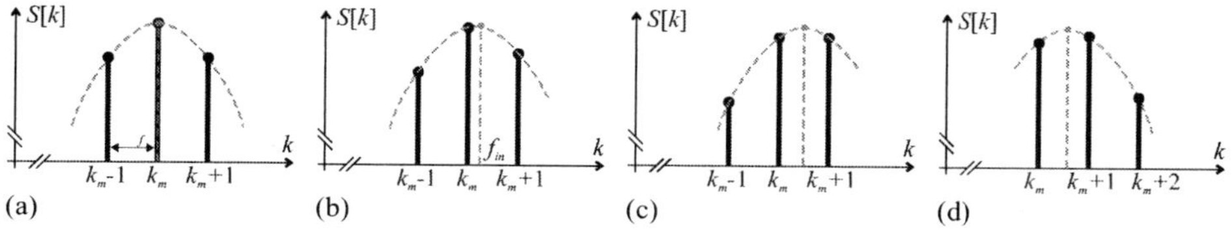

FIGURE 1. The principle of discrete spectrum interpolation.

This example illustrates the fact that continuous frequency f_{in} can be estimated, even if it is located between two bins, by calculating the maximum abscissa of an interpolation curve of the discrete spectrum peak. This maximum can be located between $k_m - 1/2$ and $k_m + 1/2$, where k_m is the index of the biggest bin within the range of interest. If one needs to resolve the cases presented in Fig. 1 and symmetrical ones, it is necessary for the interpolation to have at least three node points. As seen in Fig. 1(c), for efficient interpolation the minimal width of the spectral peak is 3 bins.

If in the continuous-time Fourier Transform (FT)

$$S(f) = \int_{-\infty}^{\infty} s(t)\exp(-\mathrm{j}2\pi f t)dt \tag{4}$$

one assumes time domain sampling $t=nT_s$ and frequency domain sampling $f=k\Delta_f$, then the integral can be replaced by a sum with N elements and FT (4) becomes DFT as in (1) when taking T_s and Δ_f as units in the discrete time and frequency domains. For this reason discrete spectra in the paper are considered as continuous spectrum samples taken at multiples of Δ_f, allowing examination of discrete spectra between the bins by means of the FT (4). Replacing the ordinary frequency f in (4) by

$$\varphi = \frac{f}{\Delta_f} = f\frac{N}{f_s} = fNT_s = fL \,, \tag{5}$$

referred to as the normalized frequency, adjusts continuous spectra to the scale of the discrete spectrum bin indexes. It allows expressing discrete and continuous spectra in one convenient scale (e.g. φ = 10.5 for frequency lying exactly between bin 10 and 11).

Interpolated spectral peaks have shapes corresponding to the spectrum of the window function, since windowing can be considered as a modulation applied to each sinusoidal component of the analyzed signal. Consequently, the goal of the interpolation methods discussed hereafter is to find the maximum abscissa of such a shape while knowing only its three discrete spectrum samples.

Results of the methods are presented for weighted cosine windows with fast decaying small sidelobes, described by Nuttall [1]:

$$w_c(t) = w_r(t)\sum_{i=0}^{3} c_i \cos\left(\frac{2\pi i}{L}t\right) \,, \tag{6}$$

where $w_r(t)$ is the rectangular window with duration from $-L/2$ to $L/2$. Basic properties and coefficients of the weighted cosine windows are listed in Table 1. In the window names of the form aTb, a stands for the number of terms on which the main lobe width depends, and b for the order of the highest continuous derivative, determining the sidelobe asymptotic fall-off. As an example, 4T1 is a four-term window with the main lobe full width of 8 bins ($2a$) and continuous first derivative, resulting in the sidelobe decay of 18 dB/octave ($6b$+12).

Three Gaussian windows are also considered. They are defined as

$$w_g(t) = w_r(t)\exp\left(-r^2\frac{t^2}{2L^2}\right) \,, \tag{7}$$

where r is the ratio between the window length L and its standard deviation σ, i.e. the Gaussian shape is truncated at $\pm\,\sigma r/2$.

The window choice is a compromise between a few parameters, three of which are listed in Table 1. The main lobe width determines the minimal distance between two spectral peaks, which still can be resolved. The highest sidelobe and sidelobe fall-off characterize the spectral leakage to the component of interest from respectively close and far interference. As seen in Table 1 and Fig. 2, showing an example of magnitude spectra of four windows, spectral properties of Gaussian windows are inferior to those of the weighted cosine ones. This paper deals only with interpolation on idealized (i.e. not perturbed by noise nor interference) spectra and in this case only the shape of the main lobe is important, exclusively within the interpolation range of $\pm$ 1.5 φ units. This range of Fig. 2 is magnified in Fig. 3.

TABLE 1. Parameters of windows used in the paper.

Window	Main lobe -6 dB width [bin]	Highest sidelobe [dB]	Sidelobe fall-off [dB/oct]	Window coefficients			
				c_0	c_1	c_2	c_3
Hanning	2.00	-31.5	18	1/2	1/2	0	0
Blackman	2.30	-58.1	18	0.42	0.50	0.08	0
3T1	2.36	-64.2	18	0.40897	0.5	0.09103	0
3T3	2.59	-46.7	30	3/8	1/2	1/8	0
4T1	2.69	-93.3	18	0.355768	0.487396	0.144232	0.012604
4T3	2.83	-82.6	30	0.338946	0.481973	0.161054	0.018027
4T5	3.07	-60.9	42	10/32	15/32	6/32	1/32
Gaussian r=6	2.26	-56.1	6	-	-	-	-
Gaussian r=7	2.62	-71.0	6	-	-	-	-
Gaussian r=8	3.00	-87.6	6	-	-	-	-

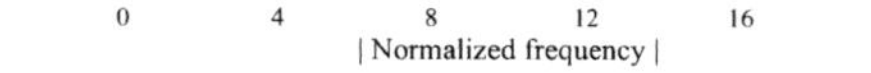

FIGURE 2. Window magnitude spectra.

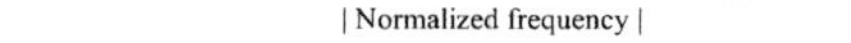

FIGURE 3. Magnified main lobes of Fig. 2.

PARABOLIC INTERPOLATION

As explained before, the goal of the interpolation process is to find the abscissa of the spectral peak maximum, knowing only three peak samples in the form of three bins of the discrete magnitude spectrum. The peak continuous shape, i.e. the window spectrum main lobe, does not need to be accurately reproduced as long as the maximum abscissa of the interpolation shape follows the measured frequency. This is why a simple parabolic interpolation (PI) can improve the discrete spectrum frequency resolution by an important factor even when, as shown in Fig. 3, the shape of the window spectrum main lobe is quite far from a parabola. To quantify this deviation, a window spectrum magnitude $W(\varphi)$ can be expanded into a (normalized to the coefficient upon φ^2) Maclaurin series of the form

$$W(\varphi) \cong a_0 + \varphi^2 + a_4\varphi^4 . \tag{8}$$

Coefficient a_4, listed in Table 2, can be used as a measure of this deviation.

Let $S[k]$ be the discrete magnitude spectrum of N samples of a signal $s(t)$ containing a sinusoidal component of frequency $\varphi_{in} = f_{in}L$, and k_m be the index of the biggest bin of the corresponding discrete spectrum peak. Index k_m can be found if the bin constitutes a local maximum within a given range. Fitting a parabola

$$S_p(\varphi) = a(\varphi - \varphi_m)^2 + h \tag{9}$$

through interpolation nodes $S[k_m-1]$, $S[k_m]$, $S[k_m+1]$ and finding the abscissa of the interpolation maximum φ_m, gives

$$\varphi_{in} \cong \varphi_m = k_m + \Delta_m = k_m + \frac{S[k_m+1] - S[k_m-1]}{2\left(2S[k_m] - S[k_m+1] - S[k_m-1]\right)} \tag{10}$$

under the condition $2S[k_m] > S[k_m+1] + S[k_m-1]$.

The quantity Δ_m in (10) is the abscissa correction of the discrete spectrum maximum. It is a real number linking both, the discrete and continuous spectra, ranging from $-1/2$ when $S[k_m-1] = S[k_m]$, to $1/2$ for $S[k_m+1] = S[k_m]$.

The shape of the interpolated magnitude spectrum peak corresponds to the spectrum of the windowing function applied to the signal samples. If window $w(t)$ with magnitude spectrum $W(\varphi)$ is used, then the interpolation error $E(\varphi_d) = \varphi_m - \varphi_{in}$ is [1]

$$E(\varphi_d) = -\frac{W(\varphi_d+1) - W(\varphi_d-1)}{2(2W(\varphi_d) - W(\varphi_d+1) - W(\varphi_d-1))} - \varphi_d , \tag{11}$$

where $\varphi_d = \varphi_{in} - k_m$. The interpolation error corresponding to four windows is shown in Fig. 4 and is given in units of Δ_f. For non-perturbed spectra the error is the same around each discrete spectrum bin, i.e., is periodic with a period of Δ_f. The interpolation errors for other windows have similar shapes and can be characterized by the error maximum $E_{max} = \max(|E(\varphi_d)|)$ and its abscissa. They are listed in Table 2.

Performance of an interpolation method can be characterized by the interpolation gain, defined as the ratio of the FFT frequency resolution (3) and the method maximum error

$$G = \frac{\varepsilon}{E_{max}} = \frac{\Delta_f}{2E_{max}} . \tag{12}$$

As listed in Table 2, the PI can increase the resolution of discrete spectra by more than one order of magnitude.

GAUSSIAN INTERPOLATION

The interpolation gain can be significantly improved by fitting a Gaussian shape to find the abscissa of the spectral peak maximum located between two discrete spectrum bins. Since a Gaussian curve

$$S_g(\varphi) = \exp\left(a'(\varphi - \varphi_m)^2 + h'\right) \tag{13}$$

is a parabola in the logarithmic scale, the Gaussian interpolation (GI) reduces to the PI on the natural logarithm of the magnitude spectrum.

Analogically to PI, the window spectrum main lobe deviation from a Gaussian shape can be quantified by coefficient b_4 of a normalized Maclaurin series expansion:

$$\ln W(\varphi) \cong b_0 + \varphi^2 + b_4\varphi^4 . \tag{14}$$

The coefficient is listed in Table 2. Its value is much smaller than the corresponding a_4 of (8), especially for Gaussian windows.

The GI can be derived from (10) using logarithmic spectrum values, i.e. the interpolation nodes $S[k_m - 1]$, $S[k_m]$, $S[k_m + 1]$ are replaced by natural logarithms $\ln(S[k_m - 1])$, $\ln(S[k_m])$ and $\ln(S[k_m + 1])$. Thus, after logarithm grouping, (10) becomes

$$\varphi_{in} \cong \varphi_m = k_m + \Delta_m = k_m + \frac{\ln\left(\dfrac{S[k_m+1]}{S[k_m-1]}\right)}{2\ln\left(\dfrac{S[k_m]^2}{S[k_m+1]\, S[k_m-1]}\right)} . \tag{15}$$

Similarly, the GI error can be derived directly from (11):

$$E(\varphi_d) = -\frac{\ln W(\varphi_d + 1) - \ln W(\varphi_d - 1)}{2\left(2\ln W(\varphi_d) - \ln W(\varphi_d + 1) - \ln W(\varphi_d - 1)\right)} - \varphi_d . \tag{16}$$

It is plotted in Fig. 5 for four windows. The errors for other windows have similar shapes and are characterized by the error maximum and its abscissa, as listed in Table 2. The interpolation gains are about two orders of magnitude for cosine weighted windows and well beyond three orders of magnitude for the Gaussian window of $r = 8$.

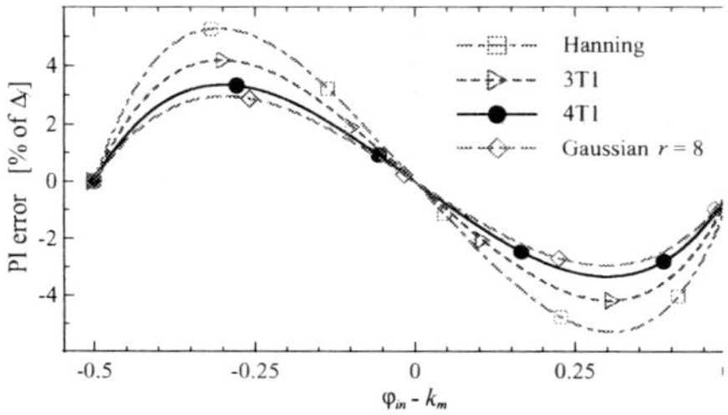

FIGURE 4. Parabolic interpolation errors.

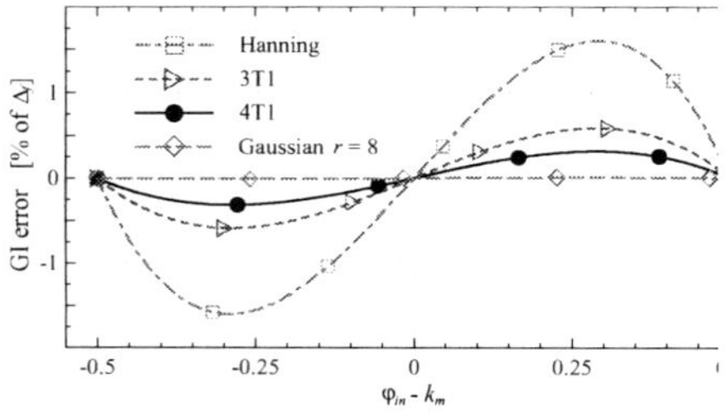

FIGURE 5. Gaussian interpolation errors.

TABLE 2. Parabolic and Gaussian interpolation performance.

	Parabolic interpolation				Gaussian interpolation			
Window	a_4 in (8)	E_{max} [% of Δ_f]	$\lvert\varphi_{in} - k_m\rvert$ @ E_{max}	Gain	b_4 in (14)	E_{max} [% of Δ_f]	$\lvert\varphi_{in} - k_m\rvert$ @ E_{max}	Gain
Hanning	0.259	5.28	0.307	9.5	0.064	1.60	0.291	31.2
Blackman	0.220	4.38	0.303	11.4	0.031	0.66	0.289	75.3
3T1	0.211	4.18	0.303	11.9	0.028	0.59	0.289	84.7
3T3	0.172	3.40	0.300	14.7	0.025	0.53	0.289	93.7
4T1	0.171	3.34	0.300	15.0	0.016	0.31	0.289	159
4T3	0.153	2.99	0.299	16.7	0.015	0.31	0.289	163
4T5	0.129	2.51	0.297	19.9	0.013	0.27	0.289	187
Gaussian $r=6$	0.252	4.95	0.305	10.1	0.015	0.24	0.281	208
Gaussian $r=7$	0.196	3.80	0.301	13.2	0.0038	0.052	0.279	970
Gaussian $r=8$	0.153	2.95	0.298	17.0	0.0007	0.0087	0.278	5756

MEASUREMENTS

The parabolic and Gaussian interpolation methods were examined with a laboratory setup shown schematically in Fig. 6, based on a tune measurement development system [1]. It consisted of a channel with a 14-bit analog to digital converter (ADC) preceded by an antialias low-pass filter (LPF), a memory for fast ADC sample storage, and a board with a floating-point digital signal processor (DSP). A PC was used to prepare processing software and to download it to the DSP board. To achieve input and clock frequencies of sufficient phase stability, the frequencies were generated by two Direct Digital Synthesizers (DDS), driven by the reference source of a frequency meter. During each acquisition, 2048 ADC samples of the sine wave input signal were stored in the memory. Then, the DSP successively performed windowing, the FFT and the power spectrum calculation. Next, the spectrum bin with the biggest amplitude was found and finally the input frequency f_{in} was calculated according to (10) or (15). The measurements were done around bin k_m = 128, with the ADC clock frequency f_s = 1.25 MHz and f_{in} about 78 kHz (mid-range of the frequency span of the setup).

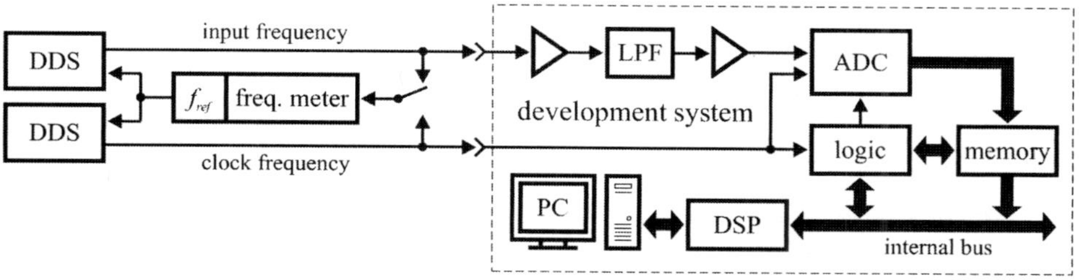

FIGURE 6. Measurement setup.

Measurement results are shown in Fig. 7. Crosses mark extreme values from 100 consecutive measurements with the same setup frequencies and dashed lines show theoretical errors as in Figs. 4 and 5. The measurement results were spread due to amplitude noise present in the analyzed spectra, which was converted during the interpolation process into a frequency jitter. This uncertainty was caused mostly by noise present in the input signal, originating in the DDS output 12-bit digital to analog converters. The quantization noise of the ADC might have contributed as well. The observed noise became more visible as the interpolation gain increased. Nevertheless, as seen in Fig. 7(d), the total of the systematic and "noise" error was 107 ppm of Δ_f, that is some 65 mHz. The interpolation gain was close to 4700 and relative measurement error of about 0.8 ppm (for f_{in} close to only $f_s/16$).

The interpolation gain of 4700, obtained at the expense of performing the Gaussian interpolation (15) within some microseconds, is equivalent to the frequency resolution of an FFT measurement without interpolation with N samples and the sampling time L multiplied by this factor. For the presented measurement it corresponds to increasing N from 2048 to almost 10^7 and L from 1.6 ms to 7.5 s. Such an amount of data would increase the FFT calculation time by a factor of 10^4, from about 2 ms to 20 s.

CONCLUSIONS

Theoretical and experimental studies have been undertaken to enhance FFT frequency measurement resolution, using parabolic or Gaussian interpolations on the discrete magnitude spectrum to find abscissa of spectral peaks maxima located between discrete spectrum bins. The interpolation yield strongly depends upon the windowing function used and, to be significant, the window spectrum main lobe should be at least 3 bins wide.

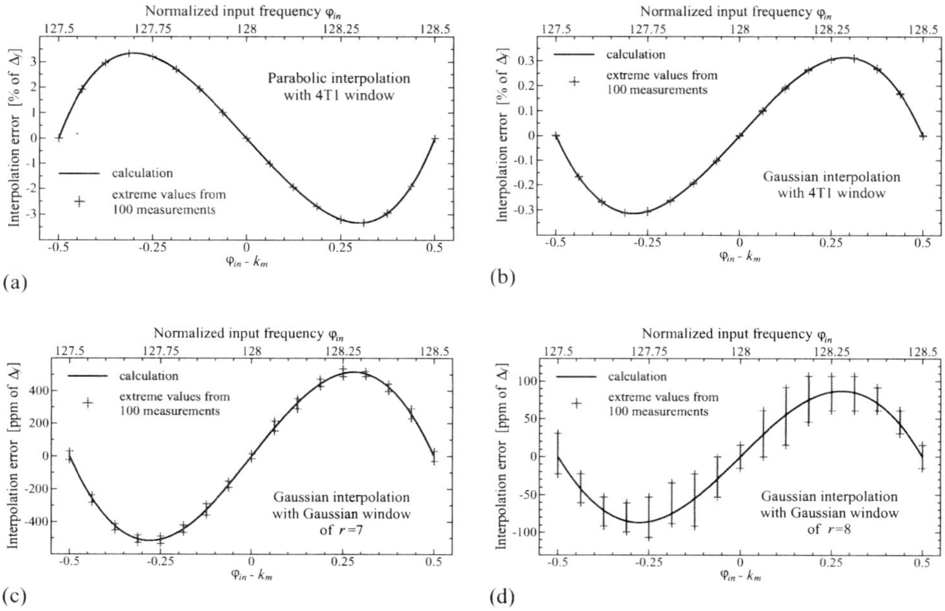

FIGURE 7. Measurement results of the parabolic and Gaussian interpolation methods.

This paper shows that the parabolic interpolation can improve the frequency resolution by more than one order of magnitude. The method computing cost is one division and one multiplication.

The frequency resolution improvement can still be better with Gaussian interpolation. A gain larger than two orders of magnitude is achievable with windows having very good spectral properties and well beyond three orders of magnitude when using steep Gaussian windows. The cost of the Gaussian interpolation is three divisions, three multiplications and calculation of two natural logarithms.

Both interpolation methods assume the measured frequency to be constant during acquisition and the spectral peak corresponding to the measured component to constitute a local maximum in a given band of the input signal spectrum. These methods do not help to resolve nearby spectral peaks and assume bin spacing to be small enough to avoid peak merging.

This paper describes systematic errors of the interpolation methods, assuming ideal discrete spectra. Soon, results will be published concerning the behavior of the methods when spectra are perturbed by noise, interference from strong components and the exponential decay of the input signal.

A direct application of these methods is FFT-based tune measurement systems. The Gaussian method with 4T1 window is used in such a system for the CERN PS Booster accelerator. In the future, similar systems will be made for the PS and LEIR machines.

Further details concerning the methods and windowing can be found in [1], containing also a list of supplemental references.

ACKNOWLEDGMENTS

The authors are indebted to all those involved in the development of the tune measurement system for the PSB accelerator. A special mention is given to E. T. d'Amico and A. Chapman-Hatchett. The authors also acknowledge the help of J. Belleman, J. Bosser, J. P. Potier and U. Raich.

REFERENCES

1. M. Gasior and J. L. Gonzalez, "Improving FFT Frequency Measurement Resolution by Parabolic and Gaussian Interpolation," *AB-Note-2004-021 BDI*, CERN, February 2004, with all references therein.

Development of Residual Gas Profile Monitors at GSI

T. Giacomini*, S. Barabin†, P. Forck*, D. Liakin†, and V. Skachkov†

*Gesellschaft für Schwerionenforschung - GSI, Darmstadt, Germany
†Institute for Theoretical and Experimental Physics - ITEP, Moscow, Russia

Abstract. Beam profile measurements at modern ion synchrotrons and storage rings require high timing performances on a turn-by-turn basis. High spatial resolutions are essential for cold beams and beamwidth measurings. The currently used RGM supported very interesting measurements and applications. Due to the readout technology the spatial and time resolution is limited. To meet the expanded demands a more comprehensive device is under development. It will be an all-purpose residual gas monitor to cover the wide range of beam currents and transversal particle distributions. Due to the fast profile detection it will operate on primary electrons after residual gas ionization. A magnetic field of 100 mT binds them to the ionization point inside 0.1-mm orbits. The high-resolution mode will be read out by a digital CCD camera with an upstream MCP-phosphor screen assembly. It is planned to read out the fast turn-by-turn mode by an array of 100 photodiodes with a resolution of 1 mm. Every photodiode is equipped with an amplifier-digitizer device providing a frame rate of ~ 10 MSamples/s.

INTRODUCTION

Residual Gas Monitors (RGMs) provide a non-destructive beam profile measurement of a circulating ion beam. Applications are: the detection of transverse beam profiles in dependence of injection matching [1], hollow beam prevention, coupling resonances, beam cooling effects, etc. These different applications call for a detector that is capable of providing a spatial resolution down to 0.1 mm for cooling and hollow beams and a frame rate up to 10 MSamples/s to determine an optimal matching condition at injection. To span the variety of applications we are developing an RGM that provides two different modes of operation realized in one device. A high-resolution readout via digital CCD camera and a high-speed readout by photodiodes are in preparation.

The beam interacts with the residual gas within the beam line and produces residual gas ions and electrons (Fig. 1a). An electrostatic field E1 accelerates the ionization products (ions or electrons) towards a Microchannel Plate (MCP). When the particles reach the MCP surface secondary electrons are produced and are accelerated into the channels by E2. Inside the channels they are multiplied by a factor of about 10^6 (Chevron configuration). A wire array behind the MCP collects the secondary electrons and is connected to an adequate electronic device (current-to-voltage converter and adc). Placing a phosphor screen behind the MCP is another readout method (Fig. 1b). The secondary electrons hit the phosphor which is mounted

CP732, *Beam Instrumentation Workshop 2004: Eleventh Workshop*,
edited by T. Shea and R. C. Sibley III

downstream of the MCP. At the impact points lightspots are produced. These spots can be observed by a CCD camera, photo-diodes or other light detectors.

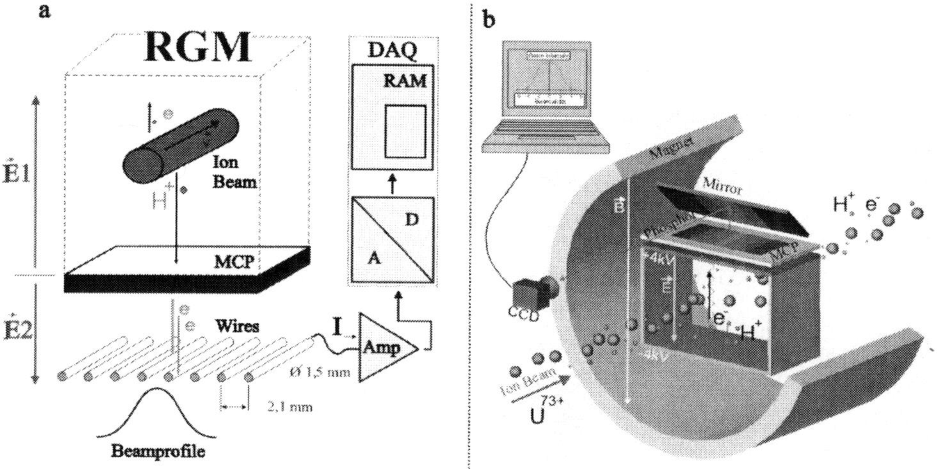

FIGURE 1. a) Function principle of the currently used RGM in the heavy ion synchrotron at GSI; b) Operational principle of the new RGM. Schematic electrostatic field box and MCP phosphor screen assembly. The beam pipe is not shown.

For precise beam observations during acceleration or cooling, a high spatial resolution down to 0.1 mm is needed. The exposure time of one beam profile will cover the 0.1- to 10-ms range. In this case the beam's image on the phosphor screen is observed with a digital CCD cam era providing the required frame rate up to 100 fps (frames per second). To realize the fast readout (one profile per synchrotron revolution) an array of photodiodes with a spatial resolution of 1 mm and a repetition rate of about 100 ns (10 MSamples/s) is foreseen. The fast readout mode necessitates the detection of electrons. To bind the residual gas electrons to the ionization point a magnetic field of about 100 mT is needed. By the way the magnetic field works against the space charge effect of cold or high intensity beams.

The currently used RGM revealed its usefulness in ion beam monitoring and enabled specific measurements; see the "Currently Used RGM" section. The low spatial and time resolution calls for a more comprehensive device, described in "New Residual Gas Monitor."

CURRENTLY USED RGM

The currently used RGM in the heavy ion synchrotron at GSI enables important measurements concerning electron cooling, injection mismatches resulting in hollow beams, acceleration and extraction of the beam. Pictures of the RGM are shown in Figs. 2a–2c. The duration of one measurement (one beam profile) can be switched between 0.5 and 5 ms. Every 10 ms a new measurement starts. The detector operates on residual gas ions which are accelerated by an electrostatic field of about 300 to 450 V/cm. The amplification is done by two MCPs in Chevron configuration (Fig. 2c). The dimensions of the MCPs are 100 mm x 30 mm and the active area is 100 mm x 26 mm. They are fixed in two sockets of glass ceramics and are fitted with two flat metal springs which are gold coated. An array of 64 wires with diameters of 1.5 mm and distances of 0.6 mm to each other is placed behind the MCPs. This wire array configuration determines the spatial resolution of 2.1 mm.

The black plates in Fig. 2a are made of glass and are covered with germanium on the inner surfaces. A voltage of 4.5 kV to 7 kV is applied to the top sheet while the lower sheet, respective to the MCP surface, is on ground potential. Difficulties with current breakthroughs via the germanium surfaces forced us to develop a new electric field box. Figures 3a–3c show the results of two different electrostatic field simulations and the final mechanical design.

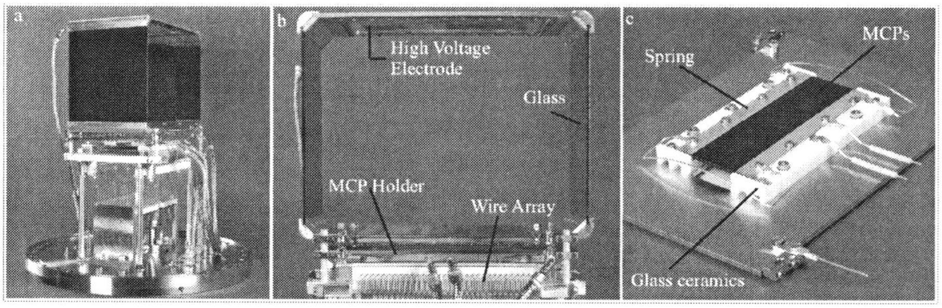

FIGURE 2. a) Picture of the currently used RGM in the heavy ion synchrotron at GSI, mounted on a flange of 300 mm diameter; b) extraction area of ionization products, MCP holder and wire array; c) MCP module in Chevron configuration, MCP size 100 mmm x 30 mm, active area 100 mmm x 26 mm, gold-covered flat springs, sockets of glass ceramics.

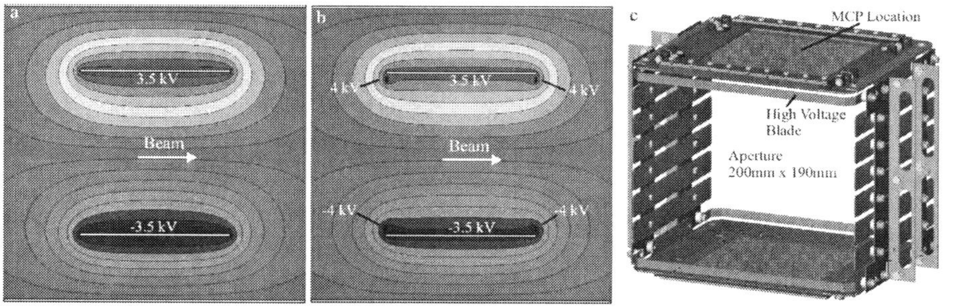

FIGURE 3. Electrostatic field box with field-forming high voltage blades. Pictures a and b show the results of simulations with and without field-forming blades for the electrostatic potential. Picture c shows the final mechanical design.

Straight electrostatic field lines between the top and bottom are important for obtaining an undistorted image of the beam. The main improvement of the new E-field box is the 4 blades at the entrance and exit of the beam path (Figs. 3b and 3c). The equipotential lines in Fig. 3a are more curved in comparison with Fig. 3b. The blades are on a higher (± 4 kV) respectively lower potential than the corresponding sheets (± 3.5 kV). The limiting factor of the length of an E-field box is usually the diameter of the flange. The blades improve the field quality significantly whereby the length of the E-field box is constant.

To test the MCPs a deuterium lamp (115- ~ 400-nm spectral distribution) is placed above the E-field box outside the vacuum chamber. It shines through a glass flange and the E-field box onto the MCPs. Thus the bottom of the box is made of a metallic grid.

Figure 4 shows the measured signals of a synchrotron cycle with 3 injections while cooling at an injection energy of 11.4 MeV/u. Every 500 ms an injection started. Finally $6 \cdot 10^7$ ions of Xe^{47+} circulated in the synchrotron and were accelerated to 600 MeV/u within 0.5 s. Every profile was measured in intervals of 10 ms with an integration time of 0.5 ms/profile. Plotting profiles on top of each other makes it easy to visualize the changes of the profiles shape during the cycle (Figs. 4a and 4d). The color alteration of the plotted profiles starts blue and turns red later in the cycle. Red profiles are not seen due to the decay of the signal at extraction. The beam latitudes of the whole cycle were plotted with 10 ms time resolution (Figs. 4b and 4e). The profiles shape gets narrow and the profiles height increases during electron cooling. The beamwidth was measured at FWHM. Due to the overlaying of the first cold beam (large narrow profile) and the new not cooled beam (flat and wide profile) the beamwidth evaluated after each following injection is smaller than after the injections before. In Figs. 4c and 4f the calculated integration (along abscissa in a, d) of the beam profiles is shown.

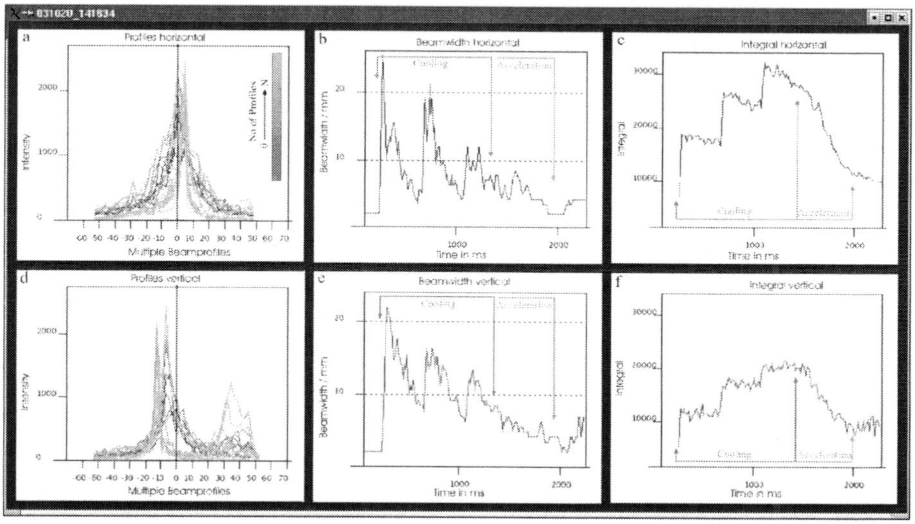

FIGURE 4. Pictures a-c show the horizontal data, d-f the vertical ones. Pictures a and d are beam profiles of one synchrotron cycle. Every 10 ms a new profile was measured with an integration time of 0.5 ms. Pictures b and e show the beamwidth at FWHM; c and f show the integral of the beam profiles.

Figure 5 shows measurements based on a misadjustment of a steerer magnet in the transfer line between UNILAC and synchrotron. A hollow beam is seen in Fig. 5a. After correction of the magnet settings profiles 5b and 5c were recorded. Due to the high data quality of the 3 measurements the corrections were completed within a few cycles.

Figure 6 shows an interesting measurement. The shapes of the horizontal and vertical profiles were exchanged between two measurements (10 ms). The basic effect could be a coupling tune resonance between horizontal and vertical phase space. The low time resolution of the detector inhibited further studies.

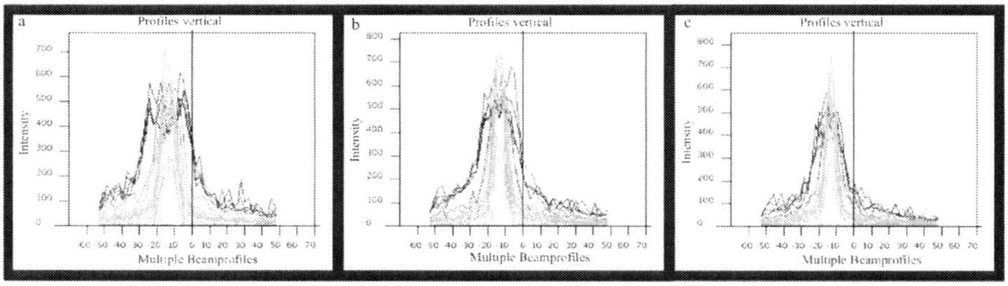

FIGURE 5. Measurement of a hollow beam in vertical direction: a shows the results of a misadjustment of a steerer magnet in the transfer line between UNILAC and synchrotron; b and c show the signal after correcting the magnets settings.

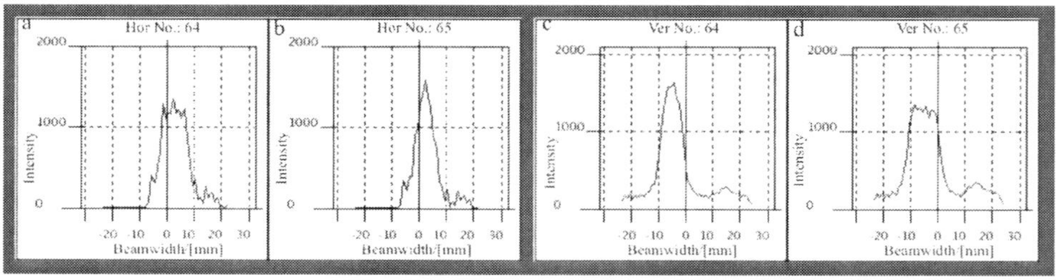

FIGURE 6. Pictures a and b show the horizontal beam profiles, c and d the vertical; between profiles 64 and 65 the profiles shape was exchanged while the position of the beam did not change.

The fast non-destructive measurement proves the big capacity of residual gas monitors for the operation and fast and exact adjustment of accelerators.

New Residual Gas Monitor

The currently used RGM showed the possible fields of applications and operations, and produced interests of more detailed beam detection that it cannot fulfil. To achieve high spatial and time resolutions a new detector will be designed for determination of beam profiles during injection, cooling, acceleration and extraction. Furthermore the device has to provide the features and properties needed for ambitious accelerator experiments like measurements of injection mismatches [1], detailed observation of coupling resonances and coherent betatron oscillations [2] on a turn-by-turn basis. High resolutions in time (100 ns/measurement) and space (0.1 mm) are the basic parameters of this development. The space resolution is limited by the MCP (Chevron configuration). To conserve the space resolution of the MCP a phosphor screen and readout by a digital CCD camera are the most reasonable solution. The optical readout condition (phosphor screen) leads to an array of photodiodes for the high-speed mode. Every photodiode is equipped with an amplifier-digitizer channel. A spatial resolution of 1 mm in the fast mode is sufficient. Furthermore the fast mode requires electron detection and a fast phosphor screen. The fast phosphor screens e.g. P47 are near the ultraviolet range (400 nm) and provide decay times of 60 ns.

During ionization the electrons are pushed out of the ionization point by the atomic scattering kinematics. To keep the electrons perpendicular to the ionization point a magnetic field is applied parallel to the external electric field. The strength of the magnetic field should be of about 0.1 T and the electric field of about 0.5 kV/cm. Within 4-7 ns the primary electrons are guided inside a cyclotron diameter of 0.1 mm towards the MCP as calculated for Fig. 7. The parameters are typical for a high current operation of the SIS synchrotron at GSI. The field uniformity from the ionization point to the MCP surface has to be in the range of 95%. In general the cyclotron radius is given by the initial electron velocity after ionization [2]. Due to the circulating electrons around the magnetic field lines, the MCPs' amplification factor decreases. Applying a higher voltage to the MCPs compensates for this effect. The magnetic field can be provided by a rod-type permanent magnet [3] or by a window frame magnet. Concerning the overall length and fringe field effects rod-type permanent magnets seem to be the best solution. Permanent magnets are fail-safe concerning the higher voltage at the MCPs. Windowframe electromagnets are another possibility, however they require special precautions due to voltage breakdowns concerning the MCPs. The advantage is the possibility to switch them off.

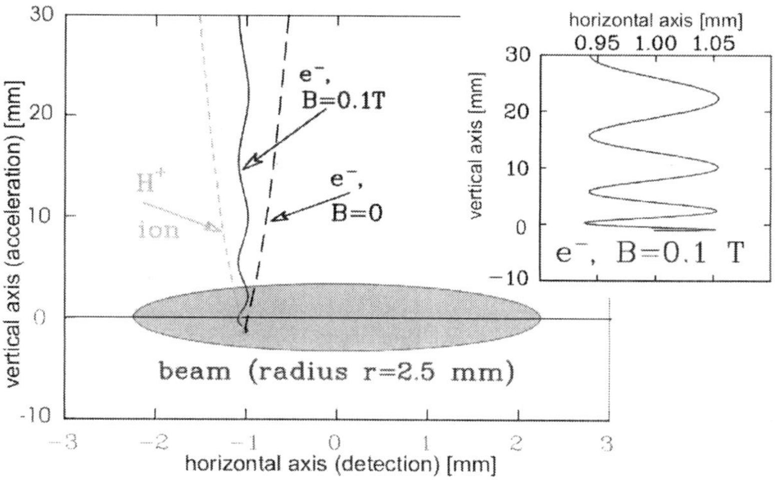

FIGURE 7. Trajectories of ionized particles with and without a magnetic field. $U^{73+}$with 10^9 ions at a bunch length of 10 m.

While the electrons are travelling towards the MCP the residual gas ions move to the opposite potential and produce secondary electrons when they hit the environmental parts. Secondary electron suppression is essential to avoid signal corruption.

The two different readouts (CCD and photodiode) get their signals from the same phosphor screen. One possible solution is the use of a mirror (Fig. 1b), which reflects the light from the phosphor either to the CCD or to the photodiodes.

The readout electronic of the photodiodes requires a data acquisition system with 10-MHz bandwidth and optimal noise and dynamic range [4]. It is separated into 4 stages. A conventional transimpedance amplifier is in the first stage to amplify the signal immediately behind the photodiode. The next stage is a variable gain amplifier that provides the connection between the first stage and the adc and adjusts the obtained signal to the adc range. The third stage is a 14-bit, 40-MSamples/s adc. A DSP with a native interface to the adc and a built-in SDRAM control the digital electronic and provide the connection to the PC. The DSP's 16-bit, 32-MB SDRAM, the 66-MHz access frequency, and the DMA (direct memory access) offer the possibility to record long time sequences.

The E-field box is placed inside the vacuum while the magnetic device due to vacuum bakeout resides outside. The MCP-phosphor assembly, the glass flange, the mirror and eventually the feedthroughs are placed between the electric field box and the magnet device. This enlarges the gap between the magnet poles. Everything inside the gap has to be designed as small as possible. The whole detector consists of 4 magnets, a horizontal and vertical and two compensation magnets. To cover a wide range of applications and locations it will be possible to detect either electrons or ions. In case of ion detection the large magnet devices can be removed and the E-field boxes are placed nearer to each other in beam direction. This allows the use of the detector in environments of less space and makes the detector all-purpose.

CONCLUSION

Important measurements have been performed with the existing residual gas monitor. Due to the chosen anode technology its spatial and time resolution is limited. The new detector will be a very flexible and modular detection system, which covers a wide range of applications. We will integrate two modes of operation: a high spatial resolution mode by a digital CCD camera and a fast turn-by-turn mode read out by a photodiode array. The detector will operate on electrons supported by a magnetic field parallel to the electrostatic extraction field. The electrons are multiplied by two MCPs and the readout is done via a phosphor screen. To enlarge the number of applications it will be possible to detect residual gas ions too. The combination of a high spatial resolution and a fast readout makes the RGM to be a multi-purpose installation, which is capable of delivering very valuable information about transverse ion beam distributions.

REFERENCES

1. G. Ferioli, C. Fischer, I. Koopman, and F. Roncarolo, "Beam Studies Made with the SPS Ionization Profile Monitor", in Proc. DIPAC-2003, Mainz, p. 116.
2. R. Connolly, P. Cameron, R. Michnoff, and S. Tepikian, "Performance of the RHIC IPM," in Proc. PAC-01, Chicago, p. 1297.
3. D. Liakin et al., "Development of a Permanent Magnet Residual Gas Profile Monitor With Fast Readout," EPAC-2004, Luzern.
4. P. Forck et al., "Advanced Residual Gas Profile Monitor for High Current Synchrotrons and Cooler Rings," in Proc. DIPAC-2003, Mainz, p. 134.

Experiences on Counter Applications and Beam Loss Measurement at the GSI Synchrotron

T. Hoffmann, D. A. Liakin*, P. Forck, and P. Moritz

Gesellschaft für Schwerionenforschung (GSI), Planckstraße 1, D-64291Darmstadt, Germany
**ITEP, Moscow, Russia*

Abstract. At the GSI Synchrotron (SIS18) and the high-energy beam transfer lines (HEBT) several particle counting detectors of different types are installed. These are connected to one dedicated 128-Channel VME based data acquisition system called "A Beam Loss Measurement and Scaling System (ABLASS)," which was recently added to the GSI control system. An overview of all its implemented features is given. Experiences gained during daily operation as well as the more sophisticated applications such as high-resolution spill structure analysis with up to 100-kHz sample rate and slow extraction optimization are presented. In addition the SIS18 is going to be the injector for the planned superconducting double ring facility SIS100/300, which is part of the GSI future project, which was recently named FAIR, the international Facility for Antiproton and Ion Research. Therefore the observation of beam losses at the SIS18 produced by the required high currents is essential to prevent activation and vacuum instabilities. The methods for the beam loss measurement are described.

INTRODUCTION

In the GSI heavy ion synchrotron SIS18 all ions can be accelerated to about 1-2 GeV/u. In the past several years many beam times for high current applications were accomplished and for medium heavy ions the space charge limit is reached routinely. For the proposed double ring facility SIS100/300 of FAIR, the international Facility for Antiproton and Ion Research, the existing SIS18 is going to serve as an injector. For the heaviest ions like Uranium the space charge limit has to be reached resulting in 10^{11} accelerated U^{28+} ions within one cycle. For this high current the beam loss during injection, acceleration, and extraction has to be minimized to prevent serious activation or damages. Charge exchange processes by residual gas collisions are lifetime limitations for ions. These stripping and e^--capture effects lead to changes of the charge states resulting in a false magnetic deflection in the bending magnets. These particles hit the chamber walls, initiating bursts of secondary particles followed by an increase in the vacuum pressure in the beam pipe, leading to additional beam loss.

With respect to this problem, A Beam Loss measurement And Scaling System (ABLASS) was developed and recently implemented into the control system. The design and the components of this system, presented in [1,2], gave an opportunity to implement several other counter applications in addition to the beam loss detectors, such as ionization chambers, scintillators and secondary electron monitors [3]. Moreover, other types of diagnostic measurements such as the dc-transformer signal,

CP732, *Beam Instrumentation Workshop 2004: Eleventh Workshop,*
edited by T. Shea and R. C. Sibley III

and machine data such as rf-amplitude, rf-frequency, and the U/f-conversion output of a Hall sensor signal installed in the reference-bending magnet can be displayed.

128 channels are permanently counted and read out by the modular VME-LynxOs system, visualized on a Linux PC. Using the standard Linux X-capability the display can be exported to the main control room or any other suitable location. The data acquisition over all channels is performed simultaneously, typically with a sampling rate of 1 kHz. To process higher rates up to 100 kHz, an expert mode has to be chosen in which the unused channels have to be disabled, e.g. for 100 kHz only 2-4 channels are available. An extraction cycle at the SIS18 generally lasts a few seconds, whereas the pause between two cycles may take only 200 ms. Even under these conditions the data acquisition runs stably, thus the choice of the selected components and the system principle itself is validated.

Since the introduction of ABLASS into the daily operation many constructive comments from the users were given and usability was increased. The fact that the operating system (OS) of ABLASS is Linux and the OS of the accelerator control system is VMS causes no problems. The graphical user interface (GUI) shows no difference, and the connection to the VMS database, which is relevant for all the beam and beamline parameters, is organized by ORACLE software interfaces.

HARDWARE OVERVIEW

A general overview of the system setup and the interfacing is presented in Fig. 1. The hardware of this data acquisition system consists of the following components [4]:

- Standard NIM electronics (Discriminator, Fan In/Fan Out, Logical Gates)
- VME-System including:
    - CES RIO3-8064 CPU, 128-MB DRAM, 100-MBit TCP/IP
    - CES 8004AA VME/VSB Display Unit
    - SIS3807, 4-Channel Programmable Clock
    - 4x SIS3801, 32-Ch. Multiscaler, 64-KB FIFO, 32-bit, 200-MHz Input
    - 2x CAEN V895, 16-Channel Leading Edge Discriminator
- GSI Timing Interface (TIF) and Decoder (ATMEL processor)
- RedHat Linux PC, 1-GHz, 100-MBit TCP/IP, Front-End
- Several Tektronix X-Terminals

The operating system for the CES RIO3 CPU is real time LynxOS on which the front-end software, developed in C, is permanently running. This front-end CPU sends the collected data via a 100-MBit TCP/IP connection to a RedHat 7.0 Linux PC, 1-GHz, 512-MB SDRAM, on which the data is stored in a shared memory segment. The software running on the PC was programmed in the Borland Kylix 3 development environment (Delphi for Linux). For more details see [1] and [5].

Once the data socket has established any ABLASS application program, as described in "OPERATING TOOLS," the data, which is stored in a shared memory segment, may be accessed. Even multiple instances of the same program may run simultaneously. By this method it is possible to present equal or different data in diverse locations with the same application window e.g. on X-terminals in the main control room and the experimental laboratory.

OPERATING TOOLS

ABLASS provides three main subprograms for displaying data. These are tools for crossbar functionality, trending, and spill structure analysis. Due to limited amount of channels the old hardware of an analog switching matrix from the main control room to the dedicated electronics room (ELR), where all analog signals from the SIS18 and HEBT are collected, was dismounted. ABLASS replaced this required functionality by counting the signals in the ELR and transmitting the total particle amount of one synchrotron extraction via TCP/IP to the main control room shown in Fig. 2-top. This function is important for beam optimization, target observation, and protection, and is used almost continuously.

The trending function shown in Fig. 2-middle gives information about the scalers final values. These values are stored for all 128 channels on a hard disk for the last 6 days. In Fig. 2-middle the chosen time scale is set to approx. 6 hours. This tool helps to observe online the result of beam optimization while increasing or decreasing the beam current. This function shows instabilities, beam dropouts, and changes in the beam currents, e.g. it is shown on which time an experiment was provided with beam.

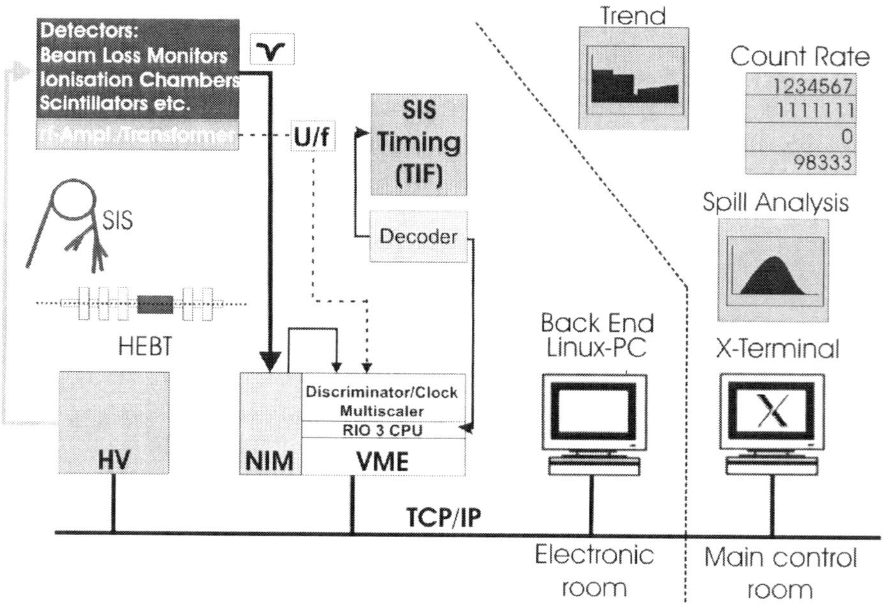

FIGURE 1. Drawing of the system overview. As a main interface between all components TCP/IP based on Ethernet is used. Beneath the front- and back-end CPUs even the high voltage power supply is connected to the network.

The spill structure analysis, Fig. 2-bottom, gives insight into the actual extraction cycle, where the count rate is shown as a function of time (time scale in seconds) and is updated at every new cycle. This tool offers the most important information about the spill quality. It allows comparison between different signals in the same time domain, up to six per window, and shows direct response to user intervention in the machine.

For this application, several useful functions that can be used in online mode were added. As on standard oscilloscopes, time-measurement between two moveable bars, zooming, binning, and flexible Y-scaling are possible. The user may accumulate data over many spills, treat data with smoothing filters or generate an FFT analysis. A calculation of the derivative and a special FM-mode, which is described as follows, are also installed.

FM-MODE: Most of the signals fed into ABLASS, e.g. produced by beam loss or generated by the HEBT counters, such as scintillators, are based on particle counting techniques and have a random pulse distribution. In contrast to this a class of processes monitored in ABLASS has a regular structure, for example the voltage output of vacuum tubes in the SIS18. To use an analog information in ABLASS as given by a time varying voltage, amplitude to frequency converters are used to adapt these signals to countable pulses, e.g. sine function in Fig.3-left A. The converted signal, shown in Fig. 3-left B, consists of countable digital pulses that are fed into the ABLASS system.

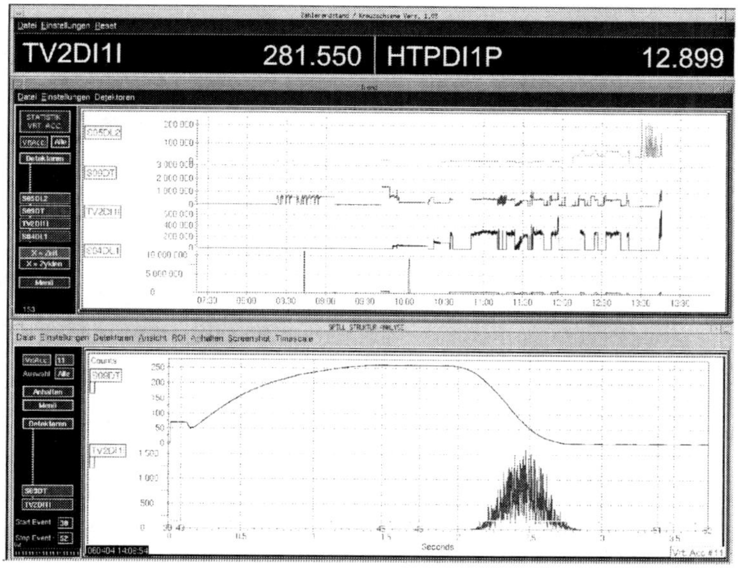

FIGURE 2. Top: Crossbar utility showing the end values of selected counters after the cycle. Middle: Trending/History utility presenting the counter end values over a period of time or cycles. Bottom: Spill structure analysis tool displays the particle distribution as a function of time within one cycle.

This type of mode should be used when the carrier frequency is much lower than the sampling frequency of the ABLASS counters and only a few pulses are occurring within a sample period. With binning and linear filtering of the modulated pulse sequence the initial shape of the signal may be reconstructed, Fig. 3-left C. An advantage for using the FM-mode is the regular pulse structure, which gives the possibility of preprocessing the input data and increasing the quality of the demodulation result.

The method for the applied evaluation is shown in the equation $A_{DM} = n/\Delta T$ where n is the number of pulses per sample period (usually this value is 1), ΔT is the time difference between two pulses of the modulation sequence and is usually much longer then the sample period. A_{DM} is the new data point for the demodulated graph C, Fig. 3-left.

The modulation noise of the signal decreases significantly, which is shown in Fig. 3-right, in which the upper graph is the FFT-spectrum of the FM-pulse sequence and the lower graph the spectrum of the evaluated signal with an obviously better signal-to-noise ratio.

BEAM LOSS MEASUREMENT

Installed devices that limit the aperture of the beam pipe can produce losses. Scrapers are used to localize beam loss and to prevent damages of sensitive injection and extraction septa. These locations are equipped with the mobile GSI type of beam loss monitors (BLMs). The BLM is well described in [1,2]. 22 monitors are read out permanently beginning with initialization of the synchrotron and ending with completion of the extraction. The main aim is to prevent permanent activation in the synchrotron tunnel in exposed areas and to increase transmission to the experimental targets and especially to the planned SIS100/300 facility.

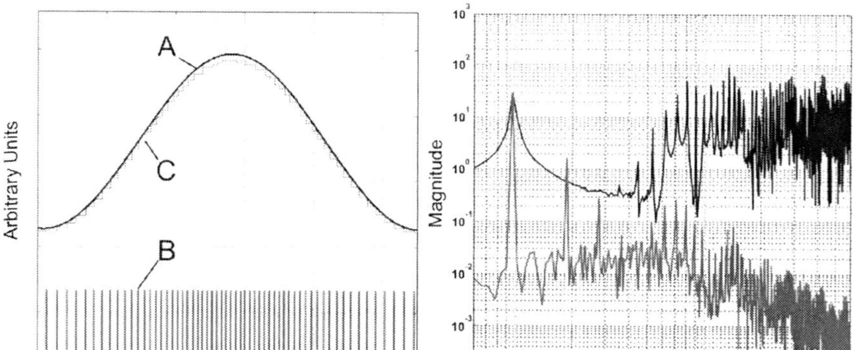

FIGURE 3. Left: Sine-function example. **A** shows the analog input signal, which is converted to digital pulses (**B**). To increase the information for an ABLASS presentation the signal is demodulated, which leads to a graph with the shape of the analog signal (**C**). Right: On the FFT-spectra the noise attenuation is shown. Upper graph: spectrum of the modulated pulse sequence, lower graph: spectrum of the demodulated output signal.

For beam loss studies some tools are implemented in ABLASS. As presented in Fig. 4 an overview of the actual BLM positions in the SIS18 tunnel is given by opening a map showing all the active devices. Due to the need for mobility of these monitors they are all mounted on movable tripods. New BLM arrangements can be easily updated in the database in this tool. A click on the monitor of interest shows a photograph of its actual position.

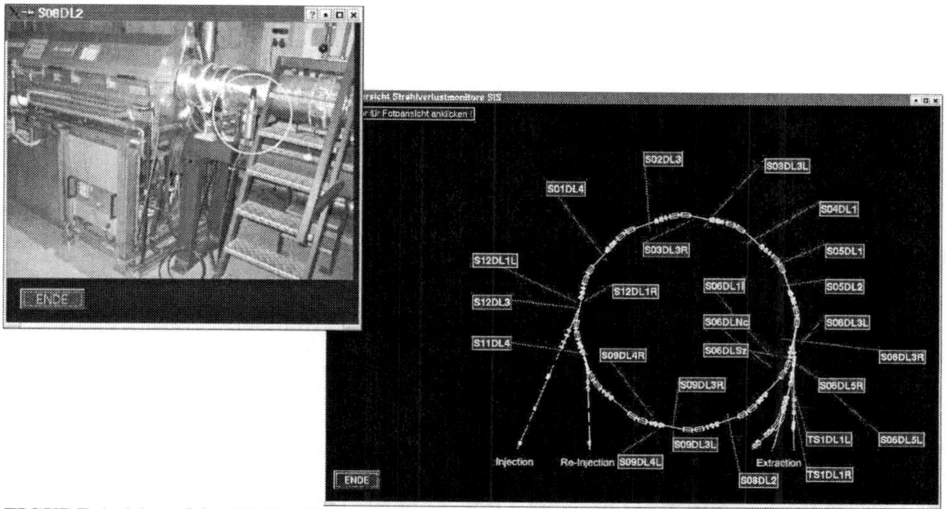

FIGURE 4. Map of the SIS18 with the positions of all available BLMs. A click on the BLM of interest shows the photograph (left) of the actual installation in the beam line.

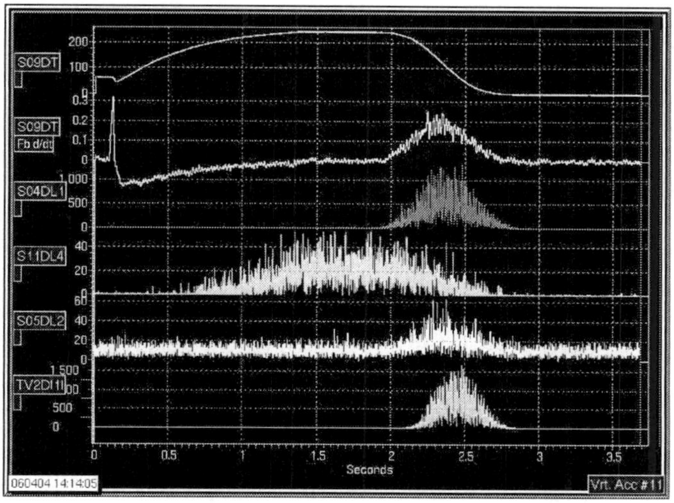

FIGURE 5. Measurement of a slowly extracted 1-GeV, $^{197}Au^{65+}$-beam. Top graph shows the dc-transformer signal in the SIS18, the second graph its derivation. Graphs 3 to 5 are generated by SIS beam loss monitor signals at different positions and the bottom curve shows the output of an ion chamber in the HEBT.

To demonstrate the usability of ABLASS for beam loss measurement, data of a slowly extracted 1-GeV, $^{197}Au^{65+}$-beam was recorded (see Fig. 5). The top graph is the output of the SIS18 dc-transformer. The injection (0-0.15 s, 11.4 MeV/u), the acceleration ramp (0.15-1.5 s), the flattop (1.5-2.0 s, 1 GeV/u), and the extraction (2.0-2.8 s) can be recognized. For the derivative of the transformer signal, shown in the second graph, smoothing finite impulse response (FIR) filters have to be applied to the raw data. The shape at the time of the extraction has to be compared with the curve generated on an experimental target, e.g., the bottom graph, which shows the output of an ionization chamber in the HEBT. For an optimized extraction efficiency the fine structure of both curves should be equal to ensure a constant angle during the complex slow extraction process. This is not perfect in the displayed case and has to be optimized by the operators. The third graph shows the beam loss signal at the electrostatic septum. This loss correlates directly to the extraction, i.e. the derivative of the transformer, as expected. At the re-injection septum, here called S11DL4, losses were detected. This loss is induced with the increasing energy during acceleration. Losses at this aperture limiting electrostatic septum can be avoided by a proper orbit correction during acceleration. S05DL2 is a BLM placed behind the kicker magnet. Losses at extraction time can be observed, but the permanent noise on the signal could be a hint to a low activation of this location. More investigations on beam loss occurrence using ABLASS have to be performed to avoid the losses and permanent activation for the high current operation.

INVESTIGATIONS ON RF-KO EXTRACTION AT THE SIS

A good example of what can be done with ABLASS as a high performance data acquisition system is this investigation of the rf–Knock-Out method, a transversal excitation on betatron frequencies for slow beam extraction [6,7].

For the GSI experimental therapy project in which cancer is treated with carbon ions, a special experiment for observing slow extractions with instantaneous spill stops and restarts within a cycle was carried out.

The intention of this experiment was to find out an adequate procedure for this application using Knock-Out extraction in cooperation with variations of the rf-amplitude in the accelerating cavity to conserve the bunched beam. The measurement of the rise and fall times of these interrupts and the optimization of the spill shape after restart of the extraction were important.

For the measurement shown in Fig. 6, a fast BC400 plastic scintillator panel of 79 x 75 x 1 mm^3 installed near the synchrotron in the HEBT was used. The displayed data belongs to a 300-MeV $^{12}C^{6+}$ beam sampled with a time resolution of 10 μs. To perform this fast data acquisition ABLASS was switched into the expert mode, in which the sampling frequency was set to 100 kHz and all channels except the one for the required counter were disabled due to the large amount of data. With these measurements it was verified that rise and fall times in the region of 300 μs and lower may be achieved. In addition, ABLASS was used for online observation of the spill shaping procedures.

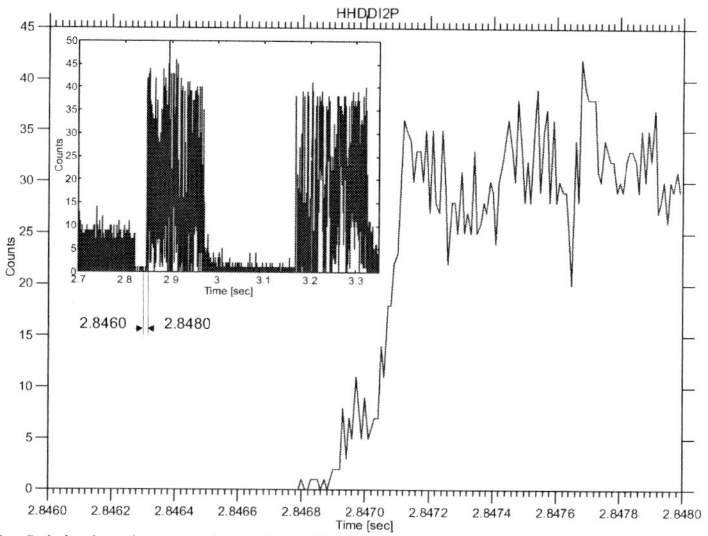

FIGURE 6. Original and zoom view of a spill stop within a cycle, sampled with high time resolution (100 kHz). It shows a rise time within approx. 300 μs (time between 2 large ticks on x-axis: 200 μs).

OUTLOOK

The development of ABLASS was started in March 2002. The final implementation of ABLASS into the control system took place in January 2004 with the aim of use for daily operation. Problems with data leaks and network performance in 2003 were solved [5]. The concept of using a Linux OS in this VMS control system has proven to be very practicable. The plan is to develop further subroutines such as a calibration tool that performs automatic series of measurements with scintillators and ionization chambers, or adding new hardware to the system like a VME-ADC to process other, not countable signals from the machine within ABLASS.

REFERENCES

1. T. Hoffmann, D. A. Liakin, and P. Forck, Proc. 10th Beam Instrumentation Workshop, BIW, Upton, New York, USA, 2002, p. 329.
2. P. Forck and T. Hoffmann, Proc. 5th European Workshop on Beam Diagnostics and Instrumentation for Particle Accelerators, DIPAC, Grenoble, France, 2001, p. 129.
3. P. Forck, T. Hoffmann, and A. Peters, Proc. 3rd European Workshop on Beam Diagnostics and Instrumentation for Particle Accelerators, DIPAC, Frascati, Italy, 1997, p. 165.
4. http://www.struck.de, www.caen.it, www.ces.ch.
5. D.A. Liakin, T. Hoffmann, and P. Forck, Proc. 6th European Workshop on Beam Diagnostics and Instrumentation for Particle Accelerators, DIPAC, Mainz, Germany, 2003, p. 164.
6. P. Forck et al., Proc. 7th European Particle Accelerator Conference, EPAC, Vienna, Austria, 2000, p. 2237.
7. T. Furukawa and K. Noda, "Fast Beam Cut-Off Method in RF-Knockout Extraction for Spot-Scanning," HIMAC-048, NIRS Chiba Japan, April 2002.

Integration of the Longitudinal Feedback System in NSRRC

K. H. Hu, C. H. Kuo, M. S. Yeh, W. K. Lau, Demi Lee, S. Y. Hsu, Teno Yang, P. J. Chou, Jenny Chen and K. T. Hsu

National Synchrotron Radiation Research Center
101 Hsin-Ann Road, Hsinchu Science Park, Hsinchu 30077, Taiwan, R.O.C.

Abstract. An operation version of longitudinal feedback baseband processing electronics was implemented. The system consists of a bunch phase detector, a 500-MS/s analog-to-digital converter and demultiplexer module (ADC/DEMUX), DSP modules, a digital-to-analog converter and multiplexer module (DAC/MUX), and an RF modulator. The ADC/DEMUX unit has a fast ADC that digitizes the bunch phase signal. The down-sampled phase error data of each bunch are then distributed to the DSP boards to perform filtering and applied control rule. In the DAC/MUX, bunch kick signals are converted into analog signals for bunch phase feedback. Design and implementation of the system will be summarized in this report.

INTRODUCTION

The storage ring of NSRRC is a 3rd generation synchrotron light source, designed to store the beam current of 200 mA at 1.5 GeV. Stable longitudinal dynamics are the most important performance issues of the storage ring. The source of the longitudinal instabilities is mainly the high order mode (HOM) of two DORIS RF cavities. Various measures were adopted to eliminate the HOM's effects during the last decade; these efforts include precision temperature control of the cavities, HOM frequency shift plunger, passive Landau cavity, etc. However, these efforts still cannot suppress longitudinal instabilities effectively. RF gap voltage modulation is used to remedy longitudinal instabilities and to improve lifetime for routine operation during the last several years. User satisfied beam is achieved but with a slight increase in energy spread for the gap voltage modulation. An active longitudinal feedback system (LFB) [1,2,4] was proposed to fight the instabilities without degrading the beam quality. The prototype LFB system has been tested and has suppressed longitudinal instability up to 80 mA in 2000 [3]. However, due to the limit of the low efficient longitudinal kicker and the capacity of the prototype electronic system, the prototype system cannot work adequately at higher current. In order to complete the system, the operation version electronics have started implementation from 2002; the system is nearly in integration phase. The commissioning is on-going. The low efficiency longitudinal kicker will be replaced by a DAFNE-like kicker; the kicker is adapted from the SLS/ELETTRA design[4]. The beam pipe is changed to an elliptical shape. Shunt impedance can be optimized to about 1 kΩ. Delivering a longitudinal feedback system to work with a

CP732, *Beam Instrumentation Workshop 2004: Eleventh Workshop*,
edited by T. Shea and R. C. Sibley III

400-mA stored beam to ensure suppression of all residual HOM for beam duct components in the SRF environment is the goal.

The project to replace two existing DORIS cavities by a single cell Cornell superconductivity RF (SRF) cavity at the storage ring will be finished in late 2004. Nearly HOM-free properties are promising for the SRF cavity. This project is also accompanied by double the stored current of the present operation (200 mA). The high order modes of the RF cavity will be almost eliminated. However, the longitudinal feedback system is still helpful to damp residues of longitudinal instabilities. The diagnostics capability of the feedback system is also useful to study the beam properties for conventional RF as well as SRF environments.

LFB HARDWARE SYSTEM OVERVIEW

A longitudinal feedback system is being developed to suppress the longitudinal coupled-bunch instabilities observed in the storage ring of NSRRC. It is a bunch-by-bunch feedback system designed for 200 bunches at 2-ns bunch interval. The longitudinal feedback system consists of major parts as shown in Fig. 1: the bunch phase detector, 500-MS/s analog-to-digital converter and de-multiplexer (ADC/DEMUX), digital signal processors array, multiplexer and 500-MS/s digital-to-analog converter (MUX/DAC), modulator, power amplifier and longitudinal kicker. Designed of the bunch phase detection circuit, the digital electronics and the 1125-MHz wide-band RF system for correction of phase errors have been tested with the beam. The advances in DSP technologies benefit us to choose a commercial off-the-shelf (COTS) DSP solution. The core of the digital electronics is a DSP array that utilizes DSP modules performing parallel processing at a data throughput of 500 Mbytes/sec. These COTS products take off the expense and risk of in-house development. Similar systems are successfully installed and operated with existing installations of LFB systems around the world.

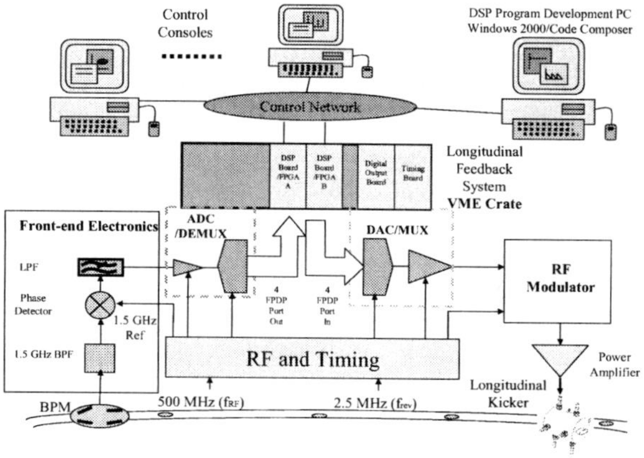

FIGURE 1. Functional block diagram of the longitudinal feedback system.

Front-end electronics include pickup BPM, bunch phase detector, and 500-MS/s ADC and de-multiplexer circuitry. The signals from BPM are sent to a bunch phase detector working at 1.5 GHz. The phase error signal of each bunch is detected by a double balanced mixer, which compares the phases of the beam signal with a 1.5-GHz reference that is phase-locked to the 500-MHz master clock. The ADC/DEMUX board digitizes the discrete phase error signal for search bunches at 500 MHz and distributes to the processors that have a slower clock rate for signal filtering. The operation of the ADC/DEMUX module is synchronized by the timing circuitry.

The filtered signals from the processors are combined by a multiplexer in correct sequence and converted into a single analog signal (the correction signal) by DAC. The kick signal is up-converted by the RF modulator, and the RF power amplifier is used to drive the longitudinal kicker that in turn kicks individual bunches every 2 nanoseconds. The RF system consists of the 1.5/3-GHz bunch phase detection unit, and the 1.25- to ~1.5-GHz (prototype kicker working at 1.125 GHz) broad band drive chain is used to drive the kicker. The new longitudinal kicker has a DAFNE-like design. The quality factor value is about 5 and shunt impedance is about 1000 Ohm.

The front-end electronics sample the bunch signal at 500 MS/sec. The decimate factor is programmable and chosen at 18 in commissioning configuration. Bunch oscillation is sampled at one turn; for the following consecution 17 turns data is neglected. The revolution period of the storage ring is 0.4 μsec. Time duration between consecutive samplings is 7.2 μsec. There are 200 buckets on the storage ring and equally distributed to 4 DSP pairs, each pair of DSP should handle 50 bunches. It needs 1 to 2 μsec to execute control rule, which is dependent upon the complexity of the control algorithm. There are more than 2 μsec that can be used for data transfer between front-end and back-end electronics to DSP modules or for housekeeping purposes. The front panel data port (FPDP) interface of the DSP board working unidirectionally was chosen for data transfer due to its easy design and implementation. The FPDP interface is 32-bits communication link capable and peak data throughput about 160 Mbytes/sec per FPDP port. Performance is satisfied with the requirement of the LFB system. The FPDP port modules are attached to a DSP board to provide eight split FPDP ports. The data of bunches from DEMUX that is prepared in advance and stored in FIFO needs about 0.5 μsec. After control rule and block filtering is executed, the DSP sends bunched error correction data to hold buffer DSRAM.

Due to the advances of the DSP technology, the numbers of DSPs in LFB systems are reduced drastically compared with similar LFB installations which are already deployed around the world. Two commercial DSP VME modules were used in the LFB system. Each VME module is equipped with quad TMS320C6202 DSP chips running in 250-MHz clock. A high MIPS integer DSP is preferred for LFB application. There are many resources on the board for various applications. Every DSP is equipped with local memory to store data and program code. Since the structure of DSP is efficient in block filtering, the control rule is applied to the 50-bunches signal in a batch way. Bi-fifo is used for communication between the DSP and the mezzanine interface. Dual port memory provides a seamless way to transfer data from DSP to VME host. Global memory provides a way for communication for VME hosts and DSPs for various operations mode. The CPLD based FPDP port interface mezzanine boards are supported by the vendor. A FPDP port interface engine provides timing, handshaking and data packing for both input and output data transfers. The FPDP port interface engine connects to the 32-bit BI-FIFO and can be individually configured for either input or output operation.

A functional block diagram of the ADC module is shown in Fig. 3. It consists of an ADC (TS8388BG), DEMUX (TS81102G0), and FPDP interface (CPLD). The DAC module function block diagram is shown in Fig. 4. This module includes FPDP interface, FPGA, timing circuit and DAC chip (RDA012M4). The function of FPGA plays to synchronize between four FPDP interfaces during data transfer from data buffers.

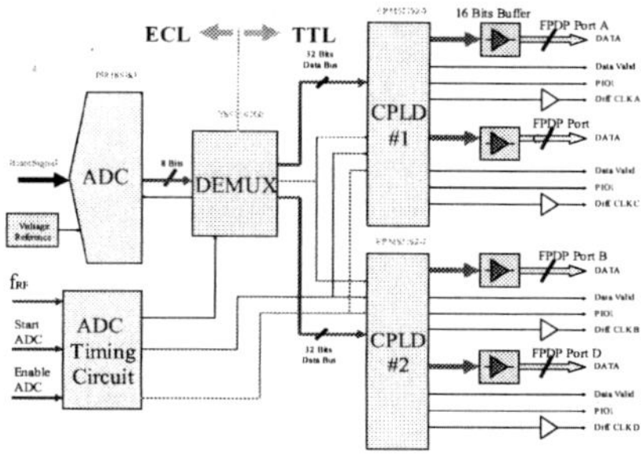

FIGURE 2. Functional block diagram of The ADC module.

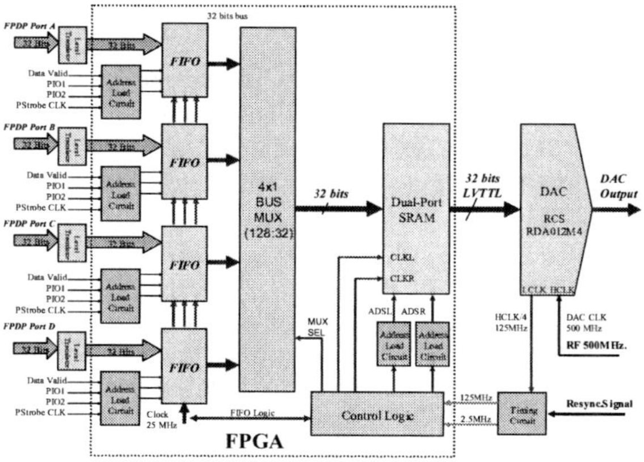

FIGURE 3. Functional block diagram of the DAC module.

SOFTWARE DEVELOPMENT

The software structure of the longitudinal multi-bunch feedback system (LFB) consists of a feedback algorithm, which is executed by a digital signal processor array and to provide raw data for beam diagnostic purposes. In order to develop, to commission and operate this system efficiently, an integrated software environment has been developed. In such an environment, we have major software components including (1) a feedback control algorithm on DSP, (2) a DSP developing environment on PC, (3) diagnostics toolkits on PC, (4) operator interface on PC, and 5) a workstation. The feedback control loop is a digital filter combined with a data input/output control. The DSP developing environment includes a code composer running on PC as well as a remote Ethernet based DSP console debugging program. The system will equip diagnostic toolkits to acquire and analyze data. The associated hardware and software are commercial products that may help to reduce the system development time. The data transportation is by the FPDP ports of the digital processing unit. There are eight input ports and eight output ports in this system.

This development system is an Ethernet-based emulator. It is able to remote control by local computer with window user interface in the Windows operation system. It supports multi-processor downloading and developing the DSP program in the same time. The code composer is a system development tool that supports an integrated development environment (IDE) for 'C6x. This environment allows the DSP code designer to edit, build, manage products, debug and profile from a single application. In addition, the IDE lets the user compile in the background as well as analyze signals graphically, perform file I/O, debug multiprocessors, and customize a C-interpretative scripting language in the IDE.

The debugging and development of DSP is to provide a simple, interrupt driven, inter-device communication protocol for distributed Digital Signal Processing applications. The protocol is based on high-level geographic addressing, in which each device is referred to by a user assigned name rather than by a range of addresses. The hardware involved in these distributed applications may utilize various interconnect schemes, such as shared memory on the VME or MIX bussed, Ethernet, etc. The debug tool provides a variety of capabilities: first, starting, stopping, or resetting the DSP; second, loading code or data into DSP, examining the registers or memory of DSP; third, hardware instruction or data-dependent breakpoints; fourth, a variety of counting capabilities including cycle-accurate profiling; and last, real-time data exchange (RTDX) between the host and the DSP. The schematic diagram of the system is shown in Fig. 4.

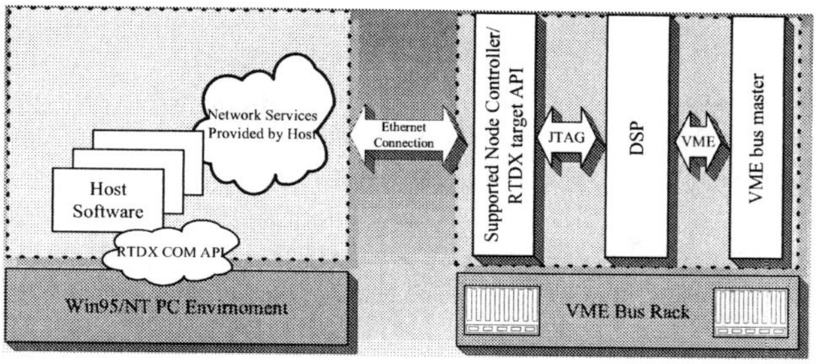

FIGURE 4. The schematic diagram of debugger and development environment.

RTDX provides continuous visibility into the way DSP applications operate in the real world, and RTDX allows system developers to transfer data between a host computer and DSP devices without stopping their target application. The system will be paused in the short time when data is transferring. Although this environment isn't enough for feedback loop of system, it can also support program debugging and development easily. The data of RTDX can be analyzed and visualized on the host using any OLE automation client. This shortens development time by giving designers a realistic representation of the way their systems actually operate.

All tasks of feedback software are optimal for communication and filtering. This task can't be swapped when a processor is accessing from FPDP port memory. The interrupt and DMA handle can avoid this problem. The phase shift, and band-pass filter tasks are embedded in the DSP module. The DSP modules include multiprocessors to process input data and to kick the system. To keep data from missing in the FPDP port communication between the MUX and DEMUX modules, there is a synchronized signal between the MUX module and DSP. The feedback procedures are idle and wait for the next working period after data I/O and signal filtering. This synchronized signal is defined to MUX module data preparation and FPDP port communication status.

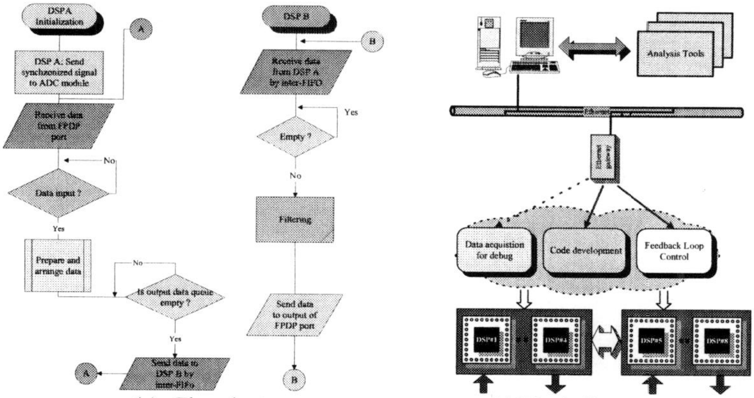

(a) Flowchart (b) Block diagram

FIGURE 5. Feedback software components.

Although the DSP debugger and development environment can establish a data exchange interface, feedback will be stopped when data are transferring. There are external diagnostic tools that can solve this problem. An integrated MATLAB simulation tool and LABVIEW are applied in console. To support various conditions in commission, adjusting phase and center frequency of the band pass filter online is efficient. Figure 5 shows the relation between various software components.

TRANSIENT DOMAIN DIAGNOSTIC SUPPORTS

The transient diagnostics have been approved as an elegant tool to aid the LFB commissioning and for various beam physics studies. It is labor intensive to integrate such functionality within the feedback electronics without supplicated design. The alternative solution was adopted to provide such function. A separated transient digitizer accompanies the feedback electronics. The diagnostics digitizer shares the feedback loop with the same bunch phase detector. The only thing is a phase adjustment to ensure that both systems read the same signal. The function block diagram of the transient domain data acquisition is shown in Fig. 6. The CompuScope 82G digitizer made by Gage Applied Technologies, Inc., was chosen because it provides SDK for LabVIEW and MATLAB. The transient digitizer access MATLAB scripts support very simple hardware access function and make it easy to integrate with MATLAB analysis scripts. The support transient domain data capture and the trigger signal can be accompanied with the feedback On/Off to capture the transient signal for 30 msec for all bunch without decimation. The acquired data might need further decimation to reduce data amounts. A typical beam pseudo-spectrum for the condition of RF gap voltage modulation On/Off is shown in Fig. 7. The sharp peak near 190 MHz is due to one of a HOM of the Doris cavity. The 2nd plunger of the RF cavity is tuned to keep the magnitude of this peak as stable as possible. The RF gap voltage modulation will further reduce the strength of the instability to provide stable beam.

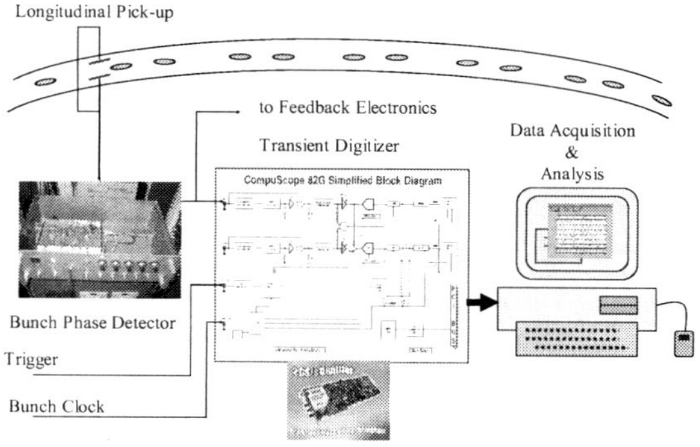

FIGURE 6. Dedicated diagnostic bunch phase data acquisition system.

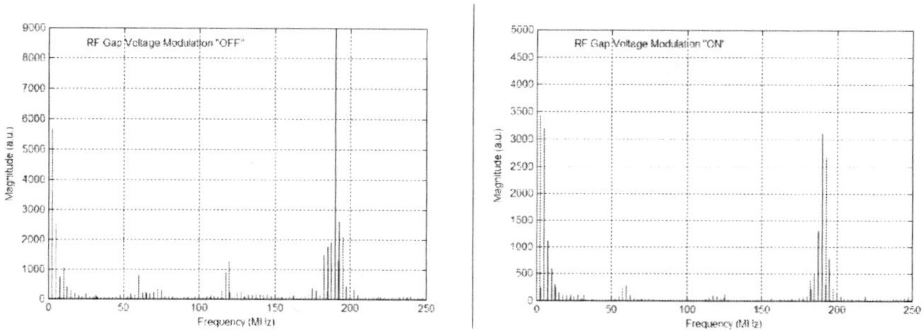

(a) RF gap voltage modulation "OFF" (b) RF gap voltage modulation "ON"

FIGURE 7. Pseudo-spectrum of the stored beam without RF gap voltage modulation.

PRESENT STATUS

The front-end and back-end electronics have been implemented and have performed intensive testing. Two DSP boards and the development environment were settled. The software for the longitudinal feedback system is under development. The prototypes of ADC/DEMUX and DAC/MUX have been implemented, and functionality is adequate. A revised version is in implementation. A dedicated transient domain data acquisition system has been set up. Necessary modification of front-end and back-end electronics is on-going to improve their performance. A prototype has demonstrated that to damp longitudinal oscillation at low current is possible. The infrastructure of the feedback software and data acquisition and analysis tools is finished. The user interface is a simple VI in LabVIEW environment; MATLAB scripts are used for complicated analysis.

ACKNOWLEDGMENTS

The authors highly appreciated the help from Dr. Micha Dehler of SLS on the kicker adaptation.

REFERENCES

1. J. D. Fox, "Programmable DSP-Based Multi-Bunch Feedback - Operational Experience from Six Installations," AIP Conference Proceedings 451, Cambridge, Massachusetts, 2000, pp. 226-233.
2. M. Tobiyama and E. Kikutani, "Development of a High-Speed Digital Signal Process System for Bunch-by-Bunch Feedback Systems," *Phys. Rev. Special Topics – Accel. and Beams*, PRST-AB 3, 012801 (2000) and reference therein.
3. W. K. Lau et al., "Preliminary Beam Test Results of the TLS Longitudinal Damper with High Performance DSP Modules," Proceedings of EPAC 2000, Vienna, pp. 1918-1920.
4. M. Dehler, "Kicker Design for the ELETTRA/SLS Longitudinal Multi-bunch Feedback," Proceedings of EPAC 2002, Paris, pp. 2070-2072.

Influences of Environmental Temperature Changes on RF Phase and Beam-Energy Drift in the PLS 2.5-GeV Linac

W. H. Hwang, H. G. Kim, W. W. Lee, S. J. Park, and Y. J. Han

Accelerator Division, Pohang Accelerator Laboratory, POSTECH, Pohang, Korea

Abstract. The change in relative phase between the accelerating rf fields and the beam bunches is the dominant factor of beam-energy change in linear accelerators. Major factors of the changing rf phase are modulator high voltage jitters in the short term and drifts in environmental conditions in the long term. We have investigated influences of cooling water and air temperature changes on the rf phase and beam energy drifts in the PLS 2.5-GeV linear accelerator. In this article, we describe how the environmental factors change the beam energy drifts.

INTRODUCTION

When the PLS linac was constructed in 1992, there were 11 klystrons operating and providing the beam energy of 2.0 GeV. Since then, the machine has fulfilled its mission of injecting beams into the PLS storage ring, the third generation synchrotron light source. In 1998, one more klystron (unit K12) and two 3-meter accelerating structures were added in order to provide more reserve in the rf power available and thus improving the beam-energy stability. In 2002, we increased the routine operational beam energy up to 2.5 GeV. The schematic diagram of the linac rf system is shown in Fig. 1[1, 2]. The PAD units shown in the figure are not yet installed. The designed beam energy spread of the linac is 0.6%. The electron beam is accelerated with pulsed rf of 2856 MHz. The rf frequency, phase, and power are very important factors in linac operations. The change of these factors gives influences on the electron beam energy and the energy spread. The magnitude of change of these factors depends on various factors such as the drive signal of klystrons, modulator beam voltage, and environmental conditions. The beam voltage of a modulator is stabilized within the design specification of ±0.5% in two stages [3]. The temperature of accelerating sections is routinely controlled within 45 ±0.2°C. The design tolerance of the drive system's phase stability is less than ±3.5° during 72 hours for the entire 145-m long drive system [4].

Until last year, the linac beam energy was changing by more than several tens of MeV in injection time and sometimes up to 200 MeV, as shown in Fig. 2. We measured beam energy variation and investigated its causes. We found it is related largely to cooling water temperatures. Also the rf phase variation by temperature

CP732, *Beam Instrumentation Workshop 2004: Eleventh Workshop*,
edited by T. Shea and R. C. Sibley III

variations affected beam energy. This paper describes measurement results of beam energy and how environment temperatures change the beam energy.

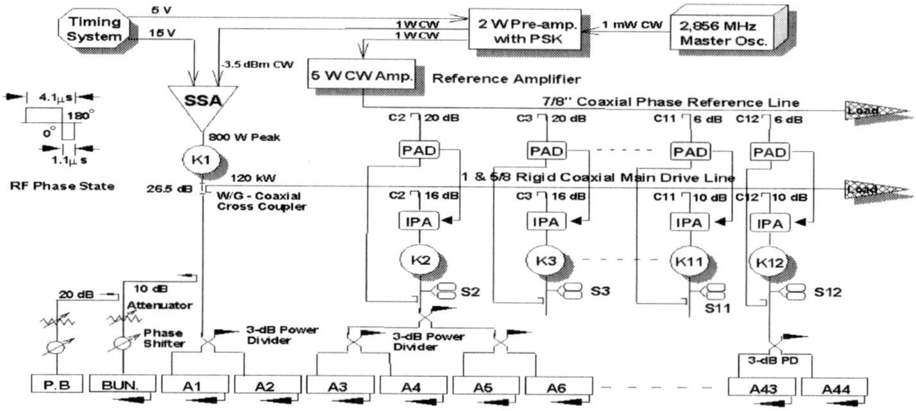

FIGURE 1. Schematic diagram of PLS linac rf sysem.

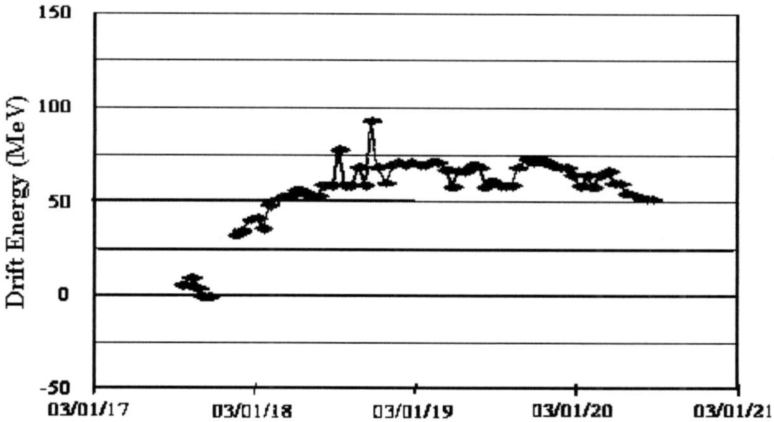

FIGURE 2. Beam energy drift in PLS linac.

BEAM ENERGY DRIFT

Energy spectrum width depends on the rf phase, and the energy spread for small angular spread is

$$\frac{\Delta E}{E} \approx \frac{1}{2}\left[\frac{\alpha}{2} + \frac{\sum \theta_n}{N}\right]^2 , \tag{1}$$

where α is bunching angular, N is the number of sections, and θ is individual rf phase error. The phase variation of accelerating column caused by the cooling temperature variation ΔT is

$$\Delta\theta = 2 \bullet \tau \bullet Q \bullet \upsilon \bullet \Delta T , \tag{2}$$

where τ is the attenuation coefficient (0.57), Q is the unloaded Q-value of accelerating column (13,000), and υ is the linear coefficient of thermal expansion (1.65 E-5). The normal cooling condition of the accelerating column in the PLS linac is controlled within $\pm 0.2^{\circ}$C and then calculated beam energy spread is about 0.5%.

Figure 3 shows the beam energy variation affected by cooling temperature of accelerating column in normal operation. The measurement is performed during 3 hours. The energy spread is about 0.9% in the variation of 0.52°C and it is more than 0.6% of designed value.

In the measurement period of 1 year, the energy variation had been more serious (>100 MeV). The variation in outside temperature caused cooling water temperature changes, and we upgraded the cooling control system to preserve the LCW temperature, as shown in Fig. 4. First, we inserted reservoir to preserve the amount of flowing water to the load. Secondly, we replaced 2-way valve to 3-way valve to control the amount of flowing water of tower and bypass. After these, the energy drift is decreased within several tens of MeV regardless of big variations in outside temperature.

PHASE MEASUREMENT

The beam energy drift is related to the rf phase by Eq. (1), and the rf phase depends on cooling temperature. So the phase measurement system is constructed to measure rf phase of the SLED output referring the SLAC phasing system [5]. We assigned the one point in the period of interaction between beam and rf pulse of 4 μsec and measured the amplitude and phase at the point. In this measurement, as shown in Fig. 5, rf amplitude coupled by a 10-dB coupler and phase information from the mixer output are sent to oscilloscope. Cooling, room, and outer temperatures are recorded by PC through a GPIB. Also, the oscilloscope and power supply are controlled by PC through a GPIB.

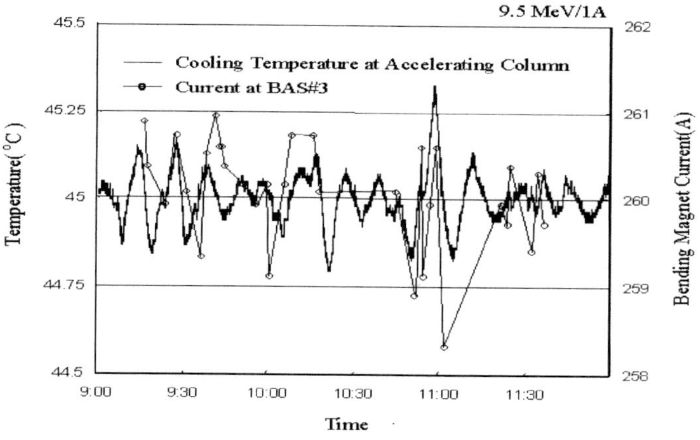

FIGURE 3. Beam energy drift caused by cooling temperature of accelerating column.

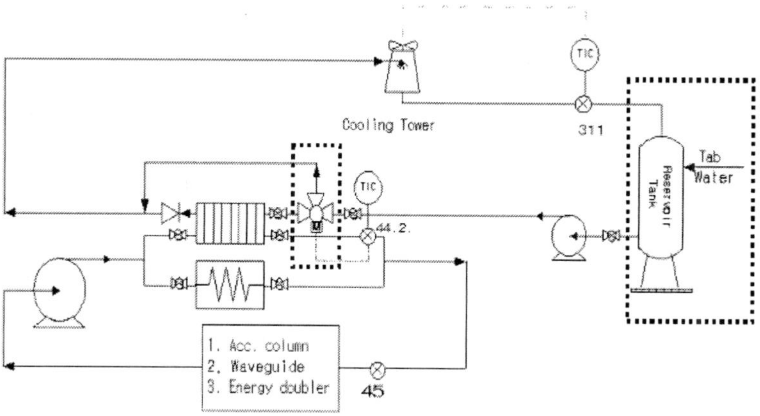

FIGURE 4. Schematic diagram of precision temperature controller.

The phase detector system consists of a bi-phase modulator, a double balanced mixer, a voltage controlled phase shifter and an amplitude detector. Amplitude-independent phase measurement is achieved by an in-situ compensation process that was done by the controller based on the LabVIEW program. The mixer output and detector output are given by the following relations:

$$V_{mix} = \alpha \left|\vec{S}\right| \cos(\phi_s - \phi_r \pm \pi / 2) + V_{offset} \quad , \tag{3}$$

and

$$V_{amp} = \beta\left|\vec{S}\right| \ . \tag{4}$$

For $\phi_s - \phi_r < 10^\circ$, the signal phase can be calculated as

$$\phi_s = \pm\frac{\beta}{\alpha}\times\frac{V_{mix} - V_{offset}}{V_{amp}} + \phi_r \ , \tag{5}$$

and each pulse phase jitter is

$$jitter = |\phi_s - <\phi_s>| \ , \tag{6}$$

where $<\phi_s>$ is the average phase. In this system, the phase was recorded by averaging 30 successive data. By high voltage modulator, maximum phase jitter of each pulse was about ± 3 degrees during 10 minutes.

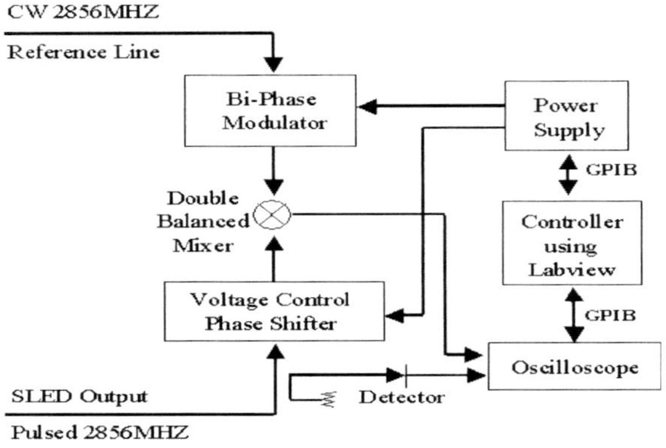

FIGURE 5. Phase measurement system.

The rf phase of the SLED output in the MK10 module was found to be changed by the room temperature variations in the MK1 module. As the temperature changes, the phase of SLED #10 is affected by the phase of klystron #1, the thermal expansion of the MDL (Main Drive Line), and reference line and master oscillator parts, etc. In this case, the rf phase is changed by 3 degrees with 1°C change in the room temperature. The room temperature of the klystron gallery is controlled to within 23 ±2°C in normal operation.

The rf phase variation of SLED output with its cooling temperature is shown in Fig. 6. The rf phase is changed by about 10 degrees per 1°C variation of the cooling water temperature. In normal operation, the temperature of the SLED is controlled to within 45 ±0.2°C.

The rf phase change of SLED output and beam energy drift, with the cooling temperature changes in klystron tubes, is shown in Fig. 7. The rf phase is changed by about 1 degree per 1°C variation of the cooling water temperature for klystron tubes. The temperature of the klystron tube is controlled within 32 ±1°C in normal operation.

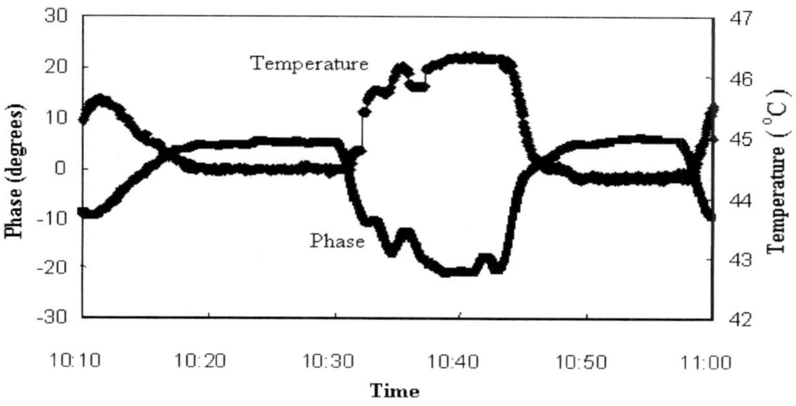

FIGURE 6. RF phase drift caused by cooling temperature of SLED.

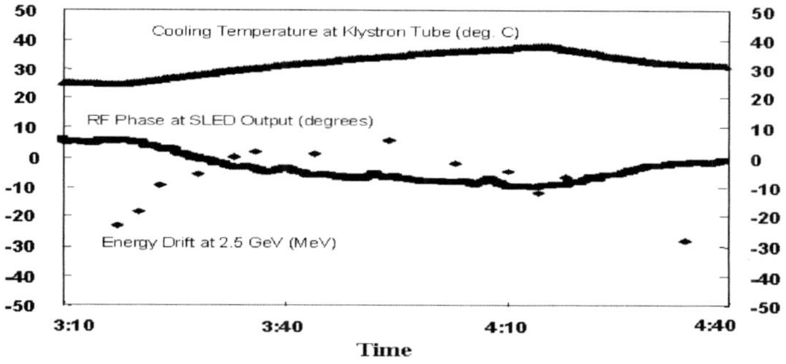

FIGURE 7. RF phase drift caused by cooling temperature of klystron tube.

SUMMARY

In the PLS 2.5-GeV linac, the beam energy is changed by cooling temperature, air condition, and modulator high voltage jitter. We found out that the rf phase variations that cause beam energy drifts were caused by environmental conditions. The SLED output phase was found to be changed by 3° per 1°C room temperature, by 10° per 1°C SLED cooling temperature, and 1° per 1°C klystron cooling temperature variations. The pulse-to-pulse phase jitter was about ±3°. Work to suppress these phase variations is under way.

REFERENCES

1. Design Report of Pohang Light Source (revised ed.), Pohang Accelerator Lab., 1992.
2. W. Namkung, "PLS 2-GeV Linac," Proc. International Linac Conf., Japan, 1994.
3. M. H. Cho et al., "High Power Microwave System for PLS 2-GeV Linac," Proc. International Linac Conf., Japan, 1994.
4. H. S. Lee et al., "Microwave System of PLS 2-GeV Linac," Particle Accelerator Conference, Dallas. Texas, 1995.
5. J. D. Fox and H. D. Schwarz, "Phase and Amplitude Detection System for the Stanford Linear Accelerator," SLAC-PUB-3071, March 1983.

DARHT AXIS II BEAM POSITION MONITORS

Jeff Johnson*, Carl Ekdahl*, and William Broste†

*Los Alamos National Laboratory, Los Alamos, NM 87545, USA
†Bechtel Nevada

Abstract. One of Los Alamos National Laboratory's (LANL's) primary responsibilities for national security is to certify the readiness of our nation's nuclear stockpile. Since the end of underground testing in 1994, LANL has used non-nuclear experiments and computational models to certify our stockpile. The Dual Axis Radiographic Hydrodynamic Test (DARHT) Facility is the next tool scientists will utilize for stockpile certification. DARHT will soon be capable of producing a three dimensional, time resolved radiographic image of a nuclear weapon pit during implosion. Data from these radiographic images will be used to validate the computational models used to study nuclear weapons. The first axis of DARHT with its single-pulse capability has been in use for about 2 years. Data returned from DARHT's first axis has been exceptional, producing the highest resolution radiographic image ever for a pit test.

The second axis of DARHT is unique in that it is designed to produce 4 time-resolved radiographic images on a hydrodynamic time scale. Four Gamma ray pulses are generated when high-energy electron pulses are kicked out of a 2-microsecond-long accelerator beam pulse. The long (2-microsecond) pulse width of DARHT's second axis accelerator presents challenging diagnostic requirements that must be achieved to produce good radiographic images. The second axis of DARHT is currently in its commissioning phase.

This paper provides an overview of the system developed to monitor the electron beam position and beam current in the DARHT second axis accelerator. The Beam Position Monitor (BPM) used on DARHT's second axis consists of an array of 4 single loop B-dot detectors mounted in a diagnostic flange. Each B-dot loop detector has a differential output allowing for common mode signal removal. The sensing loops are a Moebius type design. The Moebius loop design provides grounding at the loop, protecting recording instruments during accidental beam spill. The design of the detector body and diagnostic flange were experimentally optimized to reduce field soak. The loop construction uses a quartz dielectric insulated coax to meet the vacuum requirements of the beam line. A 25-mm square groove in the I.D. of the diagnostic flange provides additional protection from beam spill. Each loop has an area of 310 square millimeters.

Position accuracy requirements for the DARHT II BPMs are that the beam position be known to within +/- 500 microns. The inside radius of the accelerator beam pipe is nominally 5 inches (127 mm). The B-dot detector loops are located inside the flange groove at a radius of 140 mm from beam center line. Since the output signal of the b-

CP732, *Beam Instrumentation Workshop 2004: Eleventh Workshop*,
edited by T. Shea and R. C. Sibley III

dot loop will change as 1/r, when the beam centroid moves 250 microns the b-dot loop output signal will change by only 0.40%.

The ability to meet the required level of accuracy for beam position and current measurements is met through the development of split recording. Each differential output signal from the B-dot detector is split by a resistive power divider. One output of the divider is recorded on an 8-bit digitizer that captures the peaks of the derivative signals. The other output is also recorded on an 8-bit digitizer; however, this digitizer channel is set at ten times the voltage sensitivity compared to its twin. This practice will over drive the sensitive digitizer channel by a factor of ten. Laboratory experiments determined that this level of overdrive does not introduce signal distortion. After an acquisition, the data are reduced by combining the two data records. Low-amplitude data from the more sensitive digitizer channel is combined (using software) with the high-amplitude data from the other regular channel to create a single new "combined" data file. This technique of split recording gives extended dynamic range perhaps approaching 10 bits. The high-resolution data produced by split recording also allows for accurate numerical integration of the derivative B-dot signals. Split recording is used for all 8 detector outputs for each BPM during calibration and operation.

Calibration of the DARHT second axis BPMs is performed on a calibration test line. A calibration test line and pulse forming network (PFN) were designed and built that allow every BPM to be calibrated at the full operating current of the DAHRT second axis accelerator. The PFN produces a 2kA, 2μs (100kV into the 50 ohm test line) calibration pulse. The PFN output is monitored with a Pearson Current Monitor. The test line is a gas insulated (SF6) coaxial line with a 50-ohm termination resistor. The solid center conductor of the test line is moved (scanned) with micrometer stages during the calibration procedure. Calibration data are recorded and analyzed to produce calibration files for use at DARHT.

There are 21 BPMs in use on the DARHT second axis accelerator. Every BPM is recorded during each beam shot, requiring 336 digitizer channels. Recording parameters are 1 Gs/s with 15k point record length. Data acquisition chores are handled using DAAAC software from Voss Scientific Inc. The BPM design chosen for use at DARHT addresses the issues of common mode signals. However, due to the large pulse power system used to drive the accelerator induction cells, there remains a small level of background noise on the BPM recording channels. To deal with this issue, pulse power only background shots are recorded each day prior to accelerator operation. The background shot data are removed from accelerator shot data using software. After background subtraction is complete, custom IDL programs are used to determine the position and current of the beam inside the accelerator. The processed data are used to steer and focus the accelerator beam pulse during commissioning experiments and in preparation for radiographic use.

BPM System for Electron Cooling in the Fermilab Recycler Ring

Paul W. Joireman*, Jerry Cai*, Brian E. Chase* and Greg W. Saewert*

*Fermi National Accelerator Laboratory, P. O. Box 500, Batavia, IL, 60510

Abstract. We report a VXI based system used to acquire and process BPM data for the electron cooling system in the Fermilab Recycler ring. The BPM system supports acquisition of data from 19 BPM locations in five different sections of the electron cooling apparatus. Beam positions for both electrons and anti-protons can be detected simultaneously with a resolution of $\pm 50\ \mu$m. We calibrate the system independently for each beam type at each BPM location. We describe the system components, signal processing and modes of operation used in support of the electron-cooling project and present experimental results of system performance for the developmental electron cooling installation at Fermilab.

INTRODUCTION

The Fermilab Recycler ring commissioning efforts have recently provided anti-protons to establish two p–$\bar{p}$ Tevatron stores with initial average luminosities of 12.5×10^{30} cm^{-2} s^{-1} and 18.8×10^{30} cm^{-2} s^{-1}. One component of the Recycler commissioning plans involves the installation of an electron cooling system to compliment the existing stochastic cooling system. Electron cooling can reduce both longitudinal and transverse momentum spread, is independent of the stored anti-proton beam intensity and is projected to provide longitudinal cooling rates of 80 eV s / hour. [1] The electron cooling system at Fermilab will establish a DC electron beam of up to 0.5 A, direct this through a 20 m cooling section for interaction with Recycler anti-protons and then return the beam to a collector. Effective electron cooling can only occur if the angular spread of the electron beam is no greater than 0.1 mrad relative to the anti-proton or $\bar{p}$ beam. Instrumentation and diagnostics to characterize and maintain the precise alignment of electron and anti-proton beam is essential to the success of the electron cooling project.

We report a beam position monitor (BPM) system designed to measure electron and $\bar{p}$ beam positions simultaneously for the purpose of aligning the two beams. The system has an electron to anti-proton beam position resolution of $\pm 25\ \mu$m. The system has three modes of operation, measurement of DC electron beam, pulsed electron beam, and ciruculating $\bar{p}$ beam. The system makes strictly frequency domain measurements in all three modes. The DC electron beam is modulated for detection. The system can simultaneously measure both $\bar{p}$ and electron beam positons. Pulsed electron beam mode is used for diagnostics and tuning. The format of this paper is as follows. We first describe the hardware and software used to detect, acquire, process and present the raw beam signals. The next section describes operational modes of the electron cooling BPM system and presents results obtained for the electron cooling test installation at Fermilab.

CP732, *Beam Instrumentation Workshop 2004: Eleventh Workshop*,
edited by T. Shea and R. C. Sibley III

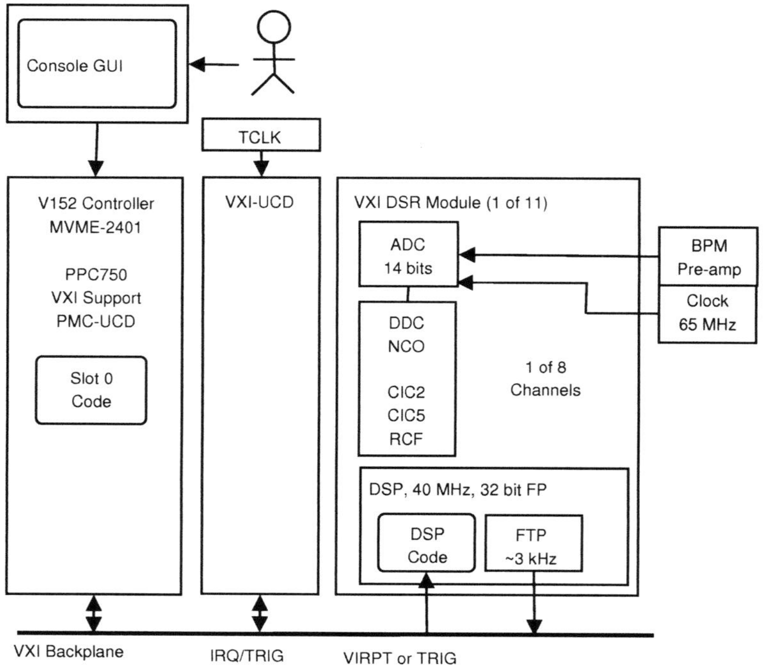

FIGURE 1. Block diagram of the electron cooling BPM system. The BPM itself is not shown.

The final section provides a discussion of the results, plans for future enhancement of the BPM system, and a brief conclusion.

EXPERIMENTAL

We define the electron cooling BPM system as all hardware and software necessary to produce a measurement of beam position. A simplified block diagram of the system is shown in Figure 1.

Hardware

The hardware can be described by following the signal path from the beam circulating in the pipe through the pre-amplifier, into the VXI DSR board and finally to the DSP, see Figure 1. The BPM detector consists of two 5 cm split plate assemblies with an inner diameter of 2 inches. One BPM assembly includes both horizontal and vertical plates. The four detector plates are connected with short lengths of coax to a four channel, low noise, charge pre-amplifier. The preamps have a 20 dB gain select on the charge integrator and a high precision resistive divider to provide a calibration signal. The output of the integrator is followed by a gain stage with a back terminated differential driver. A multi-conductor shielded twisted pair cable connects the preamps to the DSR with 15 pin D connectors. The twisted pair cable provides low loss and good channel to channel isolation in the low frequency range where processing occurs.

The digital signal receiver (DSR) has been described in detail elsewhere.[2] Briefly, the DSR is a VXI size C module that contains 8 receive channels. The channels are grouped into pairs to acquire signal from a particular set of detector plates, either horizontal or vertical. The signal chain for a single channel consists of a differential detector, 14-bit, 65 MS/s ADC (Analog Devices AD6644) followed by a **digital receive signal processor** (Analog Devices AD6620). This proceesor contains a numerically controlled oscillator (NCO) for I/Q demodulation and three stages of decimation filters. The ADC is clocked from an external low noise 65 MHz sine wave source. The filter stages consist of two cascaded integrator comb FIR filters and an FIR filter (RCF) with up to 256 user defined coefficients. The decimation rate for all three filters may be specified by the user within a fixed range prior to system operation. The NCO demodulation frequency and the gains of all filters may be changed during system operation at user request. The output of the last filter stage is latched for retrieval by an on-board 32-bit floating point DSP (Analog Devices SHARC™). The DSP provides low-level communication with the on-board hardware, enabling filter gain and receiver frequency changes, and processing resources to calculate beam positions and intensities from the filtered signal. The DSP also controls gain selection in the low noise preamp located close to the detector.

Eleven DSR modules are installed in a 13 slot VXI chassis. The first slot contains a VXI slot 0 controller card (V152 - Kinetic Systems), including Power PC 750 processor (Motorola) with 32 MB of RAM. A PMC universal clock decoder (UCD) card occupies one of the PMC slots on the controller to provide Acnet communication resources. The controller runs the application software described below and manages communication within the VXI chassis. The last slot is occupied by a VXI universal clock decoder card to provide accelerator wide timing and event decoding (Tevatron CLocK - TCLK) for the system.

Software

The system software can be broken into two categories, DSP software and slot 0 software. The DSP software is written in assembly and C using the VisualDSP++ 3.0 (Analog Devices) development environment. The slot 0 software is written in C and C++ for VxWorks, version 5.4.2 (Wind River). The DSP software is responsible for calculating beam positions and intensities, and reconfiguring the on-board hardware to fulfill operational requests. The slot 0 software is responsible for coordinating the resources with the VXI chassis to respond to end user requests for data or operational changes to the system. The end user interacts directly only with the slot 0 software through the Acnet control system interface.

The DSP software handles the raw data processing of signals received from the BPM. We consider detection of continuous or DC electron beam as a concrete example to guide our discussion. The electon beam is modulated at 32 kHz and the DSR module is tuned to this frequency. As described in previously, the incoming data, following digitization and I/Q demodulation, passes through a set of hardware filters. For DC electron beam, these filters reduce the data rate to 21.6 kHz and the data bandwidth to 800 Hz. This filtered data is then retrieved by the DSP software passed through a 100 Hz low pass IIR filter at the assembly level. These data are then used by a C routine to calculate intensity and beam positions at a rate of approximately 3 kHz. The position value x is calculated from a sum over difference formula

$$x = S\,\frac{A - kB}{A + kB} + O\,, \tag{1}$$

where A and B are the plate intensities calculated from the square root of the sum of the squares of the I/Q signals for that plate. S and O are scale and offset terms determined by the physical construction of the BPM and k is a calibration factor which accounts for small differences in signal path gain between A and B plates. The value of k is obtained from a calibration procedure which involves passing a known signal with a fixed A/B ratio through the pre-amplifier and DSR. The measured plate intensities are then used along with the known ratio to calculate a calibration constant for each BPM channel.

The C routine also filters the calculated positions by applying a 5 Hz low pass IIR filter to the position data. The two position values are referred to as 100 Hz bandwidth positions and 5 Hz bandwidth positions. These calculations are performed for all four channel pairs on each DSR module. The data is placed in a memory location that the slot 0 code maps to Acnet variables visible throughout the accelerator complex.

The slot 0 software is responsible for translating user requests for functionality or data into system action. The slot 0 software consists of the electron cooling BPM application software and several shared libraries. The shared libraries facilitate communication between the application software, the operating system, the DSR and VXI-UCD modules, and the accelerator wide control system. We will focus our discussion of the slot 0 software on those components specific to the electron cooling BPM application, the application software and the DSR shared library supporting this software.

The DSR shared library uses an object-based methodology that provides methods to create and use a DSR hardware module. The common operations that the library provides are object creation,[1] display of object information, ability to load DSP software from a specified file and slot 0 interface to a client-server protocol for requesting DSR hardware or DSP software reconfiguration.

The application software manages all resources within the VXI crate and is directly responsible for translating user requests into system functionality. At boot time, the application software creates software objects to manage the installed modules within the VXI crate and then downloads DSP code into all DSR modules using the functionality provided by the DSR shared library. The calculated positions and intensities are mapped

[1] These objects are defined as C `struct` data types not as a C++ `class`. Conventions are used to encapsulate the data related to the DSR object and C functions/methods are defined for the object by passing the object a parameter.

to a series of Acnet data variables, exposing the processed data to the accelerator wide control system. Additional system-wide control and status variables are created and mapped to Acnet variables to provide user control and diagnostics for the system. Callback functions are attached to a subset of these Acnet variables to connect user requests to application software functionality. The application software then downloads a set of configuration parameters for the installed BPM detectors to calibrate the system and enters an idle state where it waits for user request for data or operational changes.

RESULTS

The electron cooling BPM system signal conditioning and processing characteristics are, by design, dominated by the DSR VXI module. The DSR digitizes any signal present at its inputs and thus all input signals, electron and anti-proton or $\bar{p}$ are digitized simultaneously. Following digitization, the AD6620 selectively down converts either the electron beam signal at 32 kHz or the anti-proton first revolution line of 89 kHz to baseband. The in-phase (I) and quadrature (Q) signals are passed through three stages of low pass decimation filters. The signal bandwidth is reduced to 800 Hz before the IQ data is read out by the DSP at a 21 kHz rate. The DSP converts the data from 16 bit fixed point data to 32 bit floating point. A final low pass filter further reduces the signal bandwidth to 100 Hz in software prior to processing the signal to calculate beam positions.

The system has been tested in the Recycler with real $\bar{p}$-beam. Electron beam measurements, both pulsed and DC, have been made at the developmental electron cooling installation. Simultaneous measurements of the electron and $\bar{p}$ beams will not be possible until the electron cooling system is moved into the Recycler later this year. Thus, experimental data is restricted to measurements on an electron beam. Data simulating an $\bar{p}$-beam was generated for this paper using a signal generator at the Recycler revolution frequency. Two system operational modes, DC electron beam and pulsed electron beam, have been used in initial tests of the system with the electron cooling test system at Fermilab. The signal processing characteristics of DC mode have been described in the previous section. During pulsed mode operation, the electron gun emits a 2 μs pulse at a repetition rate of 1 Hz and sends a trigger indicating the pulse to the BPM system. The DSR filters broaden the detected pulse from 2 to approximately 50 μs and this data is sampled by the DSP at 1 Hz after fixed delay from the trigger pulse; the delay accounts for the processing delay through the DSR filter chain.

We initially simulated the electron and $\bar{p}$ signals using a signal generator and collected data on 100 Hz bandwidth positions and 5 Hz bandwidth positions for a representative BPM. The calculated standard deviations obtained from this data are shown in Table 1. The pre-amplifier gain was fixed to its high gain value for all data. Decreasing the pre-amplifier gain results in a corresponding increase in the calculated standard deviations by a factor of 2 to 4. Measurements made with the electron cooling developmental system on electron beam show similar noise characteristics. The 5 Hz bandwidth position data was sampled for an electron beam operating in DC mode for all installed BPM detectors. The results are shown in Figure 2 along with measurements for a calibration signal which

TABLE 1. Standard deviations for 100 Hz bandwidth position data measured for different beam types.

Beam Type	σ_X / μm	σ_Y / μm
e	5	3
$\bar{p}$	5	5

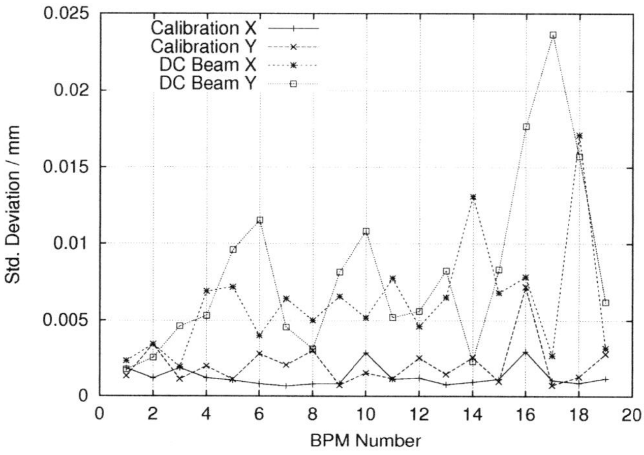

FIGURE 2. Comparison of the standard deviation σ for BPM position readings for DC electron beam relative to that obtained with a calibration signal at the same frequency. The data represent the average of twenty 5 Hz bandwidth position readings taken at 10-second intervals. Pre-amplifier gain was set to its high gain value.

simulated the electron beam signal. The average standard deviation of positions for all BPMs in DC mode is 6 μm in the horizontal (X) and 8 μm in the vertical (Y) plane. This represents a position resolution for DC mode of 9×10^{-3} % or greater than 1 part in 10,000. Standard deviations observed for the calibration signal are comparable to the bench measurements shown in Table 1. Similar measurements to those shown in Figure 2 were performed for pulsed mode operation. Due to the data collection characteristics of this mode, one sample per pulse, the calculated standard deviations are larger. Standard deviations for the pulsed beam are calculated to be 74 μm and 85 μm in horizontal and vertical planes respectively, while those calculated for a calibration source signal are 20 μm in both planes.

A simulated signal can also be used to show the stability of the calculated positions with respect to changes in detection frequency, or equivalently changes in detected beam type. This is shown in Figure 3, where position data for 2 BPM channels is shown, an artificial offset between the sets of position data has been added for clarity. The data in the figure are divided into 4 sections demarcated by the horizontal grid lines. The first section on the left shows position data for beam detected at 89 kHz, simulated $\bar{p}$-beam with the pre-amplifier gain set to its high gain value. The second section displays

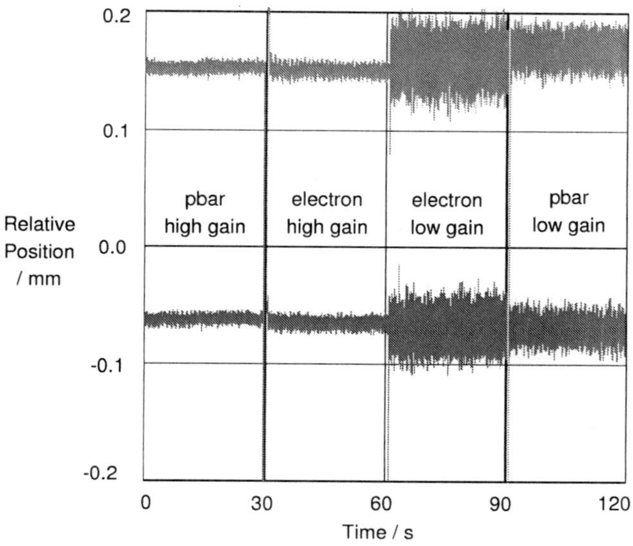

FIGURE 3. Plot of 100 Hz bandwidth position data relative to beam type and pre-amplifier gain changes. Each major horizontal division corresponds to a different beam type/pre-amplifier gain pairing, as indicated.

data for beam detected at 32 kHz, simulated electron beam, with the same pre-amplifier gain. In the last two sections the pre-amplifier gain was set to its low gain value. The third section displays electron beam position data and the fourth, $\bar{p}$-beam position data. The deviation in average position comparing all data is less than 30 μm. The BPM system has also been used to analyze oscillations in the position data readings focused on determining the source of these oscillations. A particular BPM is instructed to acquire 2048 data points at a 3.095 kHz rate. The acquired data is stored in a DSP buffer for later extraction and FFT analysis. A sample of the FFT amplitude spectrum for a given set of acquired position data is shown in Figure 4.

CONCLUSION

The electron cooling BPM system provides a frequency selective, narrow bandwidth, high precision BPM system which meets or exceeds the specified requirements of the electron cooling project. The system provides low noise position data at a resolution (DC mode) of greater than 1 part in 10,000. A series of upgrades to the system are in progress. The electron cooling project will begin moving its operational hardware to a location adjacent to the Recycler ring. The new installation will increase the number of BPMs to a total of 34 requiring an additional VXI crate and modules. The calculation of position currently relies on a characterization of the response of the BPM detector to well defined beam positions. This characterization or calibration uses a linear fit of

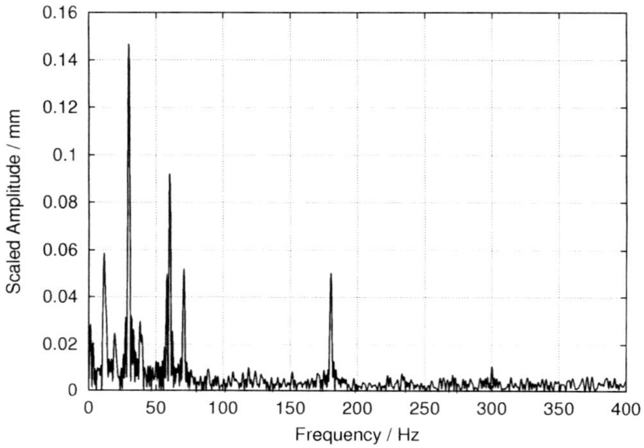

FIGURE 4. FFT amplitude spectrum of position data. No peaks of note occur above 400 Hz in the plot.

the logarithm of the plate ratio to position yields constants, which can be converted and used in Equation 1. Small non-linearities $< 20\,\mu$m have been observed for BPMs in the cooling section of the apparatus, the region of interaction between the electron beam and anti-proton beam. In order to accurately calculate the position across the BPM aperture a third order fit to the original calibration data is planned to smooth out any non-linearities in the observed position in the cooling section. The goal is to reduce the observed non-linearities in this region to $< 10\,\mu$m.

ACKNOWLEDGMENTS

We would like to acknowledge the contributions Alexander Shemyakin and the electron cooling group (operational system measurements), Keith Meisner and Jerry Cai (slot 0 software) contributions, Barry Barnes and Philip Varghese (DSP software), and Dan Klepec (DSR board construction), for their work on the electron cooling BPM system.

REFERENCES

1. Nagaitsev, S., *Electron Cooling for the Recycler Ring*, http://www-bdnew.fnal.gov/doereview03/Current/09_Electron\%20cooling\%20for\%20t-1.pdf.
2. Chase, B. E., Meisner, K. G., "A Digital Signal Receiver VXI Module for BPM and Phase Detection Processing," *Beam Instrumentation Workshop 2002: Tenth Workshop*, edited by G. A. Smith, et. al. AIP Conference Proceedings 648, 2002, pp. 368-375.

Transverse Feedback for Electron-Cooled DC-Beam at COSY

V. Kamerdzhiev and J. Dietrich

Forschungszentrum Jülich GmbH, Institut für Kernphysik, Germany

Abstract. At the cooler synchrotron COSY, high beam quality is achieved by means of beam cooling. In the case of intense electron-cooled beams, fast particle losses due to transverse coherent beam oscillations are regularly observed. To damp the instabilities a transverse feedback system was installed and successfully commissioned. Commissioning of the feedback system resulted in a significant increase of the e-cooled beam intensity by single injection and when cooling and stacking of repeated injections is applied. External experiments profit from the small diameter beams and the reduced halo. A transverse damping system utilizing a pick-up, signal processing electronics, power amplifiers, and a stripline deflector is introduced. Beam current and Schottky spectra measurements with the vertical feedback system turned on and off are presented.

INTRODUCTION

The cooler synchrotron COSY provides proton (deuteron) beams in the energy range of 45 (75) MeV to 2,7 (2,2) GeV for experiments in medium-energy physics. Electron cooling is used at injection energy to increase the space density of the beam before the acceleration. If needed, stochastic cooling is applied after acceleration at high energies. The intensity of an electron-cooled beam at injection energy is limited by transverse instabilities. Major losses are due to the vertical coherent beam oscillations.

Figure 1 illustrates the behavior of the electron-cooled proton beam at injection energy. Vertical and horizontal difference signals of a beam position monitor (BPM) together with beam current transformer (BCT) signal are demonstrated in this figure [1]. One sees that after several seconds of cooling, the beam becomes unstable in both transversal planes, first in the horizontal one and then in the vertical. However, the major particle loss (in addition to initial losses at the very beginning of the machine cycle) occurs after the vertical coherent oscillation has reached certain amplitude. This is due to the smaller vertical aperture in the arcs of the machine, the chromaticity value, strong longitudinal cooling ($\Delta p/p$ is small), and the increasing space charge impedance during the cooling process.

To cure the transverse instability chromaticity correction via sextupol magnets can be applied [2]. In this case, the enhancement of Landau damping results in higher instability threshold. Also longitudinal "heating" of the beam has a stabilizing effect. The most efficient method of curing the instability is active damping of coherent beam

CP732, *Beam Instrumentation Workshop 2004: Eleventh Workshop,*
edited by T. Shea and R. C. Sibley III

oscillation via beam feedback [3]. The goal is to damp coherent oscillations of the coasting proton beam at injection energy (β = 0,3).

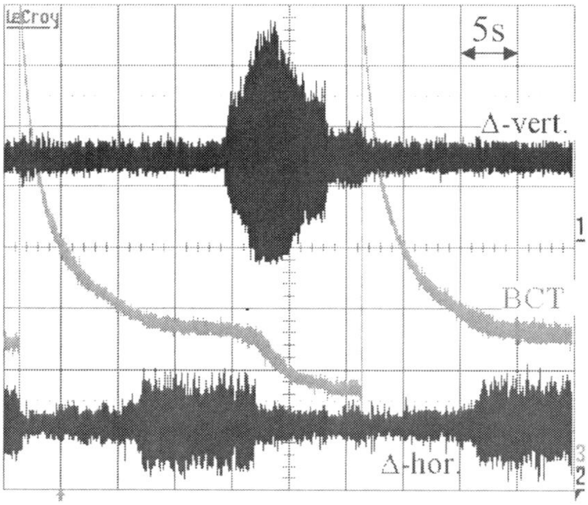

FIGURE 1. BCT signal, vertical and horizontal BPM Δ-signals in time domain.

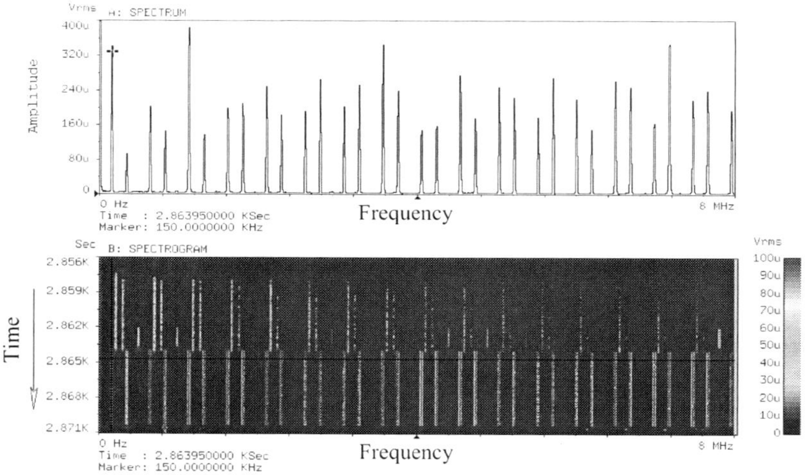

FIGURE 2. Delta signal of the vertical Schottky pick-up in frequency domain against time (spectrogram, lower image) and the frequency spectrum (upper image) corresponding to the time marked on the spectrogram by the horizontal black line.

TRANSVERSE BEAM FEEDBACK

The transverse feedback system, also called damper, typically consist of a pick-up, circuit generating the difference signal, filter rejecting the revolution frequency harmonics, delay line, power amplifiers and the kicker [4]. If at COSY the gap of the ferrite cavity is properly short circuited, the transverse frequency spectrum of the coasting beam contains only betatron sidebands and almost no revolution frequency harmonics. Figure 2 shows the delta signal of the vertical Schottky pick-up in frequency domain against time (lower image) and the frequency spectrum (upper image) corresponding to the time marked on the spectrogram by the black line. Due to the fact that the spectrum does not contain revolution frequency harmonics there is no need to use notch filters. Since the instability extends in frequency domain up to 40 MHz, a broadband feedback is necessary.

The installed feedback system uses one of the beam position monitors (electrostatic, diagonally cut) and its broadband electronics to measure the beam position deviation with 70 MHz bandwidth covering about 150 beam harmonics at injection energy. This signal is delayed by means of a very well shielded coaxial cable and fed into a 180° splitter. Two power amplifiers (commercially available) per plane drive the stripline kicker wired in differential mode. Power of up to 250 W per electrode is available. Each of the downstream ports of the kicker is terminated with 50 Ω. The betatron phase advance from the pick-up to the kicker at typical tunes is about 97°.

RESULTS

Figure 3 shows the effect of the feedback system on self-exited vertical betatron sidebands. At low frequencies the signal is noisy due to the interference on the long cable (needed for transporting the signal to the control room).

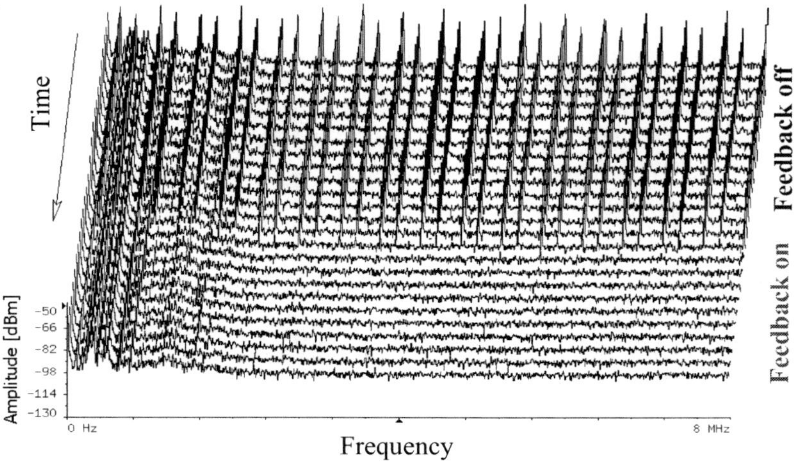

FIGURE 3. Delta signal of the vertical Schottky pick-up in frequency domain against time with vertical feedback turned off (upper part) and on (lower part).

After the feedback has been turned on, all self-exited betatron sidebands in the frequency band of the feedback system (50 kHz to 50 MHz) disappear (i.e., the coherent beam oscillation is efficiently damped). This results in significant increase of beam intensity available for the experiments after single injection and electron cooling (see Fig. 4).

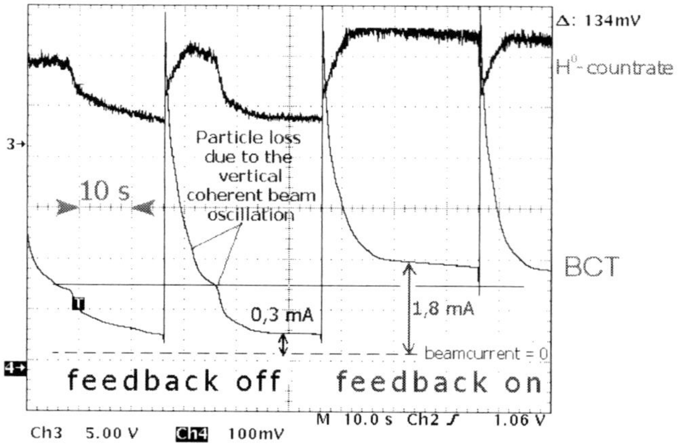

FIGURE 4. BCT signal and the H0 count rate with vertical feedback turned off and on.

One of the important applications of the feedback system is stacking of many injections in transverse phase space during electron cooling. Stacking of about 30 injections using a damper in the vertical plain allowed to reach coasting beam currents up to 9,2 mA at injection energy.

CONCLUSION

A transverse feedback system is very useful when fighting coherent beam oscillation. Since it does not affect the tune it can be turned on and off without any changes in the machine optics. Installing the vertical damper at COSY gave significant intensity increase at single injection and the possibility of stacking electron-cooled beams. To further improve the beam intensity and quality, a horizontal damper will be commissioned in several weeks.

ACKNOWLEDGMENTS

The authors are deeply grateful to H. J. Stein, I. Mohos D. Prasuhn, I. N. Meshkov, A. O. Sidorin, and the COSY team for support in experiments and useful discussions.

REFERENCES

1. I. N. Meshkov et al., Observation of Instabilities of Electron Cooled Proton Beams, IKP annual report, 2001.
2. V. Kamerdzhiev et al., Instability phenomena of electron-cooled ion beams at COSY, accepted to NIM.
3. D. Boussard, Cures of Instabilities, CERN 95-06, 22 Nov. 1995, Vol. I.
4. F. Pedersen, Feedback Systems, CERN-PS-90-49.

Ion Chambers for Monitoring the NuMI Neutrino Beam at Fermilab

Dharmaraj Indurthy[*], Ryan Keisler[*], Sacha Kopp[*,1], Steven Mendoza[*], Marek Proga[*], Zarko Pavlovich[*], Robert Zwaska[*], Deborah Harris[†], Alberto Marchionni[†], Jorge Morfin[†], Albert Erwin[‡], Huicana Ping[‡], Christos Velissaris[‡], Donna Naples[**], Dave Northacker[**], Jeff McDonald[**], Milind Diwan[††], and Brett Viren[††]

[*]*Department of Physics, University of Texas, Austin, TX 78712, USA*
[†]*Fermi National Accelerator Laboratory, Batavia, IL 60510, USA*
[‡]*Department of Physics, University of Wisconsin, Madison, WI 53706, USA*
[**]*Department of Physics, University of Pittsburgh, Pittsburgh, PA 15260, USA*
[††]*Brookhaven National Laboratory, Upton, Long Island, NY 11973, USA*

Abstract. The Neutrinos at the Main Injector (NuMI) beamline will deliver an intense muon neutrino beam by focusing a beam of mesons into a long evacuated decay volume. The beam must be steered with 1-mRad angular accuracy toward the Soudan Underground Laboratory in northern Minnesota. We have built 4 arrays of ionization chambers to monitor the neutrino beam direction and quality. The arrays are located at 4 stations downstream of the decay volume, and measure the remnant hadron beam and tertiary muons produced along with neutrinos in meson decays. We review how the monitors will be used to make beam quality measurements, and as well review chamber construction details, radiation damage testing, calibration, and test beam results.

INTRODUCTION

The NuMI beamline [1,2] at Fermilab will deliver an intense ν_μ beam to the MINOS detectors at FNAL and at the Soudan Laboratory in Minnesota. Additional experiments are foreseen. The primary beam is fast-extracted onto the NuMI pion production target in 8.56-μsec spills from the 120-GeV FNAL Main Injector. The beam line is designed to accept 4×10^{13} protons/spill. After the graphite target, two toroidal magnets called "horns" [3] sign-select and focus forward the secondary mesons (pions and kaons) into a 675-m volume evacuated to ~1 torr to reduce pion absorption, where they may decay to muons and neutrinos. The horns and target may be positioned so as to produce a variety of neutrino beam energies [4].

The instrumentation foreseen to ensure the neutrino beam quality on a spill-by-spill basis combines primary, secondary, and tertiary beam measurements. The primary beam direction is monitored by segmented foil secondary emission monitors [5] and by capacitative beam position monitors. Its intensity is measured by a beam current

[1]kopp@hep.utexas.edu.

CP732, *Beam Instrumentation Workshop 2004: Eleventh Workshop*,
edited by T. Shea and R. C. Sibley III

toroid (BCT). Additionally, the NuMI target is electrically isolated, so may be read out as a Budal-type monitor [6] of the primary beam's fraction on target. The above instrumentation measures the quality of the beam reaching the NuMI target.

The subject of this paper is the secondary and tertiary beam monitoring system, shown in Fig. 1. Its purpose is to monitor the integrity of the NuMI target and of the horns, which focus the secondary meson beam. This monitoring is accomplished by measuring the remnant hadron beam reaching the end of the decay tunnel and by measuring the intensity and lateral profile of the tertiary muon beam penetrating the beam absorber and the downstream dolomite rock. Because every muon is produced by the pion decays which produce neutrinos, the muon beam provides a good measure of the focusing quality of the neutrino beam.

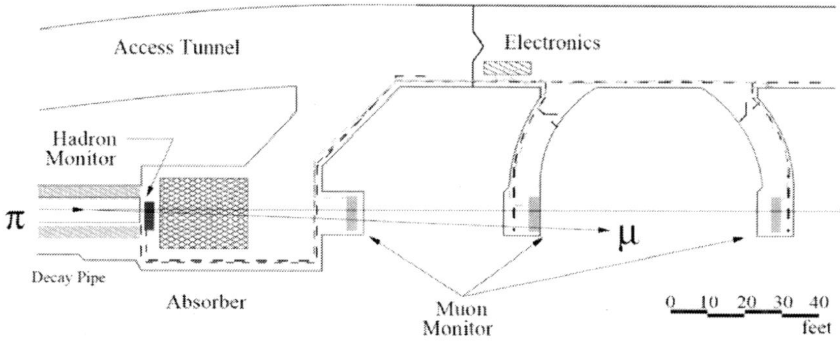

FIGURE 1. Layout of the secondary and tertiary beam monitors. Downstream of the decay volume, the hadron monitor measures flux and spatial profiles of remnant hadrons. At three stations in the downstream rock, the muon monitors measure rates and spatial profiles of the muon beam. The NuMI target and horns are 675 m upstream, to the left of this figure.

SECONDARY + TERTIARY BEAM MEASUREMENTS

The hadron monitor sits downstream of the NuMI decay volume, immediately in front of the beam absorber. It measures approximately 1 m × 1 m, and is segmented to measure the spatial profile of particles arriving at the detector. Monte Carlo calculations indicate that the charged particle rate at the center of the decay pipe is dominated by protons which did not interact in the target.

Figure 2 shows the proton beam profile at the end of the decay volume. A prominent peak is visible at beam center, due to protons which did not react in the 1-m graphite NuMI target. The 20-cm width of this peak is expected, given the mean scattering angle of 120-GeV protons passing through the target and the 725-m drift distance from the target to the hadron monitor. During normal running, this scattering width serves as a monitor that the primary beam is centered on the NuMI target. As noted in Fig. 2, it also monitors the integrity of the segmented NuMI target, since just 1 fin breaking off the target results in a more peaked proton spot. During initial beam commissioning, the hadron monitor helps align the primary proton beam to within 3 cm/725 m = 42 μRad angular precision. This compares well to the 14 μRad provided by the pre-target SEMs, which only periodically are inserted in the beam.

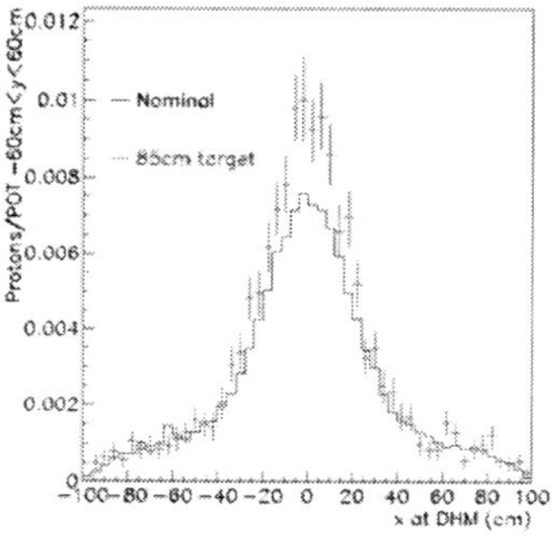

FIGURE 2. Horizontal profile of protons arriving at the downstream hadron monitor, over the band in the vertical direction of ±60 cm about beam center. The profile is shown for two cases: normal beam operations with a 1-m target (solid histogram) and the case in which the first target fin has broken off (points with errors). The error bars are due to Monte Carlo statistics.

The 3 muon monitors are located downstream of the beam absorber. Because of the steel and rock upstream of each alcove, they detect muons of increasingly higher momentum, hence detect only the decay products of higher energy parent pions. In the lowest energy ν beam, the parent pions have energies of 4-5 GeV, so many of the tertiary muons do not penetrate into Alcove 0. In the higher energy beams, the horn-focused pions have energies 12-30 GeV, so the muons penetrate into Alcoves 0, 1 and 2. Multiple scattering through the rock broadens the beam as it traverses further downstream into Alcoves 1 and 2.

Table 1 summarizes some parameters of the muon alcoves. Listed are the distance from the target, particle flux, and radiation levels. Furthermore, Table 1 indicates the minimum pion energy whose daughter muon can arrive at an alcove; for the low-energy (high-energy) ν beam, the desired pion beam energy is 4 (40) GeV. As a demonstration of the fact that the muon beam becomes more sharply peaked at high ν energy, Table 1 lists the flux on beam axis ($r = 0$) divided by the muon flux at the edge of the monitors ($r = 1$ m). Lastly, the table indicates the amount by which the muon beam centroid will shift in the scenario where horn 1 is misaligned by 3 mm. Figure 3 shows the muon beam profiles for a normal beam as well as a misaligned horn beam.

Because of the broad muon profile in the low energy beam, the muon monitors can monitor only relative neutrino event rate spill-to-spill when running in that configuration. The beam centroid shifts expected are barely measurable. Periodic short runs are therefore envisioned using the higher energy neutrino beams, which can be accomplished by moving the NuMI target remotely. Such a re-configuration thus allows a measure of the alignment of the horn optics on a periodic basis.

TABLE 1. Properties of the four ion chamber stations.

Element	Muon Monitors (Low Energy Beam)			Muon Monitors (High Energy Beam)			
	Alcove 0	Alcove 1	Alcove 2	Alcove 0	Alcove 1	Alcove 2	Hadron Monitor
Distance from Target (m)	739	751	769	739	751	769	725
Charged Particle Flux (10^7/cm^2/spill)	1.1	0.23	0.09	3.0	1.7	0.22	2500
Radiation Level (10^6 Rad/yr.)	6.6	0.7	0.15	18	10	0.7	~2000
Parent Particle Threshold (GeV)	4	10	18	4	10	18	<1
μ Beam Flux(r=0)/ Flux(r=1m)	1.3	1.1	1.1	2.1	1.5	1.5	--
μ Beam Shift for Horn 1 Offset (cm)	2.5 ±1.5	0.6 ±1.5	1.3 ±1.5	23.0 ±1.5	12.8 ±1.5	3.3 ±1.5	--

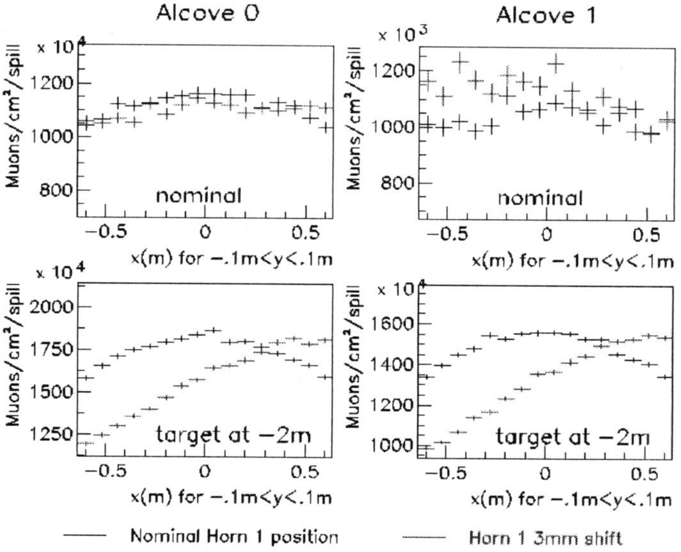

FIGURE 3. Monte Carlo calculations of muon beam profiles in Alcove 0 and Alcove 1. The upper (lower) plots are for low- (high-) energy beam. Shown are the profiles for normal beam (black points, centered distributions) and for the case in which the beam is missteered due to an offset of horn #1 (red points, offset distributions). The error bars are due to limited Monte Carlo statistics.

ION CHAMBER CONSTRUCTION

Because of the large anticipated particle fluences, signal strength will not be a significant issue. In fact, the greater consideration will be to reduce space-charge build-up in the gas volume: slow-moving positive ions will develop their own electric field which screens the applied electric field of the chamber. Such screening increases the drift time across the gas volume, increasing the probability for charge recombination in the gas. Such loss of charge manifests itself as a non-linearity in ion chamber response vs the incident particle flux. This consideration motivates the use of He gas, which produces the lowest ionization density of any of the inert gases [7,8], thereby reducing the space charge effect.

The ion chambers are parallel plates made from ceramic wafers with Ag-Pt electrodes [9]. The electrode separation is 3 mm (0.8 mm) for the muon (hadron) monitors. Because of the need to calibrate and track the relative responses of the 243 chambers in the muon arrays, each chamber is illuminated by a 1 μCi ^{241}Am α source ($E_\alpha = 5.5$ MeV), as shown in Fig. 4. The ionization current from such a source in pure He gas is typically 2-4 pA, which may be measured between spills to monitor gas quality, pressure, or temperature variations (see Fig. 4).

The hadron monitor is an array of 49 chambers mounted in a single aluminum vessel. Each chamber requires 2 ceramic feedthroughs [10], one for signal readout and one for a bias voltage (see Fig. 5). The vessel is sealed with an indium wire, whose melting point of 150°C is well above the 50-60°C temperature at which the monitor will operate. The design maximizes use of aluminum over stainless steel in order to reduce the presence of long-lived radionuclides in the detector [11]: with its current proportion of 54 lbs aluminum and 4 lbs stainless, residual activation will be 58 Rem/hr. The signal and HV lines are custom-made "cable" constructed of aluminum welding rod core, insulated by a ceramic tube, shielded by an aluminum sheath. After 1 m this cable transitions to a kapton-insulated cable [10].

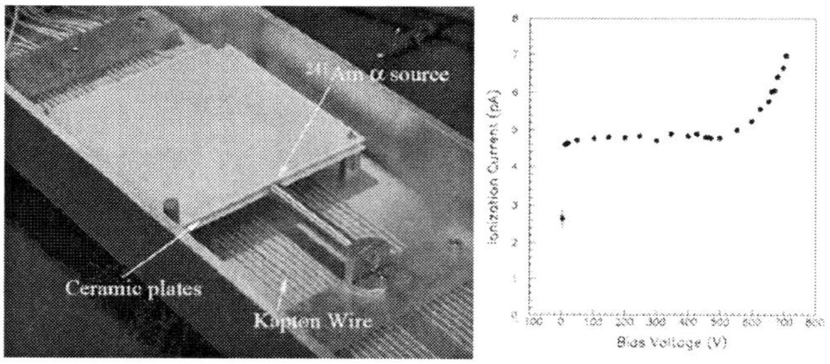

FIGURE 4. Ceramic parallel plate ion chambers. Left: Each chamber has an alpha source for calibrations. Kapton cables bring HV in and signals out. Right: Bias curve for a single chamber illuminated by the alpha source.

The muon monitor stations are each 81-chamber arrays. The layout of 9 × 9 ion chambers is achieved by mounting 9 "tubes" onto a support structure vertically. Each tube contains a tray of 9 ion chambers, as shown in Fig. 6. Because of the lower anticipated radiation levels in the muon alcoves, we routed the signals out to one end of the tube, where all feedthroughs come out one endplate of the tube. The tube is made from 6" × 2" rectangular extrusion, into which a "U"-channel "tray" is inserted. The endplates seal to the tubes using a soft aluminum wire gasket. The HV feedthroughs are custom-made using PEEK tubes and compression fittings. The signal feedthroughs are 9-pin ceramic feedthroughs [12]. The signal and HV routing within the tube is accomplished with shielded kapton-insulated cable [10]. Care was taken to minimize any exposed signal conductors to the gas volume, particularly at the feedthroughs, lest ionization in the surrounding gas collect on the conductors. Such "stray ionization" which occurs away from the ceramic chambers degrades the measurements of the muon beam's spatial profile.

Tests were performed of the radiation damage to detector components at the University of Texas 1-MW fission reactor. Samples of ceramic circuit boards with Ag-Pt electrodes, of kapton-insulated coaxial cable, of PEEK plastic and of aluminum compression fittings were exposed to ~1.2×10^{10} Rad. After exposure, the dielectrics could hold off >1000 V with <10-nA current draw. Interestingly, the kapton cable became less flexible, and the PEEK notably was of greater hardness after the test.

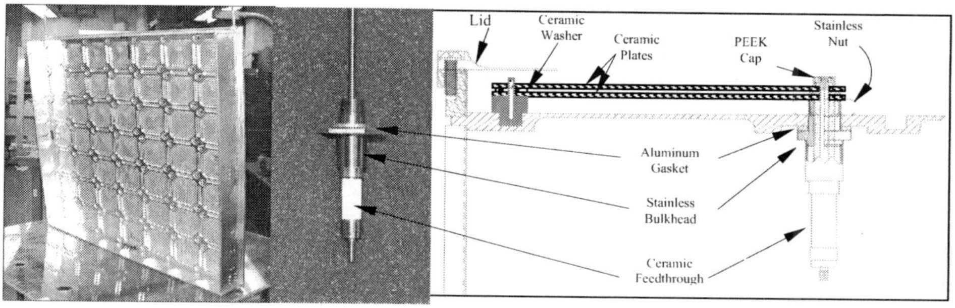

FIGURE 5. Construction details of the hadron monitor. Left: An Al vessel has penetrations through its rear plate for ceramic feedthroughs (middle). Right: Each ion chamber is supported on two of the feedthroughs. The front lid is a thin membrane welded to an Al flange, sealed to the vessel with an indium wire.

FIGURE 6. Construction details of the muon monitor "tubes." Each 6" x 2" extruded rectangular tube contains a tray to which 9 ion chambers are mounted. Each chamber has an alpha source for calibrations. Kapton cables bring HV in and signals out. All feedthroughs are at one end of the tube.

BEAM TESTS

Two tests of He-filled ion chambers have been performed, both of which are described elsewhere. The first occurred at 8-GeV FNAL Booster beam dump [13], with beam intensity varied from 10^9/spill to 4×10^{12}/spill. The second occurred at the Brookhaven ATF [14], where the beam was of 40-MeV electrons, intensities from 10^7/spill to 10^9/spill. The observed space charge effects occurred at intensities an order of magnitude larger than required for NuMI (see Table 1).

BACKGROUND RATES

The hadron monitor and the muon detectors in Alcove 0 are expected to see, by virtue of their close proximity to the beam absorber, relatively large rates of background particles, *viz.* gamma rays and low energy neutrons. Monte Carlo estimates of the neutron rates indicate as much as a factor of 10 more neutrons than charged particles reaching those detectors. Because these neutrons are produced by the proton beam hitting the absorber, they are largely uncorrelated spatially or in intensity with the neutrino beam. We attempted to measure using radioactive sources the ionization rate from neutron-induced recoils in our chambers [15]. The results indicate a background rate from neutrons of $<\frac{1}{4}$ than that expected from charged particles.

MUON MONITOR CALIBRATION

The precise measurement of muon beam profiles requires a relative chamber-to-chamber calibration within the muon monitor arrays. In fact, the uncertainties in muon beam centroid position listed in Table 1 are dominated by the chamber-to-chamber calibration uncertainty, not by the inter-chamber spacing or the number of chambers in the array. We therefore require a 1% chamber-to-chamber calibration.

The calibration procedure was to irradiate each of the detectors within the muon tubes with a source (1 Ci ^{241}Am, 30- to 60-keV γ rays). The relative ionization currents induced by this source equal the relative calibration factor to be applied to each of the 243 chambers within the muon system. All 9 chambers within a tube were read out simultaneously during one tube's calibration, so that the relative sizes of the signals from the 1-μCi α sources could be studied as well. We monitored temperature and absolute gas pressure within the tube with dedicated sensors to correct for any variations over the course of testing all the tubes. As a check of these sensors, a "reference chamber" was constructed and placed in series with the tube's gas flow. This reference chamber had a 40-μCi α source mounted inside it to act as a standard signal which should change only if gas pressure, temperature, or purity level changes.

The calibration test stand is shown in Fig. 7, and some of the results in Fig. 8. Bias voltage curves were recorded for each chamber irradiated by the 1-Ci source, and plateau ionization currents noted. As a check of the calibration procedure's precision and repeatability, as well of that of the pressure and temperature monitoring, we measured one particular chamber approximately 100 times, as shown in Fig. 8. The repeatability is 0.2 pA/123.0 pA $\approx$ 0.2%, adequate for our 1% calibration requirement. A histogram of the plateau ionization currents of all the calibrated chambers is shown in the right plot of Fig. 8. The 20-pA variation in chamber plateau currents reflects differences in the assembly of each chamber, such as the electrode spacing.

FIGURE 7. Muon monitor calibration test stand. The muon monitor tube is mounted on a fixed structure, in front of which is scanned a 1-Ci ^{241}Am γ-source.

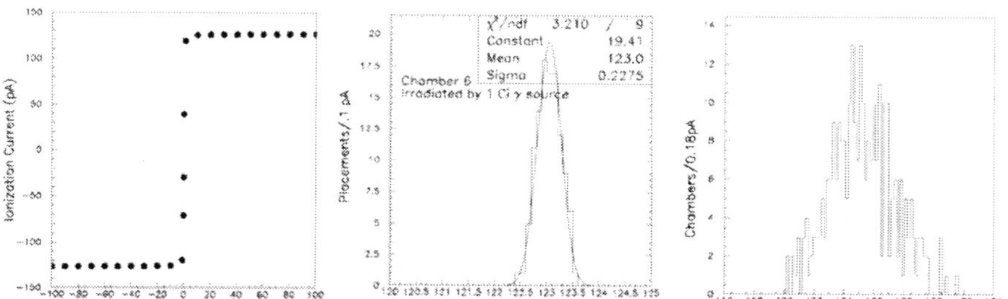

FIGURE 8. Left: Bias voltage curve of one chamber illuminated by the 1-Ci calibration source. Middle: Study of repeatability of 100 calibrations of the same chamber over the course of 8 hours. Right: Ionization currents of all 288 chambers calibrated on the test stand.

ACKNOWLEDGMENTS

It is a pleasure to thank S. O'Kelley of the University of Texas Nuclear Engineering Teaching Center, the staff of the University of Texas Physics Department Mechanical Shops, K. Kriesel of the University of Wisconsin Physical Sciences Laboratory, and

B. Baller, R. Ducar, D. Pushka, G. Tassotto, and K. Vaziri of Fermilab for valuable collaboration on this project.

REFERENCES

1. J. Hylen et al., "Conceptual Design for the Technical Components of the Neutrino Beam for the Main Injector (NuMI)," Fermilab-TM-2018, Sept. 1997.
2. S. Kopp, "The NuMI Neutrino Beam and Potential for an Off-Axis Experiment," *J. Phys.* **G29**, 1757 (2003).
3. S. van der Meer, CERN Yellow Report CERN-61-007 (1961).
4. M. Kostin, S. Kopp et al., Fermilab note NuMI-B-783 (2002).
5. D. Indurthy et al., "Profile Monitor SEMs for the NuMI Beam at FNAL," these proceedings.
6. K. Budal, *IEEE Trans. Nucl. Sci.* **14**, 1132 (1967).
7. I. Christophorou, *Atomic and Molecular Radiation Physics,* John Wiley, New York, 1971.
8. J. Dutton, "A Survey of Electron Swarm Data," *J. Chem. Ref. Data* **4**, 577-856 (1975).
9. Ceramic circuit boards fabricated by Cer-Tek, El Paso, TX.
10. Insulator Seal, Inc., Hayward, CA.
11. J. Cossair et al., "Fermilab Radiological Control Manual," http://www-esh.fnal.gov/FRCM/
12. Ceramaseal, Inc., Laurens, South Carolina.
13. R. Zwaska et al., *IEEE Trans. Nucl. Sci.* **50**, 1129 (2003).
14. J. McDonald et al., *Nucl. Instr. Meth.* **A496**, 293 (2003).
15. D. Indurthy et al., "Study of Neutron-Induced Ionization in Helium and Argon Chamber Gases." accepted for publication in *Nucl. Instr. Meth.* **A**, arXiv physics/0405016.

Profile Monitor SEMs for the NuMI Beam at Fermilab

Dharmaraj Indurthy, Sacha Kopp, Marek Proga, and Zarko Pavlovich

Department of Physics, University of Texas at Austin,
Austin, TX 78712, USA

Abstract. The Neutrinos at the Main Injector (NuMI) project will extract 120-GeV protons from the FNAL Main Injector in 8.56-μsec spills of 4×10^{13} protons every 1.9 sec. We have designed secondary emission monitor (SEM) detectors to measure beam profile and halo along the proton beam transport line. The SEMs are Ti foils 5 μm in thickness segmented in either 1-mm or 0.5-mm pitch strips, resulting in beam loss $\sim 5 \times 10^{-6}$. We discuss aspects of the mechanical design, calculations of expected beam heating, and results of a beam test at the 8-GeV transport line to MiniBoone at FNAL.

INTRODUCTION

The NuMI beamline [1,2] at Fermilab will deliver an intense muon neutrino beam to the MINOS detectors at FNAL and at the Soudan Laboratory in Minnesota. The primary beam is fast-extracted onto the NuMI pion production target in 8.56-μsec spills from the 120-GeV FNAL Main Injector. The beam line is designed to accept 4×10^{13} protons/spill. Along most of the transport line, the spot size is 1-2 mm. Because of the large 400-kW average beam power, beam loss must be kept to $<10^{-5}$ in order to keep residual activation of beam components to a manageable level and in order to prevent activation of water from aquifers near the underground NuMI tunnel.

Instrumentation along the 700-m primary beam line is to include beam profile monitors, capacitative beam position monitors (BPMs), a resistive wall monitor, and two beam current toroids (BCTs). Additionally, the NuMI target is electrically isolated, so may be read out as a Budal-type monitor [3], and an ion chamber array near the beam dump measures the remnant primary beam after a long 725-m free drift, so acts as an accurate measure of the final proton beam angle [4]. The intercepting profile monitor SEMs serve as the initial calibration of the BPMs and ion chamber.

Some of the requirements of the profile monitor SEMs are as follows: (1) the device must survive $\sim 10^{20}$ particles/cm^2 per year; (2) groundwater activation in the carrier tunnel region requires beam loss $\sim 10^{-6}$; (3) the beam centroid position accuracy must be ~100 μm along the transport line and ~ 50 μm for the pre-target SEMs; (4) the SEM chamber vacuum must be 10^{-8} Torr; (5) the measurement aperture must be ~2" and clear aperture 4"; and (6) the device must be removable from the beam without turning off the beam, and replacing the device in the beam must be achieved to 50-μm positional accuracy.

CP732, *Beam Instrumentation Workshop 2004: Eleventh Workshop*,
edited by T. Shea and R. C. Sibley III

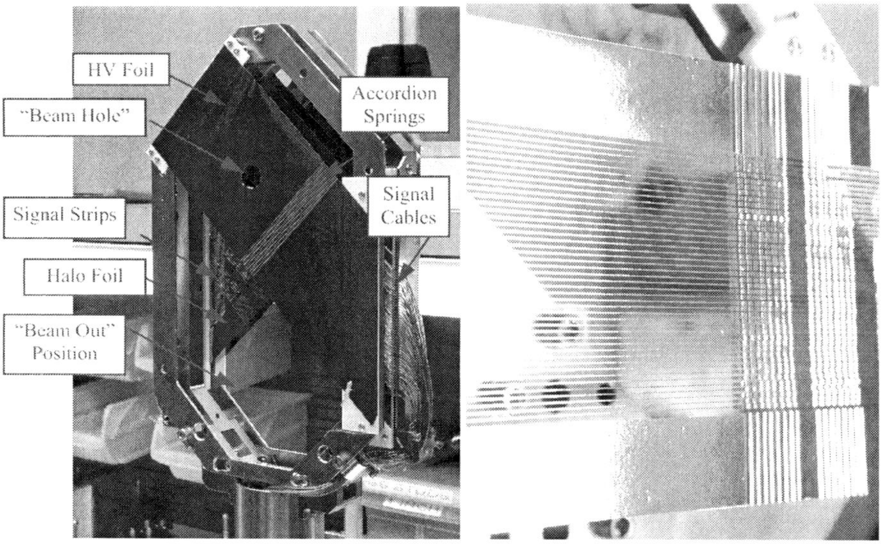

FIGURE 1. Left: Photograph of the foil paddle. Five foils are mounted, 3 bias foils and 2 segmented foils for *X* and *Y* profiles of the beam. The paddle surrounds the beam at all times, so the foils may be inserted or retracted from the beam without the paddle passing through the beam. Right: Close-up view of one segmented Ti foil. The segmented strips are at 1-mm pitch and measure out to 2.2-cm radius. A wide strip measures beam halo out to 3.8-cm radius. Accordion springs pressed into the signal and HV foils maintain tension against beam heating. The strips are 0.75 mm wide at the ends, where accordion springs are pressed in to maintain strip tension. In the area traversed by the beam, the strips are 0.15 mm wide.

PROFILE MONITOR SEM DESIGN

We have borrowed from a foil SEM at CERN [5] and designed a large-aperture SEM constructed from 5-μm Ti foils. Titanium is selected because it appears to maintain its secondary electron yield even after large particle fluences [6]. This is in contrast to the standard "multiwire" SEMs in use at FNAL which utilize Au-plated W-Rh wires, which were observed to degrade in signal yield by more than 20% over the course of fixed target operations to KTeV [7].

A view of the foil SEM paddle is shown in Fig. 1. Three planes of solid Ti foil, 2.5 μm in thickness, are interleaved with segmented foils intended for *X* and *Y* profiles. The signal foils are 5 μm thick. In this figure, the strip pitch is 1 mm. Over the 8-cm diameter area of the SEM traversed by the beam, the signal foil strips are 0.15 mm in width. A total of 44 strips are etched into the foil, along with 1.5-cm wide strips to measure beam halo out to 3.8-cm radius. The paddle completely surrounds the beam, so the foils may be inserted into the beam or retracted without the paddle frame traversing the beam. The foils are mounted on precise ceramic combs, which define the strip pitch and the foil location on the paddle [8]. A "beam hole" of 12-mm diameter in the bias foil permits most of the 1-mm beam spot to pass through without beam loss, while still permitting adequate voltage bias to maintain signal yield.

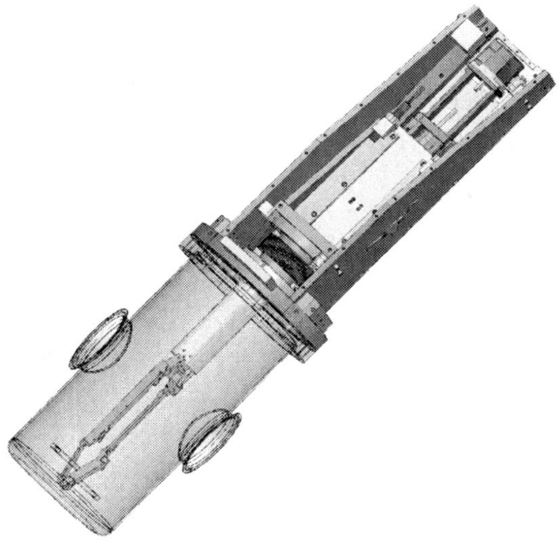

FIGURE 2. Schematic drawing of the foil SEM vacuum can and actuator assembly. The beam ports are 10-cm diameter quick-disconnect flanges. The actuator and foil paddle are mounted on a 25-cm diameter flange. The actuator consists of a stepper motor and linear stage driving a vacuum bellows to which is rigidly connected a 5-cm hollow shaft. The foil paddle is cantilevered off the shaft.

Each signal strip has 32 accordion springs pressed into its ends. The springs are elongated by approximately 4 mm prior to installation on the paddle. At this extension, our measurements indicate that a tension of ~1 g is attained on each strip. The mechanical sag δy of a strip of cross-sectional area A, length L, density ρ, may be approximated as $\delta y = g\rho AL^2/T$. This expression gives, for L = 12 cm, a tension T = 1.0 g, the density of titanium ρ = 4.5 g/cm^3, and a 5-μm × 0.15-mm strip, a sag δy = 40 μm. Digital photographs of the mounted signal foils confirm a pitch of 1.0 mm, with strip-to-strip variation <15 μm (the resolution of the photographs is 15 μm).

The foil actuating mechanism and vacuum can are shown in Fig. 2. The foil paddle is cantilevered on a 5-cm diameter hollow shaft, the other end of which is welded to a 12-cm "conflat" flange. A linear motion stage and stepper motor actuate the assembly into or out of the beam. A 9-cm outer diameter, 6.4-cm inner diameter bellows forms the vacuum seal for the actuator. Ceramic-insulated limit switches [9] halt the stepper travel at either end, while a 1.25-mm range linear variable differential transformer (LVDT) confirms the final beam "in" position of the foils with 1-μm accuracy [10]. Kapton-insulated signal cables are routed through the hollow shaft to feedthroughs at its end. The cables are bonded to the foil strips using a conducting epoxy appropriate for ultra-high vacuum [11], which is also how the strips are bonded to the ceramics. Brackets mount the linear stage to the large 25-cm diameter conflate flange on the end of the SEM's vacuum chamber. Precise dowel holes in the 25-cm flange and in the moving flange at the end of the bellows allow *in situ* optical survey of the foil position when installed in the beam line. The 16-liter vacuum chamber is a 20-cm diameter cylinder with 10-cm diameter quick-disconnect flanges at the beam ports.

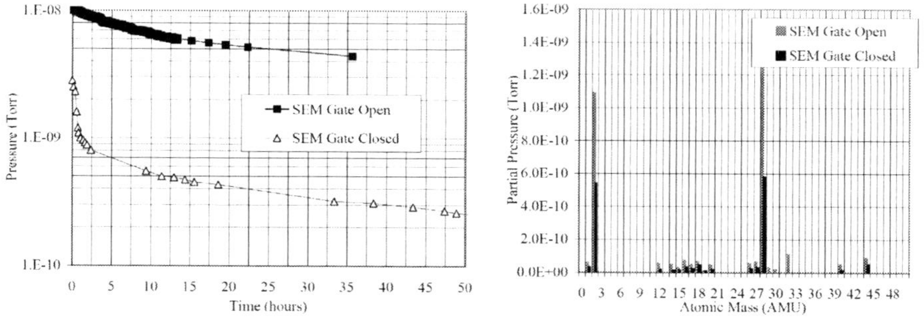

FIGURE 3. Vacuum tests of a completed SEM. Both measurements were performed on a vacuum system pumped on by a 30-liter/sec ion pump, with a gate valve which could isolate or include the SEM chamber. Left: Pump-down curves. Right: Residual gas analysis.

Tests have been performed of the linear motion actuator's repeatability. These tests include tests under vacuum load and in air. We find the repeatability is better than 15 μm over 24- to 48-hour periods, consistent with temperature variations in the laboratory and the differential thermal expansion of the materials in the actuator.

Vacuum tests were performed on a system which could isolate or include the SEM chamber via a UHV gate valve between the SEM and the rest of the system. When the SEM was isolated, a rate-of-rise measurement indicates an outgas rate of 1.2×10^{-7} Torr-liter/sec. This outgas rate, furthermore, is consistent with the ultimate pressure achieved with the SEM included in the vacuum system (see Fig. 3).

THERMAL SIMULATIONS

We have performed detailed finite element calculations of the temperature induced in the foil SEMs due to heating by the NuMI beam [12]. The heat input to the SEM comes from the energy lost by the beam, and the power dissipated by the SEM comes from blackbody radiation and from thermal conduction through the SEM material. We performed the calculations for several materials using thermal conductivity, heat capacity, and density values from Ref. [13]. We also compared the heating of wires and foils; wire SEMs cool less efficiently because blackbody radiation is proportional to the surface area of the emitter.

The temperature rise in the SEM material results from ionization energy loss by the 120-GeV protons. We use the "restricted energy loss", which accounts for ionization leading to electrons less than a cutoff energy T_{cut}: $dE/dx(T_e < T_{cut})$ [14]. For the cutoff energy T_{cut}, we use the energy at which the range of an electron is ½ the thickness of the SEM material. The range values are taken from Ref. [15]. This restricted dE/dx accounts properly for the fact that some δ-rays escape out the back of the SEM, so do not deposit their energy in the SEM. The restricted dE/dx values are given in Table 1. The effect of restricted energy loss is greater for thin foils than for wire SEMs and tends to lower the predicted energy deposited in the SEM, so this effect tends to again yield larger temperature predictions for a wire SEM as compared to a foil SEM.

TABLE 1. Results of thermal simulations for foil and wire SEMs.

Element	5-μm thick foil				50-μm diameter wire				Yield Stress (MPa)
	dE/dx **(MeV cm²/g)**	**Δ*T* (°C)**	**Δ*L* (μm)**	***S* (MPa)**	***dE/dx*** **(MeV cm²/g)**	**Δ*T* (°C)**	**Δ*L* (μm)**	***S* (MPa)**	
Be	1.14	64	11	188	1.23	69	20	202	240
C	1.26	186	1.3	19	1.37	202	3	20	469
Al	1.22	138	35	238	1.32	150	66	258	10-35
Ti	1.14	222	12	208	1.23	240	41	224	220
Ni	1.15	264	54	736	1.25	287	118	800	1580
Ag	1.05	449	66	709	1.13	484	106	763	
W	0.94	715	30	1305	1.02	775	64	1416	550
Au	0.93	729	84	807	1.01	791	154	877	205

Figure 4 shows the results of the thermal model for a Ti foil 5 μm in thickness. The left plot of Fig. 4 shows the temperature profile along the center-most strip at several times during the 1.9-sec beam cycle, after many transpired beam cycles. As seen in the graph, the beam causes a sharp rise in temperature at t = 0 sec. The cooling between spills is predominantly due to blackbody radiation. The (small) effect of thermal conduction is evident by the broadening temperature profile over the course of the cooling cycle. The right graph of Fig. 4 shows the linear expansion of the foil strip, and compares to the elongation expected for a Ti wire 50 μm in diameter.

Table 1 provides the temperature rise ΔT, maximum linear elongation ΔL, as well as dynamic stress $S = \alpha E \Delta T$, where α is the coefficient of linear expansion and E the Young's Modulus. The dynamic stress may be compared to the "yield stress" for the material, the point at which the material may deform plastically. Beryllium, carbon fiber and titanium are preferable from the point of view of long-term material damage; i.e., they have dynamic stress values below the yield stress.

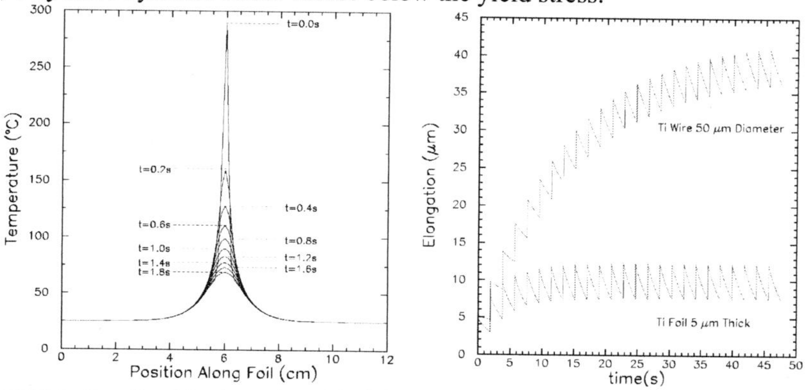

FIGURE 4. Results of the thermal model of 5-μm Ti foil in the NuMI beam. Left: Temperature along the centermost strip at several time increments through one beam cycle: at t = 0 sec the beam passes through the foil, and at t = 1.8 sec the foil has cooled to its minimum temperature, just prior to the next spill. Right: Net elongation of a 12-cm long, 5-μm thick, Ti foil strip as a function of time, showing the repeated heating and cooling of several beam cycles. The Ti foil's elongation is compared to that for a 50-μm diameter Ti wire.

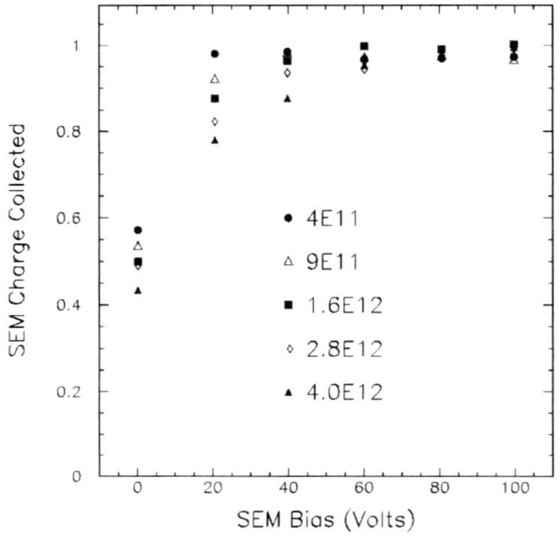

FIGURE 5. Fraction of SEM charge collected *vs* the bias voltage. The vertical scale is normalized to 1.0 for the data point at 100 V and 4×10^{11}/spill. The study is repeated for 5 beam intensities.

The Ti foil SEM experiences 12-μm elongation. Because of the 4-mm extension in the accordion springs, we note that this elongation results in <1% tension loss. Of note as well is the fact that a W-wire SEM, when strung on a frame at the maximum stress (the yield stress), stretches the wires by 160 μm; thus, beam heating will drop the tension of such a wire SEM by nearly ⅓. Such loss motivates the accordion springs for the foil SEM or individual springs on wire SEMs as has been used by previous workers [16].

BEAM TEST RESULTS

A prototype foil SEM detector was tested in the 8-GeV transfer line from the FNAL Booster to the MiniBoone experiment. The 8-GeV beam has $\sigma_{beam} \approx 3$ mm at the location of the SEM, and the spill is 1.56 μsec in duration. The beam intensity was varied from 4×10^{11}/spill to 4×10^{12}/spill. A BPM was available within 3 m of the SEM and the beam intensity was measured at Booster extraction with a BCT.

The first test was to measure the signal yield off the SEM as a function of voltage on the bias foils. The results are shown in Fig. 5 for several spill intensities. The vertical axis in Fig. 5 is the fractional charge collected, normalized to be 1.0 for the data point at 100 Volts and 4×10^{11}/spill intensity. The data suggest a need for larger applied voltage as the beam intensity increases.

The second purpose of this test was to study the SEM resolution for determining the beam centroid. This was accomplished by combining two studies, one in which beam was at a stable position, and one in which it wandered over the SEM.

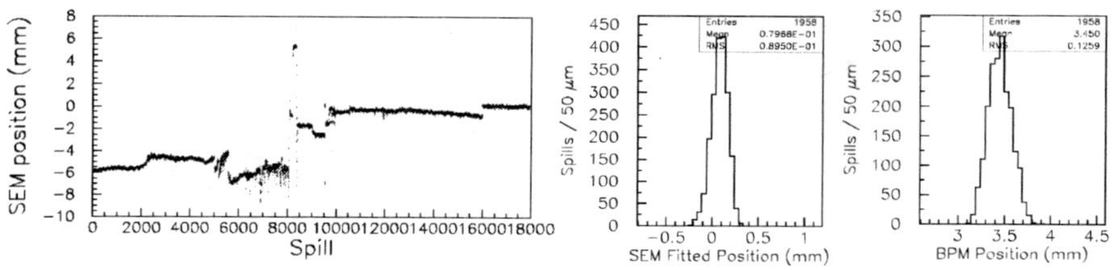

FIGURE 6. Left: Beam centroid position as measured by the SEM *vs* spill number for a several-hour period. Middle: Projection of the left plot over the spill range 16000-18000. Right: Histogram of the BPM centroid measurement for the same range of spills as the middle plot for the SEM.

The results of the first resolution study are shown in Fig. 6. The beam appears to be stable for spills 16000-18000. A histogram of the SEM and BPM centroids over this spill range are shown in the middle and right graphs of Fig. 6, showing RMS values of 90 μm and 125 μm, respectively. These magnitudes are a combination of the detector resolutions as well as the extent to which the beam wanders. We have:

$$\sigma_{SEM}^2 + \sigma_{wander}^2 = (90\mu m)^2 \quad \text{and} \quad \left(\sigma_{BPM}^2 + \sigma_{wander}^2\right)\alpha^2 = (125\mu m)^2 \ . \tag{1}$$

The factor $\alpha = 1.13$ is a scale factor between the SEM and BPM and reflects that the calibration of the two detectors may be imperfect; i.e., one millimeter in the SEM may be more than a millimeter as read out by the BPM. The value of α is obtained below. To factor out the effect of beam wandering, we take the difference of these results:

$$\sigma_{BPM}^2 - \sigma_{SEM}^2 = (125\mu m/\alpha)^2 - (90\mu m)^2 \ . \tag{2}$$

In the second resolution study, we correlated the fitted SEM centroid with the BPM. The upper two plots of Fig. 7 show the BPM and SEM position for 6499 spills histogramed over a several-hour period. During this period, the beam wandered by ~2 mm, as is seen in both detectors. These plots likewise suggest $\sigma_{SEM} < \sigma_{BPM}$.

The lower left plot of Fig. 7 shows the correlation of the BPM and SEM positions and a line fit to the data. The lower right plot of Fig. 7 shows the residuals of these data with respect to the fitted line. The RMS of the residuals, which has contributions from both the BPM and SEM resolutions, is:

$$\alpha^2\sigma_{BPM}^2 + \sigma_{SEM}^2 = (120\mu m)^2 \ . \tag{3}$$

The scale factor α here may be obtained from the slope of the scatter plot of SEM vs BPM centroid in Fig. 7; we find $\alpha = 1.13$. Combining Eqs. (2) and (3),

$$\sigma_{SEM} = 62\mu m \quad \text{and} \quad \sigma_{BPM} = 92\mu m \ . \tag{4}$$

These results, achieved with a 1-mm pitch SEM in a 3-mm beam, suggest that the SEM will be adequate to achieve 100-μm resolution in the 1- to 2-mm NuMI beam.

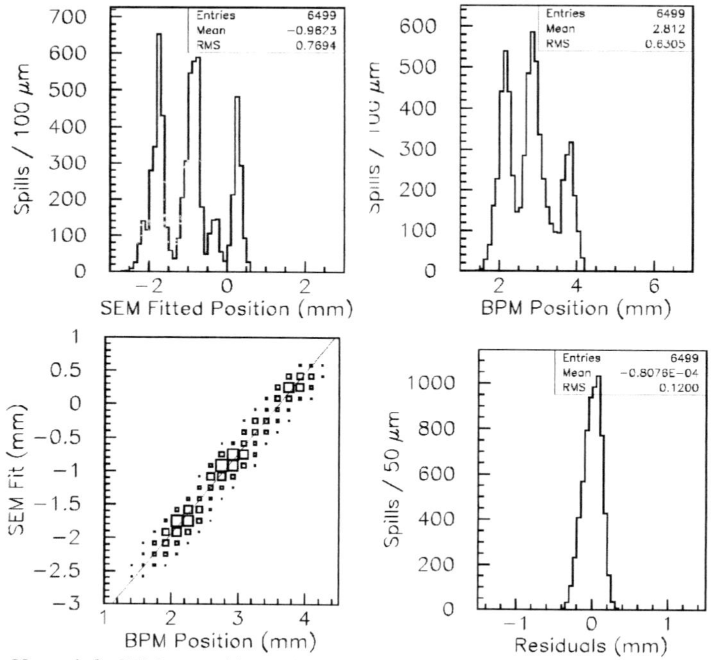

FIGURE 7. Upper left: SEM centroid position for a few hour time period in which beam center moves by nearly 3 mm. Upper right: Same info as reported by the nearby BPM. Lower left: Scatter plot of SEM and BPM centroid positions, along with a best fit line. Lower right: Residuals of the plot at left.

ACKNOWLEDGMENTS

It is a pleasure to thank M. Raizen and K. Shih, of the University of Texas; B. Baller, S. Childress, D. Harris, and C. Kendziora of FNAL; and especially G. Ferioli and J. Camas of CERN for helpful conversations. We thank R. Zwaska, T. Kobilarcik, C. Moore, and G. Tassotto for assistance with the beam test. This work was supported by the U.S. Department of Energy and the Fondren Family Foundation.

REFERENCES

1. J. Hylen et al., "Conceptual Design for the Technical Components of the Neutrino Beam for the Main Injector (NuMI)," Fermilab-TM-2018, Sept. 1997.
2. S. Kopp, "The NuMI ν Beam and Potential for an Off-Axis Experiment," *J. Phys.* **G29**, 1757 (2003).
3. K. Budal, *IEEE Trans. Nucl. Sci.* **14**, 1132 (1967).
4. D. Indurthy et al., "Ion Chambers for Monitoring the NuMI Neutrino Beam," these proceedings.
5. G. Ferioli, private communication.
6. G. Ferioli and R. Jung, CERN-SL-97-71(BI), Proceedings of DIPAC, Frascati (Rome), Italy, 1997.
7. G. Tassotto, private communication, and R. Drucker, R. Ford, and G. Tassotto, FERMILAB-Conf-98/062, Proceedings of DIPAC, Frascati (Rome), Italy, 1997.

8. Ceramics fabricated by Kadco, Inc., Easton, MA, U.S.A.
9. Honeywell ceramic-insulted switch, model 2HT1.
10. Lucas-Schaevits Sensors, HR050 mildly radiation-resistant 1.25-mm stroke LVDT.
11. Epo-Tek H27D conducting epoxy.
12. S. Kopp, T. Osiecki, and Z. Pavlovich, "Studies of Beam Heating of NuMI Proton Beam Profile Monitors," Fermilab note NuMI-B-926 (2004).
13. *Handbook of Physical Quantities*, edited by I. S. Grigoriev and E. Z. Meilikhov, CRC Press, 1997.
14. B. Rossi, *High Energy Particles*, Prentice Hall, Inc., Englewood Cliffs, 1952.
15. National Institutes of Standards and Technologies, "Stopping Powers of Electrons, Protons, and Alpha Particles," http://physics.nist.gov/PhysRefData.
16. J. Krider and C. Hojvat, *Nucl. Instr. Meth.* **A247**, 304 (1986).

Synchrotron Radiation Profile Monitor for the HERA Positron Beam

G. Kube, R. Fischer, and K. Wittenburg

Deutsches Elektronen–Synchrotron DESY, Notkestrasse 85, D-22603 Hamburg, Germany

Abstract. For the measurement of the positron beam emittance at the e–p storage ring HERA at DESY a monitor is used which is based on the direct imaging of visible synchrotron radiation from a bending magnet. While the resolution of profile measurements by synchrotron radiation is already strictly limited by inherent effects, the observation in off-axis geometry modifies the measured vertical angular intensity distribution additionally. As a result the beam image is broadened due to the increased contribution of the diffraction limited resolution. In the present article detailed calculations concerning the image resolution especially in view of off-axis observation are presented. Furthermore the calculations indicate that the influence of resolution broadening effects may be overestimated by the standard formulas used as rule of thumb.

INTRODUCTION

The precise determination of the beam emittance is essential for the understanding of luminosity in colliding beam experiments as the ones at the proton–positron storage ring HERA. While the emittance itself is no direct accessible quantity, the beam width is usually measured so that the emittance can be calculated based on the knowledge of beam opticle parameters. For the measurement of the e^+ beam size at HERA a profile monitor is used which utilizes the visible part of synchrotron radiation (SR) from a bending magnet to form an image of the beam. The image resolution of this kind of monitor is affected by inherent effects like diffraction and depth of field. However, by careful calculations of these resolution broadening contributions, the real beam size can still be derived.

The development of a model to describe these influences has turned out to be nontrivial. Attempts range from rules of thumb to quite elaborate models, see Refs. [1, 2] and the references therein. The model presented in this article follows the outlines described in the references quoted above. Special emphasis is put on the discussion of resolution broadening influences in vertical direction due to off–axis observation, i.e. if a part of the radiation used for image formation is shielded. Examples are the usage of a slotted mirror as described in Ref. [3], or moving the mirror away from the beam axis in vertical direction as it is the case for the HERA monitor. The reason for is to reduce thermal heating of the light extracting mirror due to absorption of hard X–ray radiation which would result in an image deformation.

CP732, *Beam Instrumentation Workshop 2004: Eleventh Workshop*,
edited by T. Shea and R. C. Sibley III

PROFILE MONITOR SETUP

In luminosity operation the HERA positron beam energy amounts to E = 27.6 GeV. According to the beam optical design parameters, in this mode the e^+ beam should have a horizontal (1σ) size of σ_x = 1143 μm and a vertical one of σ_y = 337 μm. Task of the optical system is to provide an image of the beam onto the chip of the CCD camera.

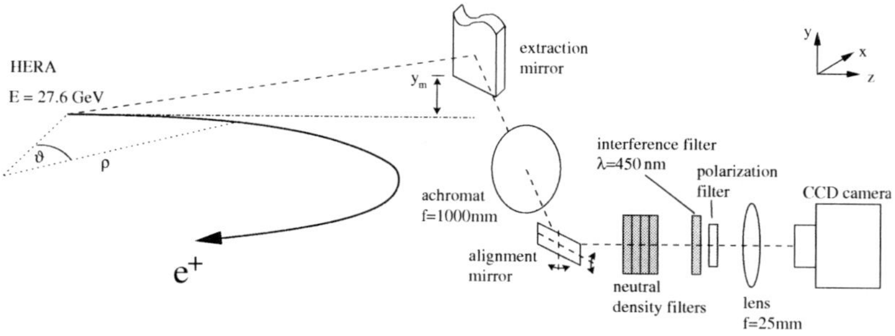

FIGURE 1. Schematic picture of the profile monitor setup.

In Fig. 1 a sketch of the monitor setup is shown. The emitted SR radiation is extracted out of the vacuum system by a mirror (thickness 2 mm, width 15 mm) which is located $\hat{p}$ = 6216 mm away from the central part of the bending magnet. The mirror surface has an inclination angle of 45° with respect to the xy plane, hence the light is reflected out perpendicular to the beam axis. An achromatic lens with focal length f = 1000 mm, located p = 6485.5 mm away from the source point, forms an image in a distance p' = 1182.3 mm. The resulting real intermediate image is then magnified by a second lens (f = 25 mm) onto the chip of the CCD camera (*JAI CV–M300E* with 768 × 494 pixels of size 11.6 × 13.5 μm^2). The total magnification factor of the optical system amounts to V = 0.55. The video output of the camera is fed to a commercially available 8 bit PCI framegrabber board (*Data Translation DT3155*) for digitalization and finally analyzed by a standard personal computer.

In order to minimize chromatic errors and to improve the diffraction limited resolution an interference filter with central wavelength λ = 450 nm is used. With the polarization filter it is possible to select the horizontally (σ) or vertically (π) polarized light component in order to investigate the influence on the diffraction limited resolution. The neutral density filters serve to adjust the incoming light intensity to avoid saturation of the camera chip. The alignment mirror which is rotateable about the x and y axis allows for adjustment of the image in the center of the CCD.

The light extracting mirror is made of beryllium which has a high thermal conductivity together with a low absorption coefficient for hard X–rays. Thereby, a large part of the emitted SR spectrum with critical energy $\hbar\omega_c$ = 76 keV will cross the mirror without absorption and reduce thermal heating. Additionally, the mirror is cooled by water. However, these precautions are not sufficient to reduce thermal heating down to a level that no mirror surface distortion occurs. The absorption of the low energetic part of the SR spectrum at the surface leads to a deformation of the beam image and even the destruction of the mirror. Therefore, in the normal operation mode of the monitor the extracting mirror bottom edge is placed y_m = 2...4 mm above the beam axis. In this geometry the X–ray part of SR which is emitted close to the beam axis will not affect the mirror, while the optical SR components which are emitted under larger angles are reflected out of the vacuum system.

DIFFRACTION LIMITED RESOLUTION

The resolution broadening calculations are based on the approach of Hofmann and Méot [1] which starts from the assumption of spherical wave propagation through obstacles under the condition of scalar Fraunhofer diffraction. Consider a point source located in the origin of the source plane (x_0, y_0) which is imaged onto the origin of the image plane (x', y'), cf. Fig. 2.

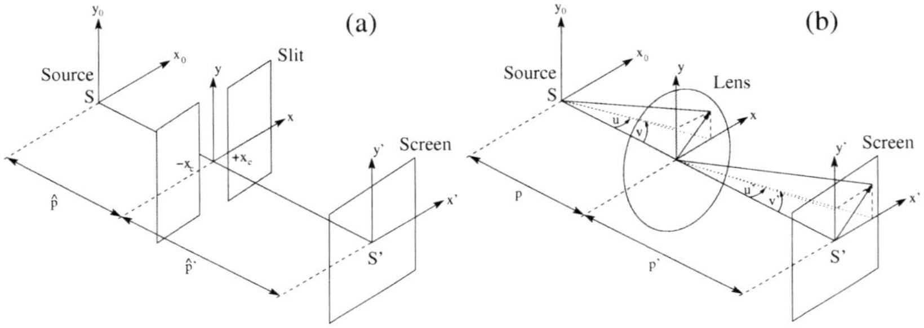

FIGURE 2. Geometry for the calculation of horizontal (a) and vertical (b) diffraction limited resolution.

The field distribution of the Fraunhofer diffraction pattern is determined by the Fourier transform of the source field distribution $f(x,y)$ over the aperture

$$F(x',y') = \iint_{aperture} f(x,y)\, e^{-ik(xx'+yy')/p'} \mathrm{d}x\,\mathrm{d}y = F_x(x') \times F_y(y')\,, \tag{1}$$

supposed the integrations in x,y over the limiting aperture can be carried out independently. $u' = x'/p'$, $v' = y'/p'$ are the emission angles in the image plane. From the amplitude distribution in the image plane the intensity distribution can be calculated by

$$I_x(x') = |F_x(x')|^2, \qquad I_y(y') = |F_y(y')|^2\,. \tag{2}$$

Horizontal Distribution. In horizontal direction the aperture is defined by the horizontal dimensions of the light extracting mirror. As indicated in Fig. 2(a) the situation is similar to the diffraction at a slit aperture, while the slit width $w = 2\,|x_c|$ is defined by the projection of the mirror width onto the beam axis. According to Refs. [1, 2], the amplitude distribution f can be considered as uniform across the slit because of the circulating motion in the horizontal plane. The evaluation of Eq.(1) in x direction results in the diffraction pattern from a single slit

$$F_x(x') = 2x_c \,\mathrm{sinc}\left(\frac{2\pi\, u_{max}}{\lambda\, \hat{V}}\, x'\right) \tag{3}$$

with the magnification $\hat{V} = \hat{p}'/\hat{p}$ and $u_{max} = x_c/\hat{p}$ the maximum angle in the horizontal plane. From Eq.(3) it is obvious that the diffraction limited resolution is mainly determined from the slit resp. the mirror width.

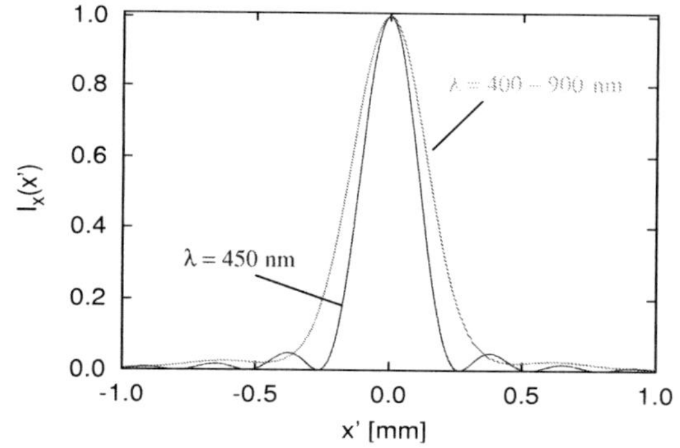

FIGURE 3. Diffraction limited normalized horizontal single electron intensity distribution $I_x(x') = |F_x(x')|^2$ of SR for the HERA profile monitor, assuming the field distribution of Eq.(3) with $\hat{V} = 1$.

Fig. 3 shows the calculated horizontal intensity distribution for the HERA profile monitor ($x_c = 5.3$ mm, $\hat{p} = 6216$ mm), assuming a magnification $\hat{V} = 1$ and a point source located in the origin of the source plane. The advantage to use monochromatic light is clear to see: the intensity distribution for the larger wavelength interval is smoothed out, leading to an increased contribution of the diffraction broadening.

In order to quantify the horizontal diffraction broadening it is possible to fit a normal distribution at the central part of the calculated intensity profiles. For $\lambda = 450$ nm this gives a value of $\sigma_{x,d} = 95.2\ \mu$m which is comparable to the result from the approximative formula given in Ref. [2] $\sigma_{x,d}^{app} = 1.39156\,\lambda\,(2\pi x_c/\hat{p}\sqrt{2\ln 2})^{-1}$, which leads to a value of 99.3 μm. At the other hand the comparison of these values with the "classical" expression of diffraction broadening from a slit aperture which is defined by the first minimum $x_{min} = \pi$ of the $\mathrm{sinc}^2 x$ distribution results in $\sigma_{x,d}^{\prime app} = \frac{1}{2}\lambda/(x_c/\hat{p}) = 249.8\ \mu$m, which is much larger than the values quoted above.

While in the fit procedure only the central part of the distribution is considered, i.e. the oscillating tails are not taken into account, the determination of the diffraction broadening based on the first order minimum is a rough estimation which overestimates the influence. Therefore in the next step the influence on the measured beam profile was investigated in order to specify a value for the resolution broadening. For this purpose the intensity distribution $I_x(x') = |F_x(x')|^2$ with $F_x(x')$ according to Eq.(3) was convoluted with the normal distributed horizontal beam profile, and the width of the resulting distribution was determined by fitting a normal distribution. From the derived width an "effective" value for the horizontal diffraction resolution can be deduced by quadratical subtraction of the beam size σ_x, leading to a value of $\sigma_{x,d} = 187.9\ \mu$m which is used to correct the contribution of the diffraction broadening.

Vertical Distribution. In vertical direction the inherent angular distribution of SR light is normally the limiting factor. The amplitude distribution of the electric field is given by

$$f_{y,\sigma}(\xi) = A_\sigma \frac{\hbar\omega}{2\hbar\omega_c}(1+\xi^2)\,K_{2/3}\Big[\frac{\hbar\omega}{2\hbar\omega_c}(1+\xi^2)^{3/2}\Big] \quad (4)$$

$$f_{y,\pi}(\xi) = A_\pi \frac{\hbar\omega}{2\hbar\omega_c}\,\xi\sqrt{1+\xi^2}\,K_{1/3}\Big[\frac{\hbar\omega}{2\hbar\omega_c}(1+\xi^2)^{3/2}\Big] \quad (5)$$

(see e.g. [4]) with $\xi = \gamma v = \gamma y/p$ the reduced vertical emission angle. $f_{y,\sigma}$ is the polarization component in the orbit plane, $f_{y,\pi}$ the perpendicular one. The amplitudes A_σ, A_π are independent on the emission angles u, v and can be disregarded for the integration in Eq.(1).

However, in the case of off–axis observation the lower integration limit in Eq.(1) is defined by the position of the light extracting mirror bottom edge y_m, i.e. $\xi_{min} = \gamma y_m/\hat{p}$, and the field distribution in the image plane is determined by

$$F_{y,\sigma/\pi}(y') = \frac{p}{\gamma}\int_{\xi_{min}}^{+\infty} d\xi\, f_{y,\sigma/\pi}(\xi)\,\exp(-i\frac{kp}{\gamma p'}\xi y')\,. \quad (6)$$

In Fig. 4 calculated vertical intensity distributions $I_y(y') = |F_y(y')|^2$ with F_y according to Eq.(6) are shown, assuming a magnification $V = p'/p = 1$. For mirror positions $y_m \leq$ -10 mm there is no more change in the shape of the distributions. This value corresponds formally to the case $\xi_{min} = -\infty$, i.e. the conventional on–axis observation which is investigated in Refs. [1, 2]: the distribution of the σ polarization component has a narrow shape which can be described by a normal distribution while the one of the π component has a minimum on–axis. If the mirror is moved upward, the width of the σ polarization component is continuously broadened. With increasing y_m the on–axis minimum of the

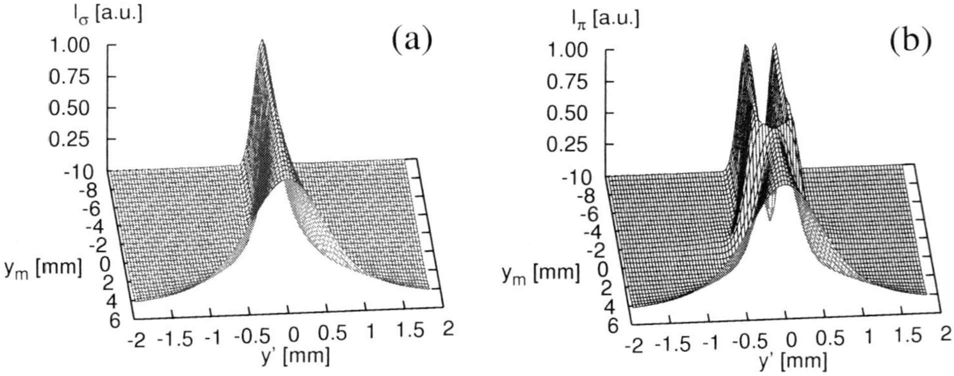

FIGURE 4. Diffraction limited vertical single electron intensity distribution in the image plane as function of the light extracting mirror bottom edge y_m in the case of σ (a) and π (b) polarization. y_m < 0: bottom edge is below the beam axis. The intensity is normalized to it's maximum value at each mirror position.

π component is smeared out and the shape of both polarization component distributions becomes more and more similar.

In order to quantify the contribution of the vertical diffraction broadening, the calculated intensity distributions $I_y(y')$ in Fig. 4 was convoluted with the vertical beam profile, assuming a normal distribution. The "effective" vertical resolution broadening was derived in a similar way than for the horizontal distribution by fitting the convoluted profile with a normal distribution with subsequent quadratical subtraction of the beam size σ_y. The results are plotted as function of y_m in Fig. 5(a) for both polarization states.

According to that, the contribution of the σ polarization component is continuously increasing with increasing y_m. Anyhow, if $y_m \leq 0$ mm the usage of this polarization component will lead to a better resolution than the usage of the π polarization. For larger mirror positions the resolution becomes independent on the polarization state.

In Fig. 5(b) the measured total width of the vertical beam profile is shown, including resolution broadening effects due to diffraction and depth of field. The measured beam size strongly reflects the influence of the vertical diffraction broadening as expected from Fig. 5(a).

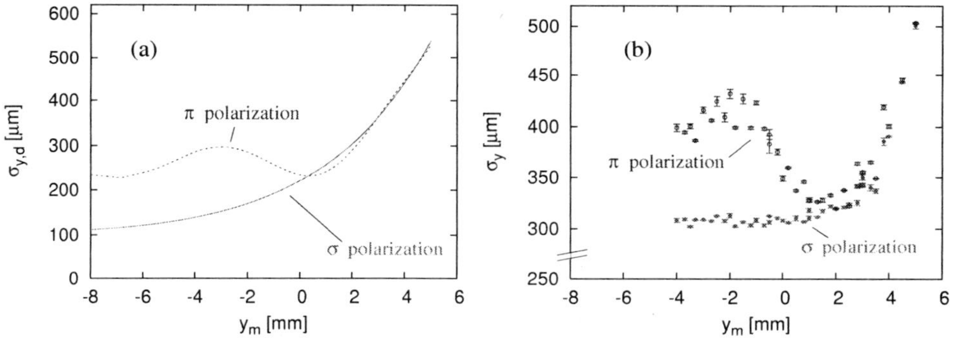

FIGURE 5. (a) Calculated contribution of the vertical diffraction broadening as function of the light extracting mirror bottom edge y_m. (b) Measurement of the vertical beam profile width as function of y_m, including resolution broadening contributions due to diffraction and depth of field.

Additionally, Fig. 6 shows screen shots of measured beam spots for two different mirror positions and both polarization states. While for y_m = -4.3 mm the measured vertical beam size for σ polarization is much smaller than for π polarization, they become independent on the polarization state for large y_m as expected from the calculations presented in Fig. 5(a).

For the HERA profile monitor the mirror position is chosen as y_m = +2 mm. This value is a compromise between a reasonable value for heat load and a sufficient vertical resolution.

The calculated resolutions in Fig. 5(a) show again larger deviations compared to the standard formula for vertical diffraction broadening, $\sigma_{y,d}^{app} = \frac{1}{2}\lambda/\Psi = (\frac{\pi}{6}\lambda^2\rho)^{1/3}$ = 400 μm, with Ψ the typical SR opening angle and ρ the magnet bending radius. Apart from the aspect of off–axis observation this formula overestimates the resolution broadening drastically. Therefore calculations based on the standard formula are more a rough estimation than a precise calculation of the SR diffraction limited resolution.

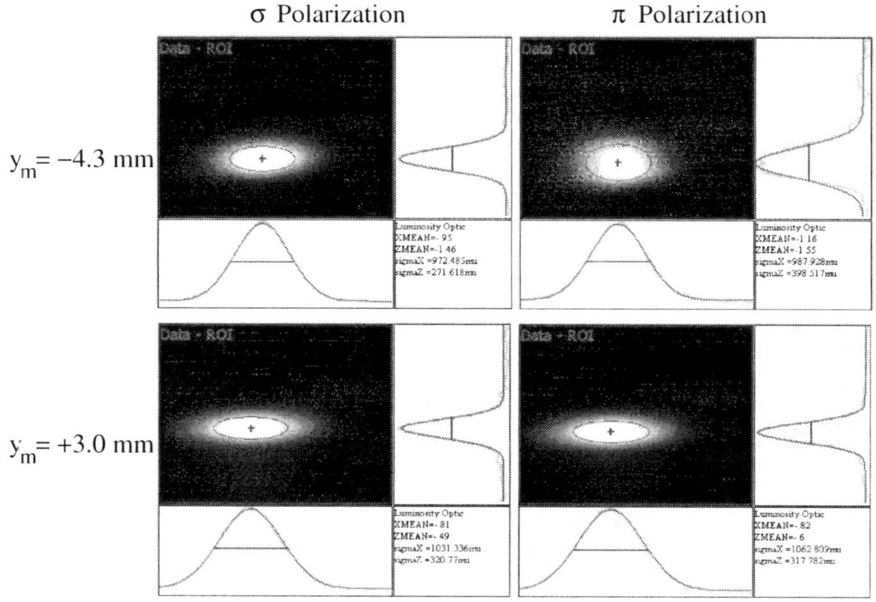

FIGURE 6. Screen shots of measured beam spots for two different mirror positions and both polarization states. The beam image is shown together with the projections onto the horizontal (x) and vertical (z) axis. By means of a fit the beam positions ($XMEAN$, $ZMEAN$) are determined together with the widths in both planes. Note that the range of interest (ROI) was chosen different for both mirror positions.

FIELD–DEPTH LIMITED RESOLUTION

The calculations for the depth of field are based on the considerations in Ref. [2]. The particle trajectory is divided in small slices which are treated as point sources, and their angular distributions are summed up as if they emit incoherently. The SR source region consists of a number of slices that are focused by the lens at different at various distances from the image plane (x', y').

In first order imaging theory, a point with emission angle α which is a distance s away from the vertical source plane y_0 (i.e. a single slice), is imaged onto a point y' in the image plane according to $y' = -\frac{p'}{p}(y_0 + s\,\alpha)$. Under the simplified assumption that y_0 and α are normal distributed in the source plane with widths σ_y resp. σ_α, the distribution of y' in the image plane is also normal distributed with width $\frac{p'}{p}\sqrt{\sigma_y^2 + (\rho\,\vartheta\sigma_\alpha)^2}$. In this relation the orbit coordinate was replaced by $s = \rho\,\vartheta$ according to Fig. 1.

Summing up the contributions along the particle trajectory results in the vertical intensity distribution in the image plane

$$I_{y,fd}(y') = I_0 \int_{-\vartheta_{lim}}^{+\vartheta_{lim}} \frac{p\rho\, h(\vartheta)\,\mathrm{d}\vartheta}{p'\sqrt{2\pi}\,\sqrt{\sigma_y^2 + (\rho\,\vartheta\,\sigma_\alpha)^2}} \exp\left[-\frac{(p y')^2}{2p'^2\{\sigma_y^2 + (\rho\,\vartheta\,\sigma_\alpha)^2\}}\right] \qquad (7)$$

with I_0 a normalization constant and $\vartheta_{lim} = x_c/\hat{p} + \Psi$ the accepted angle, taking into account the aperture limitation given by the extraction mirror together with the SR typical opening angle. The function $h(\vartheta)$ accounts for the amount of light extracted by the mirror from different trajectory slices:

$$h(\vartheta) = \frac{1}{2}\left[\mathrm{erf}\left(\frac{x_c/\hat{p} - \vartheta}{\sqrt{2}\sigma_\vartheta}\right) - \mathrm{erf}\left(-\frac{x_c/\hat{p} + \vartheta}{\sqrt{2}\sigma_\vartheta}\right)\right] \tag{8}$$

with σ_ϑ the width of the normally distributed horizontal emission angle. Eq. 7 holds also for the horizontal intensity distribution with the exception that the intensity distribution from each slice is shifted by $\Delta x = \rho\,\vartheta^2$ against the position at $s = 0$, taking into account the curvature of the trajectory which results in an additional broadening.

Calculations based on this model lead to a resolution broadening of $\sigma_{y,fd} = 154\ \mu$m for the vertical and $\sigma_{x,fd} = 292\ \mu$m for the horizontal intensity distribution. These values are again smaller than the ones according to standard approximations (see e.g. Refs. [1, 5]), $\sigma^{app}_{y,fd} = \rho\, x_c/\hat{p}\,\Psi = 185\ \mu$m, resp. $\sigma^{app}_{x,fd} = \rho\,(x_c/\hat{p})^2 = 440\ \mu$m. The deviation in horizontal direction becomes even larger because the approximative contribution due to the orbit curvature (see e.g. Ref. [5]) $\sigma_{x,c} = \rho/2\,(x_c/\hat{p})^2 = 220\ \mu$m has to be taken into account.

SUMMARY

In the present article a beam profile monitor system using visible SR radiation from a bending magnet at the HERA storage ring has been described. By use of models for diffraction and depth of field effects described above it is possible to determine the actual beam rms sizes. The calculations show large discrepancies in comparison to the ones based on standard formulas which estimate resolution broadening. Moreover special attention is drawn to the influence of off–axis observation which smears out the vertical resolution additionally. While the absolute values of the measured beam sizes deviate up to now from their design values (reason seems to be a discrepancy to the particle beam design optics), the monitor is already now a versatile tool for studies of relative changes in the beam emittances.

REFERENCES

1. A. Hofmann and F. Méot, Nucl. Instr. Meth. **203** (1982) 483.
2. Å. Andersson and Juri Tagger, Nucl. Instr. Meth. **A 364** (1995) 4.
3. A. S. Fisher *et al.*, in Proceedings of EPAC96, p.1734.
4. A. Hofmann, Proceedings of the CAS 98-04.
5. J. A. Clarke, *A Review of Optical Diagnostics Techniques for Beam Profile Measurements*, EPAC 94, www.astec.ac.uk/preprints/epac94/JAC-Profile_paper.pdf.

APS Storage Ring Monopulse RF BPM Upgrade*

R. Lill, A. Pietryla, E. Norum, and F. Lenkszus

Advanced Photon Source, Argonne National Laboratory
9700 South Cass Avenue, Argonne, IL 60439, USA

Abstract. The Advanced Photon Source (APS) is a third-generation synchrotron light source in its ninth year of operation. The storage ring monopulse radio frequency (rf) beam position monitor (BPM) was designed to measure single-turn and multi-turn beam positions for operations and machine physics studies. Many of the components used in the original design are obsolete and costly to replace. In this paper we present a proposal to upgrade the monopulse rf BPMs in which the existing system hardware is repartitioned and the aging data acquisition system is replaced. By replacing only the data acquisition system, we will demonstrate a cost-effective approach to improved beam stability, reliability, and enhanced postmortem capabilities. An eight-channel ADC/digitizer VXI board with sampling rate of up to 105 MHz (per channel) and 14-bit resolution coupled with a field-programmable gate array and embedded central processing will provide the flexibility to revitalize this system for another decade of operation. We will discuss the upgrade system specifications, design, and prototype test results.

INTRODUCTION

The Advanced Photon Source (APS) is a third-generation synchrotron x-ray source that provides intense x-rays for basic and applied research. The stability of the beam is largely dependent on the resolution and accuracy of the beam position monitor (BPM) system. The APS storage ring presently makes use of 280 monopulse radio frequency (rf) BPMs designed to measure single-turn and multi-turn beam positions. The rf BPM signal processing topology is a monopulse amplitude to phase (AM/PM) technique for measuring the beam position in the x- or y-axis [1]. A logarithmic amplifier channel measures the beam intensity. The processing electronics are located above the tunnel in 40 VXI crates with up to 9 BPMs per crate. The rf BPM system front-end and timing system was upgraded in 2001 [2,3], but the overall system performance has suffered because of an aging data acquisition system. The rf BPM data acquisition is presently being handled in a VXI-based signal conditioning and digitizing unit (SCDU) [4]. Each SCDU module provides signal conditioning for beam position in the x or y plane and sum signals. The sum signal is peak detected and the normalized position data is integrated on the SCDU. Both signals are digitized via 12-bit ADCs and stored in output registers. These registers are read at 271 kHz (revolution frequency) for up to nine BPMs by the memory scanner module in the same VXI crate via the local bus. Each VXI crate has a memory scanner and beam history module.

* Work supported by the U.S. Department of Energy, Office of Basic Energy Sciences, under contract No. W-31-109-ENG-38.

CP732, *Beam Instrumentation Workshop 2004: Eleventh Workshop*,
edited by T. Shea and R. C. Sibley III

The memory scanner provides a programmable (2-2048 turn) boxcar averager for each BPM x and y position. It also provides a high-speed fiber-optic port to stream data to the feedback system.

There are many programmable logic device chips on this module that have grown obsolete and are therefore very expensive to support; additionally there are only a limited number of spare boards available after nine years of operation.

The data from the VXI local bus are also shared with the beam history module. This module provides storage memories for each BPM for postmortem analysis. The 32-k by 32-bit memory operates as a first-in first-out (FIFO) buffer and is capable of storing approximately 50 milliseconds of storage ring BPM data. This module suffers from many of the same problems realized in the memory scanner.

RF BPM UPGRADE PROPOSAL

The upgrade approach repartitions the existing system hardware such that the rf section from the button electrodes to the receiver output remains electrically the same and the data acquisition system is replaced as illustrated in the shaded areas in Fig. 1. This approach removes the monopulse receiver from the VXI-based SCDU and locates it in a designated EMI-shielded chassis. This has many mechanical advantages such as improved cooling and ease of replacement. The receiver's center frequency remains at 352 MHz with an intermediate frequency bandwidth of 10 MHz. The receiver interface card shown in Fig. 1 will provide the control interface for self-test, gain selection, commutation, and horizontal or vertical plane measurement selection.

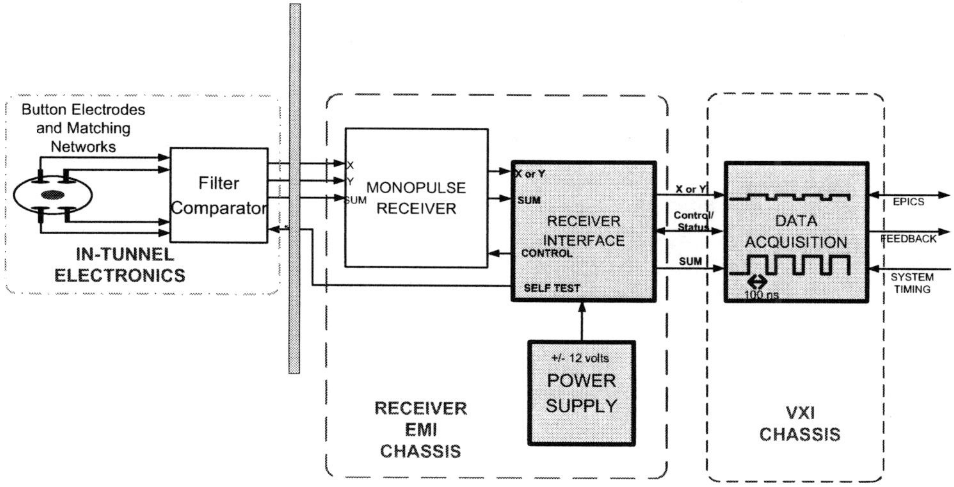

FIGURE 1. System block diagram.

DATA ACQUISITION DESIGN

The proposed data acquisition block diagram shown in Fig. 2 is an eight-channel ADC/digitizer VXI board. A single VXI board will provide the data acquisition for up to four BPMs. The gated integrator and peak detectors used in the SCDU data processing are replaced by ADCs running at a beam-synchronized frequency of 88 MHz (one quarter of the rf frequency). Presently the SCDU data acquisition only acquires data for a single 50-ns sample per turn regardless of the fill pattern. The proposed upgrade system can measure a single bunch per turn or all of the bunches in the train in a 10-MHz bandwidth. This allows for more measurements per turn, resulting in a reduction of the noise floor scaling as $\sqrt{N}$, where N is the number of samples collected. The present trend at the APS points toward increasing the number of buckets being filled. If we assume a typical fill of 24 or 324 bunches with six coherent samples per bunch, a resolution improvement of a factor of 12-18, respectively, can be implied.

The design will have many reconfigurable functions. We are presently evaluating the Lattice ispPAC ™ 80 [5] for the ADC input lowpass filter shown in Fig. 2. In-system programming allows reconfigurable gain and cutoff frequency. We are presently working on trade-off studies to determine the filter requirements to minimize bunch-pattern dependence.

This upgrade also facilitates a new feature or mode of operation called "oscilloscope," which will stream data in a 10-MHz bandwidth. This will be useful in machine studies to analyze bunch-to-bunch instabilities as higher currents are explored.

ADC Front End

At the time of selection we found two data converters that met our initial criteria of having greater than 12 bits and the ability to sample at an integral divisor of our rf frequency (351.929 MHz) for easy synchronization. The two converters we found were the Analog Devices AD6645-105 [6] and the Texas Instruments ADS5500 [7]. The Analog Devices converter was chosen for several reasons including having better published DC and AC specifications. The Analog Devices AD6645-105 is a 14-bit, 105-MSPS converter featuring a 270-MHz input bandwidth, ± 0.5 least significant bit (LSB) of differential nonlinearity (DNL), ± 1.5 LSB of integral nonlinearity (INL), signal-to-noise ratio of 74.5 dB with 30.5-MHz input and spurious free dynamic range (SFDR) of 89 dBc (70MHz $< f_{IN} <$ 105 MHz). The Texas Instruments ADS5500 is also a 14-bit converter with a maximum sampling rate of 125 MSPS with a 750-MHz analog bandwidth, ± 0.75 LSB DNL, ± 2.5 LSB INL, 71.5 dB SNR (30-MHz input), and 84-dBc SFDR (30-MHz input).

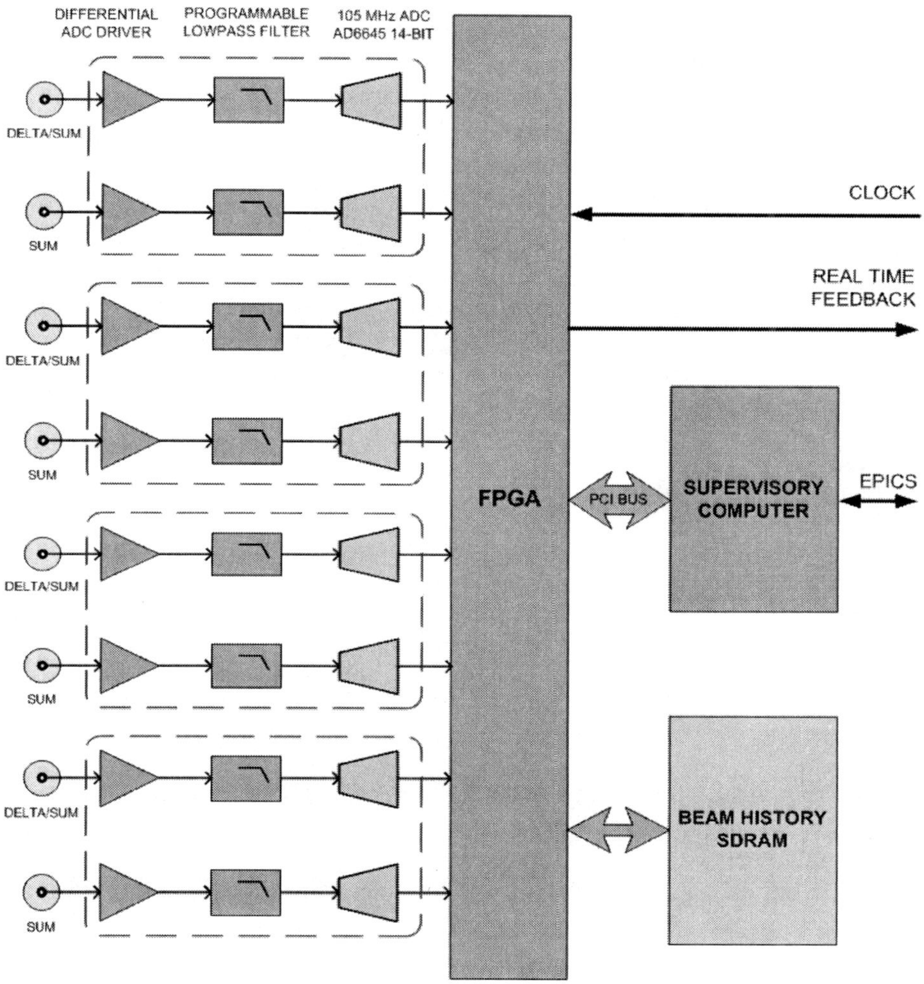

FIGURE 2. Data acquisition block diagram to process four BPMs.

Field Programmable Gate Array

Figure 3 illustrates a single channel for the Altera Stratix™ field-programmable gate array (FPGA). The clock generation and synchronization block accepts the 44-MHz clock and 'PO' marker from the facility timing system and uses these signals to synthesize an 88-MHz sampling clock to drive the analog-to-digital converters and a 'beginning of turn' marker to control the calculation of the turn-by-turn position information. This block also contains information about the fill pattern within the storage ring. This information is used to enable the summation block only when valid position signals are present. By performing the gated integration after digitization, the problems of drift and offset in the analog portion of the circuit are avoided.

The turn-by-turn beam position history is stored in a separate RAM module. A 128-Mbyte module will allow several seconds of turn-by-turn position information to be recorded.

The use of a single FPGA chip to perform the initial signal processing for all analog input channels has a number of advantages over a design using conventional digital-signal-processing devices. The FPGA chosen has several multiply/accumulate, cores, which allows the signal processing for up to eight input channels to be performed on a single chip. This provides increased reliability and reduces power consumption since almost all high-speed signals are contained within that single device and need not pass through I/O pin drivers. Altera high-density Stratix™ FPGA devices have several features that make them attractive for this application:

1. Up to 22 high-performance DSP blocks per device and dedicated multiplier, pipeline, and accumulation circuitry. These blocks will be used to perform the initial gated averaging and statistics computation as well as the low-pass filtering of the turn-by-turn position information.
2. High-speed differential I/O interface support for connection to the high-speed serial interface, which will carry the filtered position information to the orbit correction computers.
3. Support for interfaces to high-speed external memory devices, including DDR and SDR SDRAM. External memory devices will be used to store the turn-by-turn beam history for post-fault analysis or diagnostic studies.
4. Full-featured phase lock loops (PLLs) for system timing management. The on-chip PLLs will be used to synchronize the data acquisition with the system 44-MHz and P0 clocks as well as obtaining time stamp and other information from the event system.
5. Up to 7 Mbits of embedded memory. This high-speed on-chip memory will be used for buffering sample-by-sample or bunch-by-bunch information for diagnostic studies.
6. Remote system update capability, which makes it easy to modify filter parameters and other system features without removing the board from the VXI crate and without transporting programming equipment to the rack in which the crate is installed.

SPECIFICATIONS

The general system specifications are described below and in Table 1. The original specifications quoted in Table 1 are archived from the early commissioning days and are typically exceeded in routine operation at this time.

1. Measures beam position both during injection at 2 Hz and with stored beam.
2. Provides single-bunch tracking around the ring.
3. Measures position of different bunches at each BPM turn-to-turn.
4. Measures position at each turn (3.68-μs revolution period).
5. Provides average beam position for higher accuracy.
6. Provides a 32,768 sample of the beam history for each BPM.

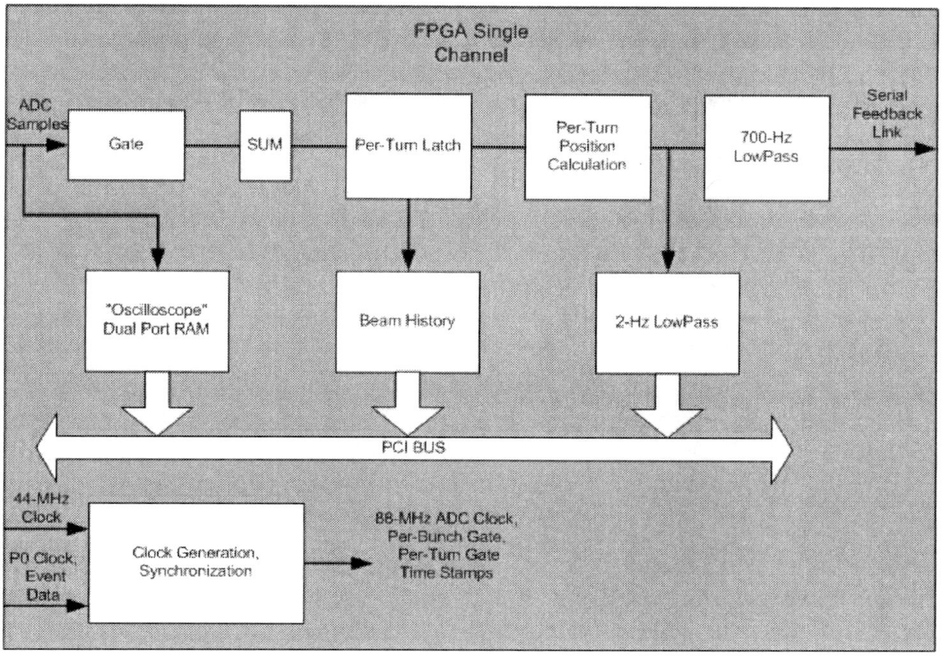

FIGURE 3. FPGA single-channel block diagram.

TABLE 1. General system specifications.

Parameter	Original Specification Limit	Specification Goal	Condition
First Turn, Single Bunch Resolution / Accuracy	< 200 μm / 500 μm rms	< 100 μm / 250 μm rms	Single Shot
Stored Beam, Single Bunch Resolution / Accuracy	< 25 μm / 200 μm rms	< 1 μm / 50 μm rms	0.1 Hz BW
Stored Beam, 24 Bunch or greater Resolution / Accuracy	< 25 μm / 200 μm rms	< 250 nm / 50 μm rms	0.1 Hz BW
Real Time Feedback, (24 Bunch or greater) Resolution	NEW	<500 nm rms	0.1-30 Hz BW
Real Time Feedback, (24 Bunch or greater) Resolution	NEW	< 2 μm rms	1-1000 Hz BW
Bunch-Bunch Oscilloscope mode (Stored Beam, Single or Multiple Bunches) Resolution / Accuracy	NEW	< 15 μm rms/ 50 μm rms	10 MHz BW
Dynamic Range, Position	± 20 mm Minimum	± 10 mm Minimum	Standard Chamber
Dynamic Range, Intensity Single Bunch 0.1-10 ma	≥ 40 dB	≥ 40 dB	200 μm deviation
Stability, Long Term	± 30 μm rms	± 1 μm rms	24 Hour +/- 0.2 Celsius

PERFORMANCE

Development on this project has just started and we are currently evaluating the Analog Device AD6645 ADC coupled to the high-speed ADC FIFO evaluation kit. The evaluation kit interfaces the ADC and provides a 256-k FIFO buffered memory, which captures the high-speed output of the AD6645. The kit also provides a software application to calculate real-time FFTs and time domain analysis. The monopulse receiver and the evaluation components were tested with 100 mA of stored beam with an equally spaced 24-bunch fill. The number of samples collected in the FIFO was varied and the noise floor relative to the ADC full-scale value was measured. A calculation based on the (receiver + ADC) measurement shown in Fig. 4 indicates that an rms system noise floor of less than 10 nm/square root Hz can be realized. Further testing will continue after the May shutdown to fully quantify ADC dynamic performance and integrate FPGA processing.

DISCUSSION

The plan is to couple the ADC evaluation board to the FPGA development board to expedite the FPGA software development. The first task for the FPGA development will be to average 5-7 samples at the 88-MHz sample rate. In the next phase we are planning to build and test a single channel from ADC through to the FPGA interface. In parallel with the FPGA development we will continue system simulations to optimize filter requirements.

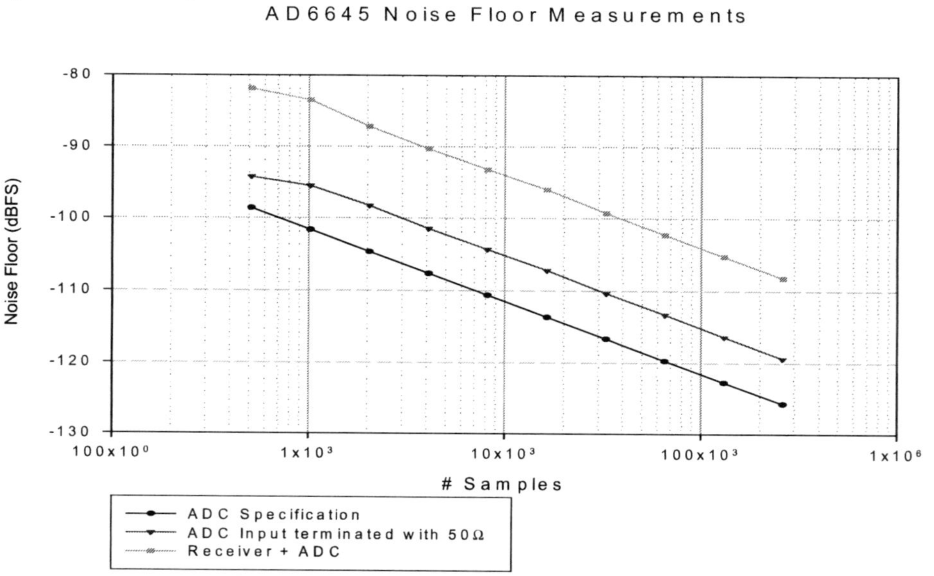

Figure 4. ADC noise floor.

ACKNOWLEDGMENTS

The authors would like to acknowledge Glenn Decker and Om Singh for many helpful discussions.

REFERENCES

1. E. Kahana, "Design of the Beam Position Monitor Electronics for the APS Storage Ring," Proceedings of the 3rd Accelerator Beam Instrumentation Workshop, AIP Conference Proceedings 252, 1992, pp. 235-240.
2. R. Lill, G. Decker, and O. Singh, "Advanced Photon Source RF Beam Position Monitor System Upgrade Design and Commissioning," Proceedings of the 5th European Workshop, DIPAC 2001 Conference Proceedings 451, Grenoble, France, 2001, pp. 180-182.
3. F. Lenkszus, R. Laird, O. Singh, R. Lill, and G. Decker, "New APS Storage Ring BPM Timing System Design," Proceedings of the 2001 Particle Accelerator Conference, Chicago, Illinois, 2001, pp. 767-769.
4. F. Lenkszus, "Beam Position Monitor Data Acquisition for the Advanced Photon Source," Proceedings of the 1993 Particle Accelerator Conference, Washington D.C., 1993, pp. 1814 -1816.
5. http://www.latticesemi.com.
6. http://www.analog.com.
7. http://www.ti.com.

Measurements of the APS Storage Ring Beam Stability at 225 mA*

A. H. Lumpkin, B. X. Yang, and M. Borland

Advanced Photon Source, Argonne National Laboratory, Argonne, Illinois 60439, USA

Abstract. The Advanced Photon Source (APS) is a third-generation hard x-ray user facility that operates nominally at 100-mA stored beam current. As part of our probing of future operation regimes we have run at 130 mA for a user run, and we are on the path to test operations at up to 300 mA. Our most recent experiments have been at 225 mA in December 2003. One sector in the ring is dedicated to diagnosing the electron beam transverse and longitudinal parameters. We use a combination of imaging tools, including an x-ray pinhole camera and optical streak camera for the bending magnet source and x-ray monochromator measurements of divergence on the insertion device source. The onset of longitudinal instabilities due to rf cavity higher-order modes (HOMs) is one of the limitations in going to higher currents. These effects are seen in the horizontal plane of the pinhole camera images since the dipole source is at a dispersive point in the lattice. We observed about a factor of two increase in horizontal beam size and integrated bunch length in the unstable condition. By correlating the size changes with the temperature readouts of the 16 rf cavity cells, we identified the source of the instability. The cavity temperature set points were adjusted to avoid an HOM.

INTRODUCTION

The Advanced Photon Source (APS) is a third-generation hard x-ray synchrotron radiation facility that was commissioned in 1996. The baseline design was to provide high-brightness x-rays from dipole sources and insertion device sources using a 7-GeV positron or electron beam with a stored beam current of 100 mA [1]. The installed rf power was designed to allow operations at up to 7.7 GeV or 300-mA current. In the last few years we have continued on the path to higher current, having operated in a user run at 130 mA and performed studies at 200 mA in September 2001. In our most recent study in December 2003, we operated at 225 mA and 7 GeV. During these studies our beam diagnostics sector instrumentation provided critical tuning information, and our insertion device beamline was the only one approved to run at this higher current.

The Sector 35 bending magnet pinhole camera views a dispersive point in the lattice so that the observed horizontal beam size provides information about changes in energy spread due to longitudinal instabilities. Optical synchrotron radiation (OSR) is also measured by a streak camera to correlate the bunch-length effects in the longitudinal dynamics. One of the main issues in these higher-current runs is control

*Work supported by the U.S. Department of Energy, Office of Basic Energy Sciences, under contract No. W-31-109-ENG-38.

CP732, *Beam Instrumentation Workshop 2004: Eleventh Workshop*,
edited by T. Shea and R. C. Sibley III

of the rf cavity higher-order modes (HOMs). As we raise the stored beam current and rf power, temperature variations in the rf cavities result in shifting resonance frequencies for HOMs in the 16 rf cells. If an HOM overlaps an upper synchrotron sideband of the beam longitudinal motion, the beam becomes unstable and may eventually be lost. By correlating the beam size changes with the rf cavity temperature, we identified which cavity temperature setpoints should be adjusted to reduce the HOMs. We subsequently were able to store the 225 mA in a reasonably stable situation. Complementary information on the bunch length and the beam divergence were also obtained and will be discussed.

Experimental Background

The injector for the storage ring includes a thermionic rf gun, an S-band linear accelerator (linac), a particle accumulator ring (PAR) operating at 325 MeV, and a booster synchrotron that ramps the energy from 325 MeV to 7 GeV. Full energy injection is done at 2 Hz into the 1104-m-circumference storage ring. One of the standard fill patterns is for 24 singlets, each separated by 54 rf buckets of the 352-MHz rf frequency. However, for the high-current runs the single-bunch current is dramatically reduced by using 500 of the 1296 rf buckets available. Even at 225 mA, this results in only about 0.42 mA per bunch.

One sector of the storage ring is dedicated to state-of-the-art beam diagnostics [2,3]. These diagnostics include an x-ray pinhole camera viewing a dipole source; a dual-sweep synchroscan streak camera imaging the OSR from the same dipole source; a cryo-cooled monochromator on the 198-period diagnostic undulator source; and another x-ray pinhole camera viewing a dipole source at a low dispersion point in the lattice. The recently installed cryo-cooling feature for the crystal in the monochromator was involved in our heat load assessments so that this was the only sector permitted to open the x-ray shutter to allow x-rays into a beamline station.

The pinhole camera uses four tungsten blades with precision stepper motors to provide remote control of the aperture size and center. Generally, we run with a 15-μm ×15-μm aperture for the horizontal and vertical planes. The source-to-aperture distance is 9.0 m, and the aperture-to-crystal distance is 8.8 m, resulting in a magnification of about 1.0. The x-ray images are converted to visible light by a YAG:Ce single crystal and imaged by a pair of achromat lenses and a CCD camera. The total system resolution is 22 μm as determined by a scan of the skew quads at low vertical coupling [2]. Normal beam sizes at the source are 95 μm (H) by 30 μm (V) for the lower emittance of 2.5 nm rad. The observed horizontal beam size (σ_x) included the dispersion-energy spread product ($\eta_x\sigma_E$) in quadrature with the betatron size and system resolution (σ_{res}) as shown in Eq. (1), where β_x and ε_x are the beta function and emittance.

$$\sigma_x = \left(\beta_x \varepsilon_x + (\eta_x\sigma_E)^2 + \sigma^2_{res}\right)^{1/2}. \quad (1)$$

The images are digitized by a dedicated Datacube MV200 video digitizer, and the beam parameters (size, emittance) are calculated and archived as process variables in the EPICS protocol.

The optical streak camera is a Hamamatsu C5680 with the vertical sweep unit being a synchroscan unit tuned to 117.3 MHz, the third subharmonic of the storage ring rf frequency. The dual-sweep plugin is installed to access the slower time scale of the ring period, 3.68 μs. Although the streak camera has ~ 0.6-ps (σ) resolution when operating on its fastest range, we use a slower sweep speed to provide the time span needed to cover bunch lengths of 20 to 80 ps. The effective resolution is then about 2 ps (σ), and this is a small effect when taken in quadrature with the bunch length. Generally, for single-bunch studies we can image 1 to 8 mA in a single bunch in a single turn. However, in these experiments the large number of bunches results in less than 0.5 mA per bunch. The synchroscan feature gives us the low jitter in the measurements so that we can synchronously sum the signal from many bunches in many turns. A newly developed interface allowed EPICS control of the streak camera and hence, remote control. The readout camera video was sent to another MV200 video digitizer, and the images saved as needed. Online profile processing was also done.

The cryo-cooled monochromator was designed in-house, and commissioning was begun in the fall of 2003. A Si crystal used in the Laue Geometry for the [220] planes is used to monochromate the x-ray beam. The bandwidth is about 1.4×10^{-4}. The monochromatic x-rays at about 22 keV are transmitted out of the vacuum chamber through a Be window strip that covers angles from 10 to 90 degrees. The x-rays are converted to visible light by a 100-μm-thick YAG:Ce single crystal, and the beam spot is imaged onto a CCD camera. Since there are over 28 insertion device beamlines in the facility, our device was used as a surrogate for the high-current test. Scaling arguments were made that with cryo-cooling, our crystal would survive up to 300 mA, and the x-ray front end and beamline were safe if the ID gap was not reduced below 17 mm. The nominal minimum gap is 12 mm with 100-mA stored beam. We were interested in how the HOMs might affect the ID beam shape. The ID pinhole camera was not used for this run.

EXPERIMENTAL RESULTS

X-ray Pinhole Camera Results and Correlation to HOMs

As mentioned previously, the horizontal beam size at the bending magnet source point shows a clear signature of size increase when the HOMs are excited and the beam is still stored. Examples of the stable and unstable conditions are shown in Figs. 1(a) and (b), respectfully.

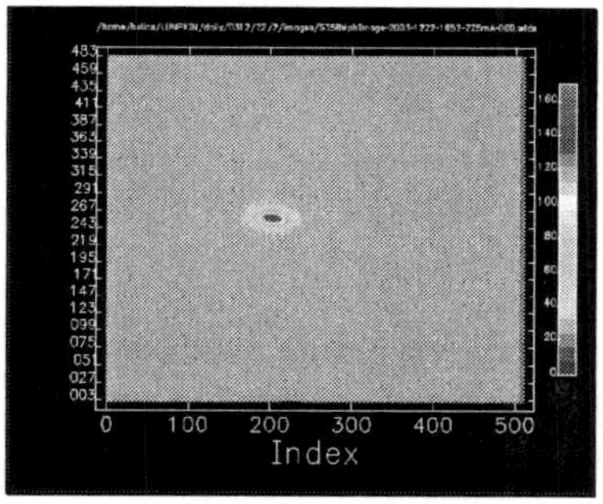

(a)

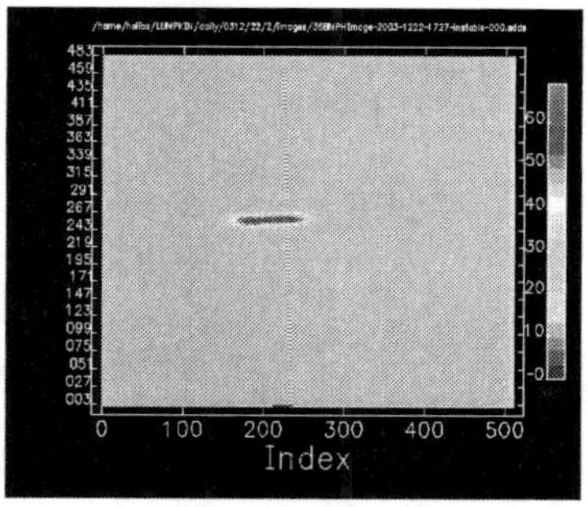

(b)

FIGURE 1. Examples of the x-ray pinhole camera image for stable (a) and unstable (b) conditions at 200 mA. The horizontal beam sizes are 90 μm and ~ 250 μm for (a) and (b), respectively.

The beam size, emittances, rf cavity temperatures, and rf reflected power are available as process variables (PVs) in EPICS. Using the SDDS toolkit [4], we were able to easily collect thousands of channels of data potentially related to the instability and to compute correlations with the horizontal emittance. Sorting on the magnitude of the correlations identified a particular cavity as having temperature readbacks highly correlated to emittance. The individual cavity temperature set point was adjusted to reduce/avoid the HOMs before more beam was stored. An example of the correlation of horizontal emittance with cavity cell temperature is shown in Fig. 2 for Sector 40 Cell 4. A strong correlation of 0.87 was calculated in this case.

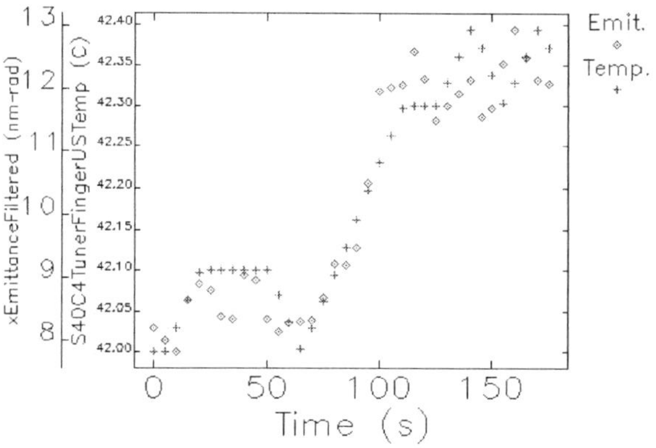

FIGURE 2. A plot showing the correlation of observed emittance and the cavity temperature. In this case, the temperature set point for Sector 40 Cell 4 was adjusted and subsequently, the emittance dropped to 2.5 nm-rad.

Optical Streak Camera Results

Complementary to the energy spread information is the actual effective bunch length. The streak camera is operated in the dual-sweep mode to provide access to the μs-scale information (one turn is 3.68 μs) and the ps-scale information for the bunch length (vertical axis). The observed bunch-length envelope in the region-of-interest (ROI) is about 46 ps as shown in Fig. 3(a). This is larger than expected for a 0.4-mA bunch and 9-MV gap voltage, but perhaps a phase oscillation is also involved, so the time averaged bunch is lengthened. Further studies are needed on this point.

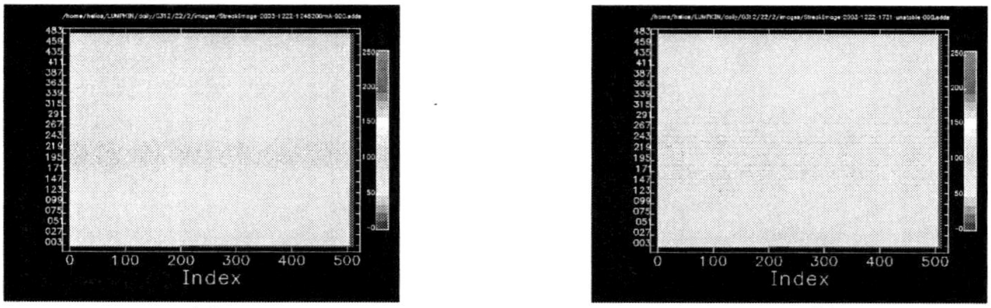

(a) (b)

FIGURE 3. Examples of the dual-streak camera image for stable (a) and unstable (b) conditions near 200 mA. The horizontal axis covers 5 μs and the vertical axis is 500 ps. The observed bunch lengths are about (a) 46 and (b) 90 ps.

In the unstable mode (also correlated with the larger horizontal beam size), the bunch- length envelope is about 90 ps (σ). We should note that for a single bunch at such low current at 9.0 MV rf gap voltage, we would expect a bunch length of 20-25 ps. The image profile from the MV200 for stable conditions is shown in Fig. 4.

ID Divergence Results

The cryo-cooled monochromator data are used to provide information about the e-beam's horizontal and vertical divergence. By using the known source distance to the monochromator and converter crystal, the projected beam image spot size at the crystal can be used to determine the beam divergence. The nominal values at 100 mA are $\sigma x' = 13$ μrad and $\sigma y' = 6.4$ μrad. For the 200-mA regime data we see slight effects, including when the HOM instability was purposely stimulated by changing the cavity temperature set points at the end of the run. The unstable condition image is shown in Fig. 5.

There is dispersion in the ID straight for this lattice; however, the effects of energy spread in the projected divergence images are masked compared to the x-ray pinhole image. There is detectable tailing in the $\sigma x'$ profile. Further studies are warranted since beam trajectory was not as well controlled for the high current run. A subset of the beam missteering interlocks was applied, and orbit feedback was off.

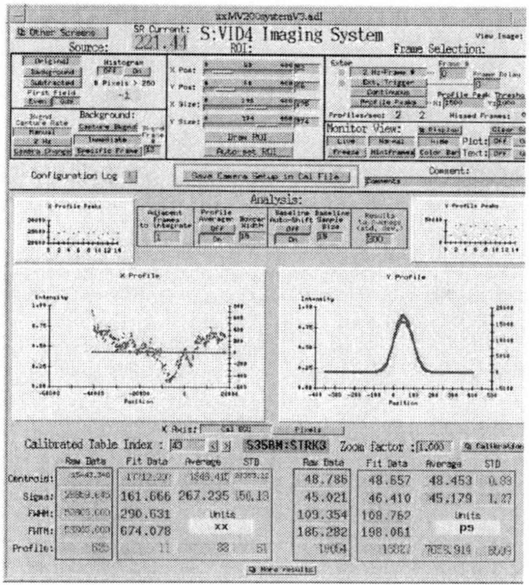

FIGURE 4. A longitudinal profile for the streak camera image of Fig. 3(a). The bunch-length profile is the y profile on the right, in this case.

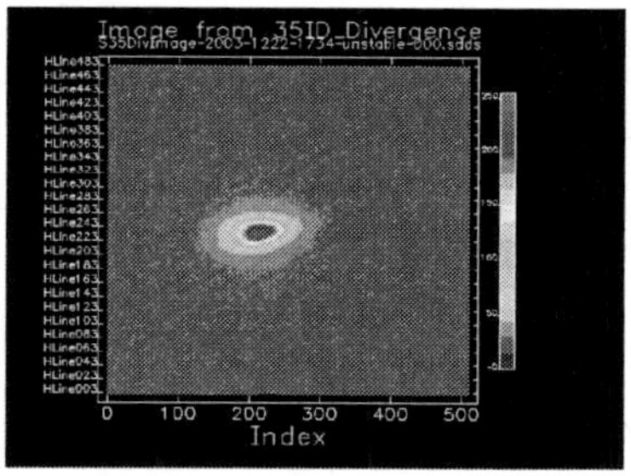

FIGURE 5. Beam divergence image from the ID source in the HOM unstable condition.

SUMMARY

In summary, the APS team successfully stored 225 mA in the ring for the first time. One of the critical aspects was the control of the HOMs. The correlation of observed horizontal beam size and cavity cell temperatures was effective. For both stable and unstable conditions the beam quality was assessed using the x-ray pinhole camera, the streak camera, and the cryo-cooled monochromator. We are developing the techniques needed to support operations at 250 to 300 mA. If the machine does operate at these elevated currents, the x-ray front ends would need upgrades for the higher heat loads.

ACKNOWLEDGEMENTS

The authors acknowledge the support of Rod Gerig (ASD) and Bill Ruzicka (AOD) for the studies and the contributions of L. Emery, K. Harkay, D. Horan, A. Nassiri, and the operations team during the test.

REFERENCES

1. J. N. Galayda, "The Advanced Photon Source," Proc. of the 1995 Particle Accelerator Conference, Dallas, Texas, May 1-5, 1995. IEEE Vol. 1, 4-8 (1996).
2. B. X. Yang, A. H. Lumpkin, L. Emery, and M. Borland, "Recent Developments in Measurement and Tracking of the APS Storage Ring Beam Emittance," BIW 2000, Cambridge, Massachusetts, May 8-11, 2000 AIP Conf. Proc. Vol. 546, 622-630 (2000).
3. B. X. Yang and A. H. Lumpkin, "Simultaneous Measurements of Electron Beam Size and Divergence with an Undulator," Proc. of the 1999 Particle Accelerator Conf., New York City, N. Y., March 29-April 2, 1999, Vol. 3, 2161-2163 (1999).
4. H. Shang, R. Soliday, L. Emery and M. Borland, "New Features in the SDDS-Compliant EPICS Toolkit," Proc. of the 2003 Particle Accelerator Conf., Portland, Oregon, May 12-16, 2003, Vol. 5, 3470-3472 (2003).

Innovative RF Design Unites Benefits of Multiplexed and Multi-Channel System

Uroš Mavrič

Instrumentation Technologies, Srebrničev Trg 4a, SI-5250 Solkan, Slovenia
uros@I-tech.si, www.i-tech.si

Abstract. The main advantages of the implemented innovative RF design are reproducibility, accuracy, low measurement uncertainty, and compact design. They were achieved by implementing the direct RF sampling and the completely new, patent pending technological approach of quasi-crossbar multiplexing and associated measurement method. The innovative quasi-crossbar switch matrix unites the benefits of both the multi-channel and the multiplexed system and at the same time compensates the disadvantages of the two. We achieve reproducibility and good "beam vs. current" dependence, which are multiplexed system characteristics and, using a multi-channel approach, we ensure a broader band of operation.

INTRODUCTION

In general, measuring linear relationships between a multitude of RF signals over a wide dynamic range, in temperature unstable environments or over a long period of time, is associated with significant errors caused by the measuring device. It is impractical, if not impossible, to build a device with parallel signal processing channels, whose transfer functions stay in constant relationship with respect to each other over a wide dynamic range, time, and temperature. Using a single-channel multiplexed system would seem to solve the problem, but it is limited to narrowband applications, has worse sensitivity because it does not process all the available signal power and suffers from aliasing problems, to name just a few problems.

The focus of this paper is on a new method for stabilization of dynamic range, time and temperature dependant parameters in a four-channel topology. The circuit presented in this is an Analog board composed of amplifiers, filters, and analog-to-digital converters, which are all susceptible to drifts.

The Analog board is shown in Fig. 1. Four SMA connectors on the left side of the board receive signals from four pick-up electrodes and connect to the quasi-crossbar switch. Each of the four parallel channels following the quasi-crossbar switch consists of a variable gain RF chain followed by the AD9433 analog-to-digital converter from Analog Devices. All the measurements presented in this article were performed on this kind of board.

CP732, *Beam Instrumentation Workshop 2004: Eleventh Workshop,*
edited by T. Shea and R. C. Sibley III

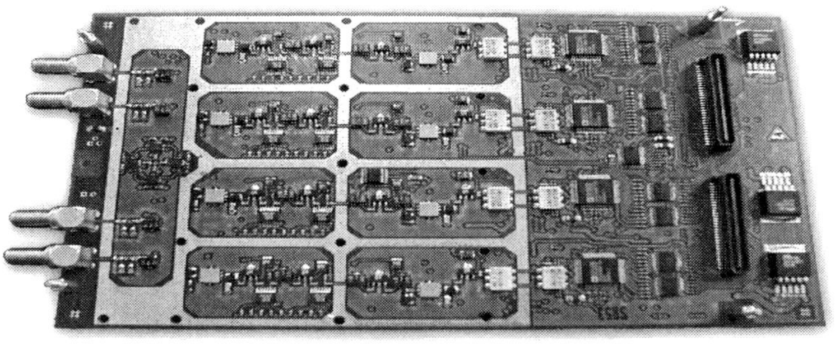

FIGURE 1: Top view of the analog board.

Theoretical Background of the RF Quasi-Crossbar Switch

Figure 2 shows a block diagram of the quasi-crossbar switch that is the key component of this new concept. The quasi-crossbar switch works as a multiplexer, which can direct any of the four inputs to any of the four RF processing chains.

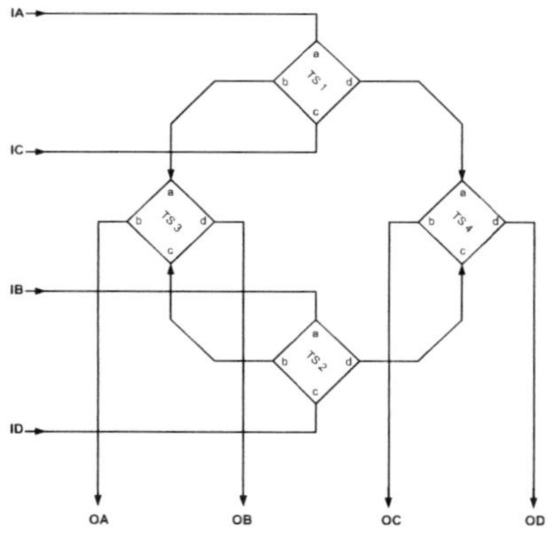

FIGURE 2: Block diagram of the quasi-crossbar switch.

Table 1 shows a multiplexing scheme, which consists of a group of 4 switch patterns out of 16 possible. Once the four combinations have been cycled through, we obtain a set of four signals on each channel output.

TABLE 1: Group of 4 switch patterns out of 16 possible.

Input/Switch Position	Position 1	Position 2	Position 3	Position 4
A	**B**	**A**	**C**	**D**
B	**A**	**B**	**D**	**C**
C	**D**	**C**	**A**	**B**
D	**C**	**D**	**B**	**A**

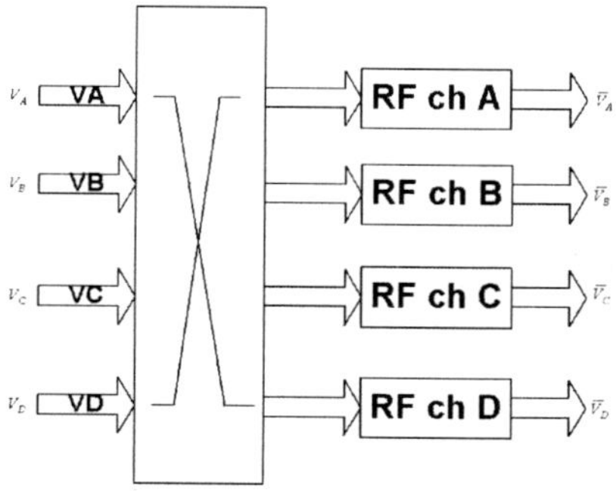

FIGURE 3: Theoretical background.

Referring to a simplified model of the Analog board, as shown in Fig. 3, the following equations can be derived from Eqs. (1) – (4). At the output of every channel we can measure four signals as follows.

On channel A output:

$$V_A \cdot G_B \cdot e^{j\varphi B}, V_A \cdot G_A \cdot e^{j\varphi A}, V_A \cdot G_C \cdot e^{j\varphi C}, V_A \cdot G_D \cdot e^{j\varphi D} \tag{1}$$

On channel B output:

$$V_B \cdot G_A \cdot e^{j\varphi A}, V_B \cdot G_B \cdot e^{j\varphi B}, V_B \cdot G_D \cdot e^{j\varphi D}, V_B \cdot G_C \cdot e^{j\varphi C} \tag{2}$$

On channel C output:

$$V_C \cdot G_D \cdot e^{j\varphi D}, V_C \cdot G_C \cdot e^{j\varphi C}, V_C \cdot G_A \cdot e^{j\varphi A}, V_C \cdot G_B \cdot e^{j\varphi B} \tag{3}$$

On channel D output:

$$V_C \cdot G_D \cdot e^{j\varphi D}, V_C \cdot G_C \cdot e^{j\varphi C}, V_C \cdot G_A \cdot e^{j\varphi A}, V_C \cdot G_B \cdot e^{j\varphi B} \tag{4}$$

A new value for each channel output, proportional to the associated input signal, can be introduced representing a weighted average of all four paths for each input signal and are derived in Eqs. (5) – (8) below. The weights are arbitrary complex gains (magnitude gain and phase shift introduced by every channel) of every channel.

$$\overline{V}_A = V_A \cdot \frac{(G_B \cdot e^{j\varphi B} + G_A \cdot e^{j\varphi A} + G_C \cdot e^{j\varphi C} + G_D \cdot e^{j\varphi D})}{4} \tag{5}$$

$$\overline{V}_B = V_B \cdot \frac{(G_A \cdot e^{j\varphi A} + G_B \cdot e^{j\varphi B} + G_D \cdot e^{j\varphi D} + G_C \cdot e^{j\varphi C})}{4} \tag{6}$$

$$\overline{V}_C = V_C \cdot \frac{(G_D \cdot e^{j\varphi D} + G_C \cdot e^{j\varphi C} + G_A \cdot e^{j\varphi A} + G_B \cdot e^{j\varphi B})}{4} \tag{7}$$

$$\overline{V}_D = V_D \cdot \frac{(G_C \cdot e^{j\varphi C} + G_D \cdot e^{j\varphi D} + G_B \cdot e^{j\varphi B} + G_A \cdot e^{j\varphi A})}{4} \tag{8}$$

From the above it can be seen that by implementing the described scheme, all the input signals are multiplied by the same complex transfer function:

$$\overline{G} = \frac{(G_B \cdot e^{j\varphi B} + G_A \cdot e^{j\varphi A} + G_C \cdot e^{j\varphi C} + G_D \cdot e^{j\varphi D})}{4}.$$

When applying the averaged values $(\overline{V}_A, \overline{V}_B, \overline{V}_C, \overline{V}_D)$ to the $\frac{\Delta}{\Sigma}$ equation the factor $\overline{G}$ can cancel out as shown in Eqs. (9) and (10):

$$x = K\frac{(\overline{V}_A + \overline{V}_D) - (\overline{V}_B + \overline{V}_C)}{(\overline{V}_A + \overline{V}_B + \overline{V}_C + \overline{V}_D)} = K\frac{(V_A + V_D) - (V_B + V_C)}{(V_A + V_B + V_C + V_D)}, \tag{9}$$

$$y = K\frac{(\overline{V}_A + \overline{V}_B) - (\overline{V}_C + \overline{V}_D)}{(\overline{V}_A + \overline{V}_B + \overline{V}_C + \overline{V}_D)} = K\frac{(V_A + V_B) - (V_C + V_D)}{(V_A + V_B + V_C + V_D)}. \tag{10}$$

The calculations show that the beam position is independent of the asymmetry of individual channel gains.

MEASUREMENTS

The following paragraph presents the effectiveness of the described method. We performed long-term stability, temperature stability, and beam vs. current dependence tests on the Analog board using the RF quasi-crossbar switch concept as described above.

Long Term Stability

The stability of position readings of the X and Y measurements during the night was measured. The estimated ambient temperature change during the experiment was 4 degrees Celsius. The plot shows the effectiveness of the quasi-crossbar switch scheme, which can be evaluated by comparing the dotted and full lines. The measurements were referenced to zero for convenience.

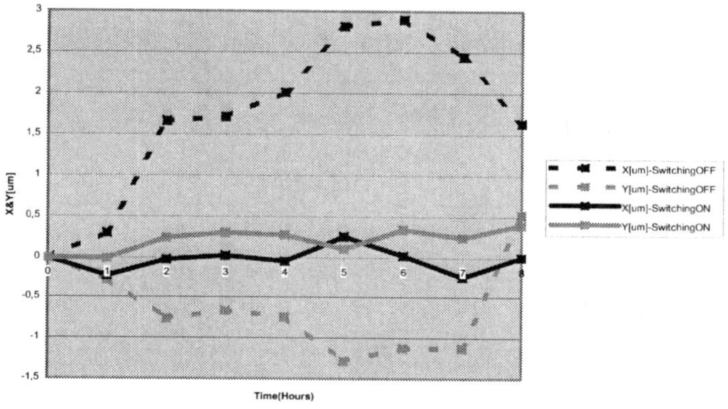

FIGURE 4: Preliminary eight-hour stability test.

Temperature Stability

The analog board was placed in a temperature-controlled chamber. X and Y were recorded as a function of temperature. The temperature was swept from 10°C to 35°C. The measurements were referenced to zero for convenience.

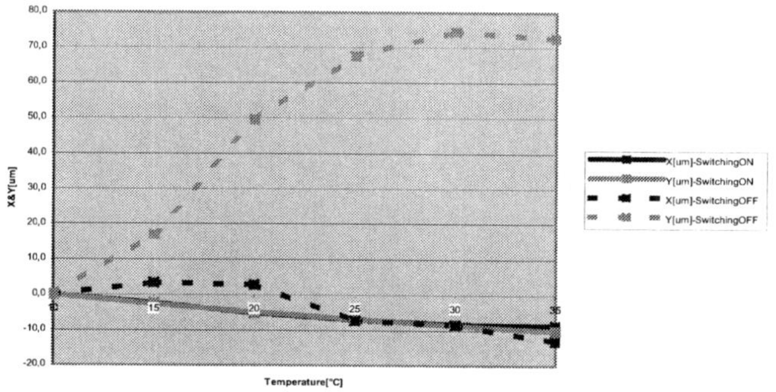

FIGURE 5: Preliminary temperature stability test.

Beam vs. Current Dependence

The plot below shows X and Y position readings versus input signal power measured. The input power to the electronics was swept from –8 dBm to –22 dBm. The measurements were referenced to zero for convenience.

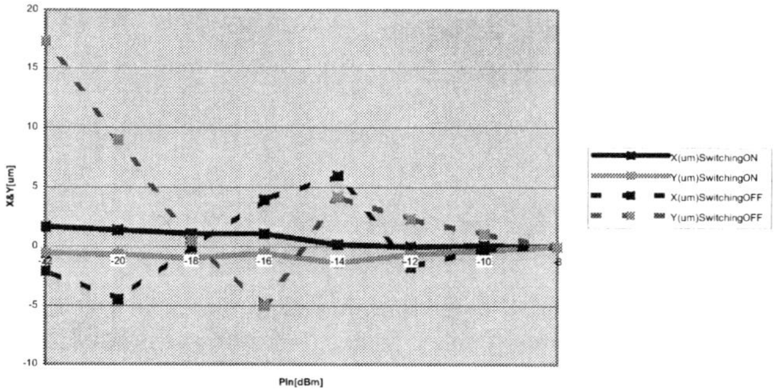

FIGURE 6: Preliminary beam current dependence test

CONCLUSIONS

In this article we demonstrated effectiveness of the quasi-crossbar switch. The equations derived show complete immunity to any asymmetry between channels after applying the quasi-crossbar switch method. The performance of the Analog board is being optimized. For this reason the following measurements have to be taken as preliminary; however, the results already show significant improvements. Last but not least, the switching frequency is also very important; if it is too low, the transfer function of individual channels may change during the process and two different inputs will not see the same gain in the same channel. On the other hand the switches have to be slow enough not to introduce interference into the system. This method and the device for carrying out the described method are patent pending.

REFERENCES

1. U. Mavrič, "An Interface for Direct RF Sampling," Thesis at Faculty for Electrical Engineering in Ljubljana, 2002.
2. R. Uršič, "Method for the Precise Measurement of Dependency on Amplitude and Phase of a Plurality of High Frequency Signals and a Device for Carrying out Said Method," patent pending.
3. www.i-tech.si.

Large-Scale PC Management and Configuration for SNS Diagnostics*

Darryl J. Murphy and J. David Purcell

ORNL, Oak Ridge, TN 37830, USA

Abstract. The Spallation Neutron Source (SNS) project's diagnostics group has begun its implementation of more than 300 PC-based Network Attached Devices (NADs) [2]. An implementation of this size creates many challenges, such as distribution of patches and software upgrades; virus/worm potentials; and the configuration management, including interaction with the SNS relational database. As part of the initial solution, a base operating system (OS) configuration has been determined and computer management software has been implemented. Each PC requires a unique configuration, but all are based on a common OS and supporting applications. The diagnostics group has started with an implementation of an XP Embedded (XPe) OS and uses Altiris® eXpress Deployment Solution™ [3]. The use of XPe and Altiris gives the diagnostics group the ability to easily configure, distribute, and manage software on a large scale. This paper describes the initial experience and discusses plans for the future.

INTRODUCTION

The SNS project is a partnership involving six U.S. Department of Energy (DOE) national laboratories: Argonne, Brookhaven, Jefferson, Los Alamos, Lawrence Berkeley, and Oak Ridge. The project is described in more detail on the SNS Web site [4].

The SNS diagnostics group is in the process of installing more than 300 PC-based NADs. Several factors were considered in determining an approach to the configuration management of the PCs. Although many of the devices are similar in application (e.g., there will be 157 Beam Position Monitors), the time scale of PC acquisition assured us of quite different base hardware configurations. Modifications for future implementations as well as deployed applications will occur as the accelerator goes through its commissioning phases. OS updates will be required. Finally, it is expected that time will affect the PC's hardware, and replacement of parts or entire new chassis will be needed.

It was obvious very early on that tight control had to be exercised to successfully manage such a large, diverse group of PCs. With a challenging objective of potential push-button PC imaging and updates, it was determined that tools like Altiris and an XPe OS would lead us to the configuration control required.

---

*Work supported by the U.S. Department of Energy under contract DE-AC05-00OR22725.

CP732, *Beam Instrumentation Workshop 2004: Eleventh Workshop*,
edited by T. Shea and R. C. Sibley III

DEPLOYMENT AND TOOLSET

Altiris Deployment Solution

Altiris Deployment Solution provides OS deployment, configuration, and software deployment across hardware platforms and OS types. We leveraged the SNS Technical Information Systems Group's experience with Altiris Deployment Solution in their environment, which includes approximately 500 desktops and laptops, and tailored the PC management solution to our PC-based NADs. Altiris Deployment Solution is a client/server PC management tool that has many features, some of which include the ability to create disk images of the PC's hard disk and to distribute these images across the network to a NAD that is configured to Wake-on-LAN and PXE boot. This can be done to one or more of these devices simultaneously. With our current OS, this process, regardless of the number of machines being configured, takes about 15 minutes. Figure 1 shows how many PCs can be stacked and imaged at one time. This software also allows us to deploy software, run specific scripts and configure each device from one console. Altiris gives us the ability to take a PC from its box and, assuming the supplier has set up the bios correctly, set it in a rack where it will boot and receive its appropriate image.

Windows Server 2003

Windows Server is required for the Altiris software. We have installed it on a dual processor 2.8-GHz server with 2-GB RAM and 140 GB of storage for disk images. This configuration exceeds initial hardware requirements and allows for future growth.

Windows XP Embedded

Windows XP Embedded is an operating system that is based on a subset of components derived from the Windows OS. Based on the same binaries as Windows XP Pro, XPe enabled us to choose the individual components needed to achieve optimum functionality. XPe was also chosen as the OS platform because of its ability to run with only the needed services and because of the small footprint on disk. Currently the embedded base OS stands at approximately 300 MB compared to approximately 1 GB for the basic XP Pro installation. This smaller image eliminates potential network security concerns as well as enables better utilization of system resources (i.e., memory). Using the Task Manager, you can see that the number of processes running and the memory required, although varying, is half that of the standard XP Pro installation. Conveniently, this lean image also speeds up the imaging process.

FIGURE 1. XPe Image Deployment Using Altiris.

Support Structure [ORACLE, Controls, or ORNL Implementation of Network Registration (NetReg)]

Network security is important to the operation of both the accelerator and to these devices. The Diagnostics group cannot allow others uncontrolled access to the network or these machines. The SNS Controls group has implemented NetReg, a process that requires all devices that use the network to go through a registration and scan procedure. The use of NetReg ensures that the network does not have devices attached to it that are not meant to be. It also scans devices prior to being attached to make sure they are virus free and properly patched.

Backup

The images are stored on a RAID 5 Storage Array and backed up to tape.

CONFIGURATION PROCESS

Determine Base Configuration

The deployment of all the diagnostic PCs is based on a standard base configuration. Instead of using XP Pro or Windows 2000 as an OS, we have used the components of XP Pro to make up our embedded OS. Determining baseline components is an ongoing process that starts with the analysis of a functioning complete OS on your planned hardware. We analyzed functioning systems using XPe tools and use only those XP Pro components that are needed. As work continues, it becomes obvious that certain excluded components are needed and that some included components can be omitted. An example of this includes the use of specific control panels. These control panels are not necessary to a running OS but are useful on our machines. At this point, we have a base list of components that provides an OS capable of not only running the required device applications but also directly supporting development.

As the base OS becomes stabilized, other standard tools, settings, or conveniences are added to the base configuration as required. These include major applications like LabVIEW, Altiris, Timbuktu, and the EPICS toolsets. Settings included are the EPICS environment variables and some basic windows settings. Other conveniences include a program that identifies which computer you are on, shortcuts to network Web pages (logbook), and a list of IP addresses for all machines that can be used with Remote Desktop or Timbuktu.

Cloning Process–Distribution

During the distribution of the devices, one has to consider the fact that the base configuration is going to be greatly distributed but that the security IDs have to be different for each implementation of XPe. One of the components included with XPe is the System Cloning Tool. This component is used to ensure that each device has a base image containing a unique computer security ID. The use of the tool gives us the ability to use Altiris to distribute images without having to touch each machine. The following image gives an example of the process.

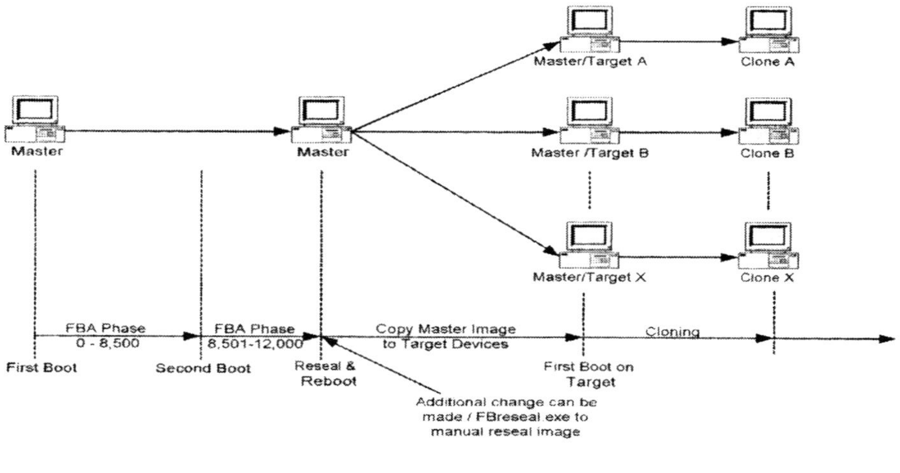

FIGURE 2. Cloning Process XP Embedded [5].

Database Interaction

One problem with deploying PCs and allowing them to boot and receive an appropriate image is that a unique security ID does not identify the device and that somehow an accelerator device ID has to be assigned to the machine. Because the PCs have to be registered with the network to function, the IP and MAC addresses are known. At boot up time, a Java application using the JDBC thin client accesses the SNS Oracle production database to retrieve the configuration file for the Input Output Controller. The application uses a global read-only account to access the IP address of the IOC making the connection. This IP is contained in the database and is used to look up the IOC record in the IOC_DVC table. The application then writes the configuration file from the IOC_DVC table into a directory and file name, which are also specified in the IOC_DVC table. No software other than the Java Runtime Environment is required to be installed on the IOC for this process.

CONTROL OF DEPLOYED DEVICES

Because the NADs are PC-based, many changes will have to be made to the systems over the life of the machine. This toolset allows us to maintain control of the devices while keeping them up to date with current applications and hardware. Because XPe is based on the Windows XP OS, it is necessary to have the ability to apply security patches to the devices in operation. To keep device downtime to a minimum, the patches will only be applied when appropriate, but we still need the ability to quickly patch our devices in the event of an emergency. Windows XP Embedded SP1 includes a method for this called Device Update Agent (DUA). There are at least a couple of ways to configure DUA. Because we have hard disks in our devices, we chose to copy the updated files to the specified directory using an Altiris Job. This "job" can be dropped on one or many systems simultaneously. Once the files are in place, DUA takes over the task of installing the patch. DUA is in a constant wait mode and does nothing until the specified directory is "loaded" with the appropriate files.

Similar to Microsoft patches, the deployed devices will also require updated applications or additional software. Again, this can be accomplished in a couple of different ways. DUA can be triggered to run an executable that has been loaded on to the machines via Altiris. Altiris can also deploy updates/additions in a manner similar to any type of file management.

Many of these deployed devices will also have to undergo hardware changes because of failures or upgrade needs. Again, Altiris allows the software changes to be handled with relative ease. Altiris can take a known configuration and monitor the changes made to it during any kind of upgrade. This is known as a Rapid Install Package (RIP). Upon completion of the upgrade, the RIP can be applied to any similarly configured devices.

Using Altiris as the management tool of the deployed images also gives us the ability to rapidly redeploy a device's current configuration to replacement hardware. As with any new configuration, applying the OS to a device takes under 20 minutes.

INITIAL EXPERIENCE AND FUTURE PLANS

Initial Experience

Our initial exposure to the use of these tools and their implementation required little learning time. This was particularly true when it came to Altiris. Previous SNS use of this application allowed us to quickly deploy it. It was not obvious to us at all during the start that XPe was going to be of any use at all. XPe was initially a very cumbersome tool. Initially, it required some patience and it took us time to ramp up and use it effectively. However, after using it for a short time, we began to see the base OS take shape and the designer tools and components have become easy to use.

Overall, the combination of these two major components have given us the ability to safely deploy an XP Pro-like image, maintain configuration control, and easily update images. The time required to deploy an image is far shorter than using standard installation methods. The OS runs just as an XP Pro image but without any unnecessary overhead. And because of this process, our standard NAD is configured in a way that allows us to comfortably continue with our installation of diagnostic systems.

Future Plans

One future idea is to move to a diskless solution. This would allow the NADs to boot up and operate from a unique disk image file located on a networked server or servers. The OS would still be Windows XP Embedded and would allow for quick (about the same as hard disk) boot times. This solution will be considered because of the gain in the ability to deploy new images to the same hardware. Before completely converting to a diskless deployment, diskless operation would provide a reliable backup system in case of hard disk failure. Currently we are evaluating a product called BXP from Venturcom [6] that can give us this ability.

The interaction with the relational database gives us the opportunity to expand the database role and use it to capture and house device images as a whole. Although Altiris provides a simple way to handle images and 300-400 images is not that great of a number overall, the relational database naturally lends itself to be the repository of all data related to a device.

Because these machines are configured with an XP Pro-like image, we have included the Remote Desktop Feature. This has given us the ability to log on to the machines over the network and has not required the use of monitors within each rack. However, XPe allows the devices to be headless (no graphical interface). Using a headless configuration may give us increased stability and/or increase the efficiency of the operating system.

REFERENCES

1. W. Blokland, T. Shea, and M. Stettler, "Network Attached Devices at SNS," DIPAC 2003.
2. http://www.sns.gov/.
3. http://www.altiris.com/products/deploymentsol/.
4. S. D. Liming, "Windows XP Embedded Advanced," RTC Books, 2003.
5. http://www.vci.com.

Developmental Status of Beam Position and Phase Monitor for PEFP Proton Linac

Sungju Park, Jangho Park*, Inha Yu, Dotae Kim, Jung-Yun Hwang, and Sanghoon Nam

Pohang Accelerator Laboratory, POSTECH, Pohang 790-784, Korea
** Physics Department, POSTECH, Pohang 790-784, Korea*

Abstract. The PEFP (Proton Engineering Frontier Project) at the KAERI (Korea Atomic Energy Research Institute) is building a high-power proton linear accelerator aiming to generate 100-MeV proton beams with 20-mA peak current. (Pulse width and max. repetition rate of 1 ms and 120 Hz respectively.) We have developed the Beam Position and Phase Monitor (BPPM) for the machine that features the button-type PU, the full-analog processing electronics, and the EPICS-based control system. The beam responses of the button-type PU have been obtained using the MAGIC (Particle-In-Cell) code. The processing electronics has been developed in collaboration with Bergoz Instrumentation. In this article, we report the present status of the system developments except the control system.

INTRODUCTION

The PEFP linac is one of the HPPAs (High-Power Proton Accelerators) that are under construction at the KAERI in Korea. Presently it aims to achieve 100-MeV beam energy with 20-mA (peak, 1-ms macro-pulse duration) beam current. Refer to Table 1 for major beam parameters of the PEFP accelerator.

TABLE 1. Beam parameters of the PEFP/KAERI Accelerator

	Operation Mode (Pulse)	note
Beam Energy	3 - 20 - 100 MeV	
β	0.08 – 0.20 – 0.43	BPM operation region
γ	1.003 – 1.021 - 1.107	
Beam Current	20 mA	Peak value of macro pulse
Macro Pulse Width	0.1 - 1 ms	
Bunch Length (FWAB)	200 ps	PARMILA simulation result

Front-ends of modern HPPAs are very complex and have tight installation spaces, especially for beam-diagnostic devices. Therefore compact beam-diagnostic devices are highly demanding for the HPPAs. PUs (Pick-Ups) for beam-position monitors of proton machines usually have been striplines because of their high sensitivity for weak-intensity proton beams. With other kinds of PUs, buttons are simple and reliable but their applications to proton machines have been limited because of their insufficient sensitivity. Modern HPPAs such as the PEFP linac are designed to have very high beam intensities, so that even the buttons could generate enough signals for

CP732, *Beam Instrumentation Workshop 2004: Eleventh Workshop*,
edited by T. Shea and R. C. Sibley III

precision beam position measurements. In this context, we have chosen the button-type PU for use in the PEFP accelerator. One of the disadvantages of the buttons is that it is difficult to predict the PU sensitivity using analytic formulas [1, 2]. In fact, the PU sensitivity for low-beta beams cannot be practically determined even by experimental methods, due to the difficulty of simulating electromagnetic fields of the low-beta beams. In this regard, we have decided to utilize the computer code for determining the sensitivity of the button-type PU. We have chosen the MAGIC code, which is a kind of PIC (Particle-In-Cell) code and can treat the particle and electromagnetic system in the full three-dimensional manner.

BPPM SYSTEM DEVELOPMENT

Numerical Modeling of PU

The theoretical estimation of the sensitivity of a stripline PU was established by R. E. Shafer [1]. It is basically 2D theory and cannot be directly applied to designing the button-type PUs, which are 3D features. There are commercial electromagnetic codes that can simulate 3D geometries, including the MAFIA T3. But the MAFIA T3 can handle only ultra-relativistic particles and cannot be applied to low-beta beams. (The port boundary of the MAFIA T3 can accept only TEM mode whose phase velocity is equal to that of the light. And low-beta beams generate fields that are not in the simple TEM mode.) Limitations of Shafer's theory and the MAFIA code have led us to consider using the MAGIC code. It is versatile and can handle arbitrary combinations of particle beams and electromagnetic structures.

Figure 1 shows the 3D model of the button PU for the MAGIC simulation and its cross-sectional view in the transverse plane. In the left figure of Fig. 1, proton bunches travel in a left-to-right direction. Beam signals are coupled to the four buttons that are installed around the circumference of the beam pipe. Buttons are connected to 50-ohm coaxial lines whose dimensions are different from those of usual fabrication. This simplified the simulation geometry without sacrificing the reliability.

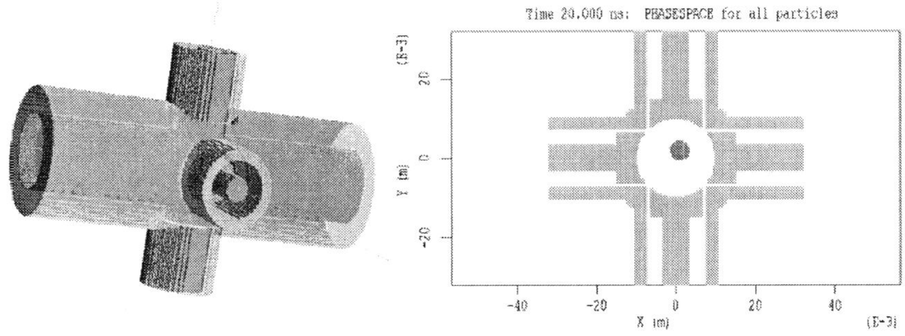

FIGURE 1. Modeling of four-button-type PU for MAGIC simulation. Left figure: 3D model, right figure: cross-sectional view. Dot inside beam pipe in right figure indicates proton beam. Diameter of beam pipe is 20 mm.

Default beam parameters used in the simulation are shown in Table 2. The left figure of Fig. 2 is the voltage waveform developed between the inner and outer conductors of the coaxial line. As expected, they are considerably longer than the beam bunch length, which is due to the axial extension of fields from the low-beta beams. (β = 0.08 for E = 3 MeV.) With increasing the beam energy, beam fields concentrate in the transverse plane, and the pulse widths of signals coupled to buttons are shortened approaching that of charge distribution in the beam. This is shown in the right figure of Fig. 2, which is the signal waveform for the beam energy of 100 MeV.

TABLE 2. Default beam parameters used in MAGIC simulation.

Beam Energy	3 MeV
Average (Peak) Beam Current	20 (570) mA
Bunching Frequency	350 MHz
Bunch Length	43 ps rms

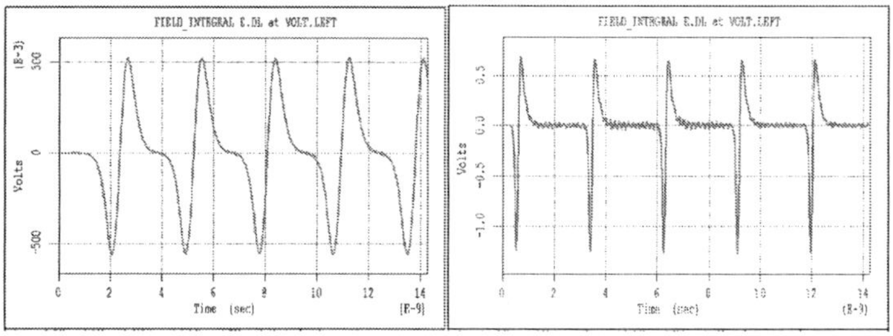

FIGURE 2. Waveforms of the voltage signal developed at the end of one of the coaxial lines for beam energies of 3 MeV (left) and 100 MeV (right).

The contour plot of the longitudinal electric field (E_z) and the vector plot of (E_x, E_z) are shown in Fig. 3. They clearly show the longitudinal extension of the field due to the low-beta beam.

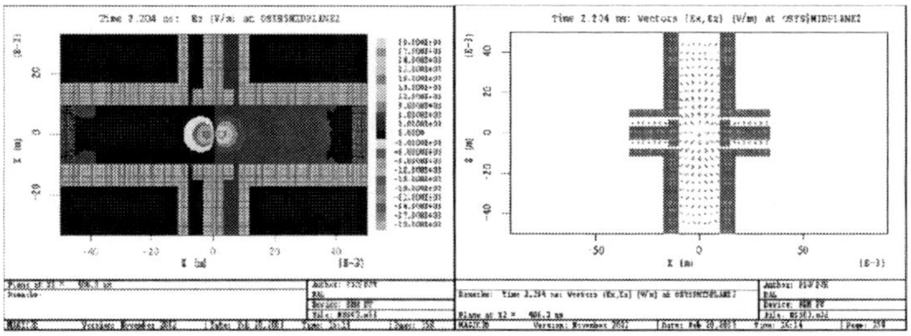

FIGURE 3. Contour (left) and vector (right) plots of electric fields showing their longitudinal extension. Right figure is rotated by 90° with respect to left one.

The dependence of the PU sensitivity on the beam energy was simulated with the result shown in Fig. 4 (left figure). As expected from Shafer's theory, the sensitivity increases dramatically with decreasing the beam energy. It also increases with increasing frequency at low energies but converges to the same value at high energies. The right figure of Fig. 4 shows the simulated dependence of signal power on beam energy. It matches well to the analytic calculation that was done by considering the effects of bunch length, cable attenuation, and PU impedance [3].

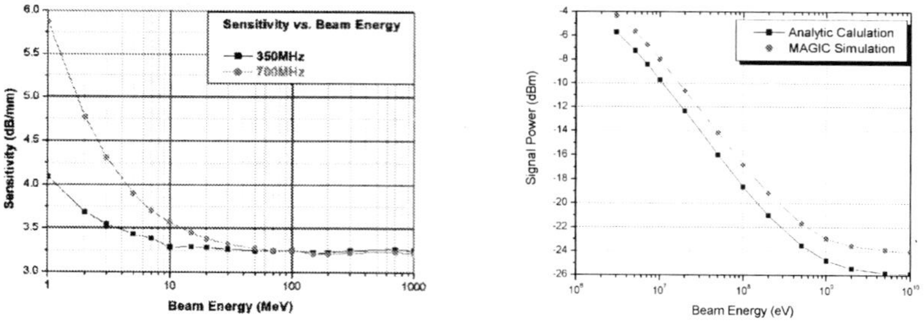

FIGURE 4. Dependence of PU sensitivity (left) and signal powers (right) on beam energy.

Sensitivity of the PU to beam position change was simulated by moving the beam in the transverse plane in the step of 1 mm. The sensitivity map was obtained by plotting the delta-over-sum values of band-pass filtered voltage signals appearing at the ends of the coaxial lines. The center frequencies and bandwidth of the band-pass filter were 350 or 700 MHz and 20 MHz respectively. See Fig. 5 for the sensitivity maps for the frequencies of 350 and 700 MHz. Note that the sensitivity for 700 MHz was about 7.5% larger than that for 350 MHz.

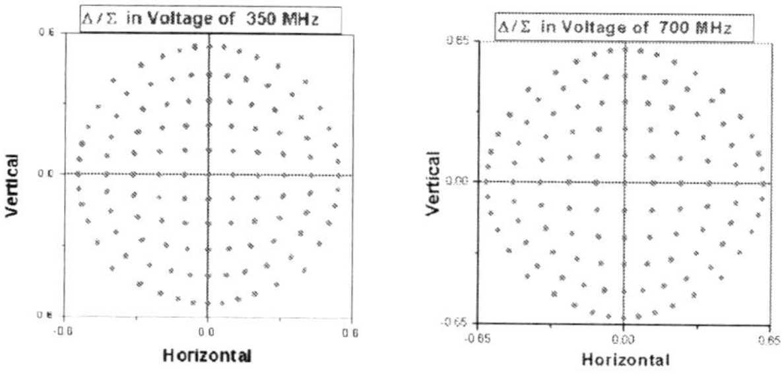

FIGURE 5. Sensitivity map of button PU at frequencies of 350 and 700 MHz. Beam pipe diameter = 20 mm. Button diameter = 12 mm. Beam energy and beta are 3 MeV and 0.08 respectively.

PU Fabrication

A prototype PU that consists of four 12-mm diameter buttons, four N-type coaxial feedthroughs, and 20-mm diameter beam pipe has been fabricated at Hitachi Electronics in Japan. The N-type feedthroughs with ceramic windows in them provide at least 2-GHz bandwidth. The PU is equipped with quick-installation type vacuum flanges, which would provide easy performance testing in the accelerator beamline. See Fig. 6.

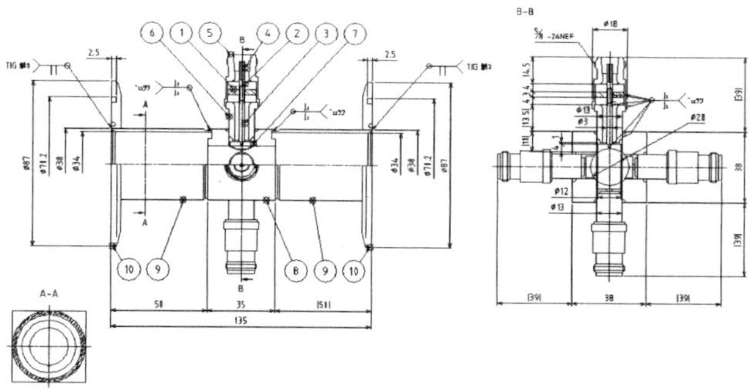

FIGURE 6. Fabrication drawing of the BPPM PU.

Development of BPPM Electronics

The BPPM electronics has been developed by Bergoz Instrumentation in accordance to our specifications. It is a full-analog processor that is based on the Bergoz Log-Ratio BPM electronics. A mezzanine board for measuring beam phase with respect to the RF reference is mounted on the LR BPM. In order to realize very low drifts in the phase detection circuit, simple low-pass filters were adopted. In addition the four button signals are combined and compared with the RF reference input, in order to minimize the dependence of the phase detection on the transverse beam position. Figure 7 is the developed BPPM electronics with the phase-detection mezzanine board shown in the upper-right corner.

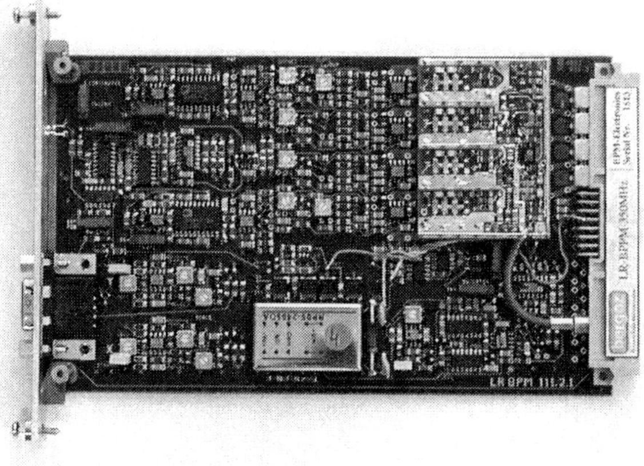

FIGURE 7. Developed BPPM electronics; phase-detection mezzanine board is shown in the upper right corner.

Required specifications for the BPPM electronics are shown in Table 3. It should provide the resolution and stability < 50 μm (at 20-mmφ beam pipe) for the position measurement and < ± 1° in the phase measurement. Since the PEFP linac was designed to operate in both CW and pulsed modes, one of the most important requirements for the BPPM electronics was that it should provide position and phase measurements for the two operation modes of the accelerator. The large bandwidth of the log-ratio position measurement scheme automatically satisfied the requirement. Transient response of the phase measurement circuit showed enough speed for the requirement.

TABLE 3. Specifications for the BPPM electronics.

	Position	Phase
Operation Frequency	350 MHz	←
Minimum Signal	-57.5 dBm @ 100 MeV	←
Powers	-4.56 dBm @ 3 MeV	←
Resolution & Stability	< 50 μm @ 20-mmφ beam pipe	< ± 1 degree
Operation Modes	CW or pulsed	←
Bandwidth of Output Signal	> 5 MHz	←

Figure 8 shows the response of the BPPM electronics to pulsed RF inputs to the position and phase circuits. The upper trace is the envelope of the pulsed RF (~ 2 ms) with its frequency of 350 MHz, the middle and lower ones are the position and phase detection signal outputs respectively. The RF reference signal input to the phase circuit is in CW. Figure 9 is the linearity performance of the phase circuit, with sinusoidal (single-frequency) signal inputs. The linearity error is less than ± 0.5°. The circuit also provides phase-detection resolution (noise rms) < 0.1° for signal input > -20 dBm. Phase error over 20-°C temperature variation is less than 0.5°.

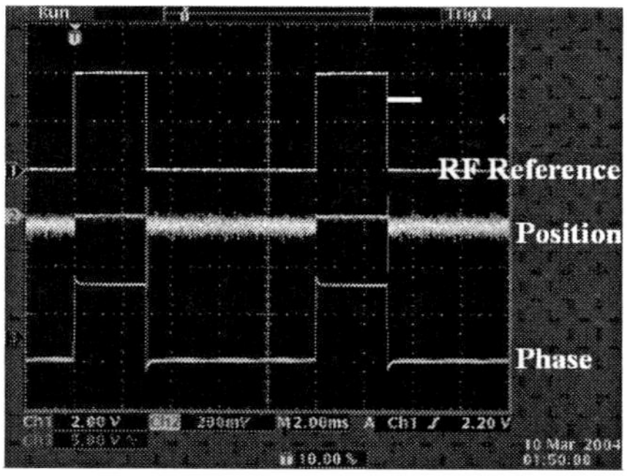

FIGURE 8. The BPPM electronics measured phase and position signal. The upper signal is a RF reference signal, the middle is a beam position signal, and the bottom is a beam phase signal.

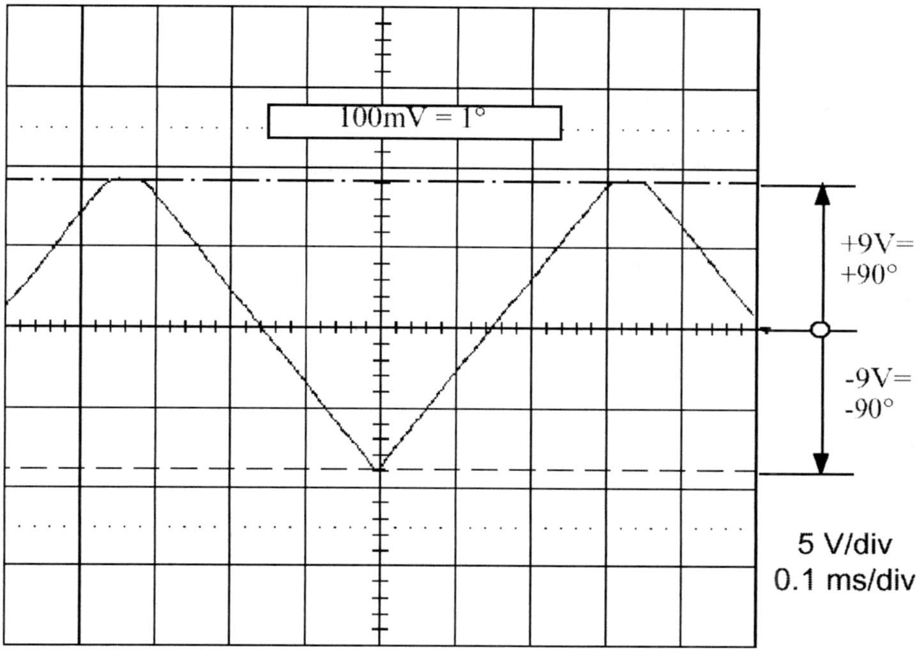

FIGURE 9. Linearity performance of the phase detection circuit for single-frequency signal inputs.

The linearity performance of the phase circuit was also tested in the PLS (Pohang Light Source), the 3rd generation light source in Pohang, Korea. The bunching frequency of the circulating beam in the PLS is 500 MHz and this is higher than the nominal operating frequency of the BPPM electronics. But since the cutoff frequency of the LPFs in the phase circuit front-end is roughly 700 MHz, the circuit operated well with the PLS beam. One uncertainty was on the bunch length. Since the PLS is an electron machine, the bunch length is very short (~ 25 ps rms). Button signals from this short bunch contain a very wide band spectrum and could cause differences in the BPPM performances to the test-bench tests with single-frequency inputs. Figure 10 is the test result of linearity error with the PLS beam signals. Note the increased linearity errors. Another test in the PEFP linac will be done in near future.

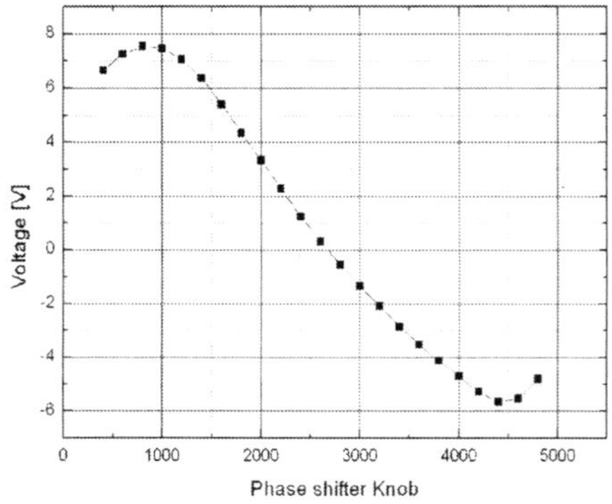

FIGURE 10. Linearity performance of the BPPM phase circuit with PLS beam signals.

CONCLUSION

We have developed the BPPM (Beam Position and Phase Monitor) system for the PEFP (Proton Engineering Frontier Project) proton linear accelerator. We conclude this article with the following summarizing remarks:

1. A button-type PU was designed utilizing the MAGIC code and a prototype one with N-type feedthroughs was fabricated.
2. Processing electronics based on the existing Log-Ratio BPM circuit was developed in collaboration with Bergoz Instrumentation. The phase detection circuit was integrated into a small mezzanine board that was mounted on the LR BPM circuit.
3. The phase circuit of the BPPM electronics can provide linearity error < ± 0.5°. Noise rms of the circuit is < 0.1° for signal input > -20 dBm. Phase error over a 20-°C temperature variation is less than 0.5°. See Ref. 4.
4. The BPPM electronics can be operated in both CW and pulsed (~ 1 ms) modes.

ACKNOWLEDGMENTS

The authors would like to appreciate the reliable cooperation by Bergoz Instrumentation in developing the full analog BPPM electronics. This work was supported by the PEFP (Proton Engineering Frontier Project), KAERI (Korea Atomic Energy Research Institute), Korea.

REFERENCES

1. R. E. Shafer, "Sensitivity of PUs for Low Beta Beams," Proc. of the BIW93.
2. S. J. Park et al., "Design of BPM PU for Low-Beta Proton Beam Using MAGIC Code," Proc. of the DIPAC03.
3. PEFP Annual Report, "Development of Beam Diagnostics for High-Power Proton Accelerator," June, 2003.
4. J. Bergoz, "Performance of BPPM Electronics Developed for PEFP Linac," Project Report, Bergoz Instrumentation, December, 2003.

Advantages of Real-Time Spectrum Analyzers in High-Energy Physics Applications

Louis Parker

Tektronix, Inc., 14200 SW Karl Braun Drive 500, Beaverton, OR 970077, USA

Abstract. Typically, particles are injected into the ring at low energy levels and then "ramped up" to higher levels. During ramping, it is important that the horizontal and vertical tune frequencies do not shift, lest they hit upon a resonant combination that causes beam instability or sudden total loss of ring beam current (beam blow up). Beam instabilities can be caused by a number of factors. Non-linearities and/or different response times of independent controls such as beam position monitor (BPM) cables and circuits, magnets for guidance and focusing of the beam, Klystrons or Tetrodes (which provide power to RF cavities that transmit energy to the beam), and vacuum pumps and monitors can all cause beam instabilities. Vibrations and lack of proper shielding are other factors. The challenge for operators and researchers is to correctly identify the factors causing beam instabilities and blow up so that costly accelerator time is not interrupted and experimental results are not compromised. The instrument often used to identify problems in particle accelerator applications is the spectrum analyzer. This paper will discuss the advantages of real time spectrum analyzers (RSA) versus swept frequency spectrum analyzers in HEP applications. The main focus will be on monitoring beam position and stability, especially during ramp-up. Also covered will be use of RSA for chromaticity measurements, Phase Locked Loop (PLL) diagnostics, and vibration analysis.

INTRODUCTION

In High-Energy Physics (HEP) applications the fundamental challenges facing operators and researchers are maintaining maximum beam power and consistent beam stability in particle accelerators. To accomplish these objectives, operators need to be able to collect information on the spectral behavior of particle beams at the time of beam ramp up as well as during various experiments.

In this application note, the causes of beam instability and "blow up"–sudden total or partial loss of beam current–will be reviewed. We'll discuss the advantages of using Real-Time Spectrum Analysis (RTSA) to analyze and solve beam loss and instability problems versus the use of Swept Spectrum Analysis (SSA). And we'll show how Tektronix Real-Time Spectrum Analyzers are used to collect, analyze, and present data in the High Energy Physics field.

Tektronix Real-Time Spectrum Analyzers have been helping HEP researchers solve beam instability issues since the widely accepted 3052 and 3054 Real Time Spectrum Analyzers were introduced over a decade ago. Today, Tektronix continues its commitment to the particle accelerator community with the latest RTSA offerings–the RSA2200 and RSA3300 series–with even more benefits for HEP research and

CP732, *Beam Instrumentation Workshop 2004: Eleventh Workshop*,
edited by T. Shea and R. C. Sibley III

maintenance. These RSA products combine the advantages of vector signal analysis (VSA), time domain analysis, and spectrum analysis in a single instrument.

CAUSES OF BEAM INSTABILITY AND "BLOW UP"

In a storage ring, particles (protons, electrons, or ions) are injected into a stable orbit that allows them to circulate for many hours. This orbit of particles is known as the "beam." The beam is maintained in the storage ring at a certain velocity until such time as it is "kicked out" to the Linear Accelerator (LINAC). During the storage time, the beam's velocity, path, coherence, and chromaticity are monitored and adjusted using magnets and RF fields. It is during this portion of the beam life cycle that it is most susceptible to errors and problems.

The beam orbit in the storage ring has design values called horizontal, vertical, and longitudinal "tune" frequencies, which are directly related to the beam's position in the ring. The electrical characteristics of the focusing magnets and RF cavities around the ring determine these values.

Typically, particles are injected into the ring at low energy levels then "ramped up" to higher levels. During ramping, it is important that the horizontal and vertical tune frequencies do not shift, lest they hit upon a resonant combination that causes beam instability or sudden total loss of ring beam current (beam blow up).

Beam instabilities can be caused by a number of factors. Non-linearities and/or different response times of independent controls such as beam position monitor (BPM) cables and circuits, magnets for guidance and focusing of the beam, Klystrons or Tetrodes (which provide power to RF cavities that transmit energy to the beam), and vacuum pumps and monitors can all cause beam instabilities. Vibrations and lack of proper shielding are other factors.

The challenge for operators and researchers is to correctly identify the factors causing beam instabilities and blow up so that costly accelerator time is not interrupted and experimental results are not compromised. The instrument most often used to identify problems in particle accelerator applications is the spectrum analyzer. In the section that follows, we discuss the advantages of real time versus swept frequency spectrum analyzers in HEP applications.

USING SPECTRUM ANALYZERS–SWEPT FREQUENCY VS REAL TIME

In swept frequency spectrum analyzers, signals are observed as a function of frequency–an approach that loses most, if not all, temporal (time relative) information. Real time spectrum analyzers allow the user to view and measure each portion of the input signal's frequency spectrum as a function of time. The differences, as they pertain to HEP applications, are as follows.

Swept Frequency Spectrum Analyzers

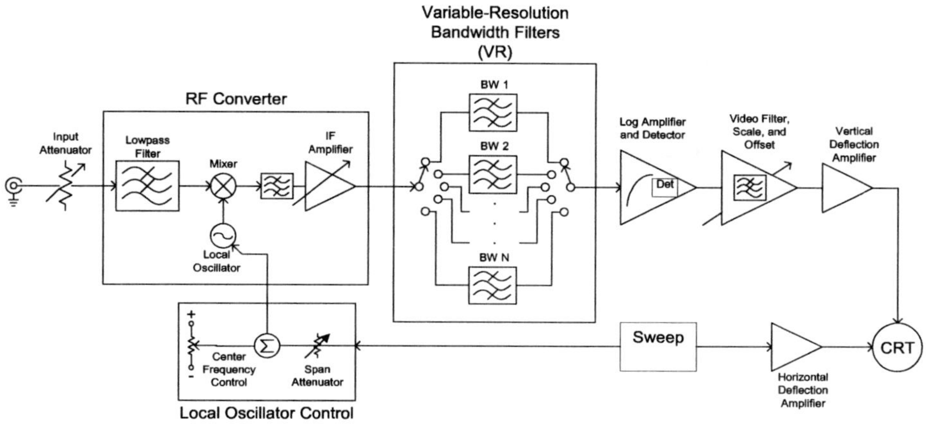

FIGURE 1. Swept frequency spectrum analyzer block diagram.

Figure 1 is a simplified block diagram of the swept frequency spectrum analyzer architecture. The local voltage-controlled oscillator continuously sweeps through a range of frequencies that are combined in a mixer stage with the input signal (a process known as heterodyning).

The mixer is a nonlinear device that produces the sum and difference of the two signals, along with the originals and their harmonics. The resulting signal components are sent through a resolution filter that extracts the desired mixing product. This signal product is then detected and sent to the Y-axis of the display (amplitude). At the same time the voltage-controlled oscillator signal is processed and sent to the display to create the X-axis sweep (frequency).

The traditional swept spectrum analyzer makes amplitude vs frequency measurements by sweeping a resolution bandwidth (RBW) filter over the frequencies of interest and recording the amplitude at each frequency point. While this method provides superior dynamic range, its disadvantage is that it only records the amplitude data in one frequency at a time. Sweeping the RBW filter over a span of frequencies takes time—n the order of seconds in some cases. A relatively stable, unchanging input signal is required.

If there is a rapid change in the signal, it is statistically probable that the change will be missed. As shown in Fig. 2, the sweep is looking at frequency segment F_a while a momentary aberration is occurring at F_b. By the time the sweep arrives at segment F_b, the error has vanished. It does not get detected. There is no way to trigger on defined signal characteristics, nor is there a way to accumulate a record of longer-term signal behavior.

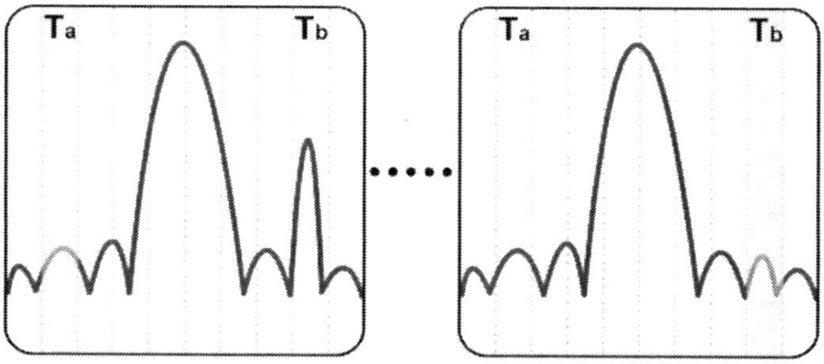

FIGURE 2: The swept spectrum analyzer steps across a series of frequency segments, often missing important transient events that occur outside the current sweep band.

The response, or settling time, of the resolution filter limits the sweep rate of the analyzer. In addition, the sweep rate is proportional to the square of the bandwidth—the narrower the resolution bandwidth, the slower the sweep rate. If the analyzer sweeps too quickly, the displayed amplitude will be lower than the actual amplitude and the apparent frequency will be shifted to the right.

Theoretically, the sweep rate should be infinitely slow, but that is not realistic, so some amount of acceptable error must be specified. Increased sweep times demand that the input signal remain stationary for even longer periods—usually disqualifying swept analyzers for the measurement of intermittent or periodic signals.

Traditionally the major advantages of swept-frequency analyzers over real-time instruments have been their dynamic range and frequency range. Recent advances in high-speed analog-to-digital (A/D) converter technology, however, have narrowed those advantages, as new real-time analyzers feature 70 dB or more of dynamic range and coverage from DC to 8 GHz—more than adequate for most HEP applications.

Real-Time Spectrum Analyzers: Storing and Analyzing Frequency and Time

As time-varying signals become more common in HEP applications, the need for an alternative approach to RF acquisition and analysis becomes more urgent. The real-time spectrum analyzer has emerged to solve this tough measurement problem. Alone among the spectrum analyzer architectures, the RTSA can trigger on a frequency domain event and then capture and analyze any passband signal that falls within its real-time bandwidth.

Figure 3 depicts the RTSA architecture. An integrated down-converter positions the real-time bandwidth on any passband up to the analyzer's upper limit. After filtering, the down-converted signal goes through an ADC that digitizes it, allowing triggering, capturing, and analysis of the signal all at once.

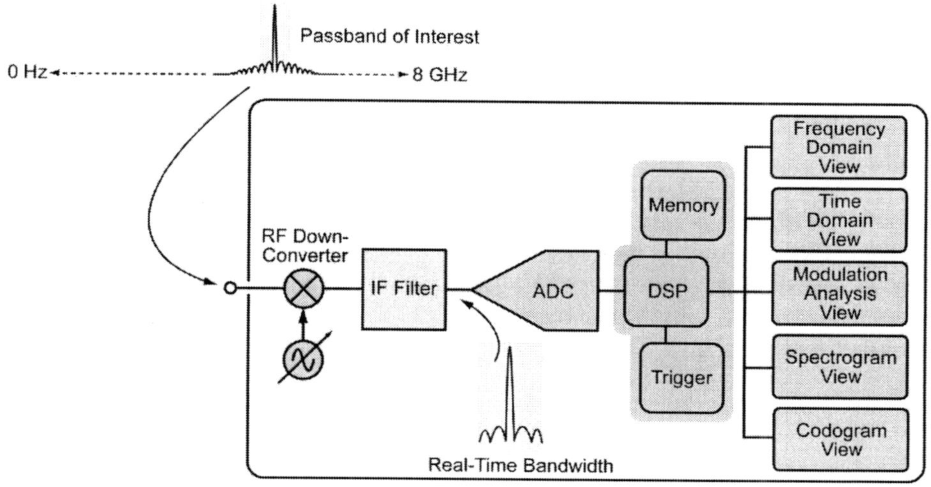

FIGURE 3: Real-time spectrum analyzer architecture. The instrument captures a full passband of frequencies all at once, and the DSP supports flexible triggering and analysis features.

Since the basic process is not one of sweeping across the RF input signal and building an image from serially acquired frequency steps, the RTSA's digital IF architecture allows a continuous capture of "snapshots" known as frames. These frames accumulate in the memory as a seamless, continuous record of time.

The memory supports a variety of display and analysis tools, including the spectrogram, which plots an entire series of frames to reveal signal changes over time. Thus the RTSA is the only RF signal analyzer optimized to produce a three-dimensional display: frequency, power (amplitude), and time.

If the passband of interest exceeds the real-time bandwidth of the RTSA, the RTSA can step through a series of frequency segments, just like the swept analyzer. In doing so, each sweep captures a band of frequencies equivalent to the RTSA's real-time bandwidth. Then the instrument concatenates the frequency bands and presents a conventional frequency-domain display.

RTSA MEASUREMENTS IN PARTICLE ACCELERATOR APPLICATIONS

Tektronix RSA2200A and RSA3300A series Real Time Spectrum Analyzers are used in particle accelerator applications to monitor beam ramping and measure horizontal and vertical tune frequencies during ramp up; measure chromaticity; and monitor amplitude, frequency, and phase modulations. In addition, the RSA's allow the display of power vs time information and provide a frequency mask trigger for monitoring beam stability.

Monitoring Beam Ramping [1]

In the following example, we will monitor a beam ramp up that is typically 160 ms in duration and compare that with the commissioning phase during which the beam may be ramped up at a rate of 1 second. We will also review a few common frequency spans settings: 20, 50, 100, and 200 KHz.

The RSA is able to capture and store up to 25 spectral waveforms in the frequency domain during the typical 160 ms ramp time (156 during the 1 second ramp time). These waveforms can be viewed in either a traditional frequency domain display (as would be seen on a Swept Frequency Analyzer) or in a Spectrogram mode where the X-axis is frequency, Y-axis is time, and power is given a color representation.

The RSA provides the ability to recall these data and step back in time through the captured waveforms in 6.4-ms intervals.

Table 1 shows the relationship of frequency span, frame rate, time resolution, and frequency resolution as well as memory length (length of events that can be stored in the RSA). The time resolution is the interval between each of the 1024 samples that makes up the frame.

TABLE 1. Relationship of span, frame time, time resolution, frequency resolution, and memory length.

Span	Frame Rate	Time Resolution	Frequency Resolution	Memory Length RTSA3300A (64 MB)	Memory Length RTSA2200A (2 MB)
200 KHz	3.2 ms	3.125 us	312.5 Hz	51.2 sec	1.6 sec
100 KHz	6.4 ms	6.25 us	156.25 Hz	102.4 sec	3.2 sec
50 KHz	12.8 ms	12.5 us	78.125 Hz	204.8 sec	6.4 sec
20 KHz	32 ms	31.25 us	31.25 Hz	512.0 sec	16.0 sec

The RSA can also display the information in a Power vs Time format with time resolution to 40 nanoseconds and dynamic range superior to an oscilloscope (see Fig. 4).

How does frequency and time resolution compare with a Swept Spectrum Analyzer sweep time? Assuming the typical 501 data points of resolution on the SSA display, the answer is shown in Table 2. Here we see the analog and FFT modes offered by some analyzers compared with the Real Time mode of the Tektronix RSA instruments at narrower resolution bandwidth (RBW) settings.

In a swept analyzer, any event that occurs at a particular frequency between the 200 msec it takes to sweep that point will not be observed. The same is true of any event occuring <15 ms in our example FFT instrument of Table 3. The real time capabilities of the Tektronix RSA, however, will allow the user to view the entire span in 6.4 ms time slices. In addition, the RSA will allow the user to store and recall these signals for post-capture analysis, a feature not offered in swept spectrum analyzers. The user can recall these signals in “frequency vs time” or “power vs time” mode with even better time resolution (6.25 us).

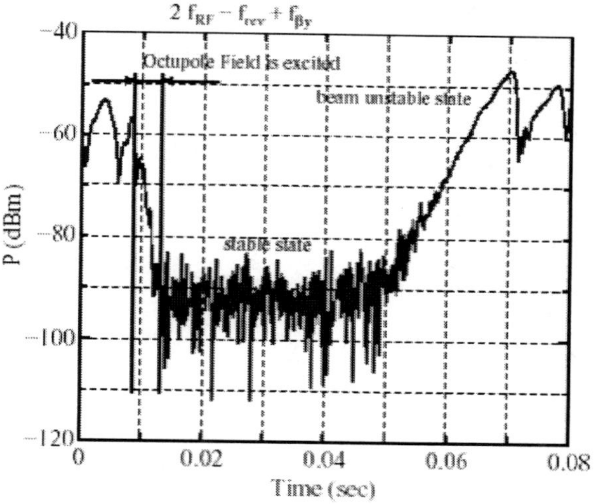

FIGURE 4. This power vs time display is used to measure power of the vertical instability at 2*f*rev - *f*rf + *fåy*. A pulse octupole magnet (used to suppress the instability) was excited at 100 A from t = 8 msec to 12.8 msec [2 ©IEEE 2003]. Note the dynamic range.

TABLE 2. Comparison of time resolution at 100 KHz frequency span between an analog swept spectrum analyzer, an instrument with FFT mode (offered in some analyzers at narrower resolution bandwidth (RBW) settings) and the RSA using Real Time technology. Note the much-improved default RBW setting using Real Time Analyzer.

Analyzer Type	**Swept**	**FFT**	**Real TIme**
RBW (default)	1 KHz	1 KHz	266.96 Hz
Sweep Time (Coupled Mode)/Time Resolution	0.2 sec	15ms	6.4 ms (frame time) 6.25 us (time resolution)

The unique architecture and approach offered in the Tektronix RSA2200A and RSA3300A series analyzers provide a definite advantage when measuring transient signals and often with improved frequency resolution. With fast frame rates, Tektronix real time spectrum analyzers allow for a more dynamic measurement of the chromaticity and application of chromatic correction.

Chromaticity Measurements

Chromaticity describes the change in betatron resonant frequency (tune) as a function of beam momentum. Chromaticity compensation is required in high-energy storage rings to avoid instability due to the head-tail (H-T) effect. We define the chromaticity of a lattice as $Q' = dQ/d\delta$, where Q is the betatron tune and δ is the fractional deviation from nominal momentum.

One approach is to measure the coherent damping rates by shock exciting the beam with a fast kicker, then observing the response of beam-position monitor signals on the

RTSA (see Fig. 5). The analyzer displays the damping envelope logarithmically, allowing, after subtraction of the incoherent radiation-damping rate, the determination of the coherent damping rate.

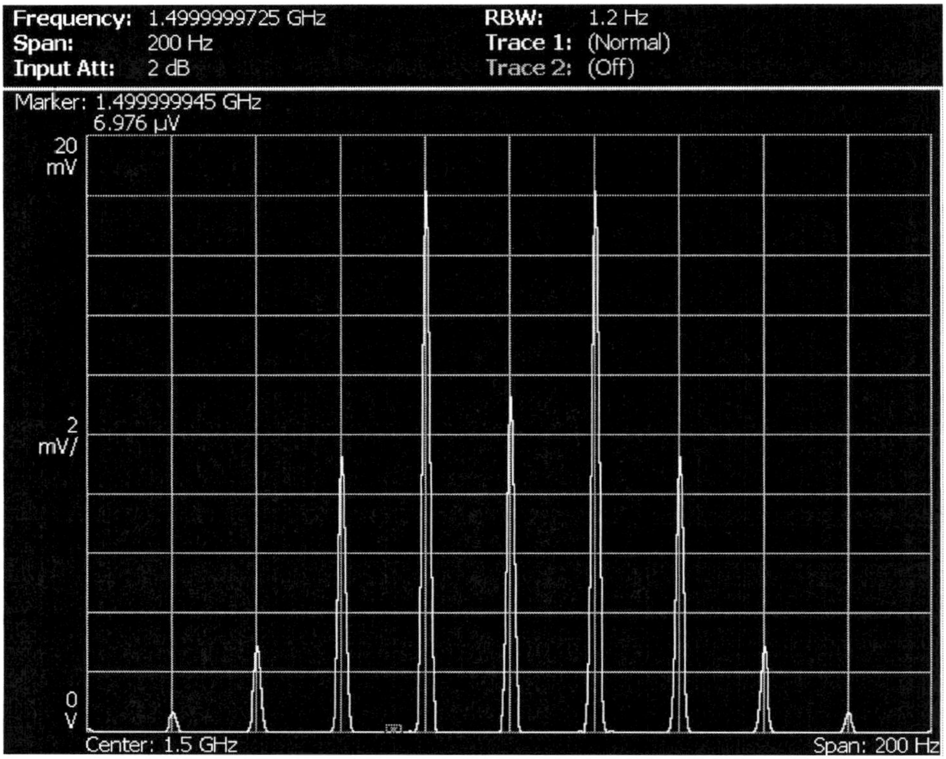

FIGURE 5a. Spectrum of a phase-modulated signal that has a 1.5-GHz carrier modulated at a rate of 20 Hz and phase modulation amplitude of 1.75 radians.

In positron or proton storage rings with many closely spaced bunches, an electron cloud can build up in the vacuum chamber due to photoemission or secondary emission. There then exists the possibility of a single-bunch two-stream instability driven by this electron cloud. Depending on the strength of the beam-electron interaction, the chromaticity, and the synchrotron oscillation frequency, this instability either resembles a linac beam breakup or a head-tail instability.

Most tune measurements in the storage ring are made at a single time in the acceleration cycle. However, if the tunes are measured over the entire acceleration cycle, as required by swept frequency analyzers, the process requires many beam cycles making data acquisition extremely laborious and grinding HEP experiments to a halt. The RTSA can acquire and retrieve tune data for the complete acceleration cycle in one beam cycle. It can also be desirable to utilize Phase Modulated RF to measure the chromaticity [3, 4]. The RSA products allow direct Phase Modulation

(PM) measurement for faster, more accurate chromaticity calculations (see Figs. 5a and 5b).

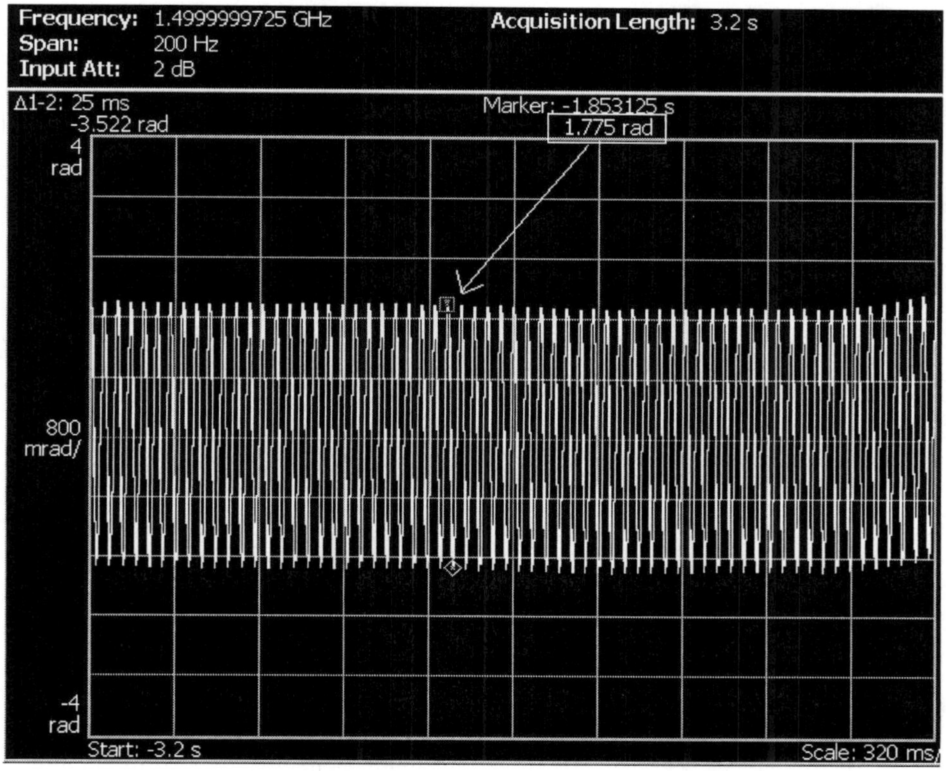

FIGURE 5b. Demodulated view of the spectrum shown in Fig 5a. Note that marker functions can now be used do directly read the phase modulation and rate.

Analog Demodulation with Multi-Domain Display

The RSA offers amplitude, frequency, and phase demodulation capabilities in multi-domain displays. By using marker functions, the RSA makes it possible to easily read out modulation directly on the display (see Fig. 6).

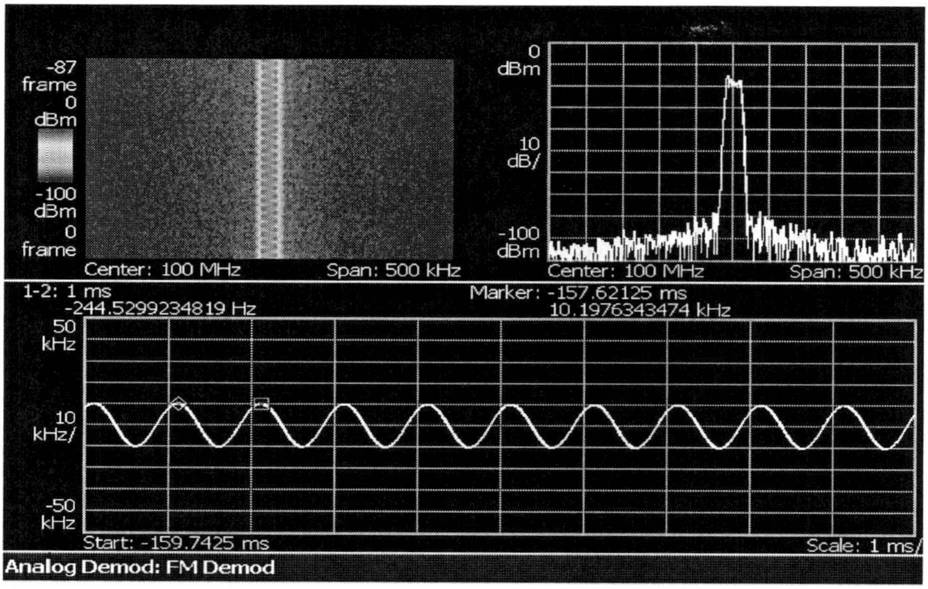

FIGURE 6. A direct readout of an FM signal. A peak marker is used to read out the peak FM deviation (around 10 kHz) and delta marker to read out FM rate (1 ms or 1 kHz). In spectrogram mode (upper left), X-axis is frequency, Y-axis is time, and a color scale represents power.

FM demodulation is useful for a number of HEP applications. In the event of beam instability, for example, it is possible to analyze the variation with high frequency and time resolution (see Fig. 7). It is also useful to monitor Betatron signals and for cases where a known amount of external FM is injected into the RF system.

In Fig. 7 we see severe signal instability in a multi-mode view. Recall from Table 1 that with a 100-kHz span the Spectrogram (upper left) and Frequency Domain (upper right) views provide time resolution of 6.4 ms and frequency resolution of 156.25 Hz. Using the FM demod mode (bottom), the time resolution (X-axis) can be set to 6.25 us and frequency resolution (Y-axis) to 16 mHz per division, allowing extremely transient and small frequency variations to be observed.

Phase modulation is especially useful to for monitoring phase stability of the beam, for analyzing and adjusting the many Phase Lock Loops (PLL) found in a typical storage ring, and for calculating chromaticity (see Fig. 8). Amplitude modulation is useful for Synchrotron measurements.

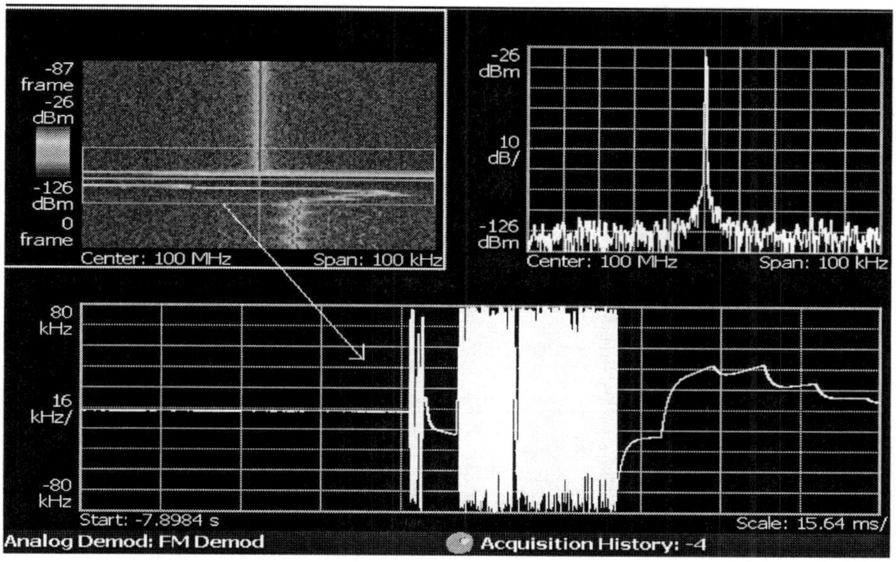

FIGURE 7. Multi-mode view of signal instability.

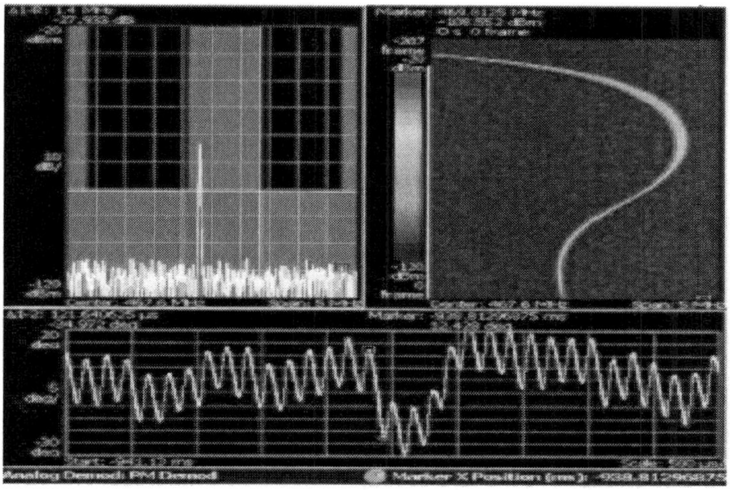

FIGURE 8. PLL responses, upper left, show frequency domain (with Frequency Mask Trigger enabled). The upper right shows Spectrogram; the frequency swing and then settling is clearly visible. Phase change vs time (PM Demod) is seen at the bottom.

Monitoring Beam Stability with Frequency Mask Trigger

Beam instability appears in the frequency domain as either signals that drift in frequency or as distortion products (see Fig. 9).

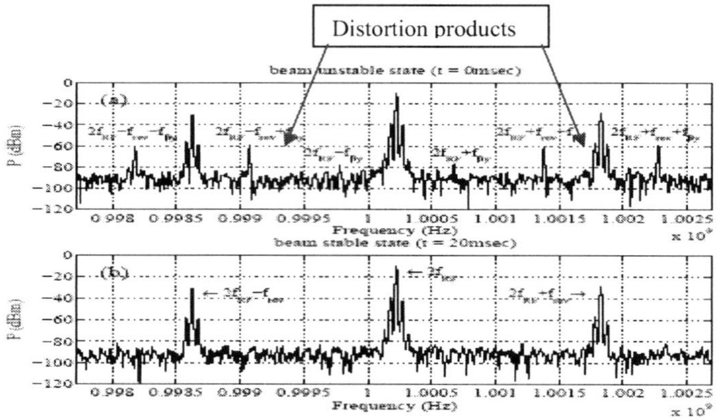

FIGURE 9. Beam instability appears as transient distortion products. An unstable beam (top) is compared to a stable beam (bottom). Note the appearance of distortion products when unstable.

By using its unique Frequency Mask Trigger (Fig. 10), the Tektronix RSA can be set to trigger should either of these events occur. A mask can be "drawn" on the display using a mouse to within one pixel and 0.1 dB of resolution.

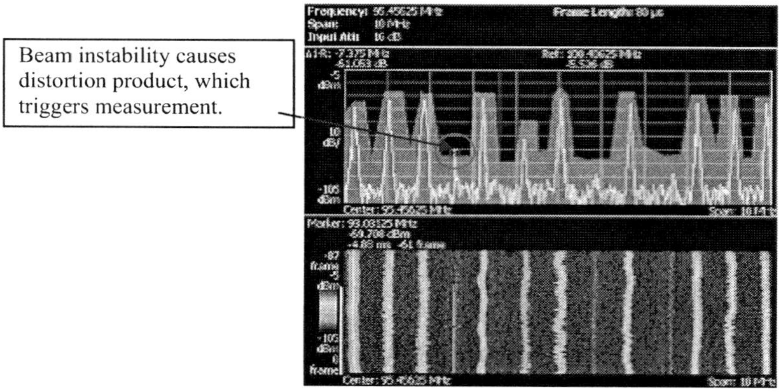

FIGURE 10. Use of Frequency Mask Trigger.

Once the RSA is triggered, it can begin recording data for later analysis, it can also output a trigger to an external device, such as an alarm. Additionally, the RTSA offers a pre-trigger setting, so it is possible to recall and analyze data just prior to the triggering event.

This is a major advantage over the traditional trigger offerings found in swept frequency analyzers, which were limited to either a TTL input or an IF level trigger. Swept frequency analyzers monitor the power level of all signals in the IF passband and, consequently, would be constantly triggered on the signal regardless of the distortion product. The RSA3300A offers theses traditional trigger inputs with the additional, unique option of a Frequency Mask Trigger.

ADDITIONAL FEATURES AND APPLICATIONS

Amplitude Correction

Additional RTSA features that can be useful in HEP applications include the ability to input an amplitude correction file with up to 3000 points of frequency response data (see Fig. 11). This can be useful for correcting for loss of BPM cables or other losses prior to the RTSA input.

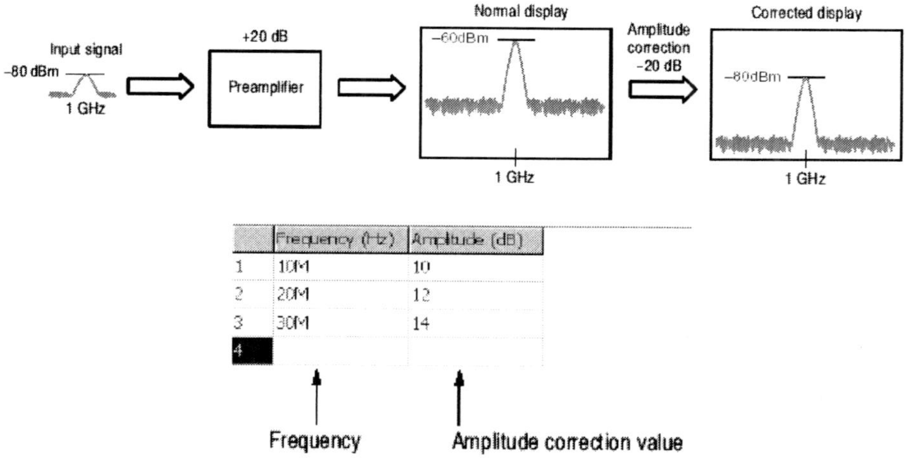

FIGURE 11. Amplitude correction.

Vibration Analysis

The RTSA can be used to measure low frequency vibrations (to DC) with excellent frequency resolution and much better update display rates than traditional swept frequency analyzers. The importance of vibration problems increases with the smaller emittance of a storage ring for synchrotron radiation source. Vertical emittance of a storage ring can be so small that the effects of vibration on the electron beam cannot be neglected. Operation of air conditioning, water-cooling, and vacuum pumps can cause vibration of the magnets, which then affects beam stability. These vibrations can be analyzed using accelerometers, the output of which can be amplified and fed to the

RTSA for FFT processing. In addition, intermittent vibration spikes can be captured reliably with the real-time capability of the RTSA.

CONCLUSION

Tektronix Real Time Spectrum Analyzers continue to lead the field in HEP applications by offering a single instrument with capabilities not found in a swept spectrum analyzer:

- Updated display of the spectrum with better frequency resolution.
- Storage of long records of seamless data (including pre-trigger data).
- A multi-domain view allowing simultaneous display of any three modes: spectrogram, power vs time, frequency domain, or demodulation data (AM, FM, and PM).
- Frequency Mask Trigger and Trigger output that allows the user to trigger on a tune frequency at the instant it goes unstable while sending a TTL signal to other instruments (i.e., alarm device) ***and*** capturing pre-trigger data.
- Data recall for postprocessing and detailed analysis.

Tektronix RTSAs deliver unique solutions and advantages for operators and researchers in the High Energy Physics field. Only Tektronix RTSAs offer the triggering, capture, and analysis features needed to reveal the true nature of particle accelerator anomalies. The RTSA is the solution that can go forward with demanding requirements of particle research.

Please contact your local Tektronix representative to request for more information and to arrange for a demonstration of these truly unique instruments.

REFERENCES

1. S. Bassanese, M. Ferianis, and F. Iazzourene"Fast Tune Measurement System for the Elettra Booster," Sincrotrone Proceedings DIPAC 2003–GSI, Mainz, Germany.
2. Tsukasa Miyajima, Yukinori Kobayashi, and Shinya Nagahashi, "Pulse Octupole Magnet System at the Photon Factory Storage Ring,"
3. Dave McGinnis, PBAR Note 656 "Chromaticity Measurements Using Phase Modulated RF," Fermi National Laboratory.
4. Oliver Bruning, "Chromaticity Measurement via RF Phase Modulation," BI Forum, Feb. 26, 2002, CERN.
5. K. Tsumaki and N. Kumagai, "Vibration Measurement of the Spring-8 Storage Ring," Proceedings of the 2001 Particle Accelerator Conference, Chicago, SPring-8, Hyogo, Japan.

Acquisition of Diagnostic Screen and Synchrotron Radiation Images Using IEEE1394 Digital Cameras

G. Rehm

Diamond Light Source Ltd, Rutherford Appleton Lab, Chilton, Didcot, OX11 9JS, UK

Abstract. In the LINAC, booster synchrotron and transfer lines of DIAMOND[1] a number of screens (YAG:Ce and OTR) as well as synchrotron radiation ports will be used to acquire information about the transverse beam distribution. Digital IEEE1394 cameras have been selected for their range of sensor sizes and resolutions available, their easy triggering to single events, and their noise-free transmission of the images into the control system. Their suitability for use under influence of high-energy radiation has been verified. Images from preliminary tests at the SRS Daresbury are presented.

INTRODUCTION

Video cameras have proven a valuable tool for many applications in accelerator installations, especially when combined with frame grabbers to gather data that can then be processed and analysed. With the introduction of CCD higher resolution have become available and camera's physical sizes could be reduced to allow use in space constrained locations.

However, analogue CCD cameras are limited in their flexibility by the frame grabber or display device they are used with. In most cases maximum integration time is limited to 20 ms. Triggering an analogue CCD camera and frame grabber to take a frame, while not impossible, required additional electronics to generate the sync signals externally. Changing the camera resolution or area of interest (AOI) during operation proves virtually impossible with this combination.

All these limitation can be overcome through the use of digital IEEE1394 cameras, where the images are digitised internally and transmitted through a standardised high speed serials bus (400 MBits/s). Control, data and power are transmitted through a single six wire cable to the camera, which can be connected to a computer using the any of the widely available and inexpensive IEEE1394 cards.

Features like the ability to completely remote control all properties of the camera (exposure time, gain, gamma, AOI etc.), a wide range of available resolutions and the easy triggering facility make these cameras a versatile new addition to the range of diagnostics.

However, this relatively new technology of cameras has only little reported record of usage and durability in an accelerator environment [2, 3]. Thus we had to evaluate their suitability before considering it for general use at DIAMOND.

CP732, *Beam Instrumentation Workshop 2004: Eleventh Workshop*,
edited by T. Shea and R. C. Sibley III

SELECTION OF CAMERAS

Suitable cameras should be B/W image, have a wide gain range, be small and affordable (<US$1500). Samples of four cameras fitting these criteria have been ordered from four manufacturers. These cameras are targeted at industrial rather than scientific use, so performance under the conditions specific for use in beam diagnostics had to be assessed (short exposure, high gain, weak signal).

TABLE 1. Data of the selected four cameras.

camera	resolution	sensor size	pixel size	gain [dB]	power [W]	ADC [Bits]	min. exp.
A	1032x778	1/3"	4.65μm	-6 to 29.6	2	12	20μs
B	1032x778	1/3"	4.65μm	0 to 24	3	10	25μs
C	1032x778	1/3"	4.65μm	0 to 26	1.5	10	60μs
D	782x582	1/2"	9.9μm	4.5 to 34.5	6	8	20μs

The mixed signal electronics of these cameras need to have been carefully designed to prevent analogue/digital interference. To check this, an image under homogeneous illumination was taken with 20 dB gain and 1 ms exposure on each camera and analysed for mean signal and noise. While the calculated overall S/N ratio was comparable, FFT analysis shows major differences (see Fig. 1).

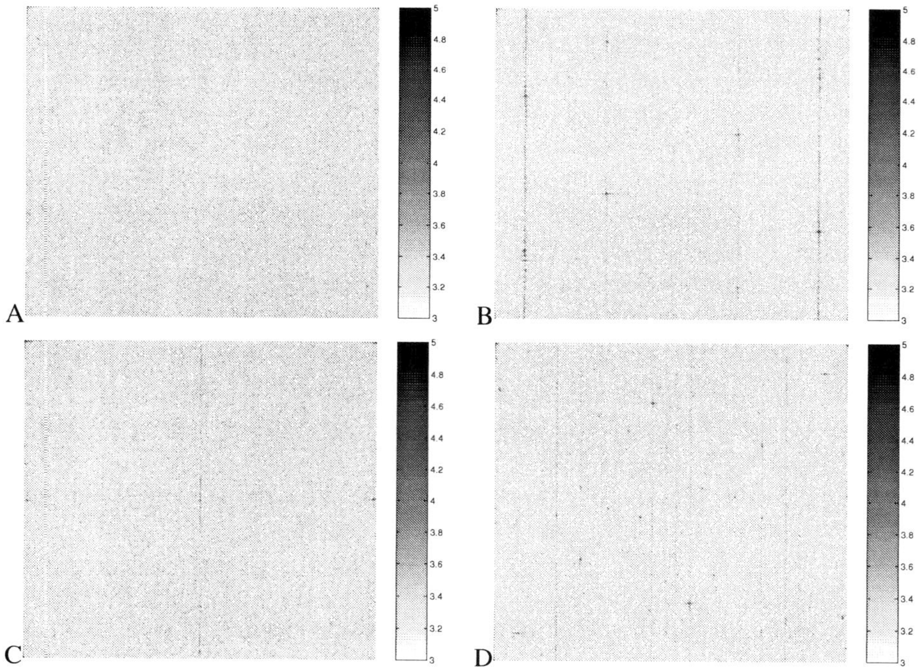

FIGURE 1. Logarithmic 2D FFT of the test images (grey levels are powers of 10).

From the FFT data it can be concluded that some digital signals couple into the analogue signal path and show up with various multiples of the pixel and line clock. This kind of interference is a disadvantage as it reduces the effectiveness of pixel binning or filtering of the image. It can also add up when summing rows or columns as often used to integrate over one axis.

At this stage it was concluded that camera A, the *PointGrey HiRes Flea*[4], meets the requirements best. Further tests for use in an accelerator environment were only carried out with this model.

CAMERA TEST IN AN ACCELERATOR

The *Flea* was installed in the booster to storage ring transfer path of the SRS in Daresbury (UK) to test it in a real accelerator application and environment. The camera was pointing at a phosphor screen in air where the 600 MeV electron bunch trains can be imaged. No shielding was placed in front of or around the camera to create worst case conditions. The booster synchrotron cycles at 5 Hz and delivers a charge of some nC in every bunch train. The camera was triggered using a connection to the booster extraction septum, so every frame taken included an image of the electron beam on the screen (no flickering or half images etc.). A snapshot from the camera taken with a long exposure time to see the setup is shown in Fig. 2.

As the triggering of flawless and precise, the exposure time could be reduced to 20 μs. This change did not decrease the brightness of the beam image at all, however, the influence of background light could be minimized this way. To reduce the amount of data which has to be transmitted and stored, an area of interest was selected which captured only the beam image. Due the the high resolution of the camera this function effectively can be used like pan and zoom with the resulting partial images still offering sufficient resolution for processing.

FIGURE 2. Image taken with the camera looking at the phosphor screen next to the electron exit window.

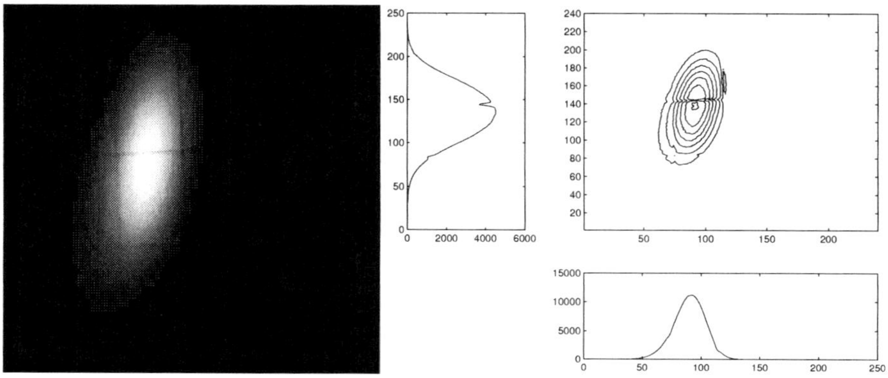

FIGURE 3. Beam image selected using AOI (left) and analyzed in MATLAB (right).

As an example the 240 by 240 pixel image in figure 3 has been loaded into MATLAB where it can be conveniently processed and displayed. The lines in the centre of the beam are actually a black line which are drawn on the phosphor screen as part of a grid used for calibration with the video camera in day to day operations.

The camera worked without any noticeable influence of the radiation bursts. Only single pixels flashed up white from time to time, indicating the impact of an ionising particle. After 4 hours of continuous operation under radiation 70 pixels showed an increased dark current. However, this was only detectable at very long integration times (>100ms), so this degradation will have no influence on the normal mode of operation. The data transmission using two 11m long cables with one repeater also was reliable at all times. As an advantage of the digital protocol the picture quality is not influenced by the cable length or the harsh EM environment (kicker pulses, etc.).

CAMERA DAMAGE DUE TO RADIATION

As the test in an accelerator could not provide any data on long time effects of larger amounts of integrated absorbed radiation doses, the *Flea* and an IEEE1394 repeater were also tested under the influence of strong γ radiation. The UK Medical Research Council was kind enough to allow use of their irradiation facilities for this purpose.

A remotely movable Co^{60} source and a shelf at a set distance provided a dose rate of 2.04 Gy/min which could be switched on and off. The camera and the repeater continued to function even in the presence of this extreme amount of ionising radiation (5 Gy is generally seen a lethal dose for humans).

To assess the damage to the camera, the average grey level of frames taken with maximum exposure time (533ms), maximum gain (29dB) and no light into the camera was recorded. However, to take these measurements the radiation sources had to be removed, as their presence led to a total "white out" of the CCD. A sequence of 5 minute irradiations provided measurements at intervals of 10.2 Gy. The result of the processed images is shown in Fig. 4.

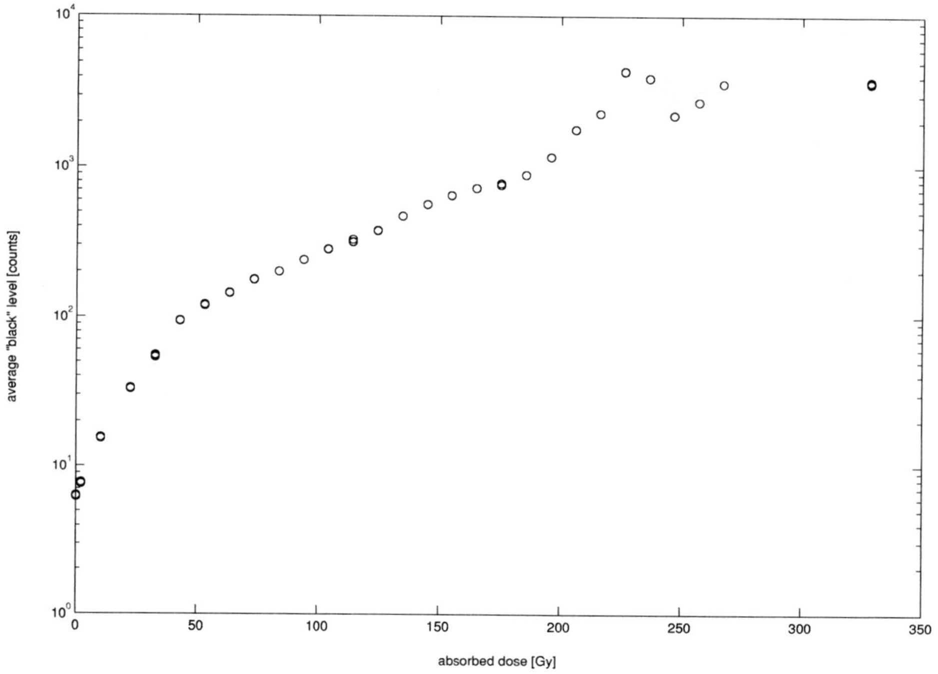

FIGURE 4. Increase of "black level" with absorbed dose.

This graph illustrates how the dark current of the CCD increases due to the radiation damage. While this effect on the image sensor was already clearly visible at low doses, the digital electronics worked flawless up to 150 Gy. At this point frame dropouts started to appear and increase with continued irradiation up to a level of 10 dropouts per minute (compared to 900 frames per minute). Above 200 Gy the observed radiation damage is no longer correlated with the measured grey level, probably other parts of the analogue electronics started to suffer damages.

After a longer irradiation to 330 Gy the digital electronics of the camera failed, i.e. the camera could not longer be detected on the IEEE1394 bus. Even disconnecting and reconnecting the camera did not change this situation. We assume that the FLASH-memory which stores the boot information of the camera had been at least partially erased as a result of the radiation.

Nevertheless, we have been very satisfied with this performance as this test proved, that digital CCD cameras suffer from no other damages as analogue CCD cameras, namely an increase of dark current. Keeping in mind that this strong dose rate leads to additional effects like accumulation of trapped charges the graph almost certainly over-estimates the damage caused by a long term exposure at low dose rates. Additionally, the ability to assess camera damage even while it is still small will allow to provide additional shielding where required.

The repeater continued to operate and was left for additional irradiation. It stopped operating correctly after an integrated dose of more the 6 kGy. However, a few days after it had been removed from the irradiation it actually started to operate again and is now fully functional again. In this case the temporary fault could have been a result of charge build-up.

ACKNOWLEDGMENTS

The author would like to thank Rob Smith from SRS Daresbury (UK) and David Kestell from the Medical Research Council (UK) for their contribution to enable the measurements presented in this paper.

REFERENCES

1. Diamond Light Source (2004), http://www.diamond.ac.uk/.
2. K. Brown, B. Frak, D. Gassner, L. Hoff, R. Olsen, T. Satogata, and S. Tepikian, "IEEE 1394 Camera Imaging System For Brookhaven's Booster Application Facility Beam Diagnostics," in *Proceedings of EPAC 2002, Paris, France*, 2002, pp. 551–553.
3. K. Honkavaara, A. Brenger, R. Fischer, D. Noelle, K. Rehlich, L. Cacciotti, M. Castellano, G. DiPirro, M. Raparelli, and R. Sorchetti, "Design of OTR Beam Profile Monitors for the Tesla Test Facility, Phase 2 (TTF2)," in *Proceedings of the 2003 Particle Accelerator Conference*, 2003, pp. 2476–2478.
4. PointGrey (2004), http://www.ptgrey.com/products/flea/index.html.

Proposed OTR Measurements of 120-GeV Protons and Antiprotons at FNAL

V. E. Scarpine*, G. R. Tassotto*, and A. H. Lumpkin†

*Fermi National Accelerator Laboratory, Batavia, IL 60510, USA
†Argonne National Laboratory, Argonne, IL 60439, USA

Abstract. Fermi National Accelerator Laboratory (FNAL) is developing optical transition radiation (OTR) detectors for beam diagnostics for their 120-GeV proton and antiproton transfer lines. As part of a collaboration to enhance the luminosity for the FNAL collider RUN II program, the quality of the proton and antiproton beams, as they are transported from the main injector (MI) to the Tevatron, will be characterized using OTR imaging techniques. A prototype detector in air has already successfully acquired OTR images of 120-GeV protons upstream of the antiproton production target. This result demonstrates that (i) the Ti and Al thin foil screens survive the 5 x 10^{12} proton beam spills, (ii) OTR is sufficient to image lower intensity antiproton beams, and (iii) the images provide two-dimensional information and higher resolution than the present multi-wire profile monitors in the transport lines. Beam bombardment effects on the Al screen and radiation effects on the lenses, filters and cameras have been evaluated for the prototype system for over 1 x 10^{19} 120-GeV protons and will also be presented. An in-vacuum OTR station is being designed for the transport lines with adjustments to the optical components as warranted by the beam characteristics and anticipated radiation environment.

INTRODUCTION

Particle-beam diagnostic techniques based on optical transition radiation (OTR) have been demonstrated at a number of facilities over a wide range of beam energy (or Lorentz factor, γ) [1,2,3]. FNAL is pursuing a number of projects where either beam luminosity for the Tevatron collider (Run II) or proton intensity for neutrino experiments require careful tracking of beam properties, such as transverse profile and shape (tilt), transverse position, emittance and intensity, throughout the facility.

Optical Transition Radiation

Optical transition radiation is generated when a charged particle transits the interface between two media with different dielectric constants [2]. The effect is a surface phenomenon that can be understood as the collapsing of the electric dipole formed by the approaching beam charge and its image charge in the dielectric at the surface. As the fields readjust, a pulse of radiation is emitted. This radiation is emitted over the visible spectrum so optical imaging techniques can be used to acquire the OTR signal and then reconstruct beam size and position at the dielectric surface. Since OTR is a surface phenomenon, thin foils are used as the converter to reduce beam scattering and minimize heat deposition.

CP732, *Beam Instrumentation Workshop 2004: Eleventh Workshop*,
edited by T. Shea and R. C. Sibley III

PROTOTYPE FNAL OTR DETECTOR

We have constructed and installed a prototype OTR detector at FNAL [4,5]. The prototype is located in an air gap in the AP-1 transport line upstream of the antiproton production target. This beamline transports 120-GeV protons with intensities up to 4.7 x 10^{12} particles in a 1.6-μs spill at a rate of one spill every 2 seconds. At this location the rms transverse beam size is a few millimeters and the OTR angular distribution ($1/\gamma$) is ~ 8 milliradians. In addition, the long-term survival of two titanium vacuum windows on either end of the air gap establishes that an OTR foil will survive the beam intensities at this location. The detector uses either a 12-μm titanium or 20-μm aluminum foil in the air gap at 45° with respect to the beam that serves as the OTR generator screen.

Prototype OTR Results

Figure 1 (left) is a false-color OTR image from 4.5 x 10^{12} 120-GeV protons through 12-μm titanium. The OTR is attenuated by a factor of 200 in order to avoid camera pixel saturation. This image shows a clear indication that the transverse beam shape is tilted at this location in the AP-1 transfer line. In Fig. 1 (right) are vertical and horizontal line profiles of light intensity relative to the peak intensity position, X_O, after correction for image distortion. These profiles indicate a full-width at half-maximum of ~12 mm in x and ~8 mm in y.

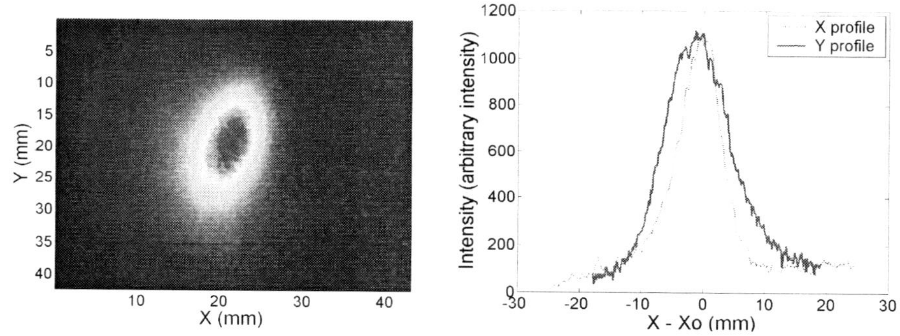

FIGURE 1. The left figure is an OTR image of 4.5 x 10^{12} 120-GeV protons through 12-μm titanium foil. The right figure is the vertical and horizontal profiles through the peak of the OTR.

Figure 2 is an image of OTR produced by 120-GeV protons using a 20-μm aluminum foil. The beam intensity is 4.7 x 10^{12} particles in a single 1.6-ms spill. The OTR intensity is reduced by 1000 in order to prevent camera saturation. Some structure is observed but it is not apparent if it is beam or foil induced.

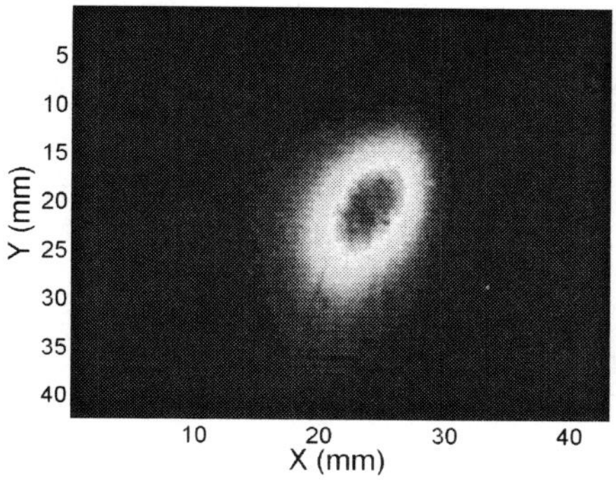

FIGURE 2. OTR image of 4.5 x 10^{12} 120-GeV protons through 20-μm aluminum foil.

PROPOSED OTR DETECTORS FOR FERMILAB

After the installation and operation of our prototype OTR detector, FNAL is in the process of developing a standard OTR detector. The primary use of this standard OTR will be as a real-time emittance monitor in the A1 transport line between the Main Injector and the Tevatron to measure both protons and antiprotons. Antiproton beam consists of four bunches separated by 396 ns of ~5 x 10^{10} particles per bunch at 150 GeV with a 95% beam spot size of 2 to 5 mm. Proton beams are used in this line to tune the beamline before operation with antiprotons. Typical proton intensities are from ~2 x 10^{10} up to 5 x 10^{11} particles.

In addition to the A1 line transport line OTR detectors, a single standard OTR detector will also be used as a beam profiling station for the high-intensity 120-GeV transport lines for NuMI. This OTR will see ~3 x 10^{13} 120-GeV protons every 2 seconds with a one sigma beam spot size of ~1 mm. Because of the high-intensity beam, issues of foil discoloration and radiation damage due to beam scatter must be addressed. For this reason we will investigate the use of very thin (~ few μm) low-density materials such as carbon or beryllium.

A single OTR detector will also be installed as a minimally intercepting OTR foil for use in the Tevatron for injection studies adjacent to the new Tevatron Ionization Profile Monitor (IPM). This OTR will primarily be used in conjunction with the IPM as a calibration tool. Typical injection studies beam consists of single bunch protons with intensities of up to 5 x 10^{11} particles. Since the Tevatron is a circulating storage ring, studies with an OTR detector would consist of a single or few-turn injection at 150 GeV before beam distortion from the OTR foil would effect the measurement or losses would produce a quench of the superconducting magnets.

Standard OTR Detector Design

The design of this standard OTR detector consists of a vacuum can containing two 3.5-inch clear aperture foils that can be moved in at 45 degrees to the beam or positioned completely out of the beam. The foils have the mounting option to measure beam in both directions using a single optical system. Design of the foil mounts allows in-situ quick foil replacement or the option of upgrading different foil material.

OTR from a foil will be passed through a quartz vacuum window to an external optics and camera acquisition system. A lens system collects the OTR generated in a cone of radius $2/\gamma$ about the optical axis and focuses it on to a Charge Injection Device (CID) camera. The CID camera has been shown to operate beyond a total dose over 1 MRad [6]. To maintain focus over the foil, the camera is mounted at an angle to satisfy the Scheimpflug condition [7]. Images of OTR are acquired and saved to a computer using a standard frame grabber card. Camera light levels are controlled by the use of selectable neutral density filters.

The acquisition camera will use a standard 30-ms frame time but future upgrades may utilize a gated image intensifier to operate with shorter integration times allowing for single bunch measurements.

OTR Detector Issues for High-Intensity Proton Beams

In order to operate OTR detectors in proton transport lines, beam and radiation effects must be considered. Beam bombardment effects on foils can cause a change in the foil surface reflectivity. Radiation effects can cause damage to lenses, filters and cameras. Our prototype detector has operated in a high radiation environment with over ~1 x 10^{19} 120-GeV protons passing through a 20-μm Al foil.

While our prototype detector remained in the AP-1 line, it was exposed with up to several hundred kRads of radiation mostly produced by scatter from the surrounding transport line vacuum windows and from missteering of the 120-GeV protons. The exposure to this radiation produced a darkening of our optical achromatic lens and also damage to our CID camera. Darkening of the lens can be compensated for by a change of neutral density filters but this will increase the minimum beam intensity required to detect OTR. Darkening of optical lenses can be drastically reduced by using only quartz lenses, which are several orders of magnitude more radiation-resistant than standard optical glass. However the use of only quartz can require a more complicated lens design in order to reduce optical aberrations and achieve the optimal image resolution.

Figure 3 is a standard resolution target taken with our CID camera after operation in our prototype OTR detector. The image shows a marked reduction in image quality but no dead or hot pixels. Figure 4 (left) is a false color image of a dark field image from this camera while Fig. 4 (right) is a horizontal cross section of this dark field image. These images show radiation damage to semiconductor material in the form of increased dark current and reduction of the CID camera charge transfer efficiency. Correction for these effects can still allow for OTR detection but with reduced camera dynamic range.

FIGURE 3. Resolution target image taken with CID camera irradiated with a few hundred kRad.

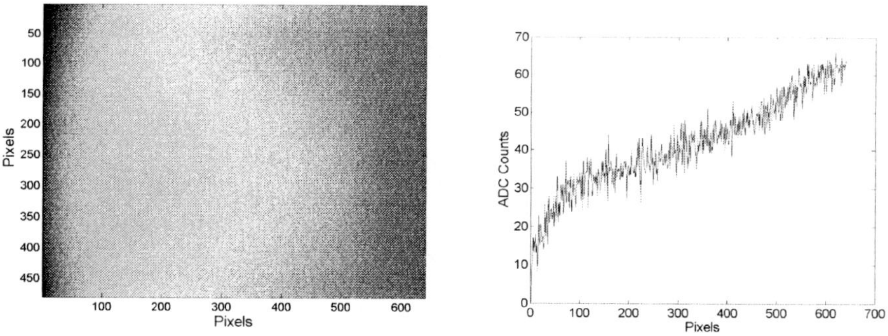

FIGURE 4. The left figure is a false-color dark image of our irradiated CID camera. The right figure is a horizontal cross section of the dark image.

Long term exposure of titanium vacuum windows in transfer lines to high-intensity proton beams has shown discoloration and changes in foil reflectivity. Figure 5 (left) is an image of a 3-mil thick titanium vacuum window with over 10^{20} 120-GeV protons and a one sigma spot size of ~0.4 mm. Figure 5 (center) is for a similar window that has seen only about 3 x 10^{18} protons but with a one sigma spot size of only 0.15 mm. Darkening of these vacuum windows occurs on both the air and vacuum side and appears darker on the vacuum side. Figure 5 (right) is of our 20-μm aluminum OTR foil that has seen ~10^{19} 120-GeV protons with an rms beam spot size of a few mm. This foil shows no change in surface features or reflectivity.

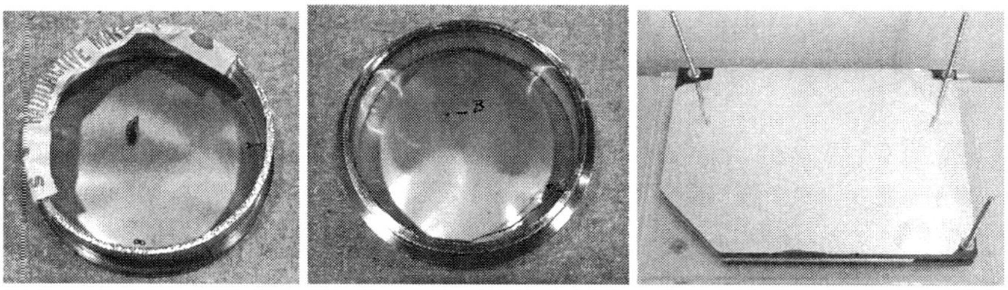

FIGURE 5. The left image is of a 3-mil thick titanium vacuum window exposed to over 10^{20} 120-GeV protons. The center image is a similar vacuum window exposed to ~3 x 10^{18} 120-GeV protons but with a smaller beam spot size. The right image is of our prototype OTR 20-µm aluminum foil exposed to ~10^{19} 120 protons with a larger beam spot size.

CONCLUSION

We have developed and installed a prototype OTR detector using 12-µm titanium and 20-µm aluminum foils. OTR light levels are large enough to allow imaging at particle intensities as low as ~1 x 10^{10}. A single aluminum foil shows no artifacts after ~10^{19} protons however thicker titanium vacuum windows show considerable discoloration after long term exposure to intensity proton beams. High radiation exposure has reduced light transmission of our optical glass and degraded our CID camera image quality but acquisition of OTR is still possible.

Performance of our prototype detector shows that a standard OTR detector can be developed for FNAL. Applications include developing (1) a real-time emittance monitor in the A1 transport line between the Main Injector and the Tevatron to measure both protons and antiprotons, (2) a beam profiling station for the high-intensity 120-GeV transport lines for NuMI, and (3) a minimally intercepting OTR foil for use in the Tevatron for injection studies adjacent to the new Tevatron IPM.

ACKNOWLEDGMENTS

We thank our colleagues in the FNAL Beams Division for their contributions: Eugene Lorman, Jim Morgan, Tony Leveling and Elvin Harms. In addition, we wish to thank Carl Lindenmeyer, John Korienek, Ron Miksa, Karen Kephart and Wanda Newby of FNAL Particle Physics Division for their outstanding work in the mechanical design and construction of the detector. The authors also acknowledge the support of Gianfranco Ferioli (CERN) for conversations on OTR detectors. We would also like to thank Stephen Pordes (FNAL) and Bob Webber (FNAL) for their enthusiasm and encouragement for pursuing OTR detectors.

REFERENCES

1. P. Goldsmith and J. V. Jelley, "Optical Transition Radiation from Protons Entering Metal Surfaces," *Phil. Mag.* **4**, 836-844 (1959).
2. J. Bosser, J. Mann, G. Ferioli, and L. Wartski, "Optical Transition Radiation Proton Beam Profile Monitor," CERN/SPS 84-17.
3. A. H. Lumpkin, B. X. Yang, W. J. Berg, M. White, J. W. Lewellen, and S. V. Milton, "Optical Techniques for Electron-Beam Characterizations on the APS SASE FEL Project," Nucl. Instrum. Methods, vol. **A429**, 336-340 (1999).
4. A. H. Lumpkin and V. Scarpine, "The Feasibility of OTR Imaging of High-Intensity Proton Beams at FNAL," presented at the Particle Accelerator Conference, Portland, Oregon, May 12-16, 2003.
5. V. E. Scarpine, A. H. Lumpkin. W, Schappert, and G. R. Tassotto, "Optical Transition Radiation Imaging of Intense Proton Beams at FNAL," submitted to IEEE Trans. Nucl. Sci.
6. J. Carbone, J. Zarnowski, M. Pace, S. Czebiniak, and R. Carta, "Megarad and Scientific CIDs," Proceedings of SPIE, **2654**, 131-138 (1996).
7. W. J. Smith, "Modern Optical Engineering," 2nd ed., Boston, MA: McGraw-Hill, 1990, p. 52.

New Acquisition System for the PSR Beam Pulse Charge Monitor*

William C. Sellyey and Paul S. Lewis

Los Alamos National Laboratory (LANL), Los Alamos, NM 87545, USA

Abstract. A Pearson 1010 current monitor toroid has been in use for many years to measure the charge per bunch being delivered from the LANSCE Proton Storage Ring (PSR) to the Lujan Center's spallation neutron source. Improved electronics have been developed to process the toroid's signal. The new system generates a calibrated measurement of charge per pulse and is network-enabled to provide remote access to charge, current and other data via EPICS. It is experimentally demonstrated that accurate charge measurements can be made on calibration pulses that contain frequency components well above what is contained in a typical beam pulse. The new electronics consists of a National Instruments (NI) PXI-1002 chassis that contains a PXI-8176 controller, a PXI-5112 100-MS/s digitizer, and a PXI-6602 scalar and digital I/O module. The 8176 runs under the NI Real Time operating system and was programmed to integrate proton pulse waveforms acquired by the 5112 digitizer. For each beam pulse a 50-kHz pulse stream proportional to the pulse charge is generated by the 6602 and this real time information is distributed to all experimental areas.

THE PICKUP

The PSR accumulates 800-MeV protons from a LINAC and these are routed to the spallation target with a 20-Hz repetition rate. The typical pulse width and charge per bunch are 300 ns and 5 μC. The pickup is a few meters from the target. It consists of a Pearson 1010 current monitor inside a cylindrical electro-magnetic shielding container. The 1010 has an inside diameter of 10.75 inches and is mounted around a ceramic break in the 4-inch diameter beam tube. The 1010 consists of a steel tape wound toroid around which are wrapped 500 evenly spaced output turns of copper wire. A 2-Ω resistor is connected across each group of 20 turns for a total output resistance of 50 Ω. The output is accessible through an N-type connector. A single turn calibration loop was added, both ends of which are available through a BNC connector. Normally one end of this coil is connected to ground through 50 Ω and a calibration signal can be injected through the other. The device is located in a permanently high radiation area so the calibration input and signal output are usually connected to instrumentation only through cables that are 6.7 m and 48.8 m respectively. The ends of these cables are located in completely different locations and it was impractical to install an additional cable so both signals can be accessed from the same place. Thus all calibration measurements were made in the time domain. A spare unit is available and its transfer function was measured using a network

*This work was sponsored by the U.S. Department of Energy.

CP732, *Beam Instrumentation Workshop 2004: Eleventh Workshop,*
edited by T. Shea and R. C. Sibley III

analyzer. The transfer function was also measured by injecting a square wave into the calibration input and by observing the output on an oscilloscope. Results are show in Fig. 1.

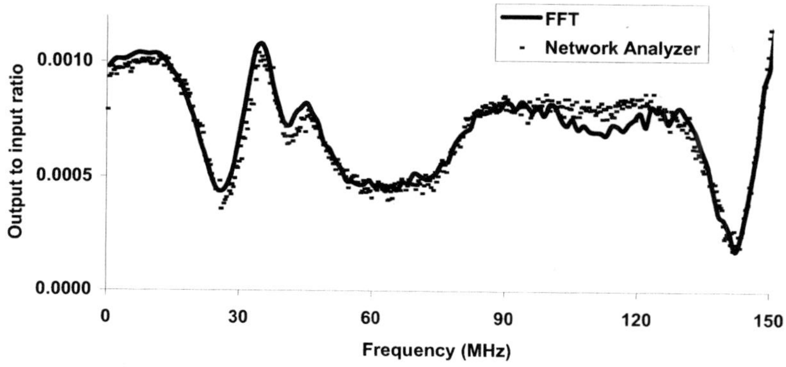

FIGURE 1. FFT output to input amplitude ratio compared to Network Analyzer transfer function.

A MODEL OF THE PICKUP

It can be shown that for sufficiently low frequencies

$$i_b - \frac{N}{R_1} v_o + i_c = \frac{2\pi \int_{t_1}^{t} v_o dt}{\mu N W ln(r_2/r_1)} + \mathrm{C}. \tag{1}$$

Here i_b is the beam current, $N = 500$ is the number of output turns on the toroid, v_o is the output voltage, R_1 is the impedance seen by the output coil, i_c is the calibration current, t is the observation time, t_1 is a time well before the onset of a current pulse, μ is the toroid permeability, W is the toriod width, r_1 and r_2 are the inside and outside radii of the toroid and C is an integration constant. In the derivation of this formula it is assumed that μ is the same throughout the toroid at any one time and at low frequencies this is a good approximation. The permeability μ can and will vary with time. According to the manufacturers' specifications, the transformer 3-db low frequency cutoff is $f_o = 400$ Hz and the output at frequencies well above this cutoff is 0.1 V/A = R_1/N. Substituting an e(jωt) expression for the current and voltage and applying the above specifications (also dropping all integration constants) one obtains

$$\frac{2\pi}{\mu N W \, ln(r_2/r_1)} = \omega_o \frac{N}{R_1} , \tag{2}$$

where $\omega_o = 2\pi f_o$. Note that ω_o is a function of μ. Rewriting Eq. 1 with i representing one of i_b or i_c and that i and v_o were 0 at t_1:

$$i = \frac{N}{R_l}\left[v_o + \omega_o \int_{t_1}^{t} v_o dt \right] . \tag{3}$$

Evaluating this from a time t_- well before a beam pulse to t_+ well after a beam pulse results in the droop voltage

$$v_o(t_+) = -\omega_o \int_{t_-}^{t_+} v_o dt . \tag{4}$$

Figure 2 shows a typical beam pulse. From this one can see that the output signal v_o is unipolar, lasts about 300 ns and is roughly triangular. Call the peak value of $v_o = V_p$ then, assuming ω_o has its low frequency value, $v_o(t_+) \approx -3.8*10^{-4}\ V_p$. This implies that when Eq. 3 is integrated to get the beam charge, the contribution from the last term is three orders of magnitude less than the contribution of the first term. Thus the term $\omega_o v_o$ is negligible for the 300-ns beam pulses from the PSR and Eq. 1 becomes

$$v_o = iR_l/N . \tag{5}$$

For the beam signals neither μ or ω_o are necessarily constant. However it will be experimentally demonstrated that the right side of Eq. 1, the source of the droop term, is in fact negligible for beam signals.

The current i can be either the beam pulse or a comparable calibration pulse. Using Eq. 5 it can be shown that the impedance of the calibration loop is

$$Z_c = R_l/N^2 = 10^{-4}\,\Omega , \tag{6}$$

for $R_l = 25\ \Omega$. If a calibration voltage is applied directly to the calibration input or through a 50-Ω cable, the input current is given by dividing the input voltage by the termination resistor. Input and output voltages were measured to get Fig. 1.

Assuming μ is constant, both Eqs. 1 and 5 are inconsistent with Fig. 1 because they both predict a constant value for the transfer function at high frequency. However it is assumed that Eq. 1 with a constant permeability is valid for the frequency range of interest. Calibration measurements with pulse widths from 300 ns to 300 μs were performed and the calibration constant (to be defined below) has the same value within a fraction of a percent for all pulse widths. The relative frequency content of the shortest calibration pulse is greater at high frequencies than for a typical beam pulse and thus it is concluded that the permeability for the beam pulse can be taken as constant at its DC value. As a consequence the estimate of the droop voltage using Eq. 4 is verified and Eq. 5 is valid for beam pulses.

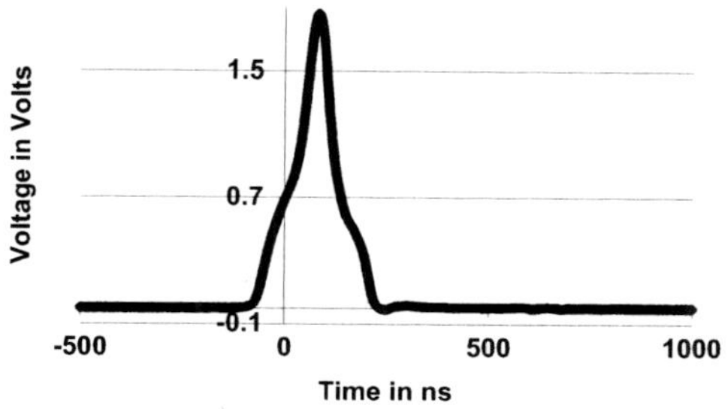

FIGURE 2. A typical beam pulse averaged over 512 pulses.

SIGNAL PROCESSING AND CALIBRATION

All measurements were done with a digitizer, either a digital oscilloscope or the PXI-5112 100-MS/s digitizer. Thus it is useful to use the discrete Fourier Transform pair [1]

$$H_n = \sum_{k=0}^{L-1} h_k e^{\frac{2\pi jkn}{L}} \text{ and } h_k = \frac{1}{L}\sum_{n=0}^{L-1} H_n e^{\frac{-2\pi jkn}{L}} , \tag{7}$$

where L is the number of samples, h corresponds to the time domain sample set and H is the discrete Fourier Transform set. Taking $n = 0$ in the first of the two and applying it to input current and the output voltage

$$I_0 = \sum i_k \text{ and } V_0 = \sum v_k , \tag{8}$$

where I_0 and V_0 are the zero frequency discrete Fourier components for the measurement interval under consideration. The i_k's and v_k's can contain signals from high frequencies but these all average out to zero when summed over the entire sampling interval. Note that I_n and V_n are frequency domain components with I_0 and V_0 representing the zero frequency or DC part. Also note that the results do not depend either on the observation interval or the sampling rate. Integrating Eq. 3 for the charge in a beam or calibration pulse results in

$$Q = \frac{N}{R_1}\int_{t_-}^{t_+} v_o dt + \omega_o \frac{N}{R_1}\int_{t_-}^{t_+} dt \int_{t_-}^{t} v_o dt' \tag{9}$$

(ω_o was taken outside the integral despite the fact that it almost certainly varies with time for short time intervals). As has already been explained the last term will be negligible for beam pulse but it will play a significant role in the calibration process. The corresponding expression in the digital picture is

$$\sum_{k=0}^{L-1} i_k = \frac{N}{R_1}\sum_{k=0}^{L-1} v_k + \frac{N}{R_1}\omega_0 \Delta t \sum_{k=0}^{L-1} \sum_{n=0}^{n=k} v_n \tag{10}$$

Multiplying the whole expression by Δt (= sample time) results in Q the input charge.

For an accurate calibration it was essential to measure the value of ω_o (at low frequencies). This was done by applying a step current input to the calibration winding and plotting $\log_e$ of the decay part of the output voltage as a function of time. This results in ω_o = 806.7/s for 25-ohm output.

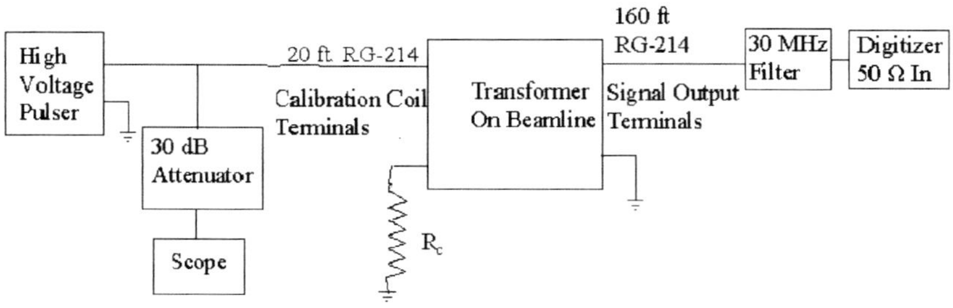

FIGURE 3. Calibration setup.

Figure 3 shows the setup for doing the calibration of the current monitor. Approximately rectangular pulses were injected into the 20-ft. calibration cable ranging in width from 300 ns to 300 μs. The input and output for a 300-ns pulse are shown in Fig. 4.

The digitally sampled current flowing into the toroid is given in terms of the Fourier components as

$$i_k = (1/L)\sum_{n=0}^{L-1} I_n e^{-2\pi jnk/L} \quad . \tag{11}$$

The I_n are given in terms of the U_n, the Fourier components of the input voltage, by

$$I_n = U_n/(Z_n + R_c) \ , \tag{12}$$

where Z_n is the non-resistive part of the cable impedance plus anything else which may be important and R_c is the termination resistance plus the DC cable resistance. It is possible that this expression does not include all possible impedances but that does not matter as long as $I_o = U_o/R_c$ at zero frequency. The charge passing through the transformer is obtained by summing over all i_k. Inserting (12) in (11) and summing over the current samples gives

$$\sum_{k=0}^{L-1} i_k = \frac{1}{L}\sum_{n=0}^{L-1} \frac{U_n}{Z_n + R_c} \sum_{k=0}^{L-1} e^{-2\pi jnk/L} = \frac{1}{L}\sum_{n=0}^{L-1} \frac{U_n}{Z_n + R_c} L\delta_{n,0} = \frac{U_0}{R_c} \ . \tag{13}$$

Thus only the zero frequency component or average voltage during the sampling interval and a DC resistance measurement are needed to determine the charge that passed through the transformer during the calibration pulse. We define a calibration constant K_a as

$$K_a = \frac{1}{R_c} \frac{\sum_{n=0}^{L-1} U_n}{\sum_{n=0}^{L-1} v_n + \Delta t \omega_o \sum_{n=0}^{L-1} \sum_{i=0}^{n} v_i}. \tag{14}$$

K_a was evaluated for the six calibration data sets. Their average is 19.28 ± 0.4% rms deviation of the six measurements. Thus K_a is constant over three orders magnitude of pulse width. It is important to note that for 300-ns calibration pulses using $\omega_o = 1614/s$ the ratio of the last term to the first term in the denominator is 0.0018 and for the 300-µs pulse it is 0.21.

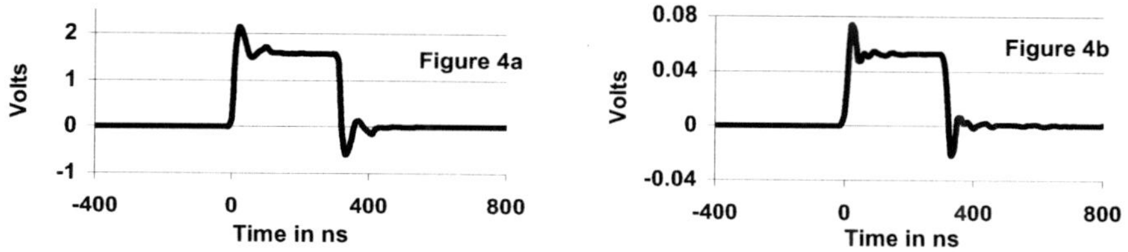

FIGURE 4. A calibration input signal (a) and output signal (b).

Figure 5 shows the Fourier Transform of the beam pulse of Fig. 2, and this is bracketed by the Fourier Transform of a 300-ns and a 30-µs calibration pulse. The amplitudes were adjusted so they are the same at the lowest non-zero frequency. Since the two calibration pulses result in the same value for K_a this demonstrates that this K_a is also applicable to the beam pulse. K_a is calculated assuming μ as constant. The fact that K_a is independent of pulse width and thus the frequency spectrum, demonstrates that it is acceptable to assume μ constant. This also demonstrates the validity of Eq. 5 for beam pulses.

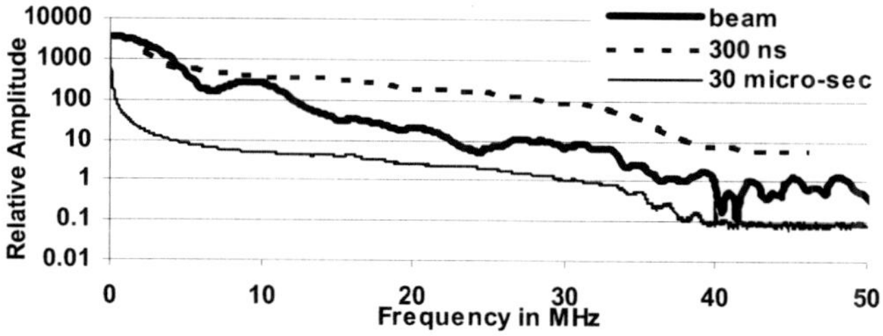

FIGURE 5. Comparison of beam pulse and calibration pulse frequency content.

To obtain measurements whose values are traceable to NBS standards, the instruments used in the calibrations were checked against a voltage reference. The 500-MHz digital oscilloscope used in the measurements was first calibrated with a DC voltage standard traceable to NBS. This was done by comparing the averaged offset corrected displacement of the oscilloscope trace on 7 ranges for 25 voltage standard output values. The standard to oscilloscope ratio was found to be 1.012 ± 0.2%. The digitizer has 42 voltage ranges and a nearly unlimited number of adjustable offset voltages. Only two ranges and offsets were used during the calibration process and these were calibrated by simultaneously applying 40-μs wide square pulses to the digitizer and the oscilloscope. The digitizer to scope ratio was 1.0068 ± 0.54% using 21 different voltages. The ranges that are to be used for beam charge measurements resulted in a digitizer to scope ratio of 0.986 ± 0.3%. The 30-dB attenuator shown in Fig. 3 was calibrated using the DC standard and the oscilloscope. The attenuation was measured as 30.105 ± 0.006 dB. The DC resistance of R_c in series with the 20-ft. cable was measured with an ohmmeter and the meter was checked with a precision Network Analyzer 50-Ω standard. This was found to be 47.68 Ω ± 0.1%. Combining all this results in K_a = 19.65 V/A ± 0.8% and for beam pulses

$$Q = K_a \Delta t \sum v_n \quad . \tag{15}$$

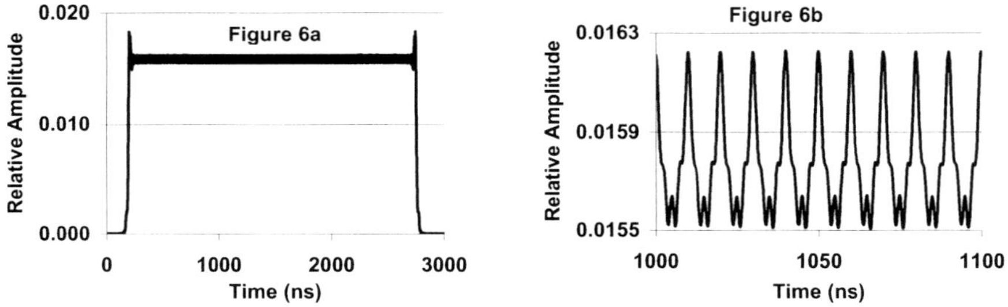

FIGURE 6. Convolution of a beam pulse with a 100-MS/s pulse train.

The 100-MS/s PXI-5112 sampling clock is not synchronized to the beam and thus different results can be obtained for identical beam pulses. To estimate the effect of this the data for Fig. 2, taken at 5 GS/s, were convoluted with a 256-long impulse train where every fiftieth sample is one and the rest are zero. The train represents the 256 beam samples per beam pulse. The result is shown in Figs. 6a and 6b. Figure 6b is an expanded version for a part of 6a. The rms deviation for the data of 6b is 1.3%. This will tend to average to zero for many beam pulses.

SOFTWARE, DIGITAL HARDWARE, AND SOME RESULTS

The primary objective of this system is to measure the beam charge and provide a 50-kHz pulse stream to a fiber-optic distribution system for each beam pulse. The number of pulses is obtained by dividing the beam charge by 20 nC. Additionally a timing pulse is generated about 60 ns after the onset of the beam pulse. Also the charge on various time scales is computed and displayed. A subset of these charge sums is distributed over the Net to EPICS.

Figure 5 represents the hardware. It is centered on a National Instruments Real Time system running under LabVIEW RT. The hardware consists of a PXI-1002 chassis that contains a PXI-8176 controller, a PXI-5112 100-MS/s digitizer, and a PXI-6602 scalar and digital I/O modules. A TB-2715 interface module plugs into the output of the PXI-6602 and the pulse stream can be accessed on a terminal block in the TB-2715. The 8176 controller runs the real time operating system under which LabVIEW programs can execute in guaranteed lengths of time without interference from other processes. The controller communicates to other devices through the Internet. LabVIEW 7.0 and LabVIEW RT 7.0 reside in the PC and constitute the development system. Programs are written on the PC and downloaded to the 8176 controller. The front panel of all processes that are downloaded to the controller can run on the PC and simultaneously on one other computer on the Internet. Only the PC can be used for program development and downloading of software to the Controller.

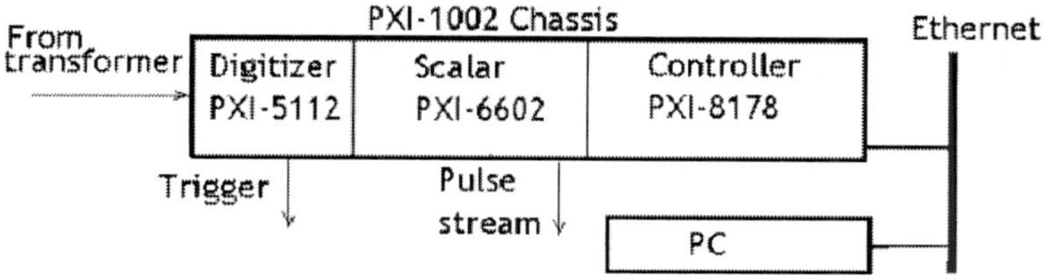

FIGURE 7. Hardware overview.

LabVIEW on the PC can be run either as conventional LabVIEW or by connecting to the Controller over the Net to operate in the Real Time (RT) mode. These two modes are mutually exclusive.

The software consists of three main parts that run on the controller with corresponding front panel displays running on another computer. The first is RTP.vi (RTP stands for Real Time Part). This needs to run at critical priority on the Controller and it generates the required pulse stream proportional to the beam charge. Its front panel contains no controls or indicators to interfere with real time operation. The second program is RTC (RTC stands for Real Time Control). This is the user interface with a complicated front panel. It also calculates most of the charge sums for display. The third part sends a subset of charge sums and time information over the Net to the PC or another computer running LabVIEW. These numbers are then communicated to EPICS by a LabVIEW program with the appropriate ActiveX interface [2]. For the PC to be the ActiveX interface it has to run LabVIEW in normal operating mode and thus it cannot act as the interface with the controller. However the Controller can run and do all its functions without any other computer running the front panel.

Once initialized RTP runs in an indefinite loop in which, for each beam pulse, it receives a 256-sample digitized array from the 5112. It finds the peak signal location in this set. The digitizer is programmed such that the peak signal is at about sample number 76. Samples 0 through about 47 are used to determine a background value. Samples 48 (about) to 230 (about) are summed and the background correction is applied. This is then multiplied by the voltage calibration, $1/K_a$ and the time between samples to determine the charge. The charge is then used to calculate the required number of pulses and this is used to program a hardware pulse generator on the PXI-6602. The 6602 then produces the pulse stream. RTP also adds the bunch charge to a running sum of accumulated charge since run start. Things like the waveform, time and many other numbers are sent to RTC. A fourth low priority program can optionally be used to archive data to the Controller hard drive every second.

The digitizer vertical parameters can be set in two ways. One is to selecting one of 7 voltage ranges from a table. The other way is to use an automatic gain adjustment feature of the program. The program will adjust the vertical position of the digitized waveform so the highest value is between digital count 85 and 107 and the lowest value is between digital count –85 and –107.

In the most common operating mode 20, 5-μC or greater beam pulses per second are delivered to the target for an average beam current of 100 μA or greater. Figure 8 shows average current recorded every minute during parts of January 26 and 27, 2004. The beam current delivered to the target is also obtained by measuring the current before injection into the PSR and this is corrected for losses observed by beam loss monitors. The fractional difference between this and the new acquisition system are also shown in Fig. 8. The difference is usually less than 1%. Some larger differences can be seen and these are probably due to differences in how the averages are calculated combined with large variations in beam current.

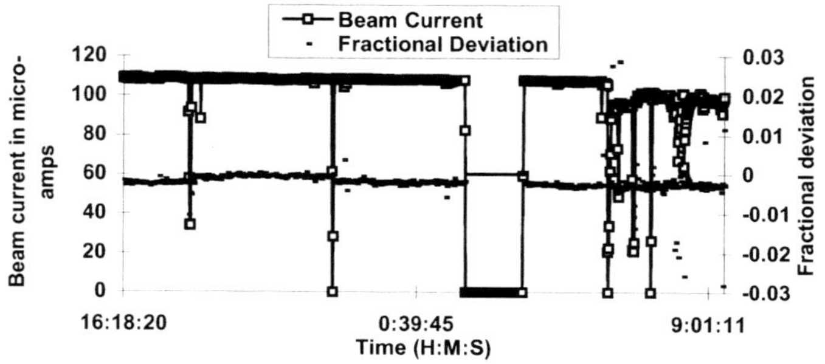

FIGURE 8. Beam current and its fractional deviation from measurements estimated by measurements by an upstream current monitor.

REFERENCES

1. W. H. Press et al., *Numerical Recipes in Fortran*, 2nd edition, New York, Cambridge University Press.
2. K.U. Kasemir, M. Pieck, and L.R. Dalesio, "Integrating LabVIEW into a Distributed Computing Environment," 8th International Conference on Accelerator and Large Experimental Physics Control Systems, San Jose, California, USA, November 27-30, 2001, LANL: LA-UR-01-5905. Also see Kasemir's online info about LabVIEW EPICS at http://lansce.lanl.gov/lansce8/Epics/ActiveXCAServer/default.htm.

Optical Synchrotron and Edge Radiation Diagnostics for Relativistic Electron Beams

A. G. Shkvarunets and R. B. Fiorito

Institute for Research in Electronics and Applied Physics
University of Maryland, College Park, MD 20742, USA

Abstract. In accelerators such as storage rings, energy recovery linacs (ERLs), and free electron lasers, bending magnets are used to alter the trajectory of a relativistic electron beam. Frequently the magnets are arranged in pairs in either a "U" or "S" configuration with an intervening straight section. In such configurations synchrotron radiation (SR) or edge radiation (ER) is produced in the bend and/or as the beam exits the field of each magnetic field region. Under certain conditions the radiation is produced in the visible region of the spectrum. The radiation from the second magnet can interfere with that of the first magnet, when a distance comparable to the coherence length of the radiation separates the two magnetic fields. For a beam with low energy spread, the visibility of the interference radiation depends on the beam divergence. To design a sensitive beam divergence diagnostic we have developed a computer code, which can calculate the far field radiation distribution produced by a relativistic electron moving along any trajectory in a magnetic field. The code model is based on the "colliding electron" approximation [1] and reproduces all known features of SR and ER. The code can account for the spatial field distribution of the dipole magnet, the effect of beam divergence and the bandwidth of the measurement. Results of our calculations are presented for the 1.9-GeV Cornell Electron Storage Ring and the 100-MeV ERL at the Thomas Jefferson National Laboratory Free Electron Laser.

SYNCHROTRON RADIATION–FARFIELD PATTERN

Electromagnetic (EM) energy radiated by an electron moving in vacuum can be found as a superposition of radiation from small parts of trajectory-arcs dl_i with local radii R_i. The radiation from dl_i is given by the formula [1]

$$dW_i(\omega,\theta,\varphi) = W_i(\omega,\theta,\varphi)\, d\Omega \;\; d\omega \;, \tag{1}$$

where

$$W_i(\omega,\theta,\varphi) = \frac{e^2}{4\pi^2 c}\left(\frac{\vec{\beta}_2\times\vec{n}}{1-\vec{\beta}_2\vec{n}} - \frac{\vec{\beta}_1\times\vec{n}}{1-\vec{\beta}_1\vec{n}}\right)^2 \;, \tag{2}$$

$dW(\omega,\theta,\varphi)$ is the energy radiated along direction $\vec{n}$ into solid angle $d\Omega = \sin\theta\cdot d\theta\cdot d\varphi$ in frequency interval $d\omega$, $\vec{\beta}_1 = \vec{V}_1/c$, $\vec{\beta}_2 = \vec{V}_2/c$, $\vec{V}_1$ and $\vec{V}_2$ are the beginning and the ending velocities of electrons in the arc dl_i, c is the speed of light and, e is the charge of the electron. Formula (1) is applicable in the long

CP732, *Beam Instrumentation Workshop 2004: Eleventh Workshop*,
edited by T. Shea and R. C. Sibley III

wavelength limit, i.e. $dl_i << \lambda_f \approx \gamma^2 \lambda$, where λ is the wavelength, λ_f is the coherence length, and γ is the Lorentz factor of the electron. An additional restriction is a small electron deflection angle in the interval dl, i.e. $d\alpha_i = dl_i / R << 1/\gamma$. This condition can be rewritten as $dl_i << \lambda_c \gamma^2 / 4$, where λ_c is the critical wavelength of the synchrotron radiation.

The total radiation of the electron from the trajectory $L = \sum_i^N dl_i$ is a coherent sum of elementary radiations:

$$dW_L = \left(\sum_i^N f(\psi_i) \sqrt{W_i} \right)^2 d\Omega \, d\omega, \tag{3}$$

where $f(\psi_i)$ is a term associated with the phase shift due to wave propagation along $\vec{n}$ and the propagation of the electron along trajectory L. Note that the radiation from dl_i does not depend on λ (similar to transition radiation TR), but that the radiation from L does depend on λ.

For calculations of the radiation, the length dl_i should be as large as possible because it directly affects the total number N of computational intervals and thus the code run time. For relativistic electron energies the length of the arc dl_i can be $>> \lambda$. For example, if the electron energy E_e = 1 GeV, $\gamma \approx 2000$, L = 1 m and λ = 1 nm, the total number of intervals $N > 250$. This estimate shows that the code run time can be reasonable.

We have developed a code, which calculates the angular distribution of SR in the far field. We have employed this code to SR from magnetic dipoles (bending magnets) with either sharp or smooth edges separated by a region without magnetic field. The code computes both horizontal and vertical polarizations and can accommodate finite waveband as well as take into account a finite electron beam divergence. In the latter case it is supposed that the beam electrons radiate incoherently (long bunch approximation). The code calculates the SR angular distribution in *X, Y, Z* coordinates. *XZ* is the horizontal plane (plane of the electron trajectory and the observation direction), *YZ* is the vertical, i.e., orthogonal, plane. The magnetic field is directed vertically only, i.e. $B = B_y$ and is considered to be uniform along the electron trajectory. The edge magnetic field distribution is approximated by a Gaussian distribution. The horizontal observation angle *Xo* observes the SR distribution measured from the vertical plane *YZ*, and the vertical observation angle *Yo* from the horizontal plane *XZ*. The general geometry for the calculations is shown in Fig. 1.

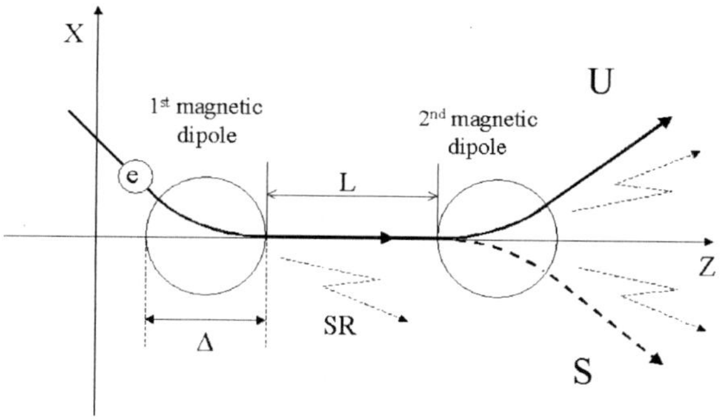

FIGURE 1. Geometry for generation of SR and ER from two magnetic dipoles in "U" or "S" configuration. Radiation is observed in the Z direction.

SR CODE VERIFICATION

In order to verify our code we have calculated the SR distribution for various conditions and compared our computational results with those published in the literature. The following computations are presented:

Standard Synchrotron Radiation–SSR

SSR is calculated from a region of homogeneous magnetic field far from the edges of the bend magnet. The well known properties of SSR are: 1) the intensity does not depend on *Xo*; 2) both polarization components depend on *Yo*; 3) the radiation is localized near the *XZ* plane – plane of trajectory of electron; and 4) the intensity is maximum for $\lambda = \lambda_c$, the critical wavelength.

Our code calculations of SSR show that intensity does not depend on *Xo*. Vertical scans of horizontal and vertical polarization components are presented in Fig. 2. These results are in very good agreement with data presented in Fig. 4-2 of ref. [2]. Maximum intensities are a function of the ratio λ / λ_c. Peak intensities calculated by our code are presented in Table 1. In this table the intensities are normalized by the peak intensity of SR calculated at $\lambda = \lambda_c$. These results are in very good agreement with the data presented in Fig. 4-1 of ref. [2].

TABLE 1. Peak intensities of synchrotron radiation.

λ / λ_c	$I_{peak}/ I_{peak}(\lambda_c)$
1000.0	0.020
100.00	0.094
10.000	0.416
1.000	1.000
0.100	0.001

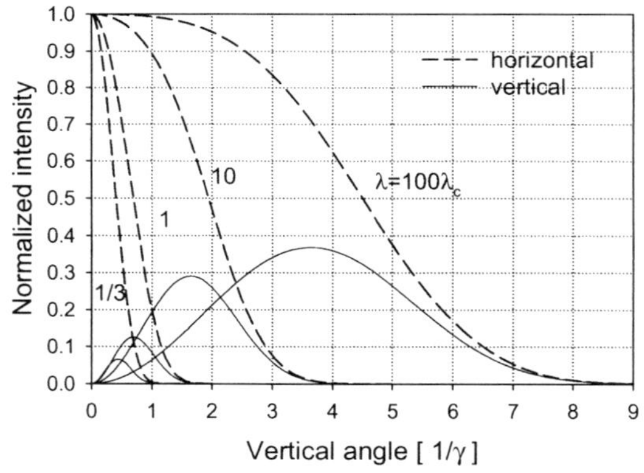

FIGURE 2. Code calculations of vertical scans of horizontal and vertical polarization components of SR normalized by corresponding horizontal components for various wavelengths in terms of the critical wavelength

Edge Radiation (ER) from a Sharp Magnetic Field Transition, $\lambda >> \lambda_c$

This case is shown in Fig. 3. The radiation pattern is almost a symmetric cone around the direction velocity of the electron exiting the magnetic field region and is very similar to the radiation pattern of transition radiation (TR), which has a symmetric cone with intensity maximum at angle $\theta_p = \gamma^{-1}$. This result is in very good agreement with the results presented in Fig. 1 of reference [3], p. 527.

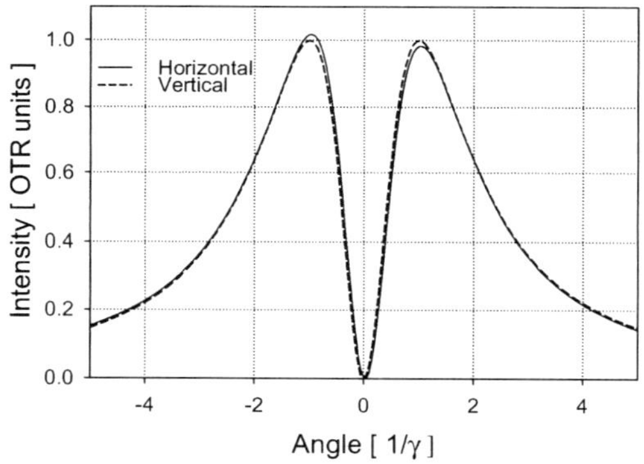

FIGURE 3. Horizontal scan of horizontal polarization component and vertical scan of vertical polarization component of magnetic edge radiation in the long wavelength limit $\lambda >> \lambda_c$.

Very Short Magnetic Field Radiation, Angle $\theta_e << \gamma^{-1}$, $\lambda >> \lambda_c$

The results of our calculation are not shown here to conserve space but have the following characteristics. The horizontal (X) polarization component is a single peak symmetric with respect to the direction of the exit velocity of the electron. In addition, 1) the horizontal width of the angular distribution of the radiation is smaller than the vertical width, 2) the peak intensity of the vertical (*Y*) polarization component is 16 times smaller than the peak of the horizontal (*X*) component, and 3) the *Y* component consists of 4 peaks symmetric with respect to the direction of the exit velocity of the electron. Here the angular coordinates of the peak intensities are (*Xo, Yo*) = (± 0.7, ± 0.7). These results are in very good agreement with those of Fig. 2 of ref. [3].

Edge Radiation Interferences (ERI)

In this calculation radiation from two magnetic field regions interact. The electron radiates coherently at each position in the magnetic field but is incoherent with respect to a neighboring electron in the bunch. The total radiation pattern is the result of superposition of many individual coherent sources – computed along the trajectory in the magnetic field. One can expect that if an electron passes by two short magnetic fields each having a width Δ separated by a distance L $>>\Delta$, then the total radiation will be modulated similar to that observed from an OTR interferometer [4]. The radiation pattern envelopes (i.e. single magnetic region or foil radiation patterns) are different for ER and TR but one expects that the fringe structure should be almost the same. This is the case. Figure 4 shows a comparison of the intensities of OTRI and ERI produced by the interaction of a 100-MeV electron with two foils separated by L = 50 mm and two short kickers with spatial extent Δ = 1 mm separated by L = 50 mm.

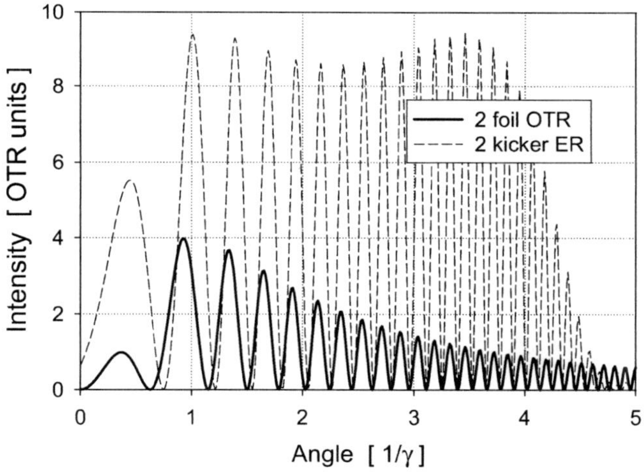

FIGURE 4. Comparison of edge radiation from two short kicker magnets (Δ = 1 mm) separated by a distance L = 50 mm, with OTR from two metallic foils separated by the same distance.

SPECIFIC CALCULATIONS FOR REAL CASE SCENARIOS

ESRI at the Cornell Electron Storage Ring (E = 1.9 GeV, $\lambda >> \lambda_c$)

At the CESR a straight section with length L = 17.944 m lies between two U-shaped bending magnetic dipole sections each with radius R = 1.406 m. The critical wavelength for each of these dipoles, λ_c = 11.45 nm. At optical wavelengths ($\lambda \approx$ 500 nm $>> \lambda_c$) the intensity of ER is comparable to OTR. The angular pattern of the radiation is an asymmetric cone with almost zero intensity in the direction of the electron velocity in the straight section. The radiations from each of the two magnetic edges interfere. Beam angular spread can affect the visibility of the fringes. Like OTRI, this effect can be used to measure the angular divergence of the beam.

One of the ways to increase sensitivity to the beam divergence is to decrease the wavelength of observation. In Fig. 5 a vertical (Y) scan of the vertical component of polarization is presented for an observed wavelength $\lambda = 200$ nm. The vertical components from each magnet overlap almost completely resulting in an interference pattern shown in Fig. 6, which is very similar to OTRI observed from two metallic foils. The effect of the divergence decreases the visibility (depth) of the fringes. The calculations shown here indicate that an rms divergence $\sigma_y \approx 2$ μrad can be measured.

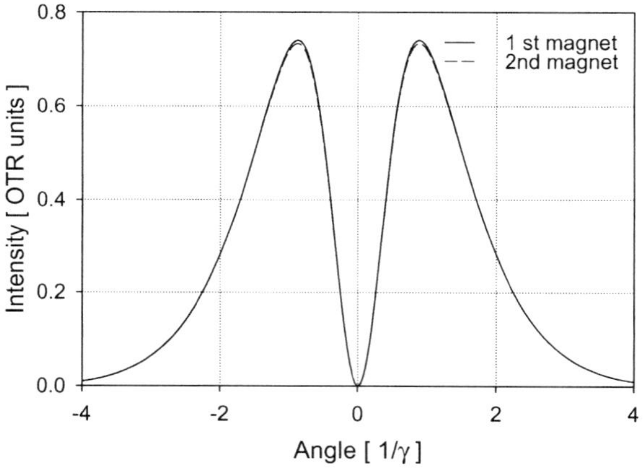

FIGURE 5. Vertical scan of vertical polarization component of edge radiation from the Cornell Electron Storage Ring (CESR), E = 1.9 GeV, λ = 200 nm, λ_c = 11.4 nm.

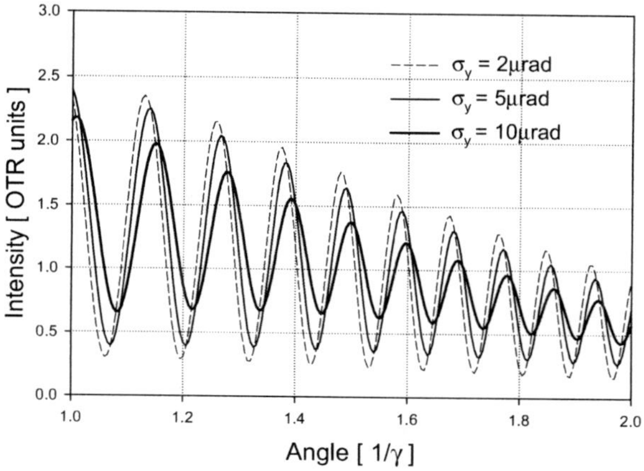

FIGURE 6. The interference intensity distribution of the components shown in Fig. 5 above, produced by a beam with three values of vertical angular spread σ_y for a beam with fixed horizontal divergence $\sigma_x = 40$ μrad.

Kicker Interferometer for the TJNAF FEL Accelerator, $\lambda \approx \lambda_c$

We demonstrate that the angular divergence of a *moderate* energy electron beam can be measured by using a SR interferometer composed of two magnetic kickers with opposite directions of magnetic field ("S" electron trajectory of Fig. 1). For example, a 100-MeV electron moving in a magnetic field with strength B$\approx$ 0.3 Tesla has a critical wavelength λ_c of SR that is in the visible region of the spectrum. The pattern of radiation in this case is very different from the edge radiation, since high radiation intensity is produced from the entire region of the kicker, not just the edge of the field. In this case the horizontal radiation pattern is proportional to the electron deflection angle when $\theta_e > \gamma^{-1}$. Overlap of the radiation from the two kickers takes place if these have opposite magnetic field directions, i.e. the "S" configuration shown in Fig. 1.

Figure 7 shows the interference of the horizontal polarization of SR from two kickers in the "S" configuration produced by electron beams with various values of angular divergence along with the radiation from each magnetic dipole.

Calculation of the vertical scan of the vertical polarized component of the interference radiation from two kickers show that the beam divergence has a similar effect on the depth of the fringes observed. The difference is that the width of the envelope of the vertically polarized intensity cannot be controlled by the horizontal beam deflection angle, as is the case for the horizontal polarization.

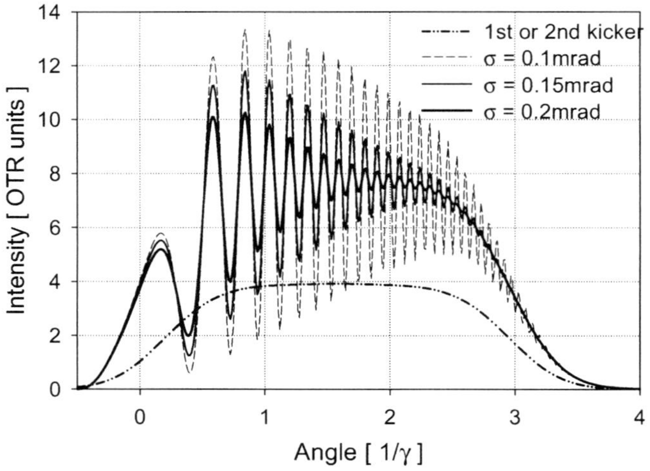

FIGURE 7. Horizontal scan of the horizontal polarization of two interfering kicker magnets, E = 100 MeV, $\lambda = \lambda_c = 600$ nm, electron deflection angle $\theta_e = 3.2/\gamma = 16.3$ mrad.

SUMMARY

We have demonstrated computationally that SR and ER interferences can be used as a sensitive diagnostic of beam divergence for high and moderate energy electron beams. Such a diagnostic can be tuned to the energy and divergence range of the beam of interest by controlling the observation wavelength, magnetic field and separation distance between the magnets.

ACKNOWLEDGMENTS

This work is supported by the Office of Naval Research and the Joint Technology Office of the U.S. Department of Defense.

REFERENCES

1. L. D. Landau and E.M. Lifshitz, *The Classical Theory of Fields,* Pergamon, Oxford, 1971.
2. X-Ray Data Booklet, Lawrence Berkeley Laboratory, University of California, 1986.
3. R. A. Bosch, *Nuclear Instr. and Meth. in Phys.* ,Research A 386, 1997.
4. R. B. Fiorito and D. W. Rule, AIP Conf. Proc. 319, R. Shafer ed., 1994, p. 187.

Toward a Cold Electron Beam in the Fermilab Electron Cooler

V. Tupikov[*,†], S. Nagaitsev[*], S. Seletskiy[‡] and A. Shemyakin[*]

[*]*Fermilab, PO Box 500, Batavia, IL 60510, USA*
[‡]*University of Rochester, Rochester, NY 14627, USA*

Abstract. Fermilab is developing a high-energy electron cooling system to cool 8.9-GeV/c antiprotons in the Recycler ring [1]. Cooling of antiprotons requires a round electron beam with a small angular spread propagating through a 20-m long cooling section with a kinetic energy of 4.3 MeV. To confine the electron beam tightly and to keep its transverse angles below 0.1 mrad, the cooling section will be immersed into a solenoidal field of 50-150 G. This paper describes the technique of measuring and adjusting the magnetic field quality in the cooling section and presents preliminary results of beam quality measurements in the cooler prototype.

INTRODUCTION

To prepare and test major elements of the electron cooling system, a full-scale prototype has been assembled and commissioned in a separate building outside the Recycler ring. The major differences are shorter beam lines, a shorter cooling section (9 solenoidal modules instead of 10), and a lower beam energy (3.5 MeV instead 4.3 MeV). One of the goals for the prototype was to develop a procedure for measuring and minimizing an effective temperature of the electron beam in the cooling section. The subject of the paper is one of the most important components of the effective temperature, a beam centroid motion in the cooling section (CS). The motion may be caused by a non-optimal matching of the beam at the entrance of the cooling section, by a discrepancy between the mechanical axis of the cooling section and the solenoid axis, and by imperfections of the solenoid magnetic fields that lead to transverse field components. While the two former can be relatively easily adjusted according to beam-based measurements, the transverse field components have to be measured directly and compensated according to the magnetic field measurements.

The specifics of the Fermilab electron cooler in comparison with existing electron cooling devices are a combination of its low magnetic field in the cooling section, 50-150 G, and its high electron energy, 4.3 MeV. As a result, the corresponding Larmour wavelength, 7-20 m, is significantly larger than a typical scale of changes in the transverse components of the magnetic field, which are about the diameter of the solenoid, 20 cm. Hence, such a short region of the cooling section solenoid can excite

[†]tupikov@fnal.gov.

CP732, *Beam Instrumentation Workshop 2004: Eleventh Workshop*,
edited by T. Shea and R. C. Sibley III

the electron transverse motion with an angle of α_x proportional to an integral of the transverse field component B_y over this region:

$$\alpha_x = \frac{1}{(B\rho)} \int_L B_y \cdot dz \,, \tag{1}$$

where $B\rho = 1.6 \cdot 10^4$ G·cm. To keep the contribution of such excitations well below the restriction for the total electron angle, 0.1 mrad, the transverse fields should be compensated down to the level of

$$\int_L B_y \cdot dz \le 0.3 \text{ G·cm}. \tag{2}$$

For this level of compensation, the magnetic measurements should have a mG-range resolving capability. The sensor used in measurements (see the next section) did have the necessary resolution but lacked a long-term stability. In the final sections of the paper we will discuss possible ways to improve the compensation of the transverse field according to beam-based measurements.

MAGNETIC MEASUREMENTS

Design of the Cooling Section

The 20-m long CS consists of ten 2-m long identical solenoid modules separated by instrumentation gaps [2]. The gaps are used for the beam diagnostics instrumentation and vacuum pump connections. The main solenoid in each module is flanked by trim solenoids to preserve the average value of Bz over the gaps. Transverse fields are compensated by 10 pairs (per module) of flexible-circuit dipole coils. A gap between 2 modules is shown in Fig. 1, and the CS parameters are listed in Table 1.

Each module is equipped with a pair of capacitive pickups used as beam position monitors (BPMs). Transverse positions of the BPMs with respect to each other were calibrated with an accuracy of 0.1 mm by using a wire stretched over the length of the CS. The measured rms BPM noise is 5 μm in the DC beam mode and 100 μm in the pulse mode. The detailed BPM description can be found in [4].

Compass Based Magnetic Measuring System

The transverse magnetic fields were measured by a compass-based sensor [3]. The schematic layout of the measuring system is shown in Fig. 2. A mirror attached to the compass reflects a laser beam to a 4-segmented photodiode. Using a pair of differential signals from X and Y photodiode segments, two identical electronic feedback systems generate currents in compensation dipole coils, wound around the compass. The fields of the coils rotate the compass until the reflected laser light comes to the center of the photodiode. The value of coil currents in the equilibrium is used as a measure of the solenoid transfer field. The compass and the coils are mounted on a cart, which is pulled through the vacuum chamber.

TABLE 1. Cooling Section Parameters

Parameter	Value
Total Cooling Section	
Total length of the CS	20 m
Number of modules	10
Gap length	8 cm
Module Solenoid	
Length / ID / OD (cm)	188.2 / 15 / 20
Magnetic field / current	40 G / 1 A
Trim Solenoid (2 per module)	
Length / ID / OD (cm)	3.5 / 15 / 20
Magnetic field / current	49 G / 1 A
Maximum field/current	0.8 G / 1 A
Dipole Corrector (10 pairs per module)	
Length trim / main	3.56 / 23.28 cm
Maximum field/current	1.0 G / 1 A

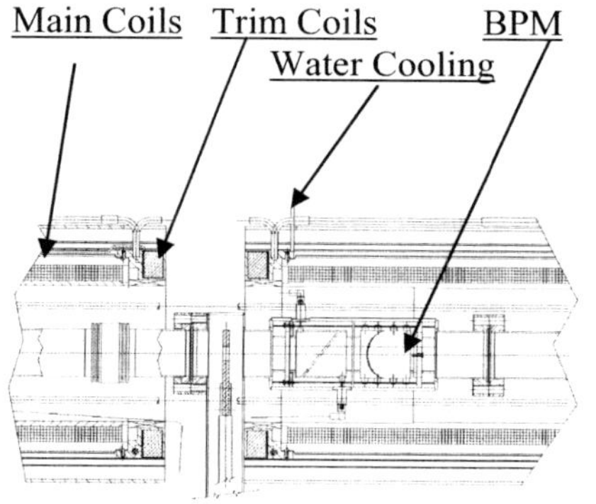

FIGURE 1. Cross-section of instrumentation gap between two solenoid modules (gap shielding is not shown).

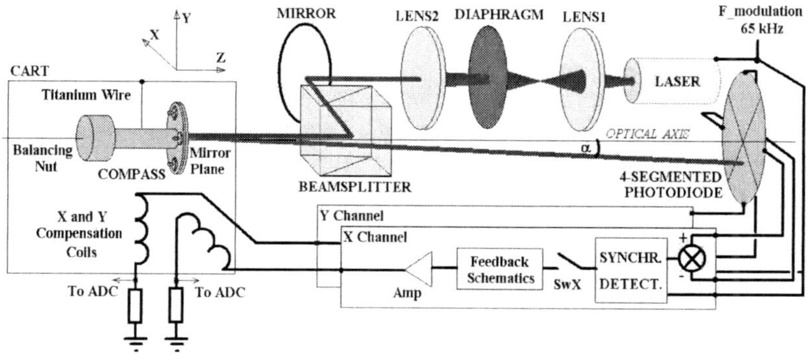

FIGURE 2. The layout of the magnetic measuring system.

Measurements have shown that the system has the resolution of 1 mG and linearity better than 1 mG in the range of up to 1 G. Its main problem is slow drifts that may be as high as 10 mG in several hours and 40 mG from day to day.

Measured Field Maps

To decrease the effect of the drifts, the CS fields were measured in two stages. First, the fields were measured with 1-2 cm steps that took a week. Second, the cart was passed through the entire CS with measurements taken only in one point per module; this measurement was performed in 2 hours. The results differed from those found in the first run by as much as 40 mG. The differences were interpreted as being caused by drifts in the first measurement and were subtracted from the first set of data in each module. The result of measurements at Bz = 100 G is shown in Fig. 3 in light colored curves. Based on these data, the dipole corrector settings for the optimum compensation of fields in the CS were calculated. The expected compensated transverse field in the CS (*Bxy_exp*) is shown in Fig. 3 in dark color. A simulation of the electron trajectory in these fields predicted the rms value of the electron angles in the CS at 0.03 mrad.

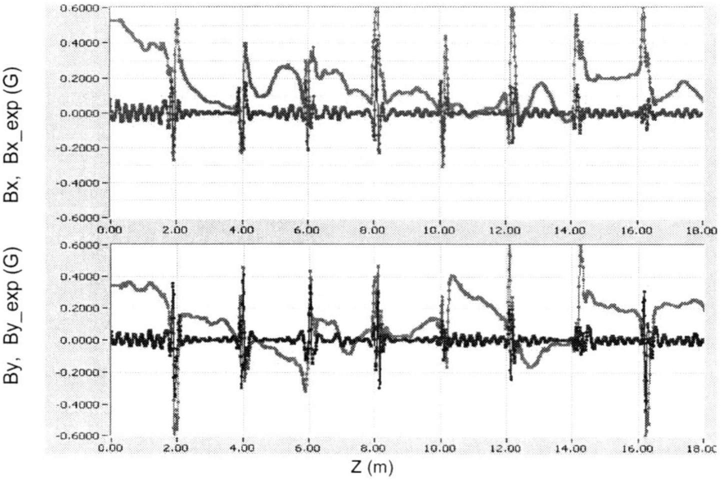

FIGURE 3. Transversal magnetic field in CS at Bz = 100 G (light color – initial with compensation coils off, dark color – after compensation by dipole correctors).

FIELD ADJUSTMENT BY BEAM TRAJECTORY

Field Compensation and Beam Angles

The transverse field components at 100 G of the longitudinal field were compensated by dipole correctors ($I'xy_{n,m}^{100G}$, n – solenoid, m – corrector). The trajectory of the beam, corresponding to initial compensation was measured by BPMs.

The trajectory could be represented as a sum of Larmour oscillations and the drift, caused by an error in determination of the average angle of cooling section solenoids. The drift was eliminated by the simultaneous change in the currents of all dipole correctors. The Larmour oscillation, caused by the beam angle at the entrance of CS, was suppressed by two sets of external (to the CS) upstream steerers.

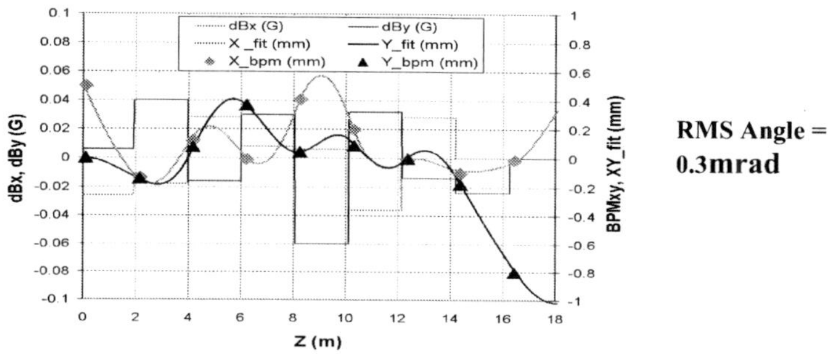

FIGURE 4. BPM readings (dots) & their fits (solid curves) at Bz = 100 G after altering the field dBx, dBy (thin lines).

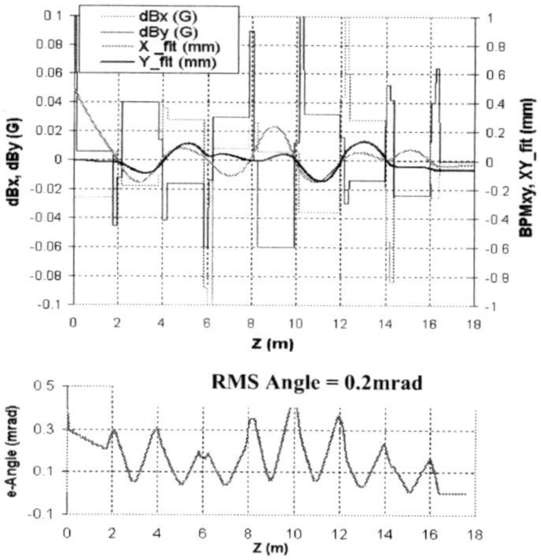

FIGURE 5. Simulation of beam trajectory and angle improvements after field correction in the gaps by trim dipoles only.

After that the trajectory looked as shown in Fig. 4. Let us assume that the modules were measured with the errors (dBx and dBy) with respect to each other. Then one can fit the measured points with the curve, simulating the electron's trajectory in the measured magnetic field (Fig. 4). The parameters of the fit are dBx and dBy. The RMS angle obtained from this simulation is 0.3 mrad.

It is clear, that applying the same additional current to each of 10 dipole correctors wound on each 2m-long module, trying at the same time to zero the closest downstream BPM reading, we could compensate the field up to the achievable beam angle limit of 0.03 mrad (for more detailed explanation see below). But even in a worst case, when further BPM zeroing is done with trim dipole correctors only (see field spikes in Fig. 5), the estimation gives 0.2 mrad for the RMS angle, i.e. improving the angles by 30%. Settings for dipole currents in the file after such adjustment can be updated as follows:

$$Ixy_{n,m}^{100G} = I'xy_{n,m}^{100G} + dIxy_{n,m}^{GAP} \quad for \;\; any \;\; solenoid \;\; n, \;\; but \;\; m = 0{,}9 \;\; only \qquad (3)$$

Compensation Procedure for Different Longitudinal Fields

We verified the assumption, that the gap regions are the main sources of constant components in transversal field, calculating the dipole settings at different longitudinal fields (the initial data used in recalculation were settings at $Bz = 100\ G$, denoted as $Ixy_{n,m}^{100G}$ [A] where n – solenoid number, m – dipole corrector number).

During the magnetic measurements in the CS we were convinced that the magnetic shield suppresses completely the external fields with magnitudes up to 2 G inside of each 2m-long module. So the only residual Bz-independent (or constant) component is present in the gaps (more details about measured components see in [3]). An indirect

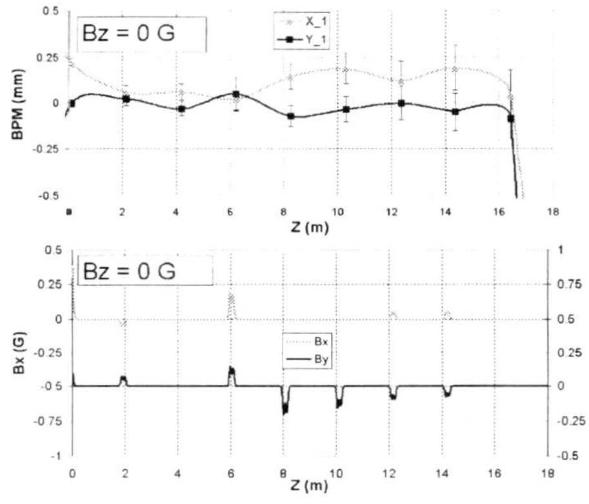

FIGURE 6. BPM readings (dots) & their fits (solid curves) after field correction made by trim dipoles Bx, By (bottom plot)

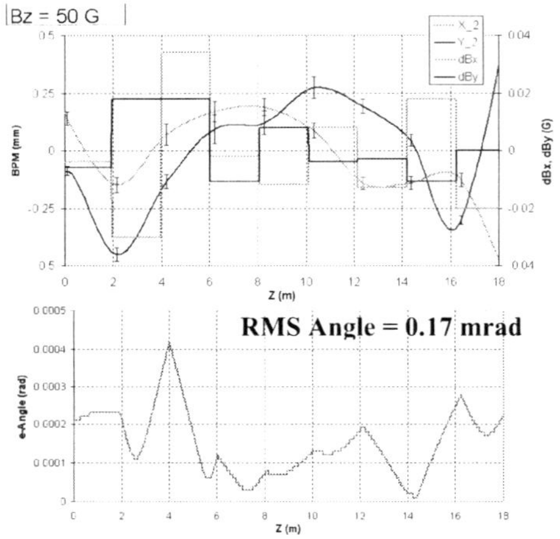

FIGURE 7. BPM readings (error bars) & their fits (solid curves) after dipole settings were recalculated for Bz = 50 G. Beam angles and offsets at the CS entrance: θx = -0.2 mrad, θy = -0.1 mrad, X0 = 0.15 mm, Y0 = -0.1 mm.

determination of this field was performed by passing the beam through the CS at Bz = 0 G. The closest downstream BPM readings were zeroed with adjustment of the currents in trim dipoles. The resultant beam trajectory and the corresponding transverse fields are shown in Fig. 6 on top and bottom plots respectively. Let's denote the corresponding dipole settings as $Ixy_{n,m}^{0G}$ [A]. Then the settings for any longitudinal field Bz can be calculated by the formula:

$$Ixy_{n,m}^{Bz} = \frac{Bz}{100G} \cdot (Ixy_{n,m}^{100G} - Ixy_{n,m}^{0G}) + Ixy_{n,m}^{0G} , \tag{4}$$

where n – solenoid and m – dipole numbers. The result for the beam trajectory recalculated at Bz = 50 G, is shown in Fig. 7. The beam's rms angle, equal to 0.17 mrad, is quite close to the original value that beam angle had at Bz = 100 G.

Estimations of the Achievable Angle in the CS

The estimation of the angle of the beam can be made from the following consideration. Suppose that the transverse field $B_{\perp}(z)$ in the cooling section was set with some error. And therefore the beam has an angle θ with respect to the axis of the CS. The trajectory of the beam is also displaced from the axis of the CS. The displacement of the beam in every BPM can be zeroed by the introduction of an additional transverse field ($B_{\perp add}$) into the cooling section. This procedure can also suppress the angle θ significantly, provided $B_{\perp}(z)$ satisfies the conditions that will be found below.

Let's consider the motion of an electron in the module consisting of a regular part of one solenoid and one gap. This module corresponds to the region between two BPMs. It can be shown that the measured longitudinal magnetic field $B_z(z)$ in CS can be substituted by its average value B_z in the equations of motion. The relative error in the calculated angle resulting from such a substitution does not exceed 1%. Thus the equations of motion are:

$$\begin{cases} r' = \theta \\ \theta' = i \cdot k \cdot (-\theta + \dfrac{B_{\perp}(z)}{B_z}) \end{cases} , \tag{5}$$

where $k = \dfrac{q \cdot B_z}{p_z \cdot c}$, $r = x + iy$, $\theta = \theta_x + i\theta_y$, $\theta_{x,y} = p_{x,y}/p_z$, $B\perp = B_x + iB_y$, q is the charge of an electron, p_x, p_y, p_z are x, y and z projections of the electron's momentum and an apostrophe denotes the differentiation over z. The solution of (5) is

$$\begin{cases} r = \dfrac{1}{B_z} \cdot \left(I_{\perp} - e^{-ikz} \cdot I\right) \\ \theta = \dfrac{i \cdot k}{B_z} \cdot e^{-ikz} \cdot I \end{cases} ; I_{\perp} = \int_0^z B_{\perp}(\xi) d\xi , \quad I = \int_0^z e^{ik\xi} B_{\perp}(\xi) d\xi . \tag{6}$$

Here an assumption that the electron has a zero displacement and zero angles at the entrance of the module was made. Suppose that r at the exit of the module is zeroed by the introduction of the transverse field, constant over the whole length (L) of the module, i.e. $B_{\perp add}(z) = const$ for $0 \leq z \leq L$:

$$\begin{cases} r(L) = 0 = \dfrac{1}{B_z} \cdot \left(I_{\perp} - e^{-ikL} \cdot I\right) - \dfrac{B_{\perp add}}{B_z} \cdot \dfrac{i(1 - e^{-ikL}) + kL}{k} \\ \theta(L) = \dfrac{i \cdot k}{B_z} \cdot e^{-ikz} \cdot I - \dfrac{B_{\perp add}}{B_z} (1 - e^{-ikL}) \end{cases} . \tag{7}$$

It follows from (7) that $\theta(L) = 0$ if integrals of the transverse field over the module satisfy the following condition (8):

$$I_{\perp} = I \cdot \frac{kL}{2} \cdot \frac{e^{-ikL} + 1}{\sin(kL)} \tag{8}$$

Of course, an ideally compensated transverse field satisfies (8). In a case if equality (8) is not fulfilled precisely it allows estimating the angles at the exit of the module as well as the angles inside the module (one has to substitute L with z for the latter).

The formula (8) replaces the estimations of the angle with the estimations of the errors in compensation of the transverse field. The upper limit of the possible error in the compensation of transverse field $\Delta B_{\perp}(z)$ in the CS was found from the analysis of magnetic measurements. This error can be separated into two parts. The first part is $\Delta B_{\perp}(z) = b_{\perp} \cdot z$, where $b_{\perp} \cdot z = 30mG$ at maximum for the module of 2 m length. The second part is due to the fact that the solenoid and gap in the same module may be compensated with different constant errors $\Delta B1_{\perp}$ and $\Delta B2_{\perp}$, with $\Delta B_{\perp} = |\Delta B1_{\perp} - \Delta B2_{\perp}| = 50mG$ at maximum. Finally, it was found that zeroing the displacement in every BPM_n by the introduction of transverse field $B_{\perp add}(z) = B_n = const \quad z \in [0, L]$ in the respective module #n would result in the rms angle $\theta \leq 0.09\ mrad$.

CONCLUSION

Transverse fields in the cooling section prototype were measured by a compass-based system, and their integrals were compensated by a set of dipole correctors according to the results of the measurements. The entrance beam angle and position were adjusted to minimize scalloping in the compensated field, caused by the entrance mismatch. The BPM readings, recorded for this optimum case, were used to simulate the beam trajectory under an assumption that the remaining beam deviations in BPMs are caused by an incomplete correction of long-term drifts in magnetic measurements. The simulation gives an estimated value for the trajectory rms angle of 0.3 mrad. Estimations and preliminary measurements show that the scalloping may be significantly decreased further by adjusting average transverse fields in separate modules according to BPM readings.

Several improvements are foreseen by the time of installing the setup in the Recycler ring in the fall of 2004. First, we hope to improve the stability of magnetic measurements. Second, the antiproton beam trajectory, which is a straight line inside the cooling section, will be used as a reference to calibrate BPM mechanical offsets with a 30-μm precision. Finally, the proposed procedure of using BPM readings for adjusting the solenoid field should provide the necessary field quality.

ACKNOWLEDGMENTS

The authors are thankful to the members of the Electron Cooling group for their help: J. Leibfritz and A. Makarov made the mechanical design of the cooling section; K. Carlson and G. Saewert provided the electrical engineering support; R. Kellet, J. Nelson, and A. Germain assembled the solenoid; C. Schmidt measured absolute BPM positions in the CS with the use of a stretched wire.

REFERENCES

1. J. A. MacLachlan, A. Burov, A. C. Crawford, T. Kroc, S. Nagaitsev, C. Schmidt, A. Sharapa, A. Shemyakin, and A. Warner, "Prospectus for an Electron Cooling System for the Recycler," Proc. of 4th Workshop on the Medium Energy Electron Cooling MEEC-98, edited by I. Meshkov, Dubna, 1999, pp. 25-113.
2. S. Nagaitsev, A. Shemyakin, and V. Vostrikov, Fermilab-FN-689, April 2000.
3. A. C. Crawford, S. Nagaitsev, A. Shemyakin, S. Seletsky and V. Tupikov, Fermilab-TM-2224, September, 2003 (http://library.fnal.gov/archive/test-tm/2000/fermilab-tm-2224.pdf).
4. P. W. Joireman and B. Chase, "BPM System for Electron Cooling in the Fermilab Recycler Ring," Proc. of this BIW-2004 workshop.

Holy Trinity of Instrumentation Development

Rok Uršič and Borut Šolar

Instrumentation Technologies, Srebrničev trg 4a, SI-5250 Solkan, Slovenia
rok@i-tech.si, www.i-tech.si

Abstract. Being user friendly should be the main guidance, beside the self-understood high performance, in today's instrumentation development. Here we identify three components of the user-friendly policy: the all-in-one concept, customization, and connectivity. All-in-one is the concept of unification of various building blocks and thus various functionalities in one product. The customization is enabled by the product's reconfigurability that allows a product to grow and support new requirements and applications without changing hardware. The consequence of the two is the capacity of the single instrument to perform a variety of tasks that before were split among different devices. The last of the three is connectivity that improves the relationship between controls and beam diagnostics, brings out-of-the-crate freedom, and opens unforeseen possibilities for intra-accelerator cooperation and remote technical support.

INTRODUCTION

Recent technology advances are enabling development of a new class of instrumentation devices. The most prominent enabling technologies are: direct digitization of RF signals; high-speed field programmable gate arrays (FPGAs) with embedded processors; high performance digital signal processing; high-speed serial interconnect technologies; and embeddable middleware and application software technologies. We show that leveraging such technologies allows a different way of thinking when it comes to developing new systems. In this article we are suggesting a shift in perspective to reap the benefits of this situation.

The need for such a shift in perspective grows from our own experience with the development and system integration of beam instrumentation devices and observing the value of similar shifts in other industries like telecommunications. We chose a practical example to explain our ideas: Libera, our new product line. The first member, the electron beam position processor, is an all-in-one, customizable product with rich connectivity options. It is intended for demanding beam position monitoring and local and global feedback applications.

Our example illustrates the result of applying these ideas to a specific application. However, there is a much broader consequence of this situation. A common, reconfigurable hardware platform allows each user to customize various attributes of the system to its needs, yet at the same time benefit from the portability of the core all-in-one encapsulated software. Flexible software architecture allows rapid adaptation to multiple applications. By leveraging "commodity" technologies, this flexibility is actually achieved at reduced cost. The use of "open standards" for intra- and inter-system communication simplifies integration within the lab, and enables a broader

CP732, *Beam Instrumentation Workshop 2004: Eleventh Workshop*,
edited by T. Shea and R. C. Sibley III

view of integration between research facilities. This allows a higher level of collaboration among laboratories and industry. Our experience shows that this is a possible, creative and productive process.

THE SIGNAL PROCESSING CHALLENGE

Direct digitization of IF or even RF signals allows functionality previously implemented in hardwired analog electronics to be implemented in digital electronics [1]. The consequence of this evolution is that the boundary between analog and digital domains moves toward the sensor. This requires tighter interactions between analog and digital parts of the system. Examples of such interactions include, but are not limited to, gain control and carrier recovery algorithms, temperature monitoring and associated systematic error corrections. We call these interactions housekeeping. Standard modular crate solutions (VME, cPCI) in general partition hardware into analog, digital, DSP and CPU boards. This partitioning requires dedicated and expensive connectors and cables, which present a likely failure point. This means lower reliability and higher cost.

Figure 1 presents a useful functional decomposition of a complex instrumentation system. The analog signal processing is followed by sampler(s) that in turn connect to the digital signal processing section. There are two "glue" functional blocks: housekeeping and timing. Timing interfaces with the external world, and in this way provides synchronization with external systems and events. It is also used for internal synchronization between different functional blocks and provides means for time-stamping of measurements. Housekeeping is a set of specific applications that are, to a great extent, transparent to the machine control system, but facilitate performance optimization and assure continuous and reliable operation.

The model of "crate thinking" was a necessary and optimal approach at one time. But technology advances have changed the rules. Where well-defined boundaries between the implementation domains were once an advantage, they have now become a limitation. The all-in-one approach allows us to "fuzzy" the boundaries between the implementation domains. This allows the designer to concentrate on a functional decomposition of the design problem with fewer constraints. It also reduces the complexity of interconnections between implementation domains, as well as enables more optimal use of advancing technology. This in turn leads to a more optimal, flexible, and cost effective implementation.

A robust, all-in-one solution, which encapsulates analog and digital hardware, sufficient processing power and fast serial interfaces seems a natural fit to this new reality [2].

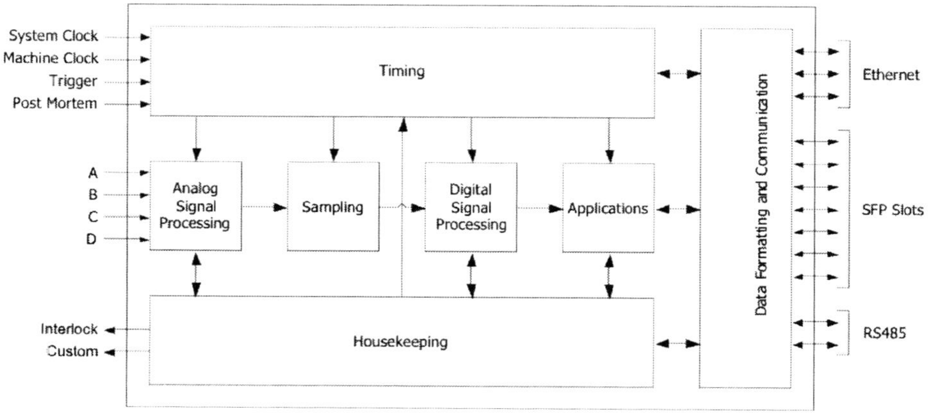

FIGURE 1. Functional decomposition of the signal processing.

THE HOLY TRINITY

All-In-One

All-In-One is the concept of unification of various building blocks and thus various functionalities in one product. The aim here is to provide an implementation domain that naturally follows the functional decomposition presented in Fig. 1. Libera is an all-in-one solution that enables accurate beam position monitoring, trouble-free commissioning, and local and global feedback building. The timing and housekeeping functions do not span across multiple boards in different racks, but are all contained in a single module.

Customization

Customization is a capability that allows adaptation of functionality of the device to specific requirements without changing hardware. For the purpose of this paper we identify three types of customization:

Parametric

Parametric customization allows adjustment of performance by simply changing parameters, without affecting architecture of the programmable hardware. Examples are analog board gain control and FIR filter bandwidth by changing tap settings.

Architectural

Architectural customization involves changes of logic configuration (FPGA firmware, DSP code and other software) to optimize performance for a particular purpose. It allows building specific logical and signal processing blocks, data-flow paths and algorithm implementations.

Communication Interfaces

The third customization involves communication interfaces that can facilitate implementation of different communication standards without changing hardware.

Connectivity

Fast serial interconnect technologies provide a brand new perspective on the process of developing high-availability and next-generation accelerator control and feedback systems.

In general, individual modules may have their own processors, operating systems and memory and can communicate independently with other modules. Because nodes are operating-system agnostic, integration is no longer required at the driver/backplane level but ascends to the network and transport layers, using standard protocols, which means significant time savings and simpler design models.

LIBERA BUILDING BLOCKS

In the following part of the paper we discuss details about the Libera electron beam position processor. A 1U 19” enclosure with a power supply accommodates analog and digital boards, which are supported by software easily adjustable to customer needs.

Analog Board

The analog board consists of a quasi-crossbar switch, four identical RF channels, four analog-to-digital converters, and an interface to the digital board.

The innovative patent-pending quasi-crossbar switch matrix unites the benefits of both the multi-channel and the multiplexed system, and at the same time compensates for the disadvantages of the two. We achieve reproducibility and good “beam vs. current” dependence, which are multiplexed system characteristics and – using a multi-channel approach – we ensure a broader band of operation. Specifically, the switch facilitates redirection of any of the four input signals to any of the four RF channels. Figure 2 shows a simplified block diagram of the analog board.

Digital Board

Core building blocks of the Digital Board are the FPGA (Field Programmable Gate Array), SBC (Single Board Computer) and Memory. The board has also a very rich set of interfaces.

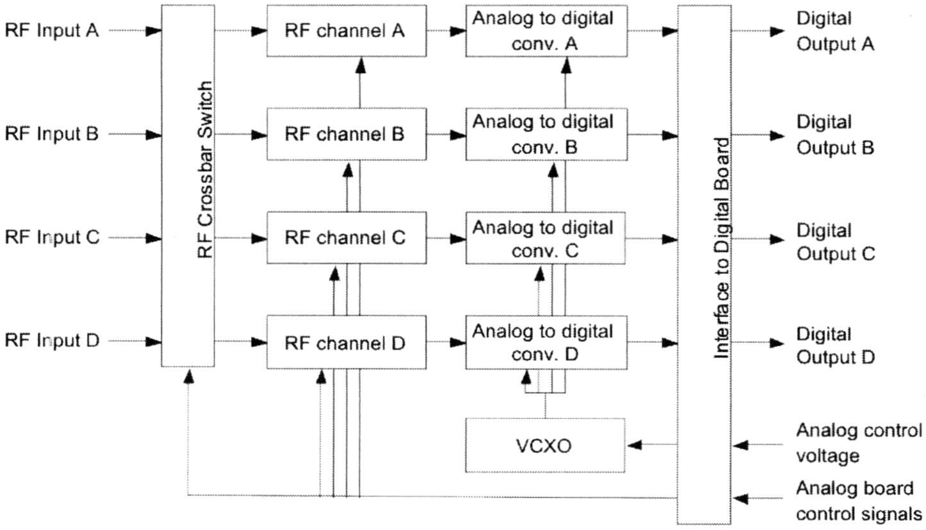

FIGURE 2. Analog board block diagram.

FPGA

The Virtex II Pro FPGA with two IBM PowerPC™ 405 processors from Xilinx is the core of the Digital Board. It is a system-on-chip (SoC) solution, which facilitates digital signal processing, communication, formatting, timing, and housekeeping. This architecture allows replacement of some hard-wired DSP functionality with more flexible procedural description of algorithms. Another powerful feature of the Virtex-II Pro FPGA is integrated 3.125-Gbps Rocket IO transceivers. The transceivers address all existing connectivity requirements as well as emerging high-speed interface standards such as Gigabit Ethernet, Fibre Channel, and Infiniband.

SBC

The Single Board Computer is mounted on a mezzanine board. It provides application software, bootstrap for the configurable FPGA logic, configuration, diagnostics, and maintenance of Libera. The Embedded Linux Operating System with networking capabilities simplifies integration into the accelerator Control System. Fast Ethernet is a native networking solution.

Memory

High capacity memory enables the storage of a large amount of data.

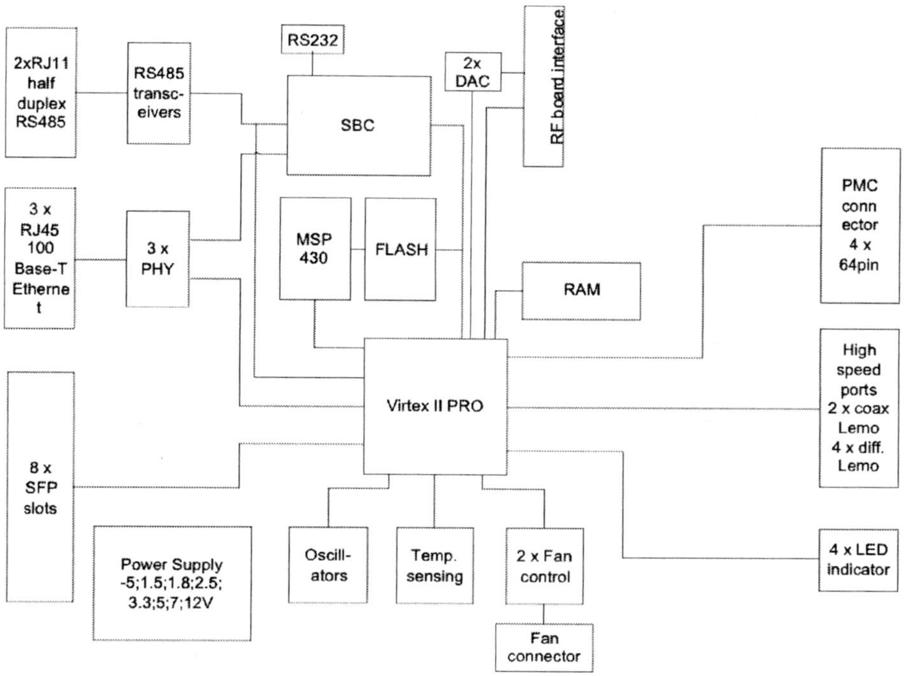

FIGURE 3. Digital Board block diagram.

SFP (Small Form Pluggable) Connectors

The SFP widely supported open standard enables hot swap of various types of fiber optic and copper-based transceivers into host equipment. This allows the user the flexibility to interchange between different protocols. Applications include: Fiber Channel, Fast/Gigabit Ethernet, ATM/SONET/SDH, and InfiniBand (IB) FDDI.

Software

The Software is divided into two groups: System Software and Application Software.

System Software

System Software is tightly incorporated into the Linux operating system running on FPGA embedded PowerPC processors and Intel Xscale/StrongArm based SBC. The software main tasks include: overall Libera control, low level communication handling, hardware control, timing and synchronization as well as data and signal processing and formatting.

Application Software

Application Software enables painless integration of Libera into the sophisticated accelerator control system. The application software can be, due to the open source nature of Linux, easily customized to the user requirements. The standard programming environment also enables easy implementation of other user packages such as EPICS, TANGO, etc. A generic sample of Application Software is provided with Libera.

LIBERA APPLICATION EXAMPLE

One of the key advantages of the Libera electron beam position processor is also that it is feedback ready. In other words, one can build local or global feedback simply by interconnecting multiple Libera devices among themselves and in this way create a network of powerful measuring devices and real-time processors. It is important to bear in mind that no additional hardware, with the exception of interconnecting cables/fibers, is needed to build such a system.

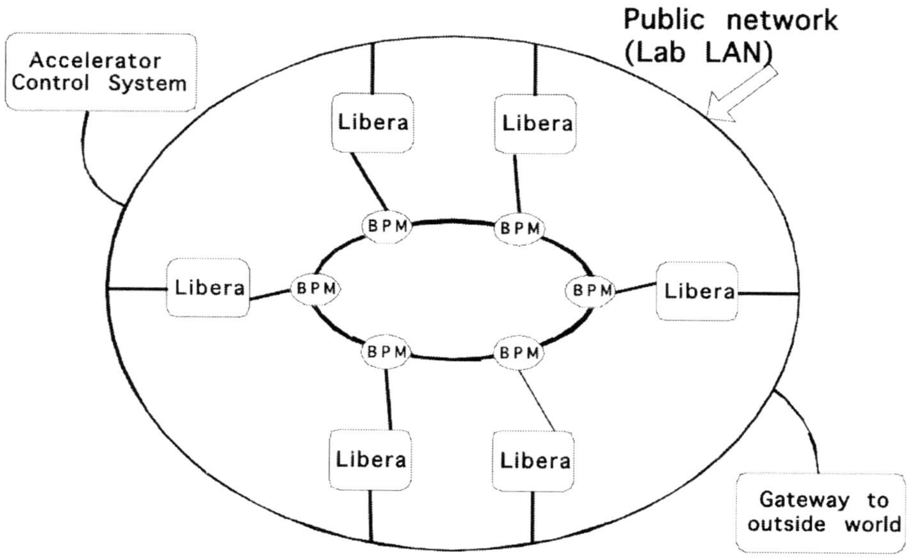

FIGURE 4. A simple BPM system implemented with Libera electron beam position processors.

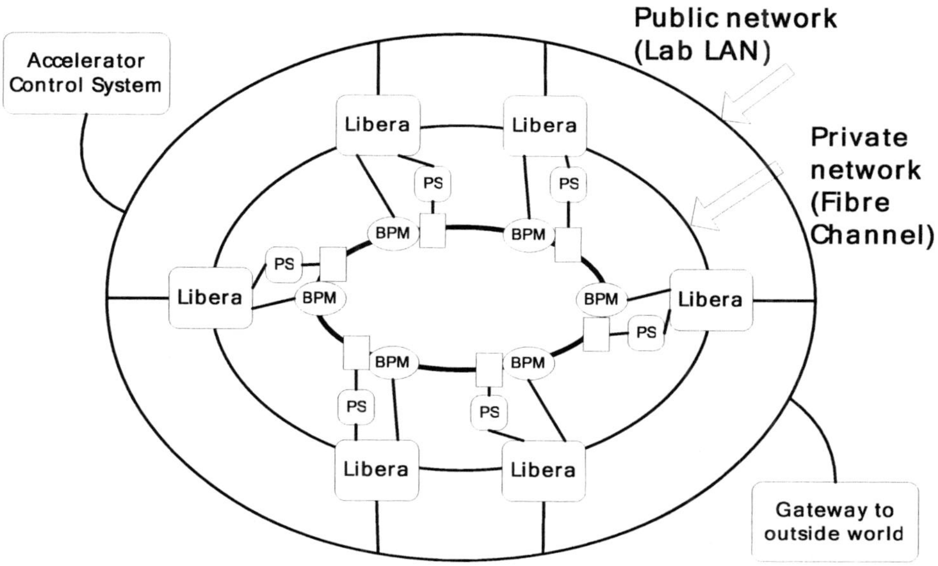

FIGURE 5. A global orbit feedback system implemented with Libera electron beam position processors.

ACKNOWLEDGMENTS

We would like to express our gratitude to Jean-Claude Denard and Lodovico Cassinari from Soleil and Guenther Rehm from Diamond. By being the launch customers for the Libera, they will have a direct influence on the design of this innovative and high performance product. We believe that this is also a good example of collaboration between industry and laboratories. All this would not have been possible without great contribution from all the members of the Instrumentation Technologies company - thank you.

REFERENCES

1. R. Ursic and R. De Monte, "Digital Receivers Offer New Solutions to Beam Instrumentation," 1999 Particle Accelerator Conference, New York, USA.
2. R. Ursic, "Reconfigurable Instrumentation Technologies, Architectures and Trends," Beam Instrumentation Workshop, 2002, Upton, USA.
3. www.i-tech.si.

RF Beam Position Monitor for the SNS Ring*

Kurt Vetter, Peter Cameron, Craig Dawson, Chris Degen,
Martin Kesselman, and Joseph Mead

Brookhaven National Laboratory, Upton, NY 11973, USA

Abstract. The Spallation Neutron Source Ring accumulates 1060 pulses of 38-mA peak current 1-GeV H-minus particles from the Linac through the HEBT line, then delivers this accumulated beam in a single pulse to a mercury target via the RTBT line. The dynamic range over the course of the accumulation cycle is 60 dB. As a result of particle energy distribution the 402.5-MHz RF bunching frequency quickly de-coheres during the first few turns. In order to measure first-turn position a dual-mode BPM has been designed to process 402.5-MHz signal energy during the first few turns then switch to a Baseband mode to process de-cohered energy in the low MHz region. The design has been implemented as a dual mother/daughter board PCI architecture. Both Baseband and RF calibration are included on the RF BPM board. A prototype system has been installed in the SNS Linac.

INTRODUCTION

Due to the large particle energy distribution the beam quickly de-coheres within a few turns resulting in a reduction of the 402.5-MHz beam structure. In order to achieve the required beam position resolution a dual-mode BPM has been designed to process 402.5-MHz beam energy during the first few turns then switch to a Baseband mode to process beam spectral energy up to 7 MHz during the remainder of the accumulation cycle. This paper will discuss only the RF BPM processing mode.

BPM SYSTEM ARCHITECTURE

In order to provide the specified beam position accuracy during the entire accumulation cycle a dual-mode architecture has been designed and implemented. The dual-mode approach is necessary in order to achieve ± 1-mm resolution due to the evolving nature of the beam spectral energy distribution. During the initial first few turns of the accumulation cycle the 402.5-MHz micro RF pulse structure is preserved resulting in significant signal energy at 402.5 MHz and very little energy at Baseband. After the first few turns the beam quickly de-coheres resulting in a spreading of the 402.5-MHz energy. In order to obtain the 30 dB of SNR required to achieve ± 1-mm resolution from the 26-cm diameter pickups during the first turn, the 402.5-MHz energy is down-converted to a 5-MHz IF for processing. A study of the frequency

*Work performed under the auspices of the U.S. Department of Energy.

CP732, *Beam Instrumentation Workshop 2004: Eleventh Workshop*,
edited by T. Shea and R. C. Sibley III

response of the open stripline detector shows that the maximum sensitivity is near 402.5 MHz and –46 dB at the Baseband frequency of 5 MHz as shown in Fig. 1. As the beam energy de-coheres signal energy in the Baseband region of 5 MHz is increased from a redistribution of the 402.5-MHz sideband structure. At a point in the accumulation cycle somewhere between three and ten turns (which will be determined somewhat empirically during Ring commissioning) the BPM will switch from RF to Baseband mode. The crossover point will be determined by examining the SNR at both 5 MHz and 402.5 MHz during the beam spectral energy evolution to determine the point in which they are equal. The pickup response will be included in this analysis by observing FFTs of the acquired data.

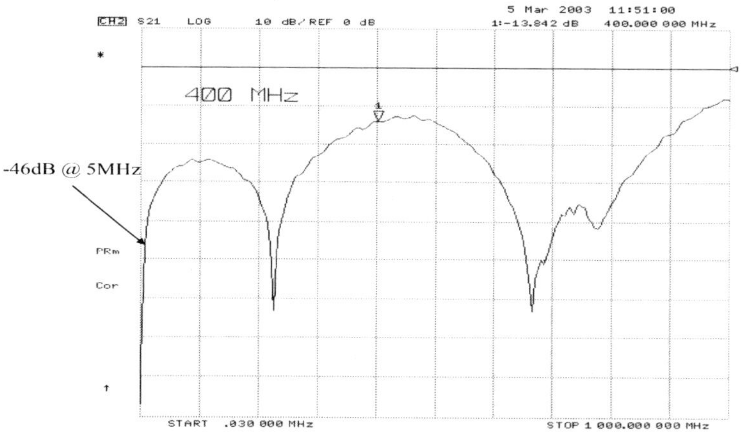

FIGURE 1. Measured S21 (H/V) 21-cm open stripline.

Listed below in Table 1 are the Accelerator Physics specifications for the Ring, HEBT, and RTBT BPMs.

TABLE 1. BPM specifications.

Parameter	Specification
Intensity	5e10 to 2e14 protons
Range	± 20 mm
Accuracy	1 mm
Resolution	1 mm
Data Structure	Average/turn-by-turn

RF BPM ARCHITECTURE AND ELECTRONICS

The two primary considerations applied to the BPM design are the time evolution of the beam spectral energy, and wide dynamic range of 60 dB. As mentioned above a dual-band approach has been implemented to achieve ± 1-mm resolution for all turns. The large dynamic range is incurred during the approximate 1000 turns of accumulation. As a result the burden to accommodate the 60-dB dynamic range is entirely within the Baseband section. Linear signal processing is used throughout both the Baseband and RF processing chains.

The RF processing architecture is simple and straightforward. Due to the coherent structure of the beam during the first few turns a large signal, approximately 0 dBm, is expected at the input of the RF electronics. Taking advantage of this large signal a completely passive single-stage heterodyne approach has been implemented as shown in Fig. 2. A front-end diplexor with a crossover frequency of 17 MHz is used to preserve a reasonable wideband match and circumvent the use of an active component at the input. A 400-MHz bandpass helical filter is used to band-limit the input signal. A level 10 mixer was chosen to preserve linearity at the mixer input to approximately +3 dBm. The receiver noise floor is -75 dBm resulting in a sensitivity of about -50 dBm to achieve a position resolution of ± 1 mm. In order to maximize flexibility and incorporate RF calibration the Silicon Labs dual-band Si4133 RF synthesizer was chosen. Only external inductors are required to set the resonant frequencies of the two integrated oscillators. Each output is programmable over ± 5% of its resonant frequency via an SPI serial interface.

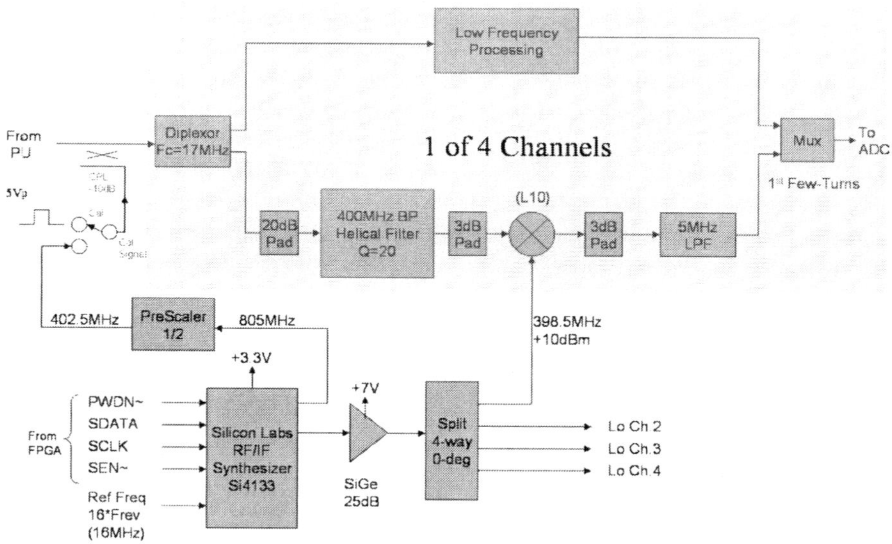

FIGURE 2. RF BPM architecture.

The IF output port of the synthesizer is set to 398.5 MHz and serves as a common LO for all four channels. The RF output of the synthesizer is set to 805 MHz. An external prescaler is used along with a GaAs switch to create a calibration burst of 402.5 MHz.

CALIBRATION

On-line calibration for both Baseband and RF is performed during a 3-ms dead time in the beam cycle. For the Baseband mode a 500-ns pulse is injected into the coupled port of a 10-dB directional coupler. The pulse is transmitted down the ¼" heliax transmission line towards the BPM electrode. When the pulse reaches the electrode it reflects from the open circuit and then propagates back toward the BPM electronics. Assuming a phase velocity of 1.4 ns/ft for Teflon it takes 700 ns to traverse 500 ft. The SNS BPM cable length has been approximated to be 500 ft so the round-trip time is 1400 ns. It is necessary to acquire data only on the calibration plane in which the calibration pulse has been activated.

RF calibration uses a 500-ns burst of 402.5-MHz CW. The reflected (and transmitted) RF calibration burst is down-converted to an IF frequency of 5 MHz for processing. Because of the low-loss plane-to-plane response [Sij] at 400 MHz it is possible to analyze all four planes of RF calibration data for each RF calibration burst. An example diagnostic would be to detect a broken cable. In Baseband mode this is not possible for two reasons: the BPM stripline is an open circuit so the pulse returns with the same polarity as injected, and the port-port coupling is very low.

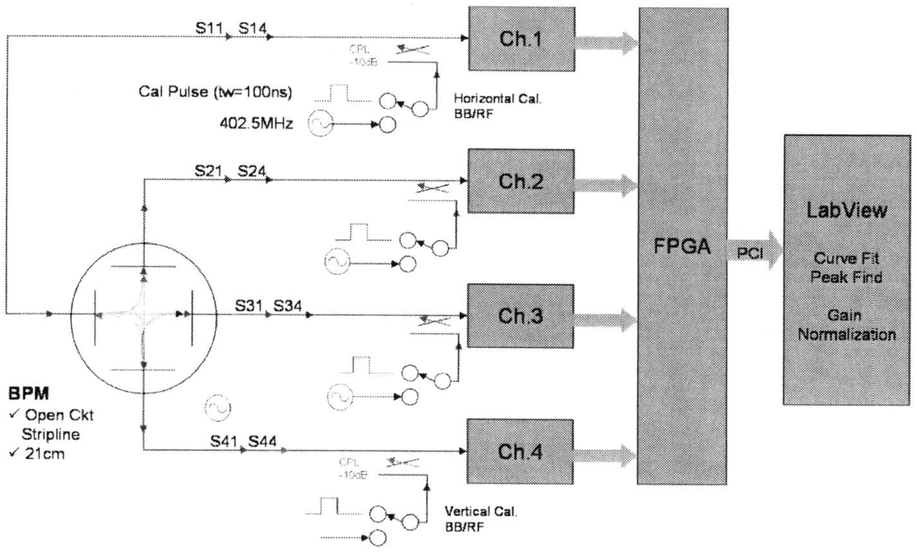

FIGURE 3. Calibration architecture.

Upon each "Start Event" a calibration pulse is initiated. Coincident at this time ADC samples are placed in memory. At the completion of the calibration period the entire data record is read by LabView. For each "Start-Event" trigger a state-machine resident in the FPGA located on the Baseband BPM PCI board sequentially adjusts the channel, gain and mode (Baseband/RF) to be calibrated. Only the 0-dB gain setting for the RF mode is calibrated. For each calibration cycle an amplitude and DC offset is calculated. A moving average filter is applied to each item of the calibration data set. Upon completion of the initial moving average filter latency calibration coefficients are updated automatically if systematic errors exceed a pre-defined bound.

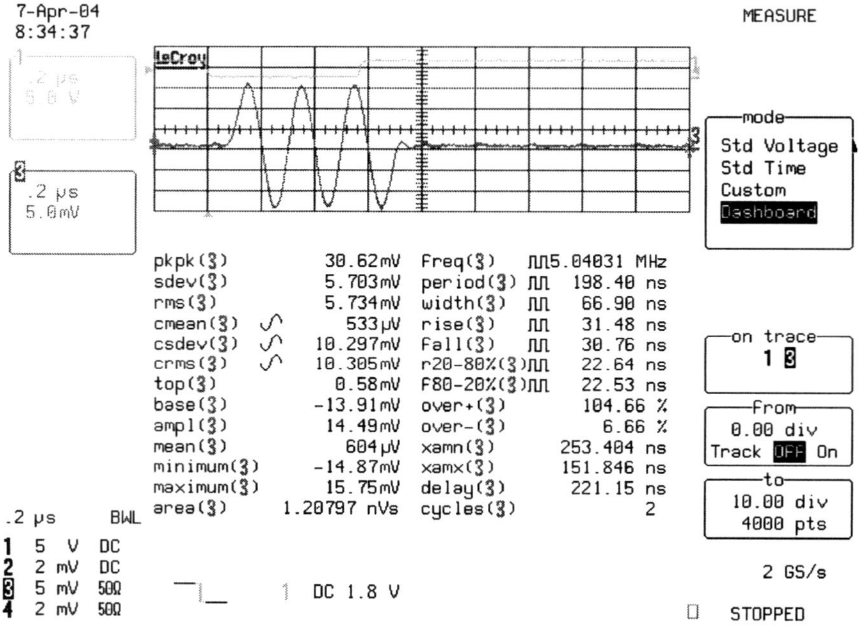

FIGURE 4. Demodulated RF calibration burst.

PROTOTYPE DEVELOPMENT AND BEAM TEST RESULTS FROM ORNL

Two completely functional BPM prototypes operating in both Baseband and RF mode have been built and tested at BNL. Shown in Fig. 5 is a test setup at BNL. The mother/daughter Baseband/RF configuration is shown on the PCI extender cards with the RF BPM visible.

FIGURE 5. RF BPM architecture.

Very encouraging results were obtained during the April 2004 Linac commissioning at ORNL. During commissioning a single BNL BPM prototype was connected immediately down-stream of DTL tank-3. Shown below is the BNL BPM response to a 20-mA, continuous 50-us 402.5-MHz RF commissioning burst. For this test the IF frequency was set to 2.5 MHz. Top trace represents the raw digitized demodulated burst. No corrections for offset have been applied at this point in the processing chain. The bottom trace is the FFT of the demodulated burst. One of the most encouraging aspects of the FFT plot is that there is no observable noise within the entire dynamic range of the system. An interesting note regarding the ORNL beam test was that the BNL BPM was able to clearly detect a 1-mm peak-peak position oscillation in tank-3 (Fig. 7).

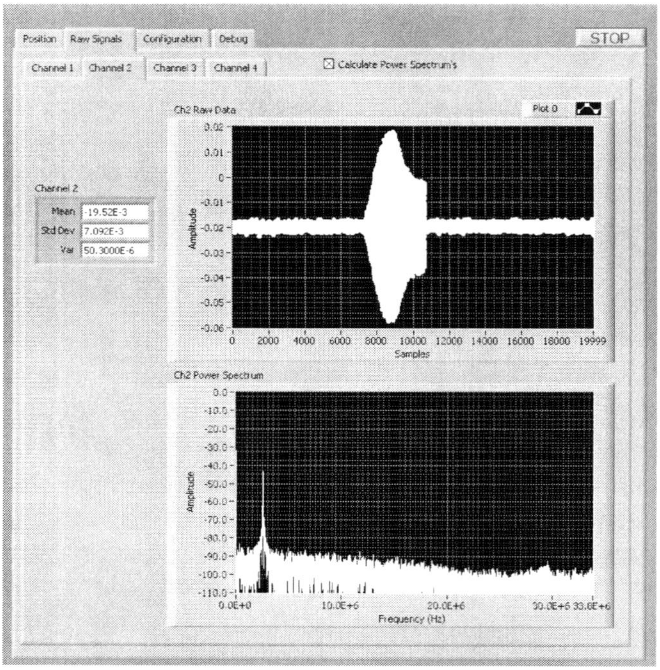

FIGURE 6. Measured beam from ORNL DTL tank-3.

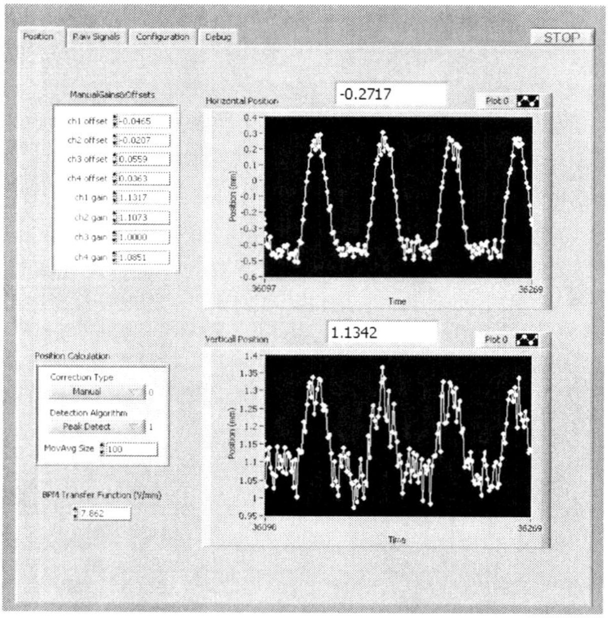

FIGURE 7. Detected DTL position oscillation.

operating and ramping energies. To provide many dynamic booster parameters, this diagnostic tool must be synchronously and automatically operated. The beam profile measurement system consists of optics, image data acquisition, and data analysis. This system is based on a new generation digital camera, which supports a wider dynamic range and faster transient image data acquisition.

FEATURES OF THE IEEE-1394 CAMERA AND REASONS FOR ITS CHOICE

Using a fully digital camera has two major advantages over the analog CCD camera. First, the A/D conversion is performed closer to the CCD sensor, keeping the amount of electronic noise to a minimum degree. Once the digitized signal is immune to noise, we can implement long haul (10 m ~ 10^3 m) applications in the accelerator research field. Various long hops solution is supported by the IEEE-1394 A/B interface. The noise immunity and isolation provided by this solution must be welcome in the accelerator environment. Secondly, unlike analog camera systems, digital systems do not suffer from pixel jitter. Each captured pixel value corresponds to a well-defined pixel on the CCD chip. In addition, high quality CCD sensors have a wider dynamic range and produce an excellent image in a low light environment. The IEEE-1394 interface is a hot pluggable and self-configuring, high performance serial bus interface that is capable of 400-Mbit/sec data transmission and will be enhanced to 3.2 Gbit/s for next generation products. The interface supports asynchronous (guaranteed delivery) and isochronous (guaranteed bandwidth and latency) data transfers. Therefore, data streams between devices are real-time with guaranteed bandwidth and no error correction needed. The IEEE-1394 complies with the memory mapped model on the IEEE1212 Command and Status Register (CSR) architecture standards with 64-bit addresses, which means standardized addressing, well-defined control and status registers standard transactions, and compatibility with other IEEE bus interfaces. Each node is re-clocked before data re-transmission. It is also a cost effective solution from the system integration point of view.

Low cost scientific IEEE-1394 general-purpose CCD cameras were chosen for this stage of upgrade. These cameras were the QICAM series of QImaging. A progressive scan interline CCD sensor gives a resolution of 1.4 million pixels in the 10/12-bit digital output. The QICAM FireWire/IEEE-1394 digital interface offers ease of use and installation with one single wire connection between the camera and computer. There is no frame grabber or additional power supply required. Frame rates of up to 100 fps can be achieved with binning and ROI selection. The QICAM is compact with a C-mount for easy mounting of various lenses. An SDK is available for integration into custom applications. The software-driven camera allows you to set camera parameters "on the fly" and ensure that the captured image is the same as the preview image. Featuring improved gain control, reduced read noise, 8 x 8 binning, and enhanced readout electronics, QImaging's line of Fast 1394 Digital cameras delivers sustained data rates of 40 megabytes/sec, full resolution 12-bit sampling at 20 MHz, and frame rates of up to 290 fps. The Intensified Retiga utilizes an 18-mm Gen III image intensifier with high-speed intensifier gate control for nanosecond exposures.

Intensifier controls provide independent trigger and gate control to allow multiple high-speed exposures within a single frame. The intensifier is fiber-optically coupled to a high resolution 1360 x 1035 progressive interline CCD. The Intensified Retiga camera is Fast 1394 interfaced for simple connectivity. External synchronization and trigger features allow for highly synchronized imaging. The camera and intensifier control are done via IEEE-1394 bus.

SYNCHROTRON RADIATION MONITOR FOR THE BOOSTER SYNCHROTRON—HARDWARE CONFIGURATION

The synchrotron radiation monitor of the booster consists of an image forming, image capture, CCD digital camera and analysis tools. This system is functioning with simple optics. A lens with focal length 500 mm is used to perform 2:1 optics. The band-pass filter is centered at 550 nm with 10-nm band-passes. The camera is working at external trigger mode. A motorized stage can move the CCD camera to accommodate various operating condition ranges. The system block is shown in Fig. 1. The optics has been optimized to minimize measured error resulting from diffraction as well as depth of field effects. Narrowed band-pass filters are used to reduce chromatic aberration and to extend the dynamic range of the CCD camera. The essentialities of the optics are to optimize the diffraction effect, depth of field and curvature. The FireWire IEEE-1394 CCD camera with 12 bits resolution was chosen to improve the functionality of the image capture and to increase dynamic range (a USB2- and Ethernet-based scientific CCD camera also emerges on the market at this moment). The camera embedded ADC on board provides exposure time control and multi-exposures. Long hauling extension of the IEEE-1394 interface is done with the aid of a fiber optical IEEE-1394 bus extender in the present system.

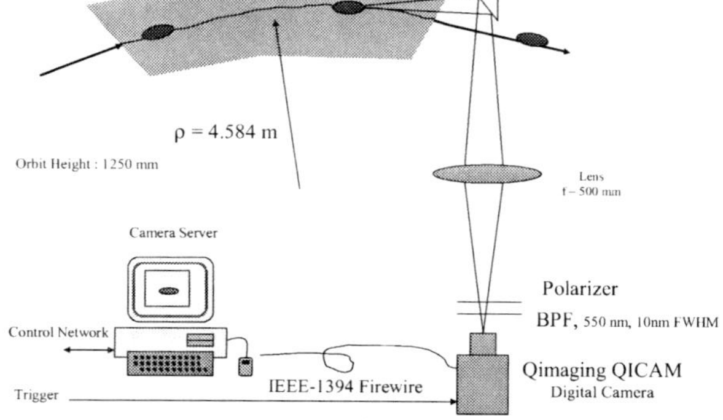

FIGURE 1. Synchrotron radiation monitor optic layout of the booster synchrotron.

SYNCHROTRON RADIATION MONITOR FOR THE STORAGE RING—HARDWARE CONFIGURATION

The synchrotron radiation monitor hardware structure of the storage ring is almost the same as that of the booster. An achromatic lens is used to perform 1:1 optics. The band-pass filter is centered at 400 nm with a bandwidth of 10 nm. The camera is working at internal trigger mode. A motorized stage can move the neutral intensity filter before the CCD camera to extend the dynamic range. The system block diagram is shown in Fig. 2. Intensified CCD camera with beam splitter is also applied to observe turn-by-turn beam profiles to support various beam dynamic studies.

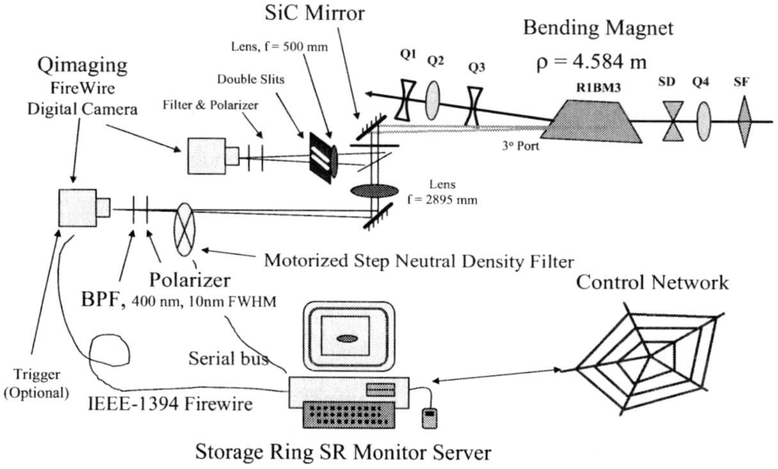

FIGURE 2. The optic layout of the storage ring synchrotron radiation monitor.

SOFTWARE ENVIRONMENT INTEGRATION

The control software is separated into two layers. One is the remote node; the other is local node. The remote node is a normal control console based on LINUX PC that includes image data access and booster parameter control (database access). These client applications run via a LabVIEW server to configure CCD parameters, acquire the image and extract feature parameters from the image. This software of the node is located on workstation/Unix and PC/Linux control consoles with commercial software MATLAB and LabVIEW installed. VI pages for routine operation are supported to operate the monitor that is useful for adjusting beam performance.

The local node is based on a windows-based PC running the LabVIEW environment; that includes 1394 camera control, network service, image processing, and data analysis. This layer is to capture image and to perform pre-processing works.

This part of the front-end PC also handles some camera parameter control including trigger sources, exposure time, times of multiple exposures, etc. A local display page must be supported at the same time for local operating and verifying.

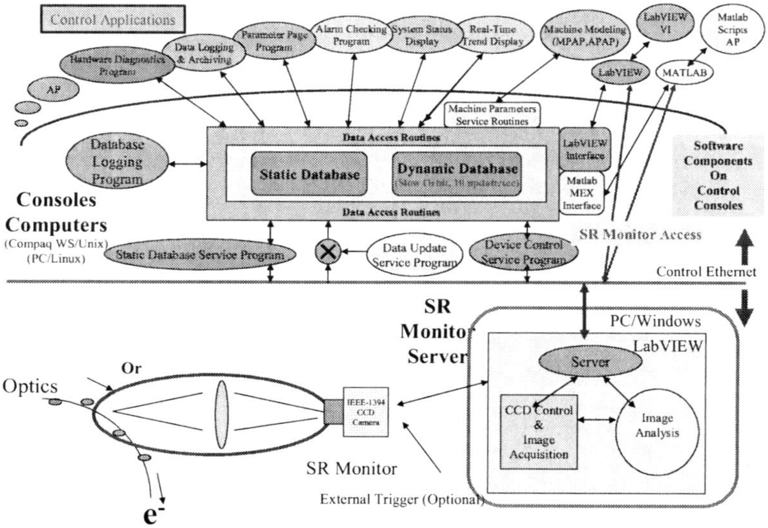

FIGURE 3. Software environment of the IEEE-1394 imaging system

The local PC is an image server to receive control commands from clients via TCP/IP protocol. This server is configured into a multithread software environment. These threads, based on the LabVIEW environment, include a digital camera driver, Ethernet communication, a database control code interface node (CIN), data processing and analysis by the Levenberg-Marquardt method CIN, and local/remote computer current status displays. These status displays include beam size, beam profile, beam image, the trigger delay relative to injection timing of the booster, and system debugging information.

Image processing and analysis include extracting the orthogonal profile and beam tilt information of the profile. The pattern reorganization of syntax approach is to identify beam object and reduce beam instability inference. A least square fit is supported to extract beam size and center position. A statistical image analysis tool to analyze spatial moment will be implemented soon to achieve better real-time performance. The display page is broadcasted to the whole institute via a CATV system, and it is web accessible. For the demand user, MATLAB scripts are also supported; the users can configure their measurement by themselves and gain the benefit of the powerful image analysis toolbox in the package. A functional block diagram of the software environment is shown in Fig. 3.

PRELIMINARY TEST RESULTS

Since the booster synchrotron is a 10-Hz machine, the injected beam is accelerated from 50 MeV to 1.5 GeV within 50 ms. Exposure time should be as short as possible for energy revolving measurements. The external trigger of the camera is synchronized to the 10-Hz trigger source. By adjusting the 10-Hz delay times, the different beam energy profile is captured from the camera. Available tools can adjust the 10-Hz delay time, camera exposure time, beam size analysis, and networking service. All information is also sent to a remote console.

The relationship of measurement timing is shown in Fig. 4. An example of the measurement is also shown in Fig. 5. The vertical beam size is reduced when the energy is increased, and it is due to synchrotron radiation damping. Multiple exposures will be used at the low energy side due to low power radiation for measurement without a scarified linearity and dynamics range.

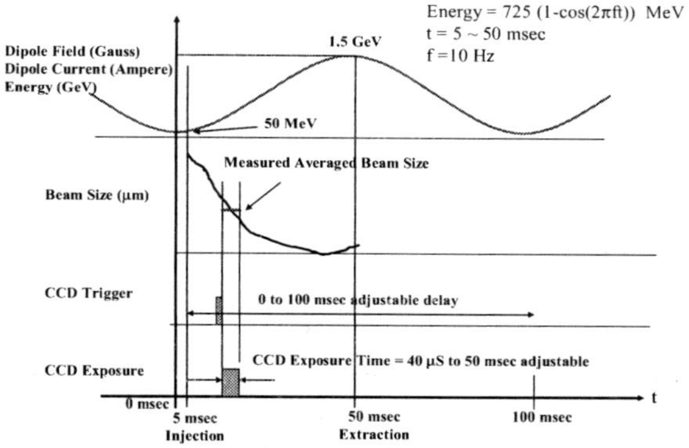

FIGURE 4. Concept of the measurement sequence during energy ramping of the booster synchrotron.

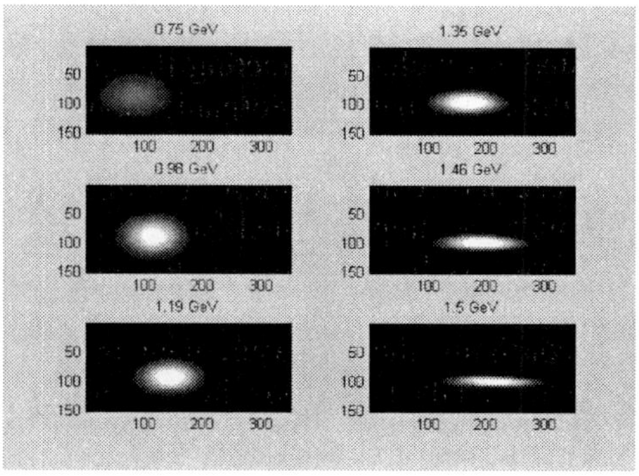

FIGURE 5. An example of observed beam profiles during ramping with various energies. One unit of the scale corresponds to 9.4 μm. The exposure time is 0.5 ms.

The graphic user interface at the remote console is shown in Fig. 6, which supports image access, profile access, exposure time of camera control, 10-Hz delay time control and booster beam current display. A similar user interface was implemented for the synchrotron radiation monitor for the storage ring as shown in Fig. 7.

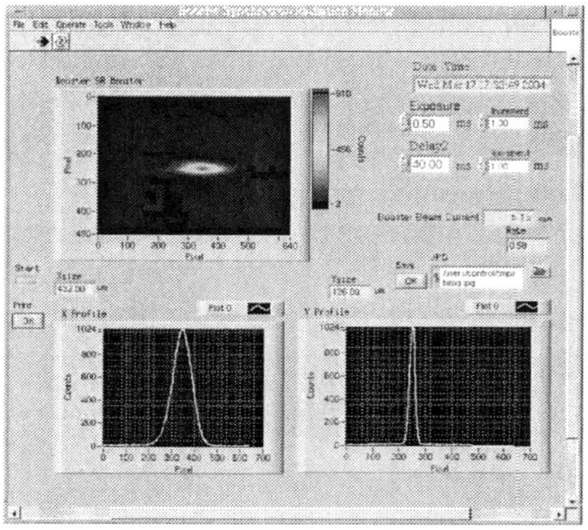

FIGURE 6. Booster synchrotron SRM user interface on control console.

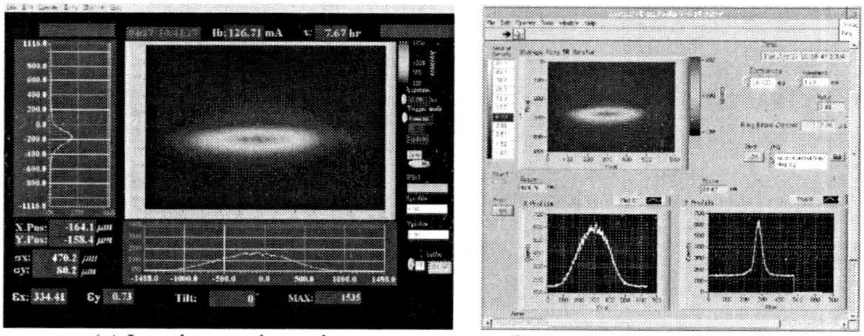

(a) Local control panel (b) Remote control panel

FIGURE 7. Synchrotron radiation monitor control panel of the storage ring at (a) local PC/Windows and at (b) remote Workstation/Unix, PC/Linux control console.

Intensified Retiga was used for low light applications. Here is an example to observe the turn-by-turn beam profile of the storage ring after setting the RF gap voltage modulation magnitude to double that of the synchrotron frequency (2 fs $\approx$ 50 kHz); that helps to relieve the effect of the high order mode (HOM) of the cavity and to stabilize the stored beam. Modulation depth is about 5% of the total 800-kV RF gap voltage. For turn-by-turn observation, the exposure time of the camera was set to 400 ns, which is the revolution time of stored beam. Trigger input to the CCD camera is synchronized with the RF gap voltage modulation source. Different delay times after trigger were observed and shown as in Fig. 8. The horizontal beam size is at minimum when the RF gap voltage is minimal, and maximal beam size occurs at maximal gap voltage setting; also, the profile period value is relative to modulation frequency. Stable horizontal beam size is obtained by the low speed imaging system that enables the integration procedure to be performed effectively.

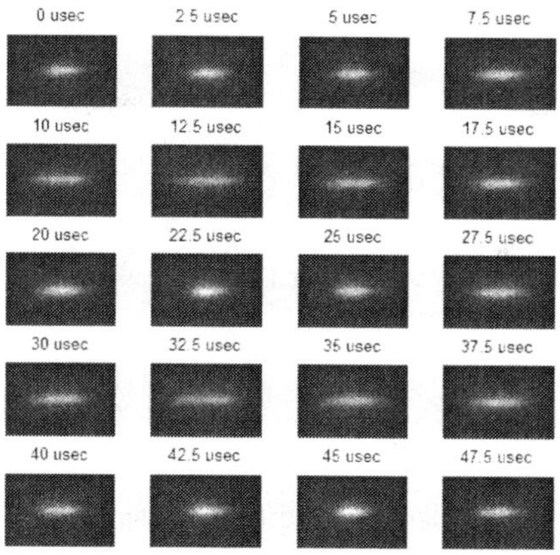

FIGURE 8. Observed RF gap voltage modulation effect corresponding to the beam profile. Period of the modulation signal is about 20 μsec.

SUMMARY

The preliminary results have shown that the purpose of wider dynamic range and better functionality of the imaging system, are all accomplished by the new generation IEEE-1394 CCD camera. Efforts put on for system integration are much lower than before. Development of various application programs to support a new imaging system is under way, and it will also provide a user-friendly interface for all kinds of users.

ACKNOWLEDGMENTS

The authors wish to thank Drs. J. Y. Hwang, K. K. Lin and C. C. Kuo for their comments and help for the work presented in this report.

REFERENCES

1. M. Ferianis, "Optical Techniques in Beam Diagnostics," Proceedings of EPAC 1998 and reference therein, pp. 159-163.
2. W. P. Leemans, "Survey Talk – New Laser and Optical Radiation Diagnostics," Proceedings of LINAC 1998, pp. 669-673.
3. K. A. Brown et al., "IEEE-1394 Camera Imaging System for Brookhaven's Booster Application Facility Beam Diagnostics," Proceedings of EPAC 2002, pp. 551-553.
4. A. Bank and P. Forck, "Residual Gas Fluorescence for Profile Measurement at the GSI UNILAC," Proceedings of DIPAC 2003, Poster Session PM18.
5. C. H. Kuo et al., "Transverse Profile Measurement System at SRRC," Proceedings of EPAC 1994, pp. 1646-1648.

ALS Control System IP I/O Module Upgrade*

J. M. Weber and M. Chin

Lawrence Berkeley National Laboratory, 1 Cyclotron Road, Berkeley, CA 94720, USA

Abstract. The Control System for the Advanced Light Source uses in-house designed IndustryPack (IP) I/O Modules in compact PCI (cPCI) chassis to control instrumentation. Each module consists of digital I/O ports and 16-bit analog I/O interfaced to instrumentation via a cPCI rear I/O card. During the past few years of installed operation, several factors have prompted investigation into the design of a new IP I/O Module. The ADC channels have significant offset drift over periods of days of initial installed operation. An in-situ calibration procedure was developed to address this problem, but it lacks speed and is inconvenient to perform. Digital I/O port limitations have led to increasing amounts of wasted I/O. Fast orbit feedback requires faster ADC sampling and better filtering than the current IP Module offers. This paper discusses the issues related to the current IP I/O Module and the design of a new Double-size IP I/O Module.

INTRODUCTION

The ALS Instrumentation Group designed custom IP Modules for the ALS Control System. Housed in cPCI crates, these modules control instruments such as magnet power supplies and beam position monitors (BPMs). Each IP Module consists of two 8-bit digital I/O ports, two 16-bit analog control channels (DACs), and four 16-bit analog monitor channels (ADCs). The modules interface to instrumentation via a custom cPCI rear I/O card. To increase channel accuracy and make modules interchangeable, each analog channel is calibrated for gain and offset compensation and the calibration data is stored in an onboard EEPROM. After several years of installed operation including changes to the control system and implementation of fast orbit feedback at the ALS, the cPCI-based hardware requires an upgrade. This upgrade will keep intact the hardware infrastructure of existing IP Carriers, cPCI chassis, and I/O Controllers (IOCs) while new versions of the IP Module and cPCI rear I/O card will replace the existing ones. The motivation for and design of the new IP Module will be presented.

MOTIVATION

Several factors motivated the IP Module upgrade. First, long term monitoring of installed IP Modules revealed that the ADC offsets drift gradually up to 25 mV in the first month of installation. The ADC data sheet specifies a maximum zero offset error of ± 10 mV with a typical drift of ± 2 ppm/°C. ALS instrumentation and control

*This work was supported by the U.S. Department of Energy under contract no. DE-AC03-76SF00098.

CP732, *Beam Instrumentation Workshop 2004: Eleventh Workshop*,
edited by T. Shea and R. C. Sibley III

requires a maximum drift of less than 10 mV/year. Since the ADC offset drift is well outside the component specifications and ALS operational requirements, an in-situ calibration procedure was developed to recalibrate installed IP Modules. The in-situ calibration compensates for the ADC offset drift, but performing the procedure is inconvenient and time-consuming.

Second, after the IP Modules were installed, developments in the fast orbit feedback required greater precision from the IP Module DACs controlling corrector magnets. To meet this requirement, a second version of the cPCI rear I/O card was developed which contains a 16-bit trim DAC divided by ten and summed with the output of each IP Module DAC for an effective resolution of 19½ bits. This version of the cPCI rear I/O card is installed only in locations used by the fast orbit feedback. As a result, there are different versions of the card in different locations, which complicates maintenance and troubleshooting of control system hardware.

Finally, upgrading or modifying the IP Module firmware requires replacing the configuration PROM for the FPGA. To replace the PROM, the IP Carrier must be removed from the cPCI chassis and the IP Module removed from the IP Carrier. The ALS Control System currently uses over one hundred IP Modules making firmware upgrades and modifications laborious. The IP2 Module uses an in-system programmable PROM that replaces the one-time programmable PROM, which greatly increases the flexibility of the FPGA firmware.

DESIGN

Physical Characteristics

IP Modules are installed in SBS Greenspring cPCI IP Carrier boards. Each IP Carrier has four IP slots and can accommodate up to four single IP Modules or two double-size IP Modules. The new Double-size IP I/O Module (IP2 Module) has twice the I/O capability of a single IP Module, maintaining the current number of analog and digital I/O channels per IP slot while freeing IP Module board space for additional features. The size of the IP2 Module (3.6" x 3.9") presented a challenge for the board layout, which resulted in a 10-layer board with components tightly packed together on both the top and bottom of the board. Figure 1 shows the IP2 Module next to its predecessor, the IP Module.

The IP2 Module has two logic connectors and two I/O connectors, one of each per IP slot. All signals on each I/O connector are pin compatible with the IP Module so that the modules may be interchanged if necessary. This allows use of the IP2 Module with existing and new cPCI rear I/O cards. One of the IP2 logic connectors is used as the IP bus interface and the other has only power and ground connections. Although the IP bus interface is the same in both modules, the IP2 Module has a modified register map requiring development of a new software driver. A block diagram of the IP2 Module design is shown in Fig. 2.

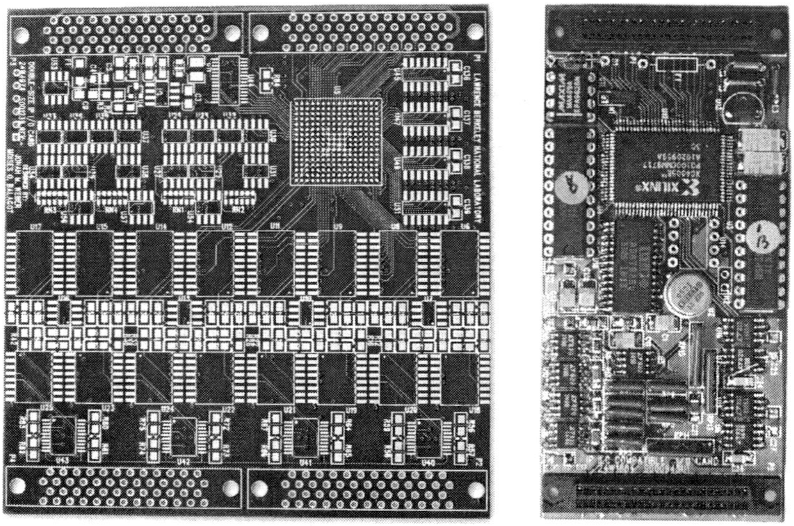

FIGURE 1. The IP2 Module top side (left) and the IP Module top side (right).

FPGA Programming

The IP Module contains an FPGA that interfaces the IP bus to I/O. Since the FPGA has volatile memory, it uses a PROM for configuration on power-up and reset. The current IP Module uses a one-time programmable configuration PROM, which must be replaced to update the FPGA logic design. To replace a PROM on an installed IP Module, the module must be removed from the cPCI chassis and the IP Carrier. As the control system expands requiring greater I/O capability, replacing every PROM is an increasingly laborious process. To allow easier reprogramming of the FPGA, the IP2 Module uses an in-system programmable PROM accessible both onboard and through a connector on the new cPCI rear I/O card.

Digital I/O

The IP Module contains 16 bits of digital I/O in two 8-bit ports. The direction of each port is controlled independently. Since many applications do not require digital I/O in 8-bit increments, some bits of the digital I/O ports are unused and inaccessible. The IP2 Module contains 32 bits of digital I/O available on eight 4-bit ports. These ports allow greater direction flexibility for the digital I/O while maintaining 16 bits of digital I/O per IP Slot.

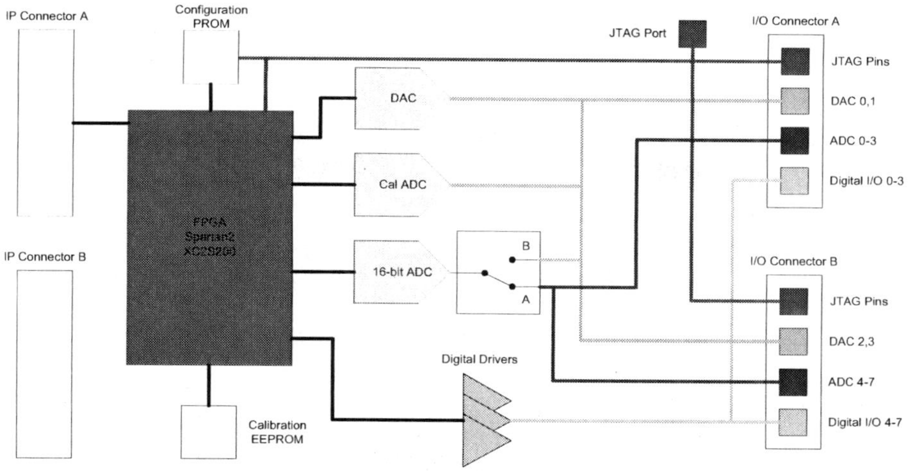

FIGURE 2. Block diagram of the IP2 Module design.

Analog Inputs

The IP Module has four 16-bit differential analog input channels on a single multiplexed ADC (Burr Brown ADS7825) sampling continuously at 10 kHz per channel. To achieve greater performance from the fast feedback, the analog inputs must sample at a higher rate for more effective filtering. Accelerator Physics also desires greater resolution from the ADC channels. The IP2 Module analog inputs must accept ±10 V differential signals. To eliminate ADC offset drift, a different ADC component and PCB assembly house are used for the IP2 Module.

The IP2 Module boasts 8 independent 16-bit analog input channels sampling at 200 kHz each with programmable gain settings of 1, 2, 4, and 8. The LTC1609A ADC used for each input accepts ±10 V differential signals. The LTC1609A noise and linearity specs are comparable to the ADS7825 used on the IP Module. The programmable gain circuit on each analog input increases the LSB resolution from 305 μV over a ±10 V range to 38 μV over a ±1.25 V range. Each channel has a 100-kHz analog anti-aliasing filter. Additional digital filtering capability is available within the FPGA to accommodate fast feedback sampling. Figure 3 shows one of the IP2 Module analog input channels.

Analog Outputs

The IP Module uses two LTC1595 16-bit DACs for analog control. Development of the fast orbit feedback required greater precision from the DACs to control corrector magnets. An additional version of the cPCI rear I/O card was designed and is currently installed to interface to instrumentation used by the fast feedback. This card includes an LTC1595 "trim" DAC divided by ten summed with the output of each IP Module DAC to create analog output. This coarse and trim DAC circuit has proven effective for fast feedback control and is used in the new IP2 Module design.

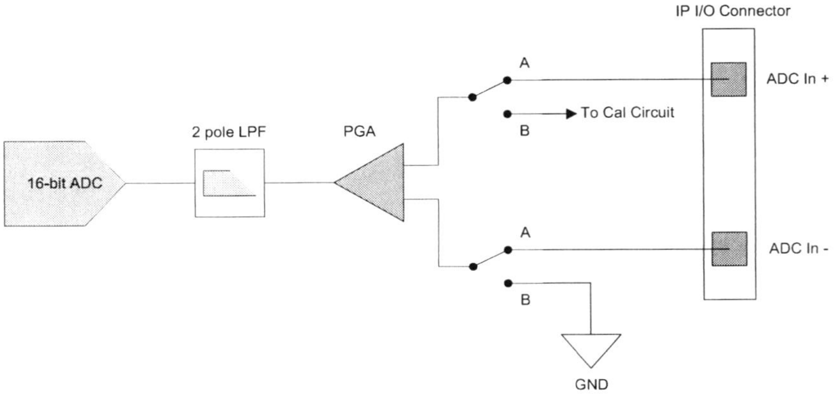

FIGURE 3. IP2 Module analog input circuit block diagram.

Each IP2 Module contains four analog output channels consisting of two LTC1595 DACs. Figure 4 shows one analog output channel. As with the cPCI rear I/O card, each coarse DAC is summed with a trim DAC divided by ten to achieve an effective 19½-bit resolution. Since the trim DACs are onboard, they can be controlled directly by the FPGA, and each DAC summing circuit requires fewer precision components to control the gain of the DAC outputs.

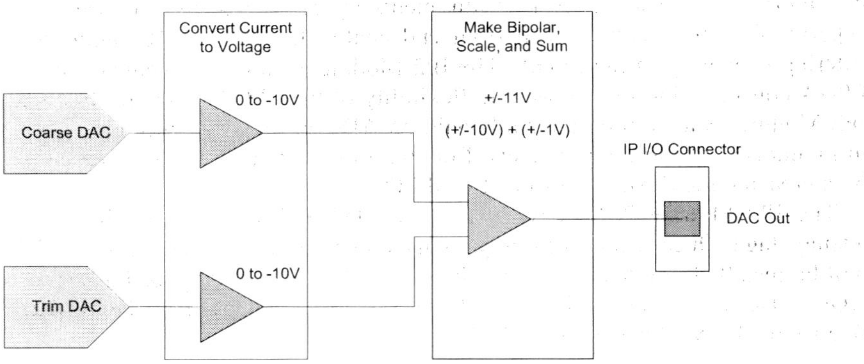

FIGURE 4. IP2 Module analog output circuit block diagram.

Analog I/O Calibration

Presently, an in-situ calibration system consisting of a laptop, voltmeter/switch unit, custom cable harness, and LabVIEW calibration software is used to calibrate analog channels. The procedure involves swapping cables and porting a laptop and voltmeter/switch unit to each cPCI chassis requiring calibration. The IP2 Module includes an onboard analog calibration circuit shown in Fig. 5. Calibration can be performed without any additional test fixtures or cable swapping by switching the analog inputs to an internal network. The control system uses calibration software to generate the calibration constants.

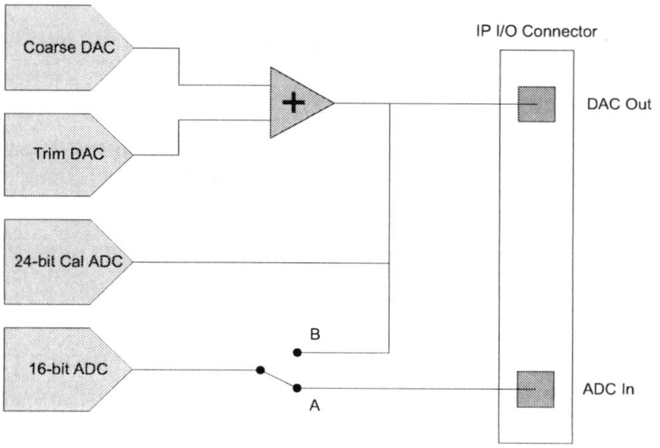

FIGURE 5. IP2 Module analog calibration circuit block diagram.

SUMMARY

The ALS Control System hardware upgrade consists of designing new versions of both the IP Module and the cPCI rear I/O card. The IP2 Module addresses the ADC offset drift with a different ADC and in-circuit calibration capability. It accommodates developments in fast feedback requirements by providing onboard trim DACs and superior filtering potential with front end analog filters on ADC inputs and digital filtering capability in the FPGA. The IP2 Module adds the convenience of in-system FPGA configuration to increase the flexibility of the FPGA program. In addition, the IP2 Module features programmable gain on ADC channels to allow greater precision in monitoring physics experiments. Finally, 4-bit digital I/O ports reduce wasted bits by increasing the flexibility of the digital I/O.

The IP2 Module PCBs have been fabricated and are currently in the prototype testing stage. Once working prototypes have been initially tested, a few IP2 Modules will be installed in a test cPCI chassis in the ALS storage ring for further testing and ADC offset drift characterization. The projected permanent installation date for the first set of IP2 Modules is early 2005.

REFERENCES

1. J. Weber and M. Chin, "In-Situ Calibration: Migrating Control System IP Module Calibration from the Bench to the Storage Ring," Proceedings of BIW 2002, Upton, New York.
2. A. Biocca et al., "The Advanced Light Source Accelerator Control System at Ten Years from Commissioning," Proceedings of ICALEPCS 2001, San Jose, California..
3. C. Steier et al., "Design of a Fast Global Orbit Feedback System for the Advanced Light Source," Proceedings of PAC 2001, Chicago, Illinois.
4. C. Steier et al., "Commissioning Results of the Fast Orbit Feedback at the ALS," Proceedings of PAC 2003, Portland, Oregon.
5. Burr Brown (Texas Instruments) ADS7825 Data Sheet, http://focus.ti.com/lit/ds/symlink/ads7825.pdf.
6. Linear Technology LTC1609 Data Sheet, http://www.linear.com/pdf/1609fa.pdf.

High-Speed, Real-Time Scope Based on Optical Fiber Delay Line Loop Technology

Yan Yin

YY Labs, Inc., P.O. Box 597, Fremont, CA 94537, USA

Abstract. Real-time oscilloscopes that are currently available go up only to 8 GHz. For transient signal measurements above 8 GHz, streak cameras have to be used. An instrument, developed by YY Labs, provides a convenient, low-cost method for such measurements. YY Labs' Single-Shot Scope captures single-shot signals in one or two channels and then regenerates 1000 copies of the one- or two-channel analogue signals to form a pulse train, with the aid of an optic-fiber delay line loop. A sampling scope recovers the original signals from the pulse train.

INTRODUCTION

For real-time measurements, currently there are two kinds of instruments. One is the regular analogue, or digitized oscilloscope, and the other is the streak camera. For the first one, the high-frequency response limit is set by the technical difficulties due to dispersion, reflections of the electrical signals at high frequencies, and semiconductor material. Therefore, the highest frequency real-time oscilloscopes only go up to 8 GHz. The streak camera has a wider frequency response range and can measure signals of a few picoseconds, but it is designed for optical signals. For measuring electrical signals, the user will have to convert them into optical signals. The setting-up procedure for streak cameras usually takes a few days or even longer.

For repetitive high-frequency signals with frequency responses above 10 GHz, a sampling scope can easily be used for the measurements.

YY Labs, therefore, has developed an instrument to capture and convert a single-shot electrical signal in one or two channels into an optical signal and then repeat the converted high-frequency single-shot optical signals 1000-fold. The original one- or two-channel single-shot signals can be recovered with a regular sampling scope.

Since the instrument works as the front-end of a sampling scope to form a real-time oscilloscope, we named the instrument **Pre-scope.**

Since the optical components have much higher frequency response limits, it is easy to reach 20 GHz or more.

The Pre-scope itself does not do signal processing; the sampling scope will do the sampling and displaying.

Various tests have been performed at SLAC with charged-particle beams and at YY Labs with a picosecond impulse signal generator.

CP732, *Beam Instrumentation Workshop 2004: Eleventh Workshop*,
edited by T. Shea and R. C. Sibley III

FIGURE 1. YY Labs' Single-Shot Pre-Scope.

Principle of the Instrument

The instrument is assembled in two boxes. The first box is named Transient Signal Capture, and the second box is named Pulse Train Generator.

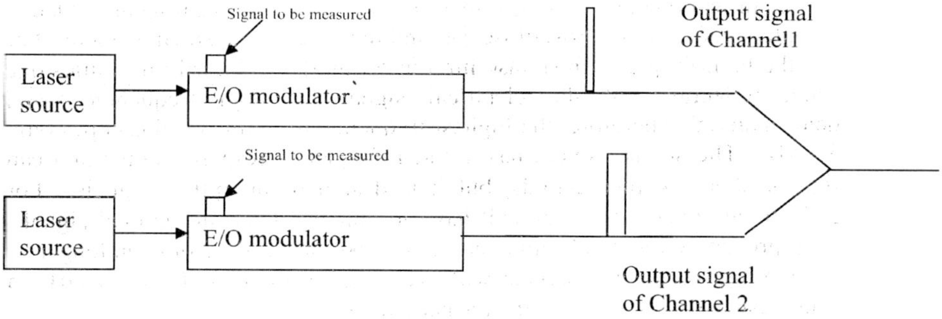

FIGURE 2. Transient signal capture.

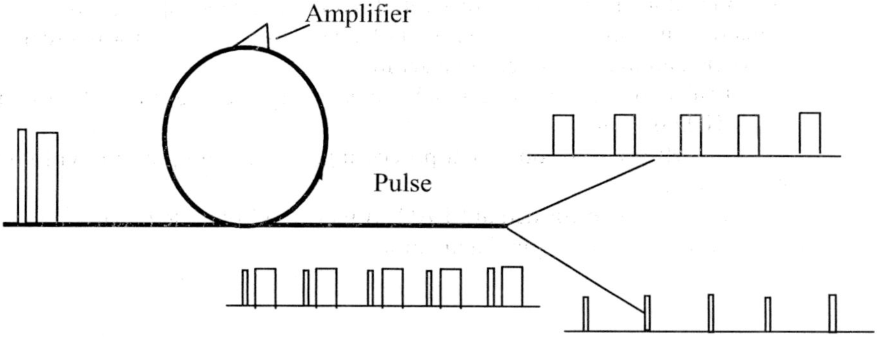

FIGURE 3. Pulse train generator.

The Transient Signal Capture box (Figure 2), which is located in a place close to the signal source, has two functions. One is to capture the transient singe-shot electrical signals, and the second is to convert them into optical transient signals, which have the same shape as the electrical signal for delivery to the Pulse Train Generator box. The Pulse Train Generator (Fig. 3), which is located in the control room, reproduces the optical signal 1000-fold to form a pulse train. Each pulse is a copy of the original signal.

The Transient Signal Capture Box has 2 electrical-signal input ports for 2 channels and one optical output port to deliver the optical signals to the Pulse Train Generator box. The distance between the two boxes is limited by the RS 485 cable, which can be as long as 1 km.

The input and output ports of the Pulse Train Generator are all optical. Two fast photo diodes are used; one is for the output signal, and the other is for the trigger signal to the sampling scope. There is a fixed delay between the two signals.

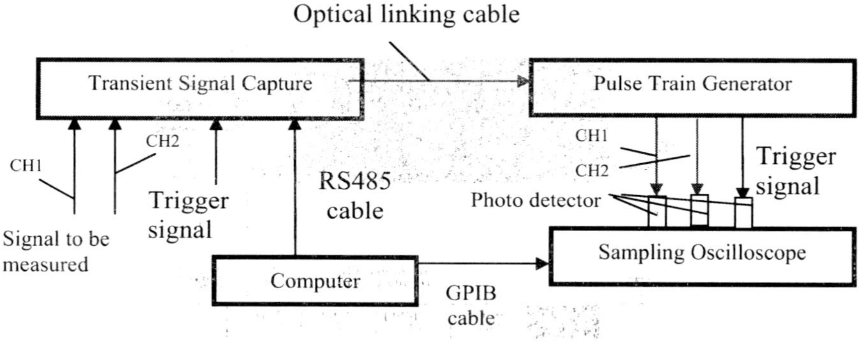

FIGURE 4. YY Labs' Single-Shot Pre-Scope arrangement.

Figure 4 illustrates the system arrangement of YY Labs' Single-Shot Pre-Scope.

Figure 5 shows the beam signals captured by the instrument.

Figure 6 is a diagrammatic representation of the circulation. Each turn takes 8 μs; 1000 turns, therefore, take 8 ms.

A sampling scope can easily recover the original signal from the 1000 pulses.

From the above description, an **equivalent sampling rate** can be calculated. If we use 20 ps/div, the sampling scope samples 512 points for the total frame. The equivalent sampling rate is $512/200 \times 10^{-12} = 2560$ GS/s, which is 100 times higher than the commercially available 20 GS/s.

The major frequency limitation of the Pre-scope resides in the E/O modulator, which converts the electrical signals into optical signals. The present model of the Pre-scope uses 20-GHz E/O modulators. 40-GHz E/O modulators are commercially available, with a penalty of cost. The photo-detectors, which convert the output optical signals into the electrical signals for the sampling scope to process, are available in 40 or even 60 GHz. The other components in this instrument are all optical fiber components, which have frequency limits above 100 GHz.

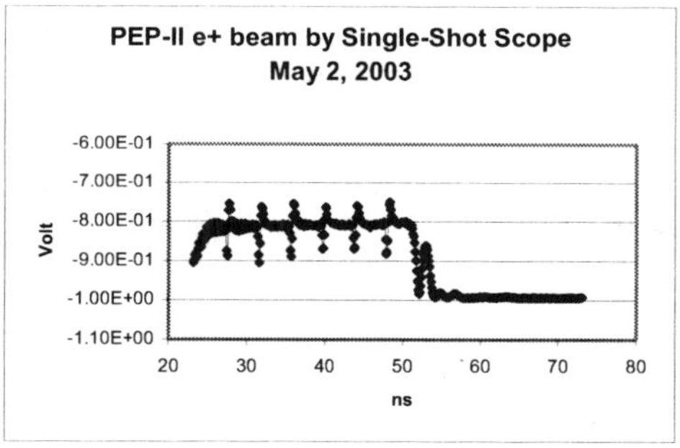

FIGURE 5. Captured beam signals.

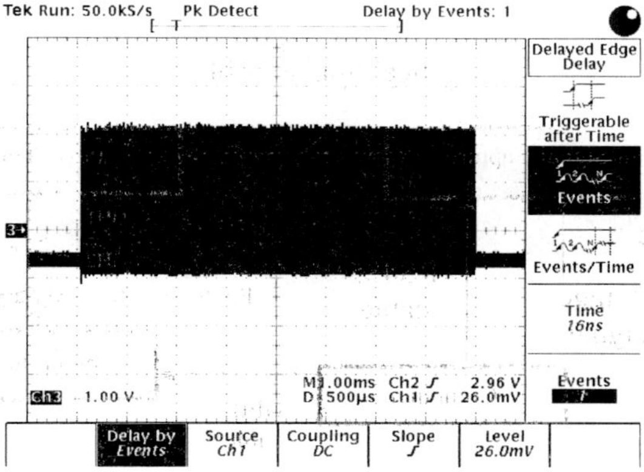

FIGURE 6. Circulation of 1000 turns.

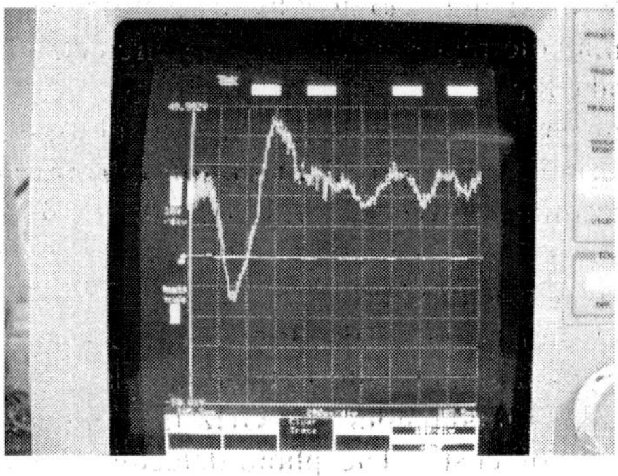

FIGURE 7. Beam signal measured with YY Labs' Single-Shot Pre-scope.

Figure 7 shows an individual bunch, as measured by YY Labs' Single-Shot Prescope, and recovered by the TEK 11801C sampling oscilloscope. The measured signal was generated by the PEP-II e+ beam.

SINGLE-SHOT 60-PICOSECOND PULSE MEASUREMENTS

An impulse signal generator manufactured by PicoSecond Pulse Lab was borrowed from SLAC. The impulse produced by this signal generator has a rising time, T_r, of 54 ps (10-90%) and a falling time, T_f, of 42 ps , as shown in Fig. 8.

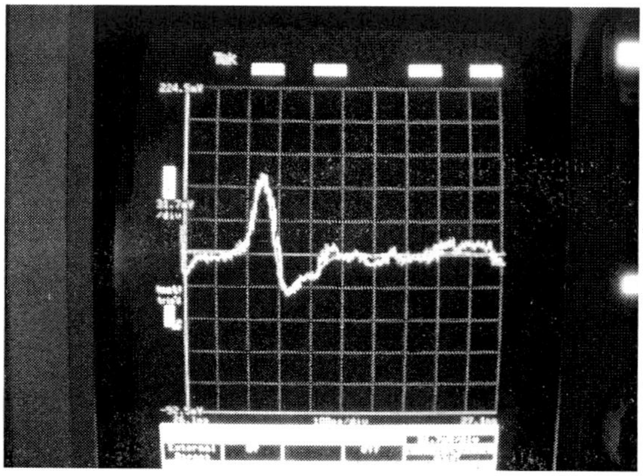

FIGURE 8. The overlapped original signal (back) and measured signal (front).

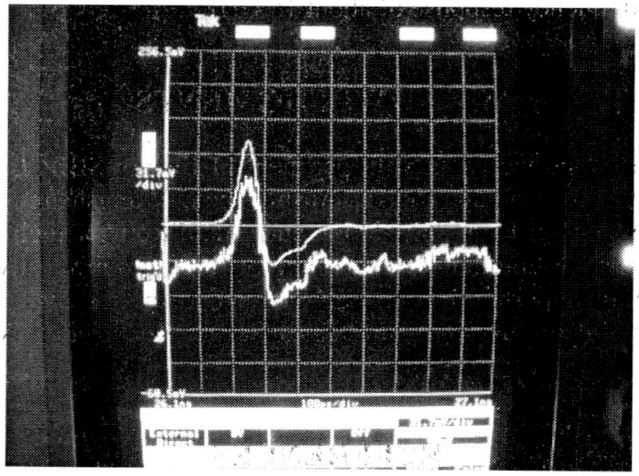

FIGURE 9. Bottom: the measured signal. Top: the original signal.

The signal was measured (a) directly with TEK 11801C with sampling head SD-24 (20-GHz frequency response), as a repetitive signal, and (b) with YY Labs' Fiber Loop Single-Shot Scope, as a single shot, in conjunction with the TEK 11801C. The accompanying pictures were taken directly from the scope. Figure 8 shows the overlapped signals from the original and from the Fiber Loop Single-Shot Scope. Figure 9 shows the corresponding separate displays. The two pictures reveal that the measured signal very closely matches the original signal.

A Fourier transform has been done to determine the frequency response of the measurement.

An impulse signal in the time domain (t) can be expressed as:

$$y = e^{-\frac{(t-t_0)^2}{2\sigma^2}}, \qquad (1)$$

where σ is the mean square root.

Its Fourier Transform in the frequency domain (ω) is:

$$F = \sqrt{2\pi}\sigma e^{-\frac{\omega^2}{2/\sigma^2}}, \qquad (2)$$

where the 1/σ is the mean square root.

When σ is 30 ps, in the frequency domain (f =ω/2π), f =1/ (2π/σ) = 5 GHz. In order to maintain the fidelity of the Gaussian pulse, the frequency response requires 3/σ, which equals 15 GHz. Therefore, the 60-ps single-shot measurement proves that the system has a frequency response of about 15 GHz.

ACKNOWLEDGMENTS

This work was supported by U.S. Department of Energy SBIR Contract No. DE-FG03-98ER82719. The author thanks Anshi Chen, Wangsheng Zhang, Guansan Chen, Xiangzhong Wang, Paul Corredoura, and Adam Reif for their contributions to this work; expresses her gratitude to SLAC and SSRL for their most valuable support in providing testing opportunities, space, and equipment for the experiments, and acknowledges helpful discussions with James Sebek, Bob Hettel, Alan Fisher, and Herman Winick.

REFERENCES

1. http://www.YYLabs.com.
2. E. Desurvire, "Erbium Doped Fiber Amplifiers: Principles and Applications," 1994, John Wiley & Sons, Inc., 499 pp. 3.
3. Yin, Yan et al., "Multichannel Single-Shot Transient Signal Measurements with a Fiber Delay Line Loop," in *Nuclear Instruments & Methods in Physics Research A*, 517 (2004) 343-348.

Appendices

11th Beam Instrumentation Workshop
May 3–6, 2004

Workshop Participants

William Accolla
Acqiris USA
234 Cromwell Hill Rd
Monroe, NY 10950
USA
Tel: 845-782-6544
Fax: 845-782-4745
E-mail: w.accolla@acqiris.com

Dean Barr
Los Alamos National Laboratory
P.O. Box 1663, MS H817
Los Alamos, NM 87545
USA
Tel: 505-667-0700
Fax: 505-665-2904
E-mail: dbarr@lanl.gov

Julien Bergoz
Bergoz Instrumentation
Espace Allondon Ouest
Saint Genis Pouilly, 01630
France
Tel: +33 450 426 642
Fax: +33 450 426 643
E-mail: bergoz@bergoz.com

Willem Blokland
Spallation Neutron Source
Oak Ridge National Laboratory
P.O. Box 2008
Oak Ridge, TN 37831
USA
Tel: 865-574-6583
Fax: 865-574-6583
E-mail: blokland@sns.gov

Mike Broadwell
Tektronix, Inc.
6333 N. State Hwy 161 #300
Irving, TX 75038
USA
Tel: 256-503-1200
Fax: 770-234-6883
E-mail: mike.broadwell@tek.com

Robert Brown
Stanford Linear Accelerator Center
2575 Sand Hill Road
Menlo Park, CA 94025
USA
Tel: 650-926-3536
Fax: 650-926-3800
E-mail: Vergne@SLAC.Stanford.edu

John Byrd
Lawrence Berkeley National Laboratory
MS 71R0259
Berkeley, CA 94720
USA
Tel: 510-486-6329
Fax: 510-486-6329
E-mail: JMByrd@lbl.gov

Lodovico Cassinari
Synchrotron SOLEIL
L'Orme des Merisiers-
BP48-Saint-Aubin
Gif sur Yvette, F-91192
France
Tel: +331 69 35 98 2
Fax: +331 69 35 94 5
E-mail: lodovico.Cassinari@synchrotron-soleil.fr

Adrian Cavalieri
University of Michigan, FOCUS Center/SPS
500 E. University
Ann Arbor, MI 48109-1120
USA
Tel: 734-764-8459
Fax: 734-764-5153
E-mail: kjrich@umich.edu

Brian Chase
Fermi National Accelerator Laboratory
P.O. Box 500, MS 308
Batavia, IL 60510
USA
Tel: 630-840-3040
E-mail: chase@fnal.gov

Daniel Cheever
MIT Bates Linac
21 Manning Rd
Middleton, MA 01949
USA
Tel: 617-253-9533
Fax: 617-253-9799
E-mail: cheever@mit.edu

Khalid Chouffani
Idaho Accelerator Research Center
Idaho State University
Research Park 150 Box 8263
Pocatello, ID 83209
USA
Tel: 208-232-0572
E-mail: khalid@physics.isu.edu

Gerald Codner
Cornell University LEPP
Wilson Synchrotron Laboratory
Ithaca, NY 14853
USA
Tel: 607-255-5749
Fax: 607-255-8062
E-mail: gcodner@lepp.cornell.edu

Stefano De Santis
Lawrence Berkeley National Laboratory
71RO259, One Cyclotron Road
Berkeley, CA 94720-8211
USA
Tel: 510-486-4726
Fax: 510-486-7981
E-mail: sdesantis@lbl.gov

Glenn Decker
10211 5th Avenue Cutoff
Countryside, IL 60525
USA
Tel: 630-252-6635
Fax: 630-252-5703
E-mail: decker@aps.anl.gov

Christophe Defrance
Bergoz Instrumentation
Espace Allondon Ouest
Saint Genis Pouilly, 01630
France
Tel: +33 450 426 642
Fax: +33 450 426 643
E-mail: defrance@bergoz.com

Al Joseph Dellapenna
Brookhaven National Laboratory
Bldg. 817
Upton, NY 11973
USA
Tel: 631-344-2175
E-mail: dellapenna@bnl.gov

Jean-Claude Denard
Synchrotron SOLEIL
L'Orme des Merisiers-
BP48-Saint-Aubin
Gif sur Yvette, 91192
France
Tel: +33 1 69 35 98
E-mail: jean-claudedenard
@synchrotron-soleil.fr

Juergen Dietrich
Forschungszentrum Jülich GmbH,
Institut für Kernphysik
Leo-Brandt Str.
Jülich, D-52425
Germany
Tel: +492461612678
Fax: +492461612670
E-mail: j.dietrich@fz-juelich.de

Rudolf Dölling
Paul Scherrer Institut
WBGA C14
Villigen PSI, 5232
Switzerland
Tel: ++41 56 310 338
Fax: ++41 56 310 338
E-mail: vera.loehle@psi.ch

Jeffrey Dooling
Argonne National Laboratory
Intense Pulsed Neutron Source
360 IPNS, 9700 S. Cass Avenue
Argonne, IL 60439
USA
Tel: 630-252-1196
Fax: 630-252-9987
E-mail: jcdooling@anl.gov

Lawrence Doolittle
Lawrence Berkeley National
Laboratory
Berkeley, CA 94720
USA
Tel: 510-486-7382
Fax: 510-486-5788
E-mail: LRDoolittle@lbl.gov

Pierre-Andre Duperrex
Paul Scherrer Institut
WBGA C14
Villigen PSI, 5232
Switzerland
Tel: 00 41 563 103 382
Fax: 00 41 563 103 383
E-mail: vera.loehle@psi.ch

Joseph Eddleman
Pulcir, Inc.
9209 Oak Ridge Highway
Oak Ridge, TN 37830
USA
Tel: 865-927-6358
Fax: 865-927-6877
E-mail: lynne@pulcir.com

Nathan Eddy
Fermi National Accelerator Laboratory
MS 308
Batavia, IL 60510
USA
Tel: 630-840-6860
E-mail: eddy@fnal.gov

Ewald Effinger
CERN Laboratory
Geneva 23
Geneva, 1211
Switzerland
Tel: +41 22 767 057
Fax: +41 22 767 774
E-mail: Ewald.Effinger@cern.ch

Joe Fabrizio
Leybold Vacuum
5700 Mellon Road
Export, PA 15632
USA
Tel: 724-325-6371
Fax: 724-325-8029
E-mail: jfabrizio@leyboldvacuum.com

Brian Fellenz
Fermi National Accelerator Laboratory
P.O. Box 500, MS 308
Batavia, IL 60510
USA
Tel: 630-840-2512
Fax: 630-840-3754
E-mail: fellenz@fnal.gov

Bibo Feng
Vanderbilt University
Box 1816, Station B
Nashville, TN 37235
USA
Tel: 615-343-6446
E-mail: bibo.feng@vanderbilt.edu

Mario Ferianis
Sincrotrone Trieste
S.S. 14 KM 163,5
Trieste, 34012
Italy
E-mail: giuliana.paoletti@elettra

Ralph Fiorito
University of Maryland
Institute for Research in Electronics
and Applied Physics, Bldg. 223
College Park, MD 20742
USA
Tel: 301-405-8484
Fax: 301-314-9437
E-mail: rfiorito@umd.edu

Claude Fischer
CERN Laboratory
Geneva 23
Geneva, 1211
Switzerland
Tel: +41 22 767 3509
Fax: +41 22 676 7740
E-mail: Claude.fischer@cern.ch

George Foster
Fermi National Accelerator Laboratory
MS 323
Batavia, IL 60510
USA
Tel: 630-840-3906
E-mail: gwf@fnal.gov

Arne Freyberger
Thomas Jefferson National Accelerator Facility
12000 Jefferson Avenue
Newport News, VA 23606
USA
Tel: 757-269-6268
E-mail: freyberg@jlab.org

Marek Gasior
CERN Laboratory
Geneva 23
Geneva, 1211
Switzerland
Tel: + 41 22 767 253
Fax: + 41 22 767 774
E-mail: Marek.Gasior@cern.ch

David Gassner
Brookhaven National Laboratory
Bldg. 911A
Upton, NY 11973
USA
Tel: 631-344-7870
Fax: 631-344-5676
E-mail: gassner@bnl.gov

Tino Giacomini
Ges. f. Schwerionenforschung mbH
Planckstr. 1
Darmstadt, D-64291
Germany
Tel: 0049 6159 71204
Fax: 0049 6159 712104
E-mail: T.Giacomini@gsi.de

J. Gilpatrick
Los Alamos National Laboratory
MS H817
Los Alamos, NM 87545
USA
Tel: 505-667-3159
E-mail: gilpatrick@lanl.gov

Gianluca Guaglio
CERN
Geneva 23
Geneva, CH 1211
Switzerland
Tel: 00 41 22 76 7807
Fax: 00 41 22 76 7956
E-mail: gianluca.guaglio@cern.ch

Robert Hardekopf
Los Alamos National Laboratory
MS H824
Los Alamos, NM 87545
USA
Tel: 505-667-1289
Fax: 505-665-2904
E-mail: hardekopf@lanl.gov

Perry Henderson
VSG-Southeast
419 The Parkway, PMB 212
Greer, SC 29650
USA
Tel: 864-801-9458
Fax: 864-801-9513
E-mail: phenderson@vsg-southeast.com

James Hinkson
GMW Associates
5805 Robin Hood Drive
El Sobrante, CA 98403
USA
E-mail: jhinkson@aol.com

Tobias Hoffmann
Ges. f. Schwerionenforschung (GSI)
Planckstrasse 1
Darmstadt, 64291
Germany
Tel: 0049-6159-71-23
E-mail: t.hoffmann@gsi.de

Kuotung Hsu
National Synchrotron Radiation Research Center
101 Hsin-Ann Road
Hsinchu Science Park
Hsinchu, 30077
Taiwan
Tel: 886 3 5780281
Fax: 886 3 5789816
E-mail: kuotung@nsrrc.org.tw

Martin Hu
Fermi National Accelerator Laboratory
P.O. Box 500
Batavia, IL 60510
USA
Tel: 630-840-6495
E-mail: martinhu@fnal.gov

Woon Ha Hwang
Pohang Accelerator Laboratory
31 San, Hyojadong Namgu
Pohang, 890-784
Korea
Tel: 82-54-279-1175
Fax: 82-54-279-1399
E-mail: hohwang@postech.ac.kr

Jeffrey Johnson
Los Alamos National Laboratory
PO Box 1663, MS P942
Los Alamos, NM 87545
USA
Tel: 505-667-9148
Fax: 505-667-7390
E-mail: johnsojb@lanl.gov

Paul Joireman
Fermi National Accelerator Laboratory
P.O. Box 500, MS 308
Batavia, IL 60510
USA
Tel: 630-840-2550
Fax: 630-840-2677
E-mail: joireman@fnal.gov

Alan Jones
Spallation Neutron Source
Oak Ridge National Laboratory
P.O. Box 2008
Oak Ridge, TN 37831
USA
Tel: 865-207-5069
Fax: 865-241-6739
E-mail: jonesaa@sns.gov

Kevin Jordan
Thomas Jefferson National Accelerator Facility
12000 Jefferson Avenue, MS 18
Newport News, VA 23606
USA
Tel: 757-269-7644
Fax: 757-269-6384
E-mail: jordan@jlab.org

Roland Jung
CERN
Geneva 23
Geneva, 1211
Switzerland
Tel: +41 22 767 329
Fax: +41 22 767 774
E-mail: Roland.Jung@cern.ch

Vsevolod Kamerdzhiev
Forschungszentrum Jülich GmbH,
Institut für Kernphysik
Leo-Brandt Str.
Jülich, D-52425
Germany
Tel: +492461616501
Fax: +492461612670
E-mail: v.kamerdjiev@fz-juelich.d

Mark Kibilko
SE Technical Sales, Inc.
P.O. Box 2760
Windermere, FL 34786
USA
Tel: 866-903-5858
Fax: 404-903-5859
E-mail: mark@setechsales.com

Sacha Kopp
University of Texas at Austin
Dept of Physics, RLM 5.208
Austin, TX 78712
USA
Tel: 512-471-9641
Fax: 512-475-7284
E-mail: kopp@hep.utexas.edu

Patrick Krejcik
Stanford Linear Accelerator Center
2575 Sand Hill Road, MS 18
Menlo Park, CA 94025
USA
Tel: 650-926-2790
Fax: 650-926-8533
E-mail: pkr@slac.stanford.edu

Gero Kube
Deutsches Elektronen-Synchrotron (DESY)
Notkestrasse 85
Hamburg, 22607
Germany
Tel: +49 40 8998 307
E-mail: gero.kube@desy.de

Jim Kurtz
Echotek Corp.
555 Sparkman Drive, Suite 400
Huntsville, AL 35816
USA
Tel: 256-721-1911
Fax: 256-721-9266
E-mail: gary@echotek.com

Jens Kuszynski
BESSY GmbH
Albert-Einstein-Strasse 15
Berlin, 12489
Germany
E-mail: jens.kuszynski@bessy.de

Thibaut Lefevre
Northwestern University
CERN-Geneva 23
Geneva, 1211
Switzerland
Tel: +41 22 767 193
Fax: +41 22 767 774
E-mail: Thibaut.Lefevre@cern.ch

Robert Lill
Argonne National Laboratory
9700 S. Cass Ave
Argonne, IL 60439
USA
Tel: 630-252-0265
Fax: 630-252-4732
E-mail: Blill@aps.anl.gv

Alex Lumpkin
Advanced Photon Source
Argonne National Laboratory
9700 S. Cass Ave., Bldg. 401,
Argonne, IL 60439
USA
Tel: 630-252-4879
Fax: 630-252-4732
E-mail: lumpkin@aps.anl.gov

Uros Mavric
Instrumentation Technologies
Srebrnicev trg 4a
Solkan, SI-5250
Slovenia
Tel: +386 5 3311700
Fax: +386 5 3311719
E-mail: uros@i-tech.si

John E. McKisson
Embry-Riddle Aeronautical Univ.
600 S. Clyde Morris Bldg.
Daytona Beach, FL 32114
USA
Tel: 386-226-7757
E-mail: mckis3ee@erau.edu

Evgeny Medvedko
Stanford Linear Accelerator Center
MS 50
Stanford, CA 94309
USA
Tel: 650 926 2685
Fax: 650 926 3800
E-mail: medvedko@slac.stanford.edu

Tony Mendez
Holifield Radioactive Ion Beam
Facility
Oak Ridge National Laboratory
P.O. Box 2008, MS 6368
Oak Ridge, TN 37831-6368
USA
Tel: 865-576-6756
Fax: 865-574-1268
E-mail: mendezajii@ornl.gov

Peter Michel
Forschungszentrum Rossendorf
Bautzner Landstraße 128
Dresden, 01328
Germany
Tel: +49 351 2603259
Fax: +49 351 2603690
E-mail: P.Michel@fz-rossendorf.de

Toshiyuki Mitsuhashi
KEK
1-1 Oho
Tsukuba, Japan
Tel: +81-298-5665
Fax: +81-298-2801
E-mail: toshiyuki.mitsuhashi@kek

Timothy Montagne
Stanford Linear Accelerator Center
2575 Sand Hill Road, MS 18
Menlo Park, CA 94025
USA
Tel: 650-926-8780
Fax: 650-926-8533
E-mail: montagne@slac.stanford.ed

Darryl Murphy
Spallation Neutron Source
Oak Ridge National Laboratory
P.O. Box 2008
Oak Ridge, TN 37831
USA
Tel: 865-300-7304
Fax: 865-241-6910
E-mail: murphydj@ornl.gov

John Musson
Thomas Jefferson National Accelerator Facility
12000 Jefferson Avenue
Newport News, VA 23606
USA
Tel: 757-269-7441
Fax: 757-269-7658
E-mail: musson@jlab.org

Jangho Park
POSTECH
31 San Hyojadong Namgu
Pohang, 790-784
Korea
Tel: 82-54-279-1184
Fax: 82-54-279-1399
E-mail: wpjho@postech.ac.kr

Stanley Pasky
Argonne National Laboratory
9700 S. Cass Avenue
Argonne, IL 60439
USA
Tel: 630-252-1768
Fax: 630-252-5703
E-mail: pasky@aps.anl.gov

Rodney Piercey
Embry-Riddle Aeronautical Univ.
600 S. Clyde-Morris Blvd.
Daytona Beach, FL 32114
USA
Tel: 386-323-8087
Fax: 386-323-8772
E-mail: Rodny.Piercey@erau.edu

Anthony Pietryla
Advanced Photon Source
Argonne National Laboratory
9700 S. Cass Ave
Argonne, IL 60439
USA
Tel: 630-252-7430
E-mail: afp@aps.anl.gov

Michael Plum
Spallation Neutron Source
Oak Ridge National Laboratory
P.O. Box 2008
Oak Ridge, TN 37831
USA
Tel: 865-576-5469
E-mail: plumma@sns.gov

James Pogge
Spallation Neutron Source
Oak Ridge National Laboratory
P.O. Box 2008
Oak Ridge, TN 37831
USA
Tel: 865-335-1437
E-mail: poggejr@sns.gov

Robert Powers
Edwards Supply Co., Inc.
315 Oak Ridge Turnpike
Oak Ridge, TN 37830
USA
Tel: 865-483-1766
Fax: 865-483-3562
E-mail: robert@edwardssupply.com

Tom Powers
Thomas Jefferson National Accelerator Facility
12000 Jefferson Ave.
Newport News, VA 23606
USA
Tel: 757-269-7660
Fax: 757-269-7658
E-mail: powers@jlab.org

Peter Prieto
Fermi National Accelerator Laboratory
P.O. Box 500
Batavia, IL 60510-0500
USA
Tel: 630-840-2509
Fax: 630-840-2677
E-mail: prieto@fnal.gov

Joseph Primeau
Acromag, Inc.
30765 S. Wixom Road
Wixom, MI 48393
USA
Tel: 248-624-1541
Fax: 248-624-9234
E-mail: jprimeau@acromag.com

Marek Proga
University of Texas at Austin
Physics Dept. RLM 5.208
1 University Station C1600
Austin, TX 78712
USA
Tel: 512-232-7189
Fax: 512-471-9637
E-mail: proga@mail.utexas.edu

Chuck Reagan
Spiricon, Inc.
2600 North Main
Logan, UT 84341
USA
Tel: 435-753-3729
Fax: 435-753-5231
E-mail: danf@spiricon.com

Guenther Rehm
Diamond Light Source Ltd
Rutherford Appleton Laboratory
Didcot, OX11 0QX
United Kingdom
Tel: ++44 1235 77807
Fax: ++44 1235 77805
E-mail: guenther.rehm@diamond.ac.uk

Andreas Ruben
Wiener
300 E. Auburn Avenue
Springfield, OH 45505
USA
Tel: 937-326-2620
Fax: 937-326-2625
E-mail: sales.us@wiener-d.com

Vadim Sajaev
Argonne National Laboratory
9700 S. Cass Ave
Argonne, IL 60439
USA
Tel: 630-252-4868
Fax: 630-252-5703
E-mail: sajaev@aps.anl.gov

Fernando Sannibale
Lawrence Berkeley National Laboratory
1 Cyclotron Rd
Berkeley, CA 94720
USA
E-mail: fsannibale@lbl.gov

Victor Scarpine
Fermi National Accelerator Laboratory
P.O Box 500
Batavia, IL 60510
USA
Tel: 630-840-2571
E-mail: Scarpine@fnal.gov

Volker Schlott
Paul Scherrer Institut
WSLA/007
Villigen, 5232
Switzerland
Tel: 0041 56 310 423
Fax: 0041 56 310 315
E-mail: volker.schlott@psi.ch

Rico Schurig
ELBE Forschungszentrum Rossendorf
Bautzner Landstr. 128
Dresden, 1328
Germany
Tel: 4.9351260209e+011
Fax: 4.9351260369e+011
E-mail: r.schurig@fz-rossendorf.de

William Sellyey
Los Alamos National Laboratory
P.O. Box 1663, MS H817
Los Alamos, NM 87545
USA
Tel: 505-665-7988
Fax: 505-665-2904
E-mail: sellyey@lanl.gov

Thomas Shea
Spallation Neutron Source
Oak Ridge National Laboratory
P.O. Box 2008
Oak Ridge, TN 37831
USA
Tel: 865-241-6087
Fax: 865-241-6739
E-mail: shea@sns.gov

Anatoly Shkvarunets
TREAP, University of Maryland
Energy Research Facility
College Park, MD 20742-3511
USA
Tel: 301-405-2311
Fax: 301-314-9437
E-mail: shkvar@umd.edu

Coles Sibley
Spallation Neutron Source
Oak Ridge National Laboratory
P.O. Box 2008
Oak Ridge, TN 37831
USA
Tel: 865-241-8055
Fax: 865-241-6739
E-mail: sibley@sns.gov

Om Singh
Advanced Photon Source
Argonne National Laboratory
9700 S. Cass Ave
Argonne, IL 60439
USA
Tel: 630-252-5409
Fax: 630-252-7432
E-mail: singh@aps.anl.gov

Steve Smith
Stanford Linear Accelerator Center
MS 50/2575 Sandhill Rd
Menlo Park, CA 94025
USA
Tel: 650-926-3916
Fax: 650-926-3800
E-mail: ssmith@slac.stanford.edu

Randall Stevens
Ludlum Measurements
c/o Pulcir, 9209 Oak Ridge Highway
Oak Ridge, TN 37830
USA
Tel: 865-927-6358
Fax: 865-927-6866
E-mail: lynne@pulcir.com

Martin Stockli
Spallation Neutron Source
Oak Ridge National Laboratory
P.O. Box 2008
Oak Ridge, TN 37831
USA
Tel: 865-241-8817
Fax: 865-241-9831
E-mail: stockli@sns.gov

Greg Stover
Lawrence Berkeley National Laboratory
1 Cyclotron Rd.
Berkeley, CA 94720
USA
Tel: 510-486-7706
Fax: 510-486-5775
E-mail: gdstover@lbl.gov

Chris Tennant
Thomas Jefferson National Accelerator Facility
12000 Jefferson Ave, MS 7A
Newport News, VA 23606
USA
Tel: 757-269-609
E-mail: tennant@jlab.org

Cyrille Thomas
Diamond Light Source
Rutherford Appleton Laboratory
Chilton, England
Tel: +441235778149
Fax: +441235778052
E-mail: cyrille.thomas@diamond.ac.uk

Michael Thuot
Los Alamos National Laboratory
LANSCE Division, MS H820
Los Alamos, NM 87545
USA
Tel: 505-667-7974
E-mail: mthuot2@comcast.net

Michael Tiefenback
Thomas Jefferson National Accelerator Facility
12000 Jefferson Avenue
Newport News, VA 23606
USA
Tel: 757-269-7430
Fax: 757-269-5024
E-mail: tiefen@jlab.org

Vitali Tupikov
Fermi National Accelerator Laboratory
P.O Box 500
Batavia, IL 60510
USA
E-mail: tupikov@bnl.gov

Klaus Unser
Bergoz Instrumentation
Espace Allondon Ouest
Saint Genis Pouilly, 01630
France
Tel: +33 450 426 642
Fax: +33 450 426 643
E-mail: unser@bergoz.com

Klaas Anko Vogel
Acqiris
P.O. Box 2203
Monroe, NY 10950
USA
Tel: 845-782-6544
Fax: 845-782-4745
E-mail: klass.vogel@acqiris.com

Jonah Weber
Advanced Light Source
Lawrence Berkeley National Laboratory
1 Cyclotron Road
Berkeley, CA 94720
USA
Tel: 510-486-5814
Fax: 510-486-5775
E-mail: jmweber@lbl.gov

Manfred Wendt
DESY
Notkestr. 85
Hamburg, 22607
Germany
Tel: +49-40-8998-251
Fax: +49-40-8998-430
E-mail: wendt@ux-bello.desy.de

Kay Wittenburg
DESY
Notkestr. 85
Hamburg, 22607
Germany
E-mail: Kay.Wittenburg@desy.de

Yan Yin
YY Labs, Inc.
P.O. Box 597
Fremont, CA 94537
USA
Tel: 510-739-6049
Fax: 510-739-6037
E-mail: yanyin@yylabs.com

Jim Zagel
Fermi National Accelerator Laboratory
MS 308, P.O. Box 500
Batavia, IL 60510
USA
Tel: 630-840-4076
E-mail: zagel@fnal.gov

Xiang-Ying Zeng
YY Labs, Inc.
P.O. Box 597
Fremont, CA 94537
USA
Tel: 510-739-6049
Fax: 510-739-6037
E-mail: yanyin@yylabs.com

Bill Zinn
Kurt J. Lesker Co.
1515 Worthington Ave.
Clairton, PA 15025
USA
Tel: 412-233-4200
Fax: 412-233-4275
E-mail: billz@lesker.com

Participating Vendors

We would like to thank the following vendors for their additional support:

- GMW Associates for hosting a special reception on Monday, May 3, 2004.
- Acquiris USA for cosponsoring the BIW04 Banquet on Wednesday, May 5, 2004.
- Bergoz Inc. for sponsoring the Faraday Cup Award.

Acqiris USA
William Accolla
234 Cromwell Hill Rd
Monroe, NY 10950
USA
Tel: 845-782-6544
Fax: 845-782-4745
w.accolla@acqiris.com
www.acqiris.com

Acromag, Inc.
Joseph Primeau
30765 S. Wixom Road
Wixom, MI 48393
USA
Tel: 248-624-1541
Fax: 248-624-9234
jprimeau@acromag.com
www.acromag.com

Drivesoft, LLC
Michael Doebelin
4813 Wayside Drive
Old Hickory, TN 37138
USA
Tel: 615-758-3718
Fax: 615-758-3973
mdoebelin@drivesoft.com
www.drivesoft.com

Echotek Corp
Jim Kurtz
555 Sparkman Drive, Suite 400
Huntsville, AL 35816
USA
Tel: 256-721-1911
Fax: 256-721-9266
gary@echotek.com
www.echotek.com

Edwards Supply Company, Inc.
Robert Powers
315 Oak Ridge Turnpike
Oak Ridge, TN 37830
USA
Tel: 865-483-1766
Fax: 865-483-3562
robert@edwardssupply.com
www.edwardssupply.com

GMW Associates (Bergoz Inc.)
955 Industrial Road
San Carlos, California 94070
USA
Tel: 650-802-8292
Fax: 650-802-8298
www.gmw.com

Instrumentation Technologies
Rok Uršič
Srebrničev Trg 4a
SI-5250 Solkan
Slovenia
Tel: +386 5 3311
Fax: +386 5 3311 719
rok@i-tech.si
www.i-tech.si

iseg-USA, LLC
Jeri L. Duncan
1655 Murfreesboro Rd., Suite A
Nashville, TN 37217
USA
Tel: 615-397-4390
Fax: 615-361-4174
jduncan@iseg-usa.com
www.iseg-usa.com

Kepco, Inc.
131-38 Sanford Ave.
Flushing, NY 11352
USA
Tel: 718-461-7000
Fax: 718-767-1102
hq@kepcopower.com
www.kepcopower.com

Kurt J. Lesker Company
Bill Zinn
1515 Worthington Ave.
Clairton, PA 15025
USA
Tel: 412.233.4200
Fax: 412.233.4275
billz@lesker.com
www.lesker.com

Leybold Vacuum
Joe. F. Fabrizio
5700 Mellon Road
Export, PA 15632
USA
Tel: 724-325-6371
Fax: 724-325-8029
jfabrizio@leyboldvacuum.com
www.leyboldvac.de

Ludlum Measurements
Randall D. Stevens
c/o Pulcir, 9209 Oak Ridge Highway
Oak Ridge, TN 37830
USA
Tel: 865-927-6358
Fax: 865-927-6866
lynne@pulcir.com
www.ludlums.com

Pulcir Inc.
Joseph D. Eddlemon
9209 Oak Ridge Highway
Oak Ridge, TN 37830
USA
Tel: 865-927-6358
Fax: 865-927-6877
lynne@pulcir.com
www.pulcir.com

SE Technical Sales, Inc.
Mark J. Kibilko
P.O. Box 2760
Windermere, FL 34786
USA
Tel: 866-903-5858
Fax: 404-903-5859
mark@setechsales.com
www.setechsales.com

Spiricon, Inc.
Chuck Reagan
2600 North Main
Logan, UT 84341
USA
Tel: 435-753-3729
Fax: 435-753-5231
danf@spiricon.com
www.spiricon.com

Tektronix, Inc.
Mike Broadwell
6333 N. State Hwy 161 #300
Irving, TX 75038
USA
Tel: 256-503-1200
Fax: 770-234-6883
mike.broadwell@tek.com
www.tek.com

Vacuum Solutions, Group-Southeast, Inc.
Perry Henderson
419 The Parkway, PMB 212
Greer, SC 29650
Tel: 864-801-9458
Fax: 864-801-9513
USA
phenderson@vsg-southeast.com
www.vsg-southeast.com

Wiener
Andreas Ruben
300 E. Auburn Avenue
Springfield, OH 45505
USA
Tel: 937-326-2620
Fax: 937-326-2625
sales.us@wiener-d.com
www.wiener-us.com

YY Labs, Inc.
Xiang-Ying Zeng
P.O. Box 597
Fremont, CA 94537
USA
Tel: 510-739-6049
Fax: 510-739-6037
yanyin@yylabs.com
www.yylabs.com

Author Index

A

B

C

D

E

F

G